Handbook of
Property Estimation Methods for Chemicals
Environmental and Health Sciences

Handbook of Property Estimation Methods for Chemicals

Environmental and Health Sciences

Robert S. Boethling
Donald Mackay

LEWIS PUBLISHERS

Boca Raton London New York Washington, D.C.

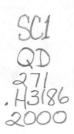

Library of Congress Cataloging-in-Publication Data

Handbook of property estimation methods for chemicals: environmental and health sciences/ edited by
Robert S. Boethling, Donald Mackay; with a foreword by Warren J. Lyman.
    p.   cm.
    Includes bibliographical references and index.
    ISBN 1-56670-456-1 (alk. paper)
    1. Organic compounds--Analysis. 2. Environmental chemistry. I. Boethling, Robert S. II. Mackay, Donald.
      QD271.H3186 2000
547′.3—dc21

for Library of Congress

99-058377
CIP

No claim to original U.S. Government works
International Standard Book Number 1-56670-456-1
Library of Congress Card Number 99-058377
Printed in the United States of America   2 3 4 5 6 7 8 9 0
Printed on acid-free paper

Foreword

In 1978, Dr. David Rosenblatt of the U.S. Army* and I had the idea to create a compilation of estimation methods for environmentally important properties of organic chemicals. This was stimulated by the problems of evaluating sites, such as Rocky Mountain Arsenal, which had many soil and groundwater contaminants about which we knew little. We sought and obtained a partnership with the U.S. Environmental Protection Agency, which had major problems evaluating new chemicals under their Premanufacture Notification Program, and the U.S. Coast Guard, which a few years earlier had been given the daunting responsibility of responding to chemical spills into water and a laundry list of potential chemicals that numbered in the hundreds. This idea culminated in the publication of the *Handbook of Chemical Property Estimation Methods* in 1982, initially published by McGraw-Hill, now by the American Chemical Society.

Considering how little guidance we had in preparing that handbook (Bob Reid's books on property estimation for the chemical process industry were our best examples**), our efforts turned out a very useful product that has endured, perhaps too long. Even as it was being published, and ever since, a continuous stream of research has led to new and better estimation methods and a correspondingly better understanding of environmental chemistry. An update of much of the handbook has been long overdue.

As noted in this book's Introduction, this new handbook was conceived 9 years ago, the objective being to provide property estimators (and those seeking general enlightenment) the needed update. I was pleased to be asked to contribute one chapter myself, and even more pleased to see the high quality of the collection of authors the editors had selected to write the other chapters. All of them are experts in the subject areas they cover. (This author selectiveness was not possible for the original handbook; we had a lot of on-the-project learning.) I was especially pleased to see this new handbook be under the creative leadership of Bob Boethling and Don Mackay, two individuals who have been true leaders in the field of environmental chemistry for the last 20 years. If you look at the Acknowledgments section of the original handbook, you will see both of them credited as reviewers.

I look forward myself to having this new handbook as a comparison to the initial one on my special bookshelf, reserved for those frequently used reliable resources, and expect that it will last at least as long a time.

Warren J. Lyman, Ph.D.
Cambridge, Massachusetts

---

* Dave Rosenblatt was employed at the time by the U.S. Army Medical Bioengineering Research and Development Laboratory, Fort Detrick. He is now an independent environmental consultant in Baltimore, Maryland.
** For example, Reid et al., *The Properties of Gases and Liquids*, 3rd ed., McGraw-Hill Book Co., New York, 1977.

Introduction

Robert S. Boethling and Donald Mackay

CONTENTS

I-1 The Need for Estimation Methods

Most industrially developed nations have government agencies that assess risk to human health and the environment resulting from the manufacture and use of chemical substances. Industry, consultants, academics, and the research community also undertake such assessments. For example, in the U.S., the Office of Prevention, Pesticides, and Toxic Substances (OPPTS) devotes considerable resources to risk analyses in its review of commercial substances that are or will be released to the environment. Canada has a similar system, and so do most nations in the European Union and Japan. As the number of potentially hazardous substances introduced into commerce grows, maximizing the efficiency of these assessments becomes increasingly important.

The ability to predict the behavior of a chemical substance in a biological or environmental system largely depends on knowledge of the physical–chemical properties and reactivity of that compound or closely related compounds. Chemical properties frequently used in environmental assessment include melting/boiling temperature, vapor pressure, various partition coefficients, water solubility, Henry's Law constant, sorption coefficient, bioconcentration factor, and diffusion properties. Reactivities by processes such as biodegradation, hydrolysis, photolysis, and oxidation/reduction are also critical determinants of environmental fate and such information may be needed for modeling. Unfortunately, measured values often are not available and, even if they are, the reported values may be inconsistent or of doubtful validity. In this situation it may be appropriate or even essential to use estimation methods.

This volume represents the culmination of a project conceived 9 years ago, whose original aim was to update the landmark *Handbook of Chemical Property Estimation Methods*

(Lyman et al., 1982) (often referred to as *Lyman's Handbook*), which was reprinted by the American Chemical Society but not updated in 1990 (ACS, 1990). *Lyman's Handbook* is still required reading for environmental scientists, but the intervening 18 years obviously have brought many advances. These are due in large measure to the synergy between basic research aimed at enhancing our understanding of environmental chemistry, and advances in computational power, computer graphics, electronic databases, and software. "Advances" doesn't quite capture the profound changes that have occurred. It is worth keeping in mind that, when *Lyman's Handbook* was published in 1982, hand calculators were still state-of-the-art technology and the personal computer was a luxury just being introduced.

Despite its age, *Lyman's Handbook* is still a valuable source of basic information on chemical properties, and many of the estimation methods contained therein remain valid. It also contains a suite of chapters on pure substance properties that normally are not of direct interest to environmental scientists; e.g., flash point and heat capacity.

Other published works relevant to estimation of properties and reactivity of chemicals may be of value to the reader. A book edited by Neely and Blau (1985) (*Environmental Exposure from Chemicals*) is less comprehensive but has several excellent chapters on basic information, such as the one on biodegradation by Klečka. Other volumes that may be useful include *The Properties of Drugs*, edited by Yalkowsky et al. (1980); and *Aqueous Solubility* by Yalkowsky and Banerjee (1991). *Chemical Exposure Predictions*, edited by Calamari (1993), is somewhat different in focus but may be useful for information on exposure modeling, especially for the soil compartment.

I-1.1   Regulatory Context

In the U.S., the safety of specific chemical substances is evaluated primarily under three statutes. Substances used as food additives, drugs, and cosmetics are regulated by the Food and Drug Administration (FDA) under the Federal Food, Drug, and Cosmetic Act (FFDCA). Chemical substances proposed for use as pesticides are regulated by the U.S. Environmental Protection Agency (EPA) under the Federal Insecticide, Fungicide, and Rodenticide Act (FIFRA), and industrial chemicals are regulated under the Toxic Substances Control Act (TSCA), whose authority is very broad and includes all commercial chemicals not addressed by other legislation. TSCA distinguishes existing chemicals and substances not yet in commerce, for which a Premanufacture Notice (PMN) must be submitted 90 days in advance of manufacture or import. New substances approved by EPA are then added to the TSCA Chemical Substance Inventory ("Inventory") and become existing chemicals as soon as EPA receives the required Notice of Commencement. EPA has received more than 30,000 valid PMNs, and submissions currently average well over 2,000 per year.

The PMN review process has evolved over time within the constraints set by TSCA. An important constraint is that submitters are required to furnish only test data already in their possession (if any) and are not required to conduct a battery of tests as a precondition for approval. This generalization holds true for basic chemical property data as well as toxicity data, and it is the main reason why TSCA has been such a powerful impetus for developing *estimation* methods for many of the parameters needed in environmental assessment. To illustrate how extreme the situation is, in one study of more than 8,000 PMNs for class 1 chemical substances (i.e., those for which a specific chemical structure can be drawn) that were received from 1979 through 1990, Lynch et al. (1991) found only 300 that contained any of the property data noted earlier as needed for environmental assessment. The U.S. is unique among industrialized nations in requiring its assessors to work in the virtual absence of test data.

In the European Economic Community (EEC), chemical control is based primarily on Directive 79/831/EEC on the "Classification, Packaging, and Labeling of Dangerous Substances," in the Sixth Amendment to the Dangerous Substances Directive (67/548/EEC). To place a new substance on the European market, a manufacturer or importer must submit a notification dossier, and, for substances whose volume is expected to exceed 1 ton/year, this includes a wide variety of test data that compose the Minimum Premarket Dataset (MPD) (Greiner, 1993). The MPD includes test requirements for melting and boiling temperature, vapor pressure, partition coefficient, water solubility, adsorption/desorption behavior, and ready biodegradability, among others. Regulation of new chemical substances under the Canadian Environmental Protection Act (CEPA) (Matheson and Atkinson, 1997) and the Japanese Chemical Substances Control Law (CSCL) (Keener, 1997) is similar.

Despite the wider availability of experimental data in these nations, risk assessments for new substances often demand estimation of environmentally important parameters. For example, degradation and partitioning processes must be considered in environmental exposure assessments, but neither Henry's Law constant nor abiotic degradation processes such as hydrolysis and photolysis are included in the MPD. Only if a substance is not readily biodegradable may requirements for abiotic degradation testing be imposed.

I-1.2   Chemical Design

The impetus to develop estimation methods also has come from the existing chemicals arena, since only a minuscule fraction of existing chemicals are blessed with a full complement of chemical and environmental data. The TSCA Inventory contains over 70,000 substances and most have not been evaluated fully. Similarly, the European Inventory of Existing Chemical Substances (EINECS) was built partially upon the TSCA Inventory and contains more than 100,000 substances (Hansen and van Leeuwen, 1995). Although it would be irresponsible to imply, as some have, that all are equally risky, an ongoing need exists to screen the Inventory and establish priorities for further review. By statute this is the responsibility of the U.S. TSCA Interagency Testing Committee (ITC), which has conducted a variety of screening exercises over the years, all of which relied heavily on estimated values of critical parameters.

The ITC is only the tip of the iceberg in chemical ranking and scoring (CRS). Virtually every EPA program now is involved in CRS in one way or another, and it is broadly used in business, government, and academia to pursue a variety of goals whose common purpose is to identify and prioritize chemicals for some subsequent action. The latter may include testing to fill data gaps for risk assessment, targeting for risk reduction through pollution prevention or other means, or evaluating competing alternatives for new product technologies, to name just a few possibilities. Besides the various ranking schemes the ITC has used over the years, examples include Reportable Quantity (RQ) adjustment methodology under the Comprehensive Environmental Response, Compensation and Liability Act (CERCLA; "Superfund"), the EPA Inerts Ranking Program for so-called "inert" ingredients of pesticide formulations, and, more recently, the Office of Solid Waste's Waste Minimization Prioritization Tool (WMPT) (U.S.EPA 1997).

A recent workshop (Swanson and Socha 1997) identified more than 100 such systems and subjected 51 of these to a comparative analysis. That analysis identified at least 12 different endpoints for persistence, 15 endpoints for mobility, partitioning, or bioaccumulation, and 34 kinds of data used to estimate exposure. It is axiomatic that the required input data simply are not available for all of the parameters and chemical substances of potential interest.

Passage of the Pollution Prevention Act (PPA) by the U.S. Congress in 1990 ushered in a new era in the philosophy of controlling risks from toxic chemicals. The PPA declared it to be the policy of the United States that pollution should be prevented at the source when possible, and it defined source reduction as any practice that:

> (i) reduces the amount of any hazardous substances ... *prior to* recycling, treatment or disposal; and (ii) reduces the hazards to public health and the environment associated with the release of such substances. (emphasis added)

Simply put, the intended message is that the best way to control risks from exposure to toxic substances is to not produce them in the first place. A direct result has been a growing emphasis on the design or redesign of commercial chemicals to make them safer for humans and the environment, without sacrificing their efficacy. Until now "safer" usually has translated into "less toxic," but there is increasing realization that environmental chemistry is important, since chemicals that persist in the environment remain available to exert toxic effects for longer periods of time than those which do not, and they may bioaccumulate. Compelling reasons that new or redesigned chemicals should reflect the principles of safe design include:

- We cannot know in advance all possible toxic effects
- Production and release may exceed initial expectations significantly if a product is particularly successful in the marketplace
- New uses may develop over time, leading to unforeseen releases and exposures
- Chemicals may be exported to nations with less stringent environmental controls
- Global distribution is possible for chemicals with certain properties.

The significance here is that, to incorporate environmental chemistry effectively into the *design* phase of chemical manufacture, reliable methods for estimating chemical properties and environmental reactivity are, and will be, of the utmost importance.

I-2 Overview of this Work

A major difference between this volume and earlier works is that this work comprehensively reviews the many recent improvements in chemical property estimation methods and focuses on those properties most critical to environmental fate assessment. Each chapter stresses practical applications of chemical property estimation, but only after thorough development of the theoretical fundamentals.

This book is intended to be a handbook to which the reader will frequently refer for details of estimation methods appropriate for the chemicals of interest. It is intended primarily for anyone involved with estimating chemical properties, but particularly those who need such data for environmental or health assessment of chemical substances. It will also be of interest to engineers, research scientists, and educators who are concerned with the fate and effects of hazardous substances.

I-2.1 Chapter Contents

In general, each property estimation chapter:

- Describes the property and its importance to chemical assessment
- Discusses the theoretical foundation of the property and how it is affected by environmental or other factors
- Presents the most important estimation methods and provides basic information on range of applicability, input requirements, and method error
- Recommends one or more methods for general use and, if feasible, provides the necessary equations, input data, and instructions to perform the calculations
- By means of sample calculations using a series of "benchmark chemicals," shows step-by-step how to use the method(s)
- Provides full reference citations for backgound literature

This book treats mostly organic compounds, not metals or inorganics. We also do not explicitly address estimation of chemical properties for complex mixtures, although some methods may be applicable to mixtures.

As in *Lyman's Handbook*, emphasis is on broadly applicable estimation methods. Given the many and varied reasons that one might be interested in chemical property estimation, we believe that most users of this book will have less interest in chemical class-specific estimation methods. Obviously such methods are reliable only for that class, which may be defined very narrowly, and they may produce substantial yet unknown error if applied to compounds that differ significantly. Many of the newer methods were developed using much larger and more varied training sets, thus are more likely to be useful for diverse and/or structurally complex compounds. Therefore, in contrast to the situation that existed in 1982 when *Lyman's Handbook* was published, current users often do not need to make decisions about which of several class-specific methods seems most applicable to the compound of interest.

Advances in hardware and software technology, together with the widespread availability of powerful yet relatively inexpensive personal computers, have diminished greatly the importance of being able to perform calculations by hand. In a way this is unfortunate because it often comes at the expense of understanding. For many users, the technical details of an estimation method are relegated to the status of mere lines of computer code, and only the final answer receives any attention. Nevertheless, we believe the benefits of current technology outweigh the risks; accordingly, widely available computer programs for chemical property estimation figure prominently in many chapters. Howard and Meylan (1997) also have summarized much of this information in a convenient series of tables.

I-2.2 Properties of Pure Substances (Chapters 1–3)

The most fundamental properties of a chemical substance are those of the substance in pure form, in most cases as a solid or liquid. Molecular mass can be deduced readily from the chemical formula or structure, although a range of values may exist for commercial mixtures. In some cases, the substance may adopt different structural (e.g., cis–trans) or enantiomeric forms, usually with relatively small physical property differences but with potentially substantial differences in ability to induce toxicity or other biological responses. The hexachlorocyclohexane isomers and enantiomers are examples, the insecticide lindane or γ HCH being the most active form.

It is important to ascertain the P-T behavior of a substance, i.e., its placement on the P-T diagram in the accompanying figure under conditions of atmospheric pressure and environmental temperature. This placement largely is determined by the triple point temperature (the unique temperature at which solid, liquid, and vapor co–exist and which is usually close to the melting point) and by the vapor pressure at typical atmospheric temperatures of 0 to 30°C, with 25°C being the temperature used most frequently when reporting data. The critical point temperature (at which liquids and vapor become indistinguishable and form a single dense fluid phase) usually exceeds environmental temperatures, except for simple gases, and is thus of limited interest environmentally. For example, for methane, ethane, propane, and benzene, the critical temperatures are 190K, 305K, 370K, and 562K, respectively. For many organic substances, decomposition occurs below the critical temperature.

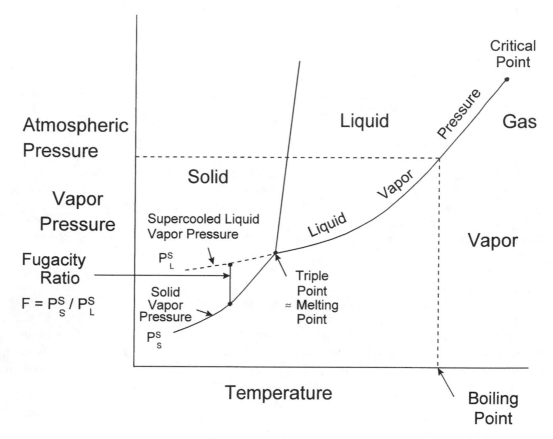

FIGURE I-1
P-T diagram of a pure substance

Of primary environmental interest are the melting point, boiling point (the temperature at which the vapor pressure equals atmospheric pressure), and related vapor pressure at environmental temperatures. Chapters 1, 2, and 3 discuss these properties. Also of interest is the super-cooled liquid vapor pressure, i.e., the vapor pressure which a solid substance would have if it were liquid at environmental temperatures. This vapor pressure, which is shown dashed in the figure, can be obtained by extrapolating the liquid's vapor pressure below the melting point. It cannot be measured directly. For example, naphthalene melts at 80°C, well above environmental temperatures. Its measured solid vapor pressure depends on the stability of the crystal structure of the pure substance, symmetrical molecules generally having greater stability, higher melting points, and lower vapor pressures (see Chapter 3). When naphthalene is in solution or sorbed to surfaces at low concentrations, there are few, if any, napththalene–naphthalene interactions or crystals; thus it behaves as if its vapor pressure is that of the liquid. In thermodynamic terms, the reference fugacity is that of the liquid state. As a result, it is necessary to estimate this hypothetical vapor pressure when predicting or relating properties in solution such as solubility in water, Henry's law constants, and sorption to organic matter and aerosols.

The ratio of the solid and liquid pressures is termed the fugacity ratio F and is usually estimated by the equation:

$$F = \exp \left(-\Delta H \left(1/T - 1/T_M\right)/R\right) \tag{1}$$

where ΔH is the enthalpy of fusion (J/mol), R is the gas constant 8.314 J/mol.K, T is the environmental temperature (K) and T_M is the melting point temperature (K). Applying Walden's Rule that the entropy of fusion ($\Delta H/T_M$) is typically 56.5 enables equation (1) to be rewritten as

$$F = \exp \left(-6.79(T_M/T-1)\right) \tag{2}$$

where 6.79 is 56.5/R. For naphthalene, since T_M is 353K, and T is 298K (i.e. 25°C), F is 0.286 and the solid vapor pressure is 28.6% of the liquid value. Likewise, the solid solubility in water is 28.6% of the liquid value.

I-2.3 Partitioning Properties (Chapters 4–11)

For most organic substances the most important phases are air, water, and organic phases such as lipids, waxes, and natural organic matter including humin, humic and fulvic acids. Partitioning to organic phases usually is quantified by measuring partitioning to n-octanol, which has the advantages that it can be obtained pure and is found to have properties similar to the glyceryl-tri fatty acid esters which constitute lipids. This gives rise to three partition coefficients or ratios of concentrations of the subject chemical, K_{OW} the octanol-water partition coefficient, K_{AW} the air-water partition coefficient, and K_{OA} the octanol-air partition coefficient. These partition coefficients are best measured under dilute conditions when activity coefficients in all three media are relatively constant and close to the "infinite dilution" values. These partition coefficients are usually expressed on a (mass/volume)/(mass/volume) basis and they are, of course, temperature dependent.

Chapters 4, 5, and 6 discuss estimation methods for these partition coefficients. The octanol-water partition coefficient K_{OW} is frequently designated P in the medicinal and biological chemistry literature.

Also of interest is the maximum capacity of each phase for chemicals, i.e., the saturation concentration above which phase separation occurs. For water, this is obviously the solubility in water. For many polar substances, the chemical and water are miscible (e.g., ethanol) and no solubility limit exists. Similarly, a solubility limit in octanol may or may not exist. For air, the solubility corresponds to the saturation vapor pressure P^S. This can be converted to a solubility in units of mol/m³ by dividing by RT, the gas constant — absolute temperature product. Chapter 7 discusses solubility in water. Solubility in octanol is not by itself of comparable interest and is not treated. Vapor pressure and solubility in water are not only of fundamental interest, but their ratio H is essentially the Henry's Law constant or air-water partition coefficient, as Chapter 4 discusses.

Partitioning between organic matter and water is of crucial importance in water bodies, sediments, and soils and is obviously related to K_{OW} as Chapter 8 discusses. It is a primary determinant of the mobility of chemicals in soils. In view of the variability in composition of organic matter, partitioning may be expressed on a basis of organic matter or organic carbon, and some variability is expected, reflecting the differing biogeneses of the organic material.

Also of crucial importance is bioaccumulation or partitioning between water and aquatic biota such as fish, as Chapter 9 discusses. Again K_{OW} plays a key predictive role, but not surprisingly, the partitioning phenomena can be quite complex and are influenced by several factors including the route of uptake (gill ventilation or diet), metabolism within the organism, growth dilution, and loss processes including fecal elimination. The extent of bioaccumulation is a critical determinant of chemical transport into aquatic and terrestrial food webs and human exposure.

Only recently has the importance of bioaccumulation from air to vegetation been fully appreciated, the literature on this subject being minuscule comparedwith bioaccumulation in aquatic biota. Accordingly, Chapter 6 treats this topic only briefly, along with K_{OA}, the octanol-air partition coefficient. In future editions of this handbook, partitioning to vegetation will be likely to merit more extensive and specific treatment.

Chapter 10 treats partitioning between the gaseous phase in the atmosphere and aerosol particles. This process is important in determining the deposition and absorption characteristics of chemicals present in the atmosphere, which can profoundly affect human uptake by respiration.

This handbook makes no attempt to describe environmental transport properties because these phenomena are specific to environmental conditions such as wind speed and water current velocity. A notable exception is the diffusion or permeation rate of a substance through biological membranes, which is the key process controlling dermal absorption and therefore a key determinant of the dose actually available to exert a toxic effect. Chapter 11 treats this topic. The reader seeking estimation methods for molecular diffusivities in air and water should consult the text by Reed et al. (1987).

I-2.4 Reactivity or Persistence (Chapters 12–16)

Obviously, chemicals which persist for prolonged periods of time in the environment are of more concern because of the potential to build up larger environmental burdens and thus higher concentrations, exposures, and potential for adverse effects. Estimating environmental rate constants of reaction or half-lives is a formidable task. These rates are not only functions of the intrinsic properties of the chemical, but they also depend on environmental variables, for example, the photolytic intensity, pH, the nature and quantity of microbial biomass present, and the presence of oxidizing and reducing chemical species. These variables can change in space and time, diurnally and seasonally. Chapters 12 through 16 discuss estimation methods for hydrolysis, atmospheric oxidation, biodegradation, and redox transformations.

A notable omission here is the estimation of acid-base dissociation that Harris and Hayes described (1990) in *Lyman's Handbook*. A fairly complete database of pKa and pKb values exists, and the reader seeking an estimation method should consult the above chapter or the definitive monograph by Perrin et al. (1981).

I-2.5 Specific Classes of Substances (Chapters 17–18)

Certain classes of substances merit special treatment because they possess unique chemical properties that generally cannot be estimated using the conventional approaches. Conventional approaches have been developed largely for "priority pollutants," which tend to be relatively non-polar organics containing primarily carbon, hydrogen, chlorine, and, less frequently oxygen, nitrogen, sulfur, and phosphorus. Two brief chapters deal with these substances, specifically surfactants (Chapter 17) and dyes and pigments (Chapter 18).

I-2.6   Benchmark Chemicals

As originally conceived, this handbook was to include worked examples of estimation methods for a group of benchmark chemicals for which reliable properties exist. The advantage of this approach is that the reader is likely to find it easier to apply the estimation methods if there are examples to follow. This proved to be more difficult than was expected and not all these benchmark chemicals are fully treated. Ideally, the estimated values should correspond closely with measured values. In some cases there are significant discrepancies, and this serves to reinforce the message that there remains a need to improve these methods in both accuracy and scope. The subject of estimating chemical properties from molecular structure and from related properties is thus a fruitful topic of research, and will remain so for many years into the future.

References

Calamari, D., Ed. 1993. *Chemical Exposure Predictions*. Lewis, Boca Raton, FL.

Greiner, P. 1993. Ecotoxicology and environmental hazard assessment in the regulatory framework of chemical control in the EEC. In *Chemical Exposure Predictions*, Ed. D. Calamari, pp 205–219. Lewis, Boca Raton, FL.

Hansen, B. and K. van Leeuwen. 1995. US-EU-Japan Collaborative Project on Fate and Effects Predictions for High Production Volume Chemicals: an Overview. *Report to the HAAB, January 1995*. HAABO195.DOC, October 1995.

Harris, J.C. and M.J. Hayes. 1990. Chapter 6 in *Handbook of Chemical Property Estimation Methods*, W.J. Lyman, W.F. Reehl, and D.H. Rosenblatt, Eds. American Chemical Society, Washington, D.C.

Howard, P.H. and W.M. Meylan. 1997. Prediction of physical properties, transport and degradation for environmental fate and exposure assessments. In *Quantitative Structure-Activity Relationships in Environmental Sciences-VII*, F. Chen and G. Schuurmann, Eds., pp 185–205. SETAC Press, Pensacola, FL.

Keener, R.L. 1997. Regulation of new polymers in the Pacific region–part II (Japan). In *Ecological Assessment of Polymers: Strategies for Product Stewardship and Regulatory Programs*, J.D. Hamilton, and R. Sutcliffe, Eds., pp 265–275. Van Nostrand Reinhold, New York.

Lyman, W.J., W.F. Reehl, and D.H. Rosenblatt, Eds. 1982. *Handbook of Chemical Property Estimation Methods: Environmental Behavior of Organic Compounds*. McGraw-Hill, New York.

Lyman, W.J., W.F. Reehl, and D.H. Rosenblatt, D.H., Eds. 1990. *Handbook of Chemical Property Estimation Methods: Environmental Behavior of Organic Compounds*. American Chemical Society, Washington, D.C.

Lynch, D.G., N.F. Tirado, R.S. Boethling, G.R. Huse, and G.C. Thom. 1991. Performance of on-line chemical property estimation methods with TSCA Premanufacture Notice chemicals. *Ecotoxicol. Environ. Saf.* 22, 240–249.

Neely, W.B. and G.E. Blau, Eds. 1985. *Environmental Exposure from Chemicals, vol. 1*. CRC, Boca Raton, FL.

Perrin, D.D., B. Dempsey, and E.P. Serjeant. 1981. *pK_a Prediction for Organic Acids and Bases*. Chapman and Hall, New York.

Reed, R.C., J.M. Prausnitz, and B.E. Poling. 1987. *The Properties of Gases and Liquids*. McGraw-Hill, New York.

Swanson, M.B. and A.C. Socha, Eds. 1997. *Chemical Ranking and Scoring. Guidelines for Relative Assessments of Chemicals.* SETAC Press, Pensacola, FL.

United States Environment Protection Agency. 1997. Waste Minimization Prioritization Tool, Beta Test Version 1.0: User's Guide and System Documentation (Draft). EPA Report No. 530-R-97-019, June 1997.

Yalkowski, S.H. and S. Banerjee. 1991. *Aqueous Solubility. Methods of Estimation for Organic Compounds.* Marcel Dekker, New York.

Yalkowsky, S.H., A.S. Sinkula, and S.C. Valvani. 1980. *Physical and Chemical Properties of Drugs.* Marcel Dekker, New York.

About the Editors

Robert S. Boethling , a native of Minneapolis, MN, has been a microbiologist at the U.S. Environmental Protection Agency (EPA) headquarters in Washington, D.C., Office of Pollution Prevention and Toxics (OPPT) since 1980 . He earned his undergraduate and Ph.D. degrees at UCLA in microbiology. He spent 2 years as a postdoctoral fellow at Cornell University, followed by 2 years in industry before coming to EPA as it began its implementation of the Toxic Substances Control Act (TSCA).

In the 1980s, in keeping with the EPA's focus under TSCA, Dr. Boethling contributed to development of guidelines for testing in the areas of chemical properties and environmental fate; estimation methods (structure/activity relationships); and test rules for chemicals recommended for EPA action by the TSCA Interagency Testing Committee. As OPPT shifted its focus in the 1990s to pollution prevention (P2) and voluntary programs, Dr. Boethling provided environmental fate expertise to Agency programs such as Design for the Environment, Environmentally Preferable Purchasing and Products, and listing/delisting of substances for the Toxics Release Inventory (TRI). In the past 3 years, he has been a major contributor to numerous EPA activities related to persistent organic pollutants and persistent, bioaccumulative, and toxic substances (PBTs). These activities include a recent *Federal Register* notice describing a PBT chemical category under TSCA's Premanufacture Notice program for new chemical substances; addition of several PBT substances and lowering of reporting thresholds for PBTs on the TRI; development of the Office of Solid Waste's Waste Minimization Prioritization Tool; and ongoing development of EPA's Voluntary P2 PBT Chemical List. He is also active in other current EPA programs including the High Production Volume Chemical testing program, an EPA–industry partnership to supply important hazard/risk assessment data for existing substances; EPA/OECD test guidelines development and harmonization; and assessment and testing of new chemicals under TSCA Section 5. He is the recipient of several EPA Bronze Medals and an EPA Science Achievement Award.

Dr. Boethling is an author of more than 50 research publications in several fields, including microbial physiology, soil biochemistry, fermentation microbiology, biodegradation, and environmental chemistry. Most of his published work is in the areas of biodegradation and design and development of environmental fate data resources and computerized estimation methods. In the latter field, he was instrumental in the development of several widely used methods for estimating chemical properties and environmental fate, such as the BIOWIN© biodegradability estimation program. He has been active in SETAC since 1983 and is on the editorial board of *Environmental Toxicology and Chemistry*. In addition to serving SETAC in this capacity, he has participated in several Society of Environmental Toxicology and Chemistry (SETAC) Pellston workshops, most recently as a steering committee member for the workshop "Evaluation of Persistence and Long-Range Transport of Organic Chemicals in the Environment" (Fairmont Hot Springs, BC; July 1998). He is an author of the popular work *Handbook of Environmental Degradation Rates* (Howard et al., 1991. Lewis Publishers, Chelsea, MI).

Donald Mackay was born and educated in Glasgow, Scotland, where he earned his Ph.D. in Chemical Engineering from what is now the University of Strathclyde. After working in a research capacity in the petrochemical industry, he joined the University of Toronto in

1967, where he taught chemical and environmental engineering until 1995. He is currently director of the Environmental Modelling Centre at Trent University in Peterborough, Ontario, Canada. This center is supported by a combination of grants and contracts with the Canadian federal government's Natural Sciences and Engineering Research Council and a consortium of chemical and petroleum companies.

Dr. Mackay's primary research interest is the development, application, validation, and dissemination of mass-balance models describing the fate of chemicals in the environment in general, and in a variety of specific environments. These models include descriptions of bioaccumulation in a variety of organisms, water-quality models of contaminant fate in lakes (notably the North American Great Lakes), rivers, sewage-treatment plants, and in soils and vegetation. He has developed a series of multimedia mass-balance models employing the fugacity concept that are widely used for assessment of chemical fate in national regions and in the global environment. A particular interest is the transport of persistent organic chemicals to cold climates such as the Canadian arctic and their accumulation and migration in arctic ecosystems.

These models require accurate data on physico-chemical properties of organic substances, which is the subject of Dr. Mackay's other interest, namely their measurement and correlation. This includes the compilation and critical review of these properties and their quantitative structure property relationships. He is co-author of the five-volume *Illustrated Handbook of Physical Chemical Properties and Environmental Fate of Organic Chemicals*, which documents data reported in the literature, and is also available in CD-ROM format from CRC Press. Dr. Mackay's hope is that a combination of the information reported in these handbooks, and the estimated data as described in the present volume, can provide a sound basis for assessment of the large and growing number of chemical substances of environmental concern.

Contributors

Roger Atkinson, Ph.D.   Air Pollution Research Center, University of California, Riverside, CA 92521

Terry F. Bidleman, Ph.D.   Atmospheric Environment Services, ARQP, 4905 Dufferin Street, Downsview, ON M3H 5T4, Canada

Robert S. Boethling, Ph.D.   Office of Pollution Prevention and Toxics, (7406) USEPA, 401 M Street, S.W., Washington, D.C. 20460

Stephen C. DeVito, Ph.D.   Office of Pesticide Programs (75096), USEPA, 401 M Street, S.W., Washington, D.C. 20460

William J. Doucette, Ph.D.   Utah State University, Utah Water Research Lab, UMC-8200, Logan, UT 84322

Frank A.P.C. Gobas, Ph.D.   School of Resource & Environmental Management, Simon Fraser University, Burnaby, BC V5A 1S6, Canada

Tom Harner, Ph.D.   Atmospheric Environment Services, ARQP, 4905 Dufferin Street, Downsview, ON M3H 5T4

Philip H. Howard, Ph.D.   Syracuse Research Corporation, 6225 Running Ridge Rd., North Syracuse, NY 13212

Peter M. Jeffers, Ph.D.   Department of Chemistry, State University of New York, Cortland, NY

Albert Leo, Ph.D.   Department of Chemistry, Pomona College, Claremont, CA 91711

Warren J. Lyman, Ph.D.   Camp Dresser and McKee Inc., 516 Cross St., Carlisle, MA 01741

David G. Lynch, Ph.D.   Office of Pollution Prevention and Toxics (7406), USEPA, 401 M Street, S.W., Washington, D.C. 20460

Kuo Chung Ma, Ph.D.   Department of Chemical Engineering and Applied Chemistry, University of Toronto, ON M5S 3E5, Canada

Donald L. Macalady, Ph.D.   Department of Chemistry and Geochemistry, Colorado School of Mines, Golden, CO 80401

Donald Mackay, Ph.D.   Environmental and Resource Studies, Trent University, Peterborough, Ontario K9J 7B8, Canada

Michael S. McLachlan, Ph.D.   Baltic Sea Research Institute, PO Box 301161, 18112 Rostock, Germany

Theodore Mill, Ph.D.   Stanford Research Institute, Chemistry Laboratory (PS-273), 333 Ravenswood Avenue, Menlo Park, CA 94025

Heather A. Morrison, Ph.D.   Aqualink, 4936 Yonge St. Suite 507, Toronto, ON M2N 6S3, Canada

Gloria W. Sage, Ph.D.   Syracuse Research Corporation, 6225 Running Ridge Rd., North Syracuse, N.Y. 13212

Martin L. Sage, Ph.D.   Department of Chemistry, Syracuse University, Syracuse, NY 13244

Wan Ying Shiu, Ph.D.   Department of Chemical Engineering and Applied Chemistry, University of Toronto, ON M5S 3E5, Canada

Dick T.H.M. Sijm, Ph.D.   Fahrenheitstraat 44, 3817 WD Amersfoort, Netherlands

Marc Tesconi, Ph.D.   Department of Pharmaceutical Sciences, College of Pharmacy, University of Arizona, Tucson, AZ 85721

Johannes Tolls, Ph.D.   Research Institute of Toxicology, Environmental Chemistry Group, University of Utrecht, Netherlands

Paul G. Tratnyek, Ph.D.   Oregon Graduate Institute of Science & Technology, Department of Environmental Sciences and Engineering, P.O. Box 91000, Portland, OR 97291-1000

N. Lee Wolfe, Ph.D.   USEPA, 960 College Station Road, Athens, GA 30605-2720

Samuel H. Yalkowsky, Ph.D.   Department of Pharmaceutical Sciences, College of Pharmacy, University of Arizona, Tucson, AZ 85721

Contents

Part IV Unusual Substances

Part I

Properties of Pure Substances

1

Melting Point

Marc Tesconi and Samuel H. Yalkowsky

CONTENTS

1-56670-456-1/00/$0.00+$.50
© 2000 by CRC Press LLC

1.1   Introduction

The melting point of a compound is the temperature at which the solid and liquid phases are in equilibrium at one atmosphere; pressure is specified because the melting process involves a change in volume and is therefore pressure dependent. Since the melting point can be determined easily experimentally, it is the most commonly reported physical property for organic compounds. However, in the absence of a rigorous theory of fusion, it is one of the most difficult to predict.

Since many solid compounds are polymorphic, they may have solid-liquid transition temperatures in addition to their melting point. For example, a compound existing as a metastable crystal may appear to melt at some temperature (e.g., 400 K) to produce a metastable liquid that can spontaneously, but perhaps very slowly, recrystallize into a more stable form. Although the compound undergoes a solid-liquid transition at 400 K, this temperature is not the true melting point. Throughout this chapter, melting point will refer only to the equilibrium melting temperature.

Because the melting point reflects the strength of a solid's intermolecular forces, it has been correlated with other physical properties and can be used in their estimation. Examples of such properties include equilibrium vapor pressure (Myrdal et al., 1994), aqueous solubility (Yalkowsky et al., 1980), fugacity ratios (see Introduction I-2.2), liquid viscosity (Przezdziechi et al., 1985), thermal conductivity (Keyes, 1959), and boiling point (Walters et al., 1995). A method for estimating melting point therefore would improve the estimation of other properties. This would make it particularly useful in the design of compounds with specific physical properties, since measurements of melting point obviously cannot be made for non-existent structures. For example, such a method could be used to develop a pesticide with a low vapor pressure to slow evaporation from foliage, or a material with specific heat transfer properties for use in conduction or insulation. Similarly, it could be used to design a low melting drug which would dissolve rapidly from a tablet in the stomach, or a high melting compound for sustained release.

This chapter deals primarily with practical applications of theoretical principles to the estimation of melting enthalpy and entropy in order to estimate melting point. As necessitated by the situations described above, the methods are based entirely upon molecular structure.

1.2   Overview of Available Estimation Methods

1.2.1   Correlation with Other Physical Properties

An excellent comprehensive review by Dearden (1999) describes how most melting point estimation methods involve correlations with other physical properties. Table 1 lists several examples of these from his review.

Table 1.1 shows that much of the work has been done on hydrocarbons and, in particular, homologous series. As a result, several very accurate methods are available for estimating the melting points of normal alkanes. For example, Broadhurst (1962) reports errors of less than 0.5°C for paraffins with chain lengths between 44 and 100 carbons. Similarly, Hanson

TABLE 1.1

Examples of Parameters Used to Estimate Melting Point.

Investigator(s)	Year	Parameter	Application To
Mills	1885	Carbon chain length	Homologous series
Longinescu	1903	Density, no. of atoms	Organic cpds.
Lindemann	1910	Atomic weight, atomic volume, vibrational freq.	Monatomic species
Prud'homme	1920	Boiling and critical temperatures	Organic cpds.
Austin	1930	Molecular weight	Long-chain homologs
Benko	1959	Molar volume at boiling point	Organic cpds.
Flory, Vrij	1963	Entropy: Rln (no. of carbons in chain)	Homologous series
Grigor'ev, Pospelov	1965	No. of axis of rotation, mechanical rigidity	Polycyclic, aromatic cpds.
Wachalewski	1970	Atomic group contribution, molecular diameter	Hydrocarbons, halogenated hydrocarbons
Eaton	1971	Solid and liquid temperature ranges	Homologous series
Syunyaeva	1981	C-C-C bond angle	Homologous series
Mackay et al.	1982	Saturated vapor pressure	Low volatility hydrocarbons, halogenated hydrocarbons
Hanson, Rouvray	1987	Topological indices: Balaban, Wiener, and carbon number	Homologous series
Tsakinikas, Yalkowsky	1988	C-C-C-C torsional angle	Aliphatic hydrocarbons
Dearden, Rahman	1988	Hydrogen bonding	Substituted anilines
Dearden	1991	Hanch hydrophobic parameter, molar refractivity	Substituted anilines

and Rouvray (1987) indicate a standard error of less than 1°C for normal alkanes between 44 and 390 carbons.

Although these methods have limited applicability, observations made during their development contributed to an understanding of the relationship between melting point and molecular structure. For example, Mills (1885) developed the following equation for the melting points of members of a homologous series:

$$T_m = \frac{\beta(x-c)}{1+\gamma(x-c)} \tag{1}$$

where x is the number of methylene groups and β, γ, and c are constants. This led to the recognition of a convergence temperature, i.e., the melting point of an alkane of infinite chain length. Similarly, Garner et al. (1931) realized that the convergence temperature is similar for all homologous series (approximately 141°C) because the head group becomes relatively insignificant as the chain increases in length. Figure 1 demonstrates this graphically for paraffins, fatty acid ethyl esters, 1-alcohols, and iso-alkyl acids. Note that, when $1/T_m$ is plotted vs. $1/n$ (n = number of chain carbons), the Y-intercept, which corresponds to infinite chain length, is approximately the same for all four series and represents the convergence temperature.

In addition to the convergence temperature, data from α and β alkylnaphthalenes and alkyl benzenes (ϕ) indicate that the location of the alkyl chain has a significant effect on the melting temperature (Figure 1.2). Also in this figure, the melting points of odd and even chained alkanes can be seen to alternate as they approach the convergence temperature. This is the result of differences in crystal packing. Both trends, however, suggest that the structure of the molecule is a determinant of crystal strength.

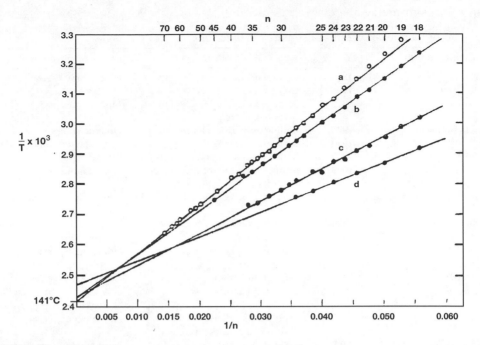

FIGURE 1.1
Reciprocal plot of temperature vs. number of chain carbons for the (a) paraffins, (b) fatty acid ethyl esters, (c) 1-alcohols, and (d) iso-alkyl acids. The intercept represents the convergence temperature.

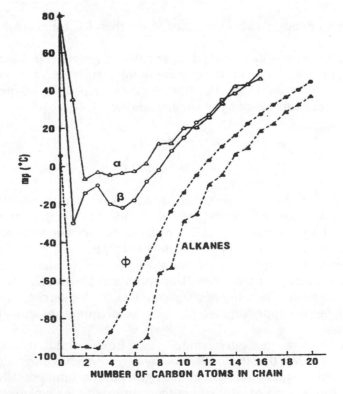

FIGURE 1.2
Melting point vs. chain length for α and β alkylnaphthalene, alkyl benzene, and paraffins.

The role of symmetry also was recognized quite early. Several investigators working with di-substituted benzenes noted that the para isomers have significantly higher melting points then the less symmetrical ortho and meta isomers (Carnelley, 1882; Beacall, 1928; Holler, 1947). It also was found that the melting points of long chain alkanes loaded with a single keto group are highest when the keto is on the first or central carbon (i.e., C1, or C9 of a C17 paraffin). That this is a function of symmetry is indicated by the lack of melting point increase when an 18 carbon chain is loaded with the keto at C9 or C10, neither being exactly central nor producing branches of equal length (Ubbelohde, 1938).

To date, no general theory of melting explains all observed phenomena. This is largely because different types of crystals — molecular, ionic, metallic, and macromolecular — give rise to different types of liquids. These liquids, or more specifically the melts (liquids near T_m), need to be characterized thoroughly in order to address the changes that occur during the transition. Since liquids and melts are not completely understood, current methods for estimating melting points are largely empirical and limited to specific types of compounds.

1.2.2 Use of Enthalpy and Entropy

The method this chapter presents is limited to organic compounds. While it is empirical, it makes use of the theoretical dependence of the enthalpy and entropy of melting upon structure.

A solid melts when its molecules obtain sufficient energy to completely destabilize the crystal lattice. When the transition temperature is reached, the free energies of the solid and liquid phases become equal. Therefore, no change in free energy is associated with the conversion from solid to liquid at the melting point, and the following equality can be written from the Gibbs equation:

$$\Delta G_m = 0 = \Delta H_m - T_m \Delta S_m \qquad (2)$$

where ΔG_m is the change in free energy upon melting, ΔH_m is the enthalpy of melting, T_m is the melting temperature, and ΔS_m is the entropy of melting. Rearrangement of Equation (2) gives an expression for the transition temperature in terms of enthalpy and entropy:

$$T_m = \frac{\Delta H_m}{\Delta S_m} \qquad (3)$$

This relationship identifies enthalpy and entropy as the thermodynamic properties that determine the value of the melting point. Because each of these properties shows a different dependence upon structure, attempts to estimate the melting point of a wide variety of organic compounds have been largely unsuccessful. The next two sections seek to correlate ΔH_m and ΔS_m with aspects of molecular structure and present methods for their estimation.

1.3 Estimating Enthalpy

Enthalpy (H) is a state variable and is defined as a compound's internal energy (U) plus the product of the system pressure (P) and volume (V) by:

$$H = U + PV \tag{4}$$

Due to the complexity of assigning absolute values of energy to a system, the *changes* in a system's enthalpy are reported rather than system enthalpy values themselves. The change in enthalpy over a phase transition at constant temperature is given by:

$$\Delta H = \Delta U + P\Delta V + V\Delta P \tag{5}$$

Since melting points are generally measured at conditions of 1.0 atm, ΔP is zero and Equation (5) reduces to:

$$\Delta H = \Delta U + P\Delta V \tag{6}$$

As a solid is heated to its melting point (T_m), the energy absorbed increases the kinetic energy of the molecules. That is, it increases their translational, rotational, and vibrational motions. This extra motion causes the molecules to be slightly farther apart from one another and thus at a higher potential energy; the latter being at an optimal minimum when the molecules are tightly packed and intermolecular attraction is maximized. When the melting point is reached, the moleculesof the solid can no longer absorb energy without causing disintegration of the crystal lattice. Therefore, when energy is added, the result is not an increase in temperature but an isothermal phase change. The enthalpy of melting is the energy required to convert the solid to liquid at the transition temperature. Its magnitude, according to Equation (6), reflects an increase in the internal energy plus the work of expansion ($P\Delta V$). The latter accounts for the increase in volume that usually occurs with melting. Note that the $P\Delta V$ contribution to the enthalpy is very small relative to the magnitude of ΔU.

This section addresses the factors that are responsible for the enthalpy of melting, i.e., the van der Waals and hydrogen bond interactions. It also provides the rationale for a group contribution scheme and presents the molecular descriptors with their group enthalpy values.

1.3.1 Hydrocarbons (Induced Dipole-Induced Dipole Forces)

For compounds composed solely of carbon and hydrogen atoms, ΔU is primarily the energy required to overcome the induced dipole-induced dipole, or dispersion interactions. The polarizability and ionization potential of the molecules determines the magnitude of dispersion interactions, and therefore of ΔH_m for hydrocarbons.

The ionization potential is the energy required to remove the most loosely held electron from the molecule. Generally, ionization potentials decrease with molecular size and the degree of unsaturation. However, they do not vary greatly for complex molecules and, to a first approximation, can be considered constant.

Polarizability is a measure of the ease with which a dipole can be induced in a molecule and is proportional to molecular volume. Since volume is a group additive function and ionization potentials are roughly constant, ΔU can be estimated by a group additive method. In addition, since $P\Delta V$ is very small compared to ΔU, the same group contributions can be used to estimate ΔH_m. This holds for the dispersion interactions in heteroatomic molecules, as well.

1.3.2 Compounds Containing Heteroatoms

For molecules containing atoms besides carbon and hydrogen, additional intermolecular forces contribute to the enthalpy of melting. These include, in order of increasing strength, dipole-induced dipole, dipole-dipole, and hydrogen bonding interactions. Ionic interactions will not be considered here, since the melting point prediction method considers only unionized species.

1.3.2.1 Dipole-Induced Dipole

Since heteroatoms are generally more electronegative than carbon, a carbon-heteroatom bond will have an electron density that is localized towards the heteroatom. The result is a permanent dipole that is analogous in its electronic properties to a bar magnet. This permanent dipole can induce a dipole in an otherwise neutral portion of another molecule in the same manner that a magnet induces a dipole in a piece of steel. The resulting attraction is referred to as a dipole-induced dipole, or Debye interaction.

The net strength of a molecule's Debye interactions is a function of its polarizability and the number and magnitude of its local dipole moments. Since induced dipoles tend to be aligned for maximum attraction, the energy of interaction resulting from Debye forces is very nearly additive.

1.3.2.2 Dipole-Dipole

A dipole-dipole interaction, or Keesom force, is analogous to the interaction between two magnets. For non-hydrogen bonding molecules with fixed dipoles, these interactions are likely to influence the orientation of the molecules in the crystal. This is because, unlike the Debye force which is always attractive, the interaction between two dipoles is only attractive if the dipoles are properly oriented with respect to one another, as is the case with magnets.

Keesom forces are a function of the number and magnitude of a molecule's local dipole moments, but since they are dependent upon the positioning of a molecule with respect to its neighbors, they may not always be strictly additive. However, since the molecules in most crystals are aligned for maximum dipolar interaction, the group interactions are often roughly additive.

1.3.2.3 Hydrogen Bonds

Many compounds containing oxygen, nitrogen, fluorine, and, to a lesser extent, phosphorus, sulfur, and chlorine, form hydrogen bonds. A hydrogen bond is the sharing of a hydrogen atom's electron by two highly electronegative atoms. This requires that the atoms entering into the bond include both a donor (A-H) and an acceptor (B). A weak hydrogen bond can also form between a donor atom and the electrons in a π orbital of an sp or sp^2 carbon.

The strength of a hydrogen bond will depend upon the electronegativities of the donor and acceptor atoms. Because they tend to be stronger than van der Waals forces, hydrogen bonds play a predominant role in the arrangement of molecules in a solid. However, if it is possible for hydrogen bonds to form within a molecule, then they may not reinforce the crystal lattice. Largely for this reason, hydrogen bonding cannot be accounted for accurately by a group contribution method. In order to account for intramolecular hydrogen bonding, non-constitutive descriptors, or correction factors, are used.

The strengths of the forces described above all depend upon intermolecular distance. Keesom forces are a function of $(distance)^{-3}$ and dispersion forces, $(distance)^{-6}$. Since hydrogen bonds require overlap of the electron clouds of H and B, they are even more sensitive to distance. Hydrogen bonds are also sensitive to the A-H-B angle, $180°$ being optimal.

1.3.3 Summation of the Intermolecular Forces

In addition to being additive, intermolecular forces can also be competitive. More specifically, the orientation of the molecules that is optimal to satisfy one type of interaction may not be ideal for another type. The result is a compromise amongst the individual attractive forces to achieve the lowest potential energy state for the crystal. Furthermore, ΔH_m is not the energy required to completely eliminate intermolecular attraction at T_m but the energy required to reduce the attraction to its level in the liquid that level being a function of intermolecular distance.

In light of these points, the enthalpy of melting is calculated from regression generated group values. A group value represents the average enthalpy of melting that results from the presence a molecular fragment in several compounds.

1.3.4 Molecular Descriptors

Calculation of a compound's enthalpy of melting can be accomplished by summing the appropriate group values, i.e.:

$$\Delta H_m = \Sigma n_i m_i \qquad (7)$$

where n_i is the number of times a particular group appears in a compound and m_i is the group's contribution to the enthalpy of melting. The molecular fragments (groups) used here are similar to those used by Simamora et al. (1994) and Krzyzaniak et al. (1995) for aromatic and non-hydrogen bonding aliphatic compounds, respectively. The m_i values are also similar but have been modified to reflect improved enthalpy of melting estimations. Note that the Y-m_i values for C, CH, and CH_2 were not determined by either of the above-mentioned authors. They were calculated using existing m_i values, literature melting points, and estimated entropies of melting. Table 1.2 presents the molecular fragments and their m_i values.

The m_i value of a particular molecular fragment is assigned according to the hybridization state of the fragment's immediate neighbors. Table 1.3 summarizes the five types of adjacent atom hybridization used. (See also Figure 1.3).

1.3.5 Assignment of m_i Values

The X-m_i value is used when a particular molecular fragment is attached only to sp^3 hybrid atoms. If the fragment instead is attached to a non-aromatic sp^2 (or sp) hybrid atom, the V-m_i value is used. Consider, for example, 2-pentene:

TABLE 1.2

Group Descriptors, Designations, and Regression Generated Values for Aliphatic, Non-Hydrogen Bonding Compounds.

Descriptor	m_i value (kJ/mol)				
	X	V	Y	YY	RYY
$-CH_3$	1.7	1.2	2.3	—	—
$-CH_2-$	2.2	1.3	−0.5	−4.6	2.0
$>CH-$	0.4	−0.5	−1.6		
$>C<$	0.1	−1.0	−3.1		
$=CH_2$	—	1.2	—	—	—
$-CH=$	2.2	2.2		—	1.7
$>C=$	1.5	1.5			0.1
$=C=$	—	1.2	—	—	—
$\equiv CH$	—	2.5	—	—	—
$-C\equiv$	1.6	1.6		—	—
$-C(=O)H$	7.1	6.4	7.0	—	—
$-C(=O)-$	6.6	4.7		2.0	3.8
$-C(=O)O-$	6.6			−0.7	8.1
$-C\equiv N$	5.9	4.3	7.4	—	—
$-C(=O)NH_2$			13.5	—	—
$-C(=O)OH$			13.2	—	—
$-F$	2.0	1.5	1.8	—	—
$-Cl$	4.0	2.2	3.1	—	—
$-Br$	4.7	3.0	3.5	—	—
$-I$	6.0	4.2	4.0	—	—
$-S-$	4.0			−3.5	2.4
$-SH$			7.0	—	—
$-SO$				0.0	6.9
$-SO_2$				1.0	9.8
$-S-C\equiv N$	8.6		4.1	—	—
$-O-$	3.2	2.7		−4.9	2.2
$-OH$			6.4	—	—
$-N$				—	3.1
$-NH$				−1.6	6.2
$-NH_2$			5.9	—	—
$-NO_2$	7.5		6.6	—	—
$-N=C=S$	9.2		7.9	—	—

Correction Factors	m_i Values (kJ/mol)	Correction Factors	m_i Values (kJ/mol)
GEM	−1.1	IHB-4	3.6
VIC	−0.2	HB-5	−1.0
RING3	−0.1	IHB-6	−1.6
RING4	−0.2	IHB-7	−2.3
RING5	−0.8		
RING6+	−0.7	CBIP	−2.0
CHbridge	1.6	OBIP	−1.1
Cbridge	6.0		

(— non-existent fragment)

$$CH_3-CH=CH-CH_2-CH_3$$
$$1 \quad 2 \quad 3 \quad 4 \quad 5$$

carbon 5 is assigned an X-m_i value while carbon 1 is assigned a V-m_i value. When an atom is joined to both sp^3 and sp^2 (or sp) atoms, it is classified according to the less saturated neighbor. Accordingly, carbons 2, 3, and 4 are given V-m_i values. If the fragment is attached

TABLE 1.3

Designation of Molecular Fragments.

m_i Designation	Description
X	Molecular fragment is attached to an sp^3 atom
V	Molecular fragment is attached to an aliphatic sp^2 or sp atom
Y	Molecular fragment is attached to an aromatic sp^2 carbon
YY	Molecular fragment is attached to two aromatic sp^2 carbons
RYY	Molecular fragment is attached to two aromatic sp^2 carbons and is part of a ring

to an sp^2 carbon that is part of an aromatic ring, it receives a Y-m_i value. Similarly, if it is attached to two sp^2 aromatic ring carbons, it receives a YY-m_i value. If the latter situation results in the fragment's inclusion in a ring, an RYY-m_i value is assigned. RYY is also the designation used for aromatic carbons and nitrogens.

1.3.6 Correction Factors

In addition to the m_i values for molecular fragments, there are 14 non-group additive m_i values or correction factors. These are employed to account for interactions between electronegative substituents, aliphatic ring and bridgehead carbons, intermolecular hydrogen bonding, aromatic carbons involved in sp^2 linkages, and biphenyl substituents which force aromatic rings out of a common plane. Table 1.4 summarizes the adjacent atom hybridizations and correction factors.

The correction factors for aliphatic compounds include geminal, vicinal, ring, and bridgehead parameters (Figure 1.3). The geminal parameter (GEM) is used to account for interactions arising from multiple electronegative substituents on a single atom. Its magnitude reflects one pair-wise interaction and when used as a correction, is multiplied by the number of such interactions. For example, 1,1,1-trichloroethane has three pair-wise interactions, which are accounted for by the addition of 3(GEM) to the sum of the group values. Similarly, the vicinal parameter (VIC) accounts for pair-wise interactions arising from electronegative substituents on adjacent atoms. Thus, perchloroethane would have a correction of 9(VIC) to account for the nine pair-wise interactions between the chlorines on adjacent carbons. As expected, the geminal parameter has a much larger (negative) value.

The RING parameters account for the fact that atoms in a ring interact differently than their non-ring counterparts. The numbers 3, 4, 5, and 6+ attached to the parameter represent the number of atoms which comprise a given ring. Note that the appropriate ring parameter is added for *each* non-bridgehead carbon in the ring. Bridgehead carbons require use of the Cbridge or CHbridge corrections.

For aromatic structures, CBIP is employed when an aromatic ring carbon is bound to an sp^2 atom of another aromatic ring. This correction factor is used in addition to the m_i value for the carbon. OBIP accounts for biphenyl substitution at the 2, 2', 6, and/or 6' positions. The IHB correction factor is used to account for intramolecular hydrogen bonding. IHB-5 through IHB-7 have negative values because intramolecular hydrogen bonding negates formation of the crystal strengthening intermolecular hydrogen bond. IHB-4, on the other hand, is positive due to the formation of reinforced intermolecular hydrogen bonds instead of a four-membered internally hydrogen bonded ring. Note that there is no correction for aromatic compounds having electronegative substituents in the ortho position.

As Table 1.2 shows, m_i values tend to increase with the ability of a group to form hydrogen bonds, its dipole moment, and its molecular size (polarizability). Thus, the amide and

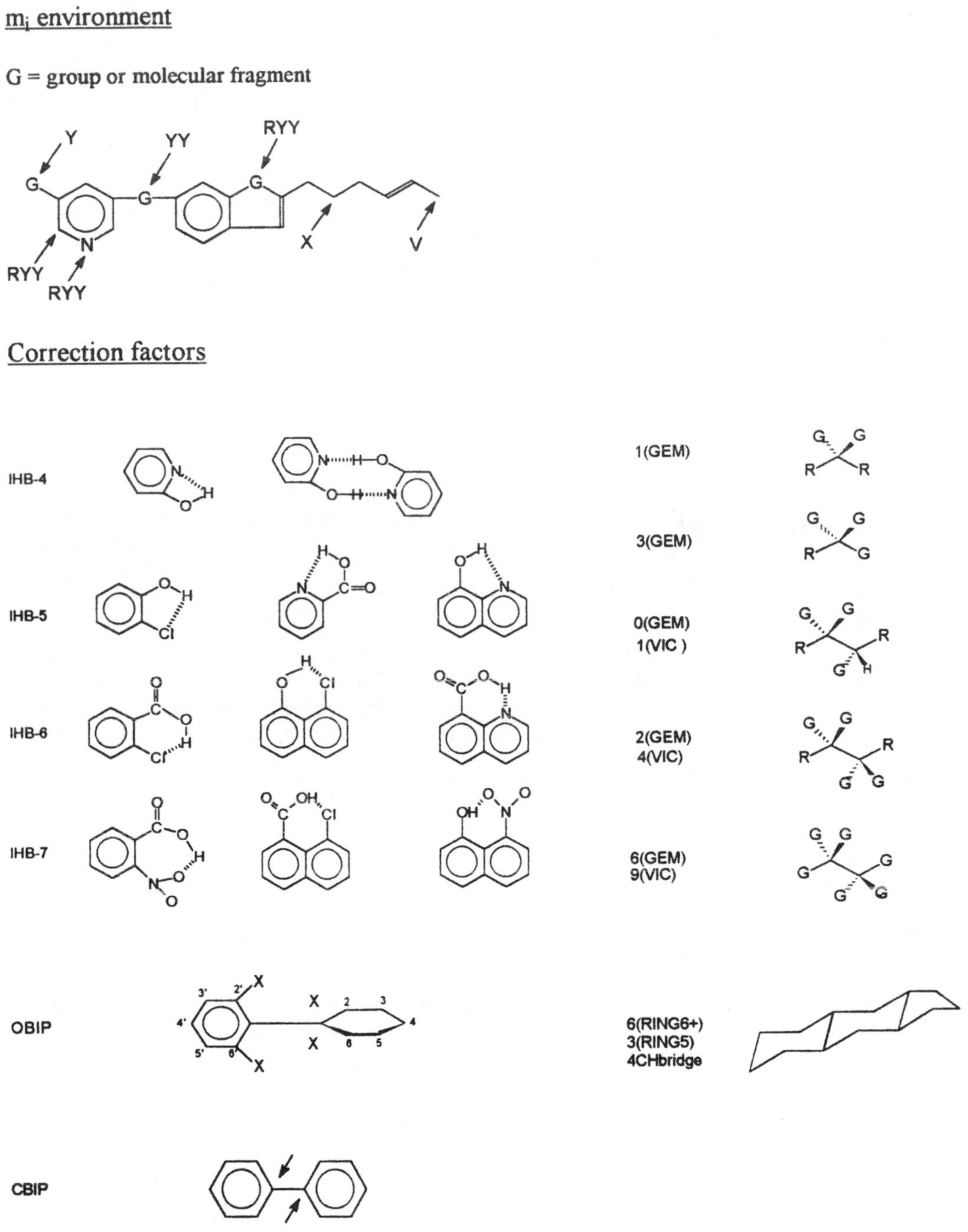

m$_i$ environment

G = group or molecular fragment

Correction factors

FIGURE 1.3
Designation of molecular fragments. (From Simamora et al., 1994)

carboxylic acid groups have the highest m$_i$ values, and of the non-hydrogen bonding groups, those containing nitrogen, sulfur, and oxygen have the highest m$_i$ values due to their larger dipole. For the halogens, which have similar dipole moments, the magnitude of m$_i$ increases with polarizability.

TABLE 1.4

Enthalpy of Melting Correction Factors.

Correction Factors	Description
GEM	Interactions between electronegative substituents on a single atom
VIC	Interactions between electronegative substituents on adjacent atoms
CHbridge	Aliphatic bridgehead carbon
Cbridge	Aliphatic bridgehead carbon
RING3	Aliphatic three-membered ring (added for each non-bridgehead carbon in ring)
RING4	Aliphatic four-membered ring (added for each non-bridgehead carbon in ring)
RING5	Aliphatic five-membered ring (added for each non-bridgehead carbon in ring)
RING6+	Aliphatic ring with six or more members (added for each non-bridgehead carbon in ring)
IHB-4	Four-membered ring intramolecular hydrogen bonding
IHB-5	Five-membered ring intramolecular hydrogen bonding
IHB-6	Six-membered ring intramolecular hydrogen bonding
IHB-7	Seven-membered ring intramolecular hydrogen bonding
CBIP	Aromatic carbon involved in sp^2 or sp linkage (used in addition to the RYY-m_i value)
OBIP	Substituent present in biphenyl at the 2, 2', 6, and/or 6' positions

1.3.7   Calculation of ΔH_m

For the calculation of ΔH_m, a molecule is first broken down into the molecular fragments from Table 1.2. Group enthalpy values are then selected from the X, V, Y, YY, and RYY columns, according to the fragments' neighbors. Finally, the appropriate correction factors are chosen and all of the m_i values summed according to Equation (7).

From Equation (7) and the values in Table 1.2, it is possible to estimate the enthalpy of melting for most pharmaceutically and environmentally relevant compounds. Although the data set used to generate these group values is quite extensive (over 2200 compounds to date), additional data is needed to obtain values for missing fragments.

1.4   Estimating Entropy

Entropy is a measure of the degree of randomness in a system. The change in entropy occurring with a phase transition is defined as the change in the system's enthalpy divided by its temperature. This thermodynamic definition, however, does not correlate entropy with molecular structure. For an interpretation of entropy at the molecular level, a statistical definition is useful. Boltzmann (1896) defined entropy in terms of the number of mechanical states that the atoms (or molecules) in a system can achieve. He combined the thermodynamic expression for a change in entropy with the expression for the distribution of energies in a system (i.e., the Boltzman distribution function). The result for one mole is:

$$S = R \ln W \tag{8}$$

where R is the molar gas constant and W represents the number of different states to which a component of a system has access. A complete derivation of this expression can be found in textbooks on physical chemistry.

By expressing entropy in terms of "accessible mechanical states," Equation (9) enables its quantification through a mechanical model. The change in entropy (ΔS_m) now can be calculated from Equation (8) by using the ratio of the number of accessible states before and after melting:

$$\Delta S_m = R \ln \frac{W_{liqid}}{W_{solid}} \qquad (9)$$

Note that since W represents all possible mechanical states, equations (8) and (9) can be interpreted as expressions of the probability of finding a molecule in a particular state.

1.4.1   Components of ΔS_m

Ubbelohde (1978) indicates that different, interrelated "modes" of entropy exist that can be treated separately and summed to give an estimate of ΔS_m. Yalkowsky (1979) recognizes four entropic components: 1) expansional—defined as the entropy resulting from translational motion; 2) positional—the entropy gained on going from an ordered crystal lattice to the disordered state of association in the liquid; 3) rotational—the entropy gained through increased rotational freedom; and 4) internal—the entropy that reflects the greater number of conformations possible for a flexible molecule in the liquid phase. Accordingly, the total entropy of fusion is expressed as:

$$\Delta S_m^{tot} = \Delta S_m^{exp} + \Delta S_m^{pos} + \Delta S_m^{rot} + \Delta S_m^{int} \qquad (10)$$

By correlating the individual components of ΔS_m with aspects of molecular structure, it is possible to estimate the entropy of fusion.

1.4.2   Rigid Compounds

For non-flexible organic compounds, only the first three entropic components must be considered. From Equation (9), the expansional entropy can be expressed as a ratio of the free volumes (V_f) to which the molecules in the solid and liquid phases have access:

$$\Delta S_m^{exp} = R \ln \frac{V_f^{liq}}{V_f^{sol}} \qquad (11)$$

The value of ΔS_m^{exp} depends on the shape of the molecule, because irregularly shaped molecules require more room to rotate then spherical molecules and because the free volume of the solid is related to packing density, which is largely a function of geometry. However, Yalkowsky (1979) showed that ΔS_m^{exp} varies only slightly for organic compounds, with values ranging from 4 to 12 J/K mol. This agrees with Bondi's (1968) statement that the packing density of spherical molecules differs by only about twenty percent from that of rod or plate-like molecules.

The positional entropy has been calculated (Hirschfelder et al., 1937; Lennard-Jones and Devonshire, 1939) to be approximately 8-12 J/K mol for most organic molecules. This number results from a mathematical expression of the probability of finding the centers of a particular number of liquid molecules arranged as they are in the solid.

The rotational entropy has been estimated to lie roughly between 25 and 40 J/K mol for rigid molecules. Yalkowsky (1979) offered an intuitive explanation for this range by comparing the three dimensional arc, through which a molecule in a solid lattice can librate or "wobble" (Figure 1.4) to the sphere through which a liquid molecule rotates.

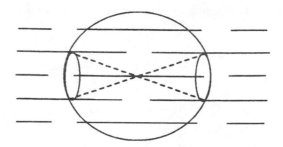

FIGURE 1.4
Area traced out by the allowed movements of a molecule in a crystal lattice.

Using Equation (9) with W_{liquid} and W_{solid} representing the area that would be traced out by maximum rotation of the liquid and solid molecules, he showed that a liquid molecule sweeps out 133 times as much area as a solid molecule that moves in a 10° arc. This gives an entropy of about R ln 133, or 40 J/K mol when the solid melts. Similarly, for a solid molecule which can rotate through a 20° arc, the rotational entropy gain upon becoming liquid is about 30 J/K mol.

An estimate of the total entropy of melting of rigid molecules is obtained by summing the averaged values of expansional, positional, and rotational entropies. For rigid molecules, this is 50 J/K mol. Note that this is consistent with the observation of Walden (1908) that the entropy of fusion for "coal-tar" derivatives is approximately constant at 56 J/K mol. Dannenfelser et al. (1993) have shown that estimates of ΔS_m for rigid molecules can be improved upon by adjusting the value of 50 J/K mol to account for differences in structure. Specifically, this is done by adjusting the rotational component of entropy.

As shown above, rotational entropy reflects differences in the extent to which the solid and liquid molecules can rotate. If the position of the molecule in the solid lattice is designated as a reference position, then it is possible to indicate the number of ways that a molecule in the liquid can be rotated to produce an indistinguishable position. For example, anthracene can be rigidly rotated into four to produce indistinguishable positions:

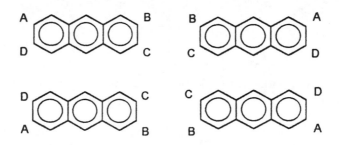

while phenanthrene can be rotated into only two indistinguishable positions:

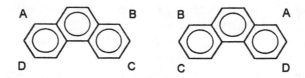

If it is assumed that being in the same orientation as the molecules of a crystal promotes incorporation into the lattice, then the value of ΔS_m^{rot} can be accounted for more accurately by subtracting from it the probability that a liquid molecule will be found in such a position, i.e.;

$$R \ln \sigma \tag{12}$$

where σ is defined by Dannenfelser et al. (1993) as the rotational symmetry number and indicates "the number of indistinguishable positions that can be obtained by rigidly rotating the molecule about its center of mass." This gives a corrected rotational entropy of:

$$\Delta S_m^{rot} = 32 - R \ln \sigma \tag{13}$$

where 32 (J/K mol) represents the rotational entropy of an unsymmetrical molecule. In calculating σ, it is assumed that groups such as CH_3, NH_2, and OH are freely and rapidly rotating so that they effectively have radial symmetry. This allows them to be treated as single atoms.

Figure 1.5 provides examples of σ for some simple molecules. For molecules which are conical (hydrogen cyanide and chloromethane), cylindrical (carbon dioxide and ethane), or spherical (neon, methane), empirically generated σ values of 10, 20, and 100 are assigned. For flexible molecules, the value of σ is unity because they cannot be rigidly rotated. Note that the σ values for spherical, cylindrical, and conical molecules are different from the values used by crystallographers.

Combining Equation (13) with the averaged values for expansional and positional entropy results in the following expression for the entropy of melting (J/K mol) for rigid molecules:

$$\Delta S_m^{tot} = 50 - R \ln \sigma \tag{14}$$

1.4.3 Flexible Molecules

For molecules whose atoms are able to assume multiple arrangements, the internal or conformational entropy component becomes a factor in calculations of ΔS_m^{tot}. This is because the molecules are restricted to a single conformation in the solid but can adopt a variety of conformations in the liquid. In the crystal of a long chain fatty acid, for example, the molecules are all in the most energetically stable, all-trans (or anti) conformation. In the liquid, however, they have sufficient energy to twist through many of the less stable conformations, including partially coiled chains. Consequently, flexible molecules have access to more mechanical (i.e., conformational) states in the liquid and experience a gain in entropy when they melt.

There are (at least) two factors which must be considered when attempting to correlate internal entropy with molecular structure: 1) for molecules with a rigid loading (or head) group, the existence of a critical length below which the chain does not contribute significantly to entropy; and 2) how to weight the more likely (i.e., more stable) conformations. Several investigators (Ubbelohde, 1965; Breusch, 1969; and Yalkowsky, 1972), have addressed the first factor the conclusion being that the chain must contain four to six atoms in order to contribute to the internal entropy. This is because shorter chains are restricted in their conformation by interactions among the rigid portions of the molecules. For purposes of calculating entropy, Yalkowsky indicates that molecules with chains less than 4-6

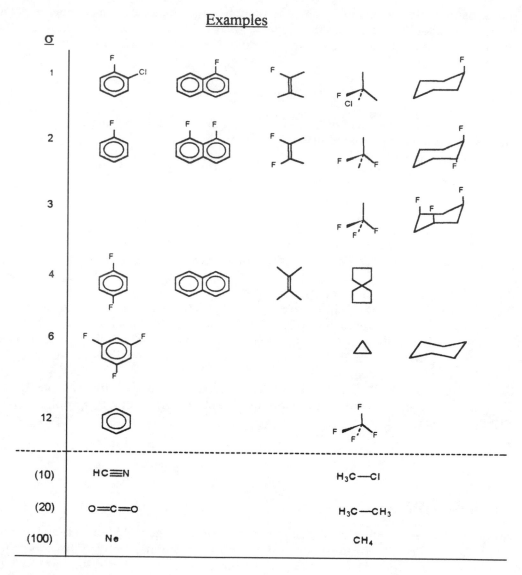

FIGURE 1.5
Rotational symmetry numbers for some simple molecules. (From Dannenfelser et al., 1996)

atoms long can be considered rigid. Note that a molecule's interactive forces determine the actual critical length.

This critical length was incorporated into a generalized parameter that could be used for all flexible molecules. Dannenfelser et al. (1995) define a variable (τ) as the number of effective torsional angles in a molecule. For a series of three bonds, a torsional angle is created by the third bond and the plane of the first two. This is analogous to the angle created by the minute and hour hands of a clock, if the connecting rod assumes the role of the middle "bond." The "effective" torsional angle extends this concept to the rigid and semi-rigid portions of a molecule by expressing their contributions to molecular flexibility as fractions of the flexibility of a normal alkane torsional angle.

The number of torsional angles is calculated by subtracting unity from the number of chain atoms (chain atoms do not include end groups such as the methyl, halogens, carbonyl oxygens which are radially symmetrical, or *tert*-butyl groups). Therefore, for saturated linear molecules, τ = SP3-1, where SP3 is the number of sp^3 atoms in the chain. Note that C, CH, CH_2, NH, N, O, and S are included as SP3.

Since there is less rotation about a double bond than a single bond, sp^2 hybridized atoms contribute less to flexibility. Dannenfelser et al. determined empirically that two sp^2 atoms result in about the same degree of flexibility as one sp^3 atom. Therefore, the number of sp^2 atoms in a molecule is multiplied by 0.5 to quantify its contribution to molecular flexibility. Note that the nitrogen, sulfur, and carbon atoms of nitro, sulfonyl, and carboxyl groups are sp^2 atoms. The sp hybrid atoms are not included in calculations of τ since they do not contribute to flexibility. In addition, all rings and fused ring systems are treated as single groups and multiplied by 0.5.

Thus, the number of effective torsional angles for any molecule is defined as:

$$\tau = SP3 + 0.5SP2 + 0.5RING - 1 \qquad (15)$$

where SP3 is the number of sp^3 chain atoms, SP2 is the number of sp^2 chain atoms and RING is the number of fused ring systems. Note that if a negative value of τ is calculated, a value of "0" is used.

Tsakanikas et al. (1988) addressed the second factor, the weight of the more probable conformations. They recognized as equivalent the three most stable conformations (i.e., the anti and two gauche) and raised this number to a power equal to the number of torsional angles in the molecule, the number of torsional angles being defined at the time only for linear alkanes. Thus, they defined a molecular flexibility number, $\phi = 3^{(n-3)}$, where n is the number of chain atoms and, since four consecutive methylene carbons defines a torsional angle, n-3 equals the number of torsional angles. Dannenfelser et al. (1995) extended this molecular flexibility number to all molecules by using their effective torsional angle parameter, τ, and accounting for the greater stability of the anticonfiguration vs. the gauche. To do this they used a value of 2.85 which is based upon an equation developed by Temperly (1956). Dannenfelser defined the molecular flexibility number as:

$$\phi = 2.85^{\tau} \qquad (16)$$

where τ is defined by Equation (15). This parameter represents the number of unique, energetically favorable conformations accessible to the molecule in the liquid. Insertion of Equation (16) into Equation (9), with the number of conformations in the solid taken to be unity, gives an expression for the internal component of the entropy of melting:

$$\Delta S_m^{int} = R \ln \phi \qquad (17)$$

Combining Equation (17) with Equation (14) gives the general equation developed by Dannenfelser et al. for estimating the entropy of fusion (J/K mol):

$$\Delta S_m^{tot} = 50 - R \ln \sigma + R \ln \phi \qquad (18)$$

Note that since σ is equal to unity for all flexible molecules and ϕ is equal to unity for all rigid molecules, one of the terms in Equation (18) will always equal zero. The two extremes of Equation (18) describe small, spherical (non-flexible) molecules and large, flexible (non-symmetrical) polymers.

1.4.4   Small Spherical Molecules

For the elements and small, roughly spherical molecules, the entropy of melting is primarily due to expansional and positional entropy. This is because no rotational entropy is gained upon melting and there is no internal entropy contribution. Using Equation (19), with a σ value of 100:

$$\Delta S_m^{tot} = 50 - R\ln(100) \qquad (19)$$

gives an estimate of 12 J/K mol for the entropy of melting. This value is in good agreement with Richards' rule that the entropy of fusion for these molecules is 10.5 J/K mol.

1.4.5   Polymers and Homologs

As the length of a polymer chain increases, τ becomes very large. At extreme chain lengths, the total entropy of fusion is almost entirely due to the internal component. Note that since the value of 50 J/K mol in Equation (19) becomes insignificant, it is possible to estimate the entropy associated with a single monomeric unit using Equation (20).

$$\Delta S_m^{tot} \rightarrow R\ln 2.85^{\tau} \qquad (20)$$

1.4.6   Hydrogen Bonding

Myrdal (1996) showed that hydrogen bonding must be accounted for in calculations of the entropy of boiling. Since hydrogen bonding is somewhat rare in gases, this finding indicates that there is some degree of association due to hydrogen bonds in the liquid phase. Bondi (1968) also concluded this. Since hydrogen bonding is known to be influential in the crystal lattice, the question arises as to whether differences in the extent of hydrogen bonding between the solid and liquid phases contribute to a gain in entropy upon melting. Yalkowsky (1979) presented strong evidence that this is not the case. He demonstrated that the entropies of melting for compounds having one or two hydrogen bonding groups did not differ significantly from those of non hydrogen bonding compounds. Therefore, no hydrogen bonding parameter is required to estimate ΔS_m^{tot}.

1.4.7   An Alternative Method for Estimating the Entropy of Melting

Recent publications by Chickos et al. (1990, 1991) present a group contribution scheme for the estimation of the entropy of fusion. They define three classes of compounds:

acyclic and aromatic molecules

$$\Delta S_m = \sum_i n_i C_i G_i + \sum_j n_j C_j G_j + \sum_k n_k C_K G_k \qquad (21)$$

cyclic molecules

$$\Delta S_m = [8.41 + 1.025(n-3)] + \sum_i n_i C_i G_i + \sum_j n_j C_j G_j + \sum_k n_k C_K G_k \qquad (22)$$

polycyclic molecules

$$\Delta S_m = \left[8.41N + 1.025(R - 3N)\right] + \sum_i n_i C_i G_i + \sum_j n_j C_j G_j + \sum_k n_k C_K G_k \tag{23}$$

where n_i is the number of identical groups in the molecule, C_i is a group contribution modifier, G_i is a molecular fragment or group, the subscript i denotes the hydrocarbon components, j denotes the carbons bearing functional group, and k denotes the functional group(s), and $K = \sum_k n_k$. For a complete listing of the individual groups, see the original research article.

Although complex, the method of Chickos et al. predicts the entropy of melting with approximately the same degree of accuracy as the method of Dannenfelser et al.; both having average errors of roughly 11 J/K mol (the former using a data set of 649 compounds and the latter, 934 compounds). For reasons of practicality, however, sample calculations of T_m using only the method of Dannenfelser et al., are provided here.

1.5 Calculating T_m

1.5.1 Recommended Method

Equation (3) can be used to calculate the melting point for most organic compounds, using the group contribution scheme based on the combined works of Simamora et al. (1994) and Krzyzaniak et al. (1995) to estimate ΔH_m and the method of Dannenfelser et al. (1996) to estimate ΔS_m:

$$T_m = \frac{\Delta H_m}{\Delta S_m^{tot}} = \frac{\sum n_i m_i}{50 - R \ln \sigma + R \ln \phi} \tag{24}$$

Since combining Equations (15) and (16) gives:

$$\phi = 2.85^{(SP3 + 0.5SP2 + 0.5RING - 1)} \tag{25}$$

Equation (24) becomes:

$$T_m = \frac{\Delta H_m}{\Delta S_m^{tot}} = \frac{\sum n_i m_i}{50 - 8.31 \ln \sigma + 8.71(SP3 + 0.5SP2 + 0.5RING - 1)} \tag{26}$$

which expresses the melting point in terms of both additive and non-additive molecular descriptors. Appendix I illustrates the application of Equation (26) to the benchmark chemicals.

TABLE 1.5

Application to Isomers.

Compound	Predicted T_m (K)	Measured T_m (K)	% Error
o-Nitroaniline	358	343[a]	4.4
m-Nitroaniline	390	387[a]	0.8
p-Nitroaniline	443	419[a]	5.7
o-Chlorotoluene	248	237[a]	4.6
m-Chlorotoluene	248	224[a]	10.7
p-Chlorotoluene	282	280[a]	0.7
1,2,3-Trichlorobenzene	334	329[a]	1.5
1,2,4-Trichlorobenzene	294	290[a]	1.4
1,3,5-Trichlorobenzene	420	336[a]	25.0
1,2,3-Triiodobenzene	395	389[b]	1.5
1,2,4-Triiodobenzene	348	365[b]	4.7
1,3,5-Triiodobenzene	497	457[b]	8.8
1,2:6,7 Dibenzanthracene	492	537[b]	8.4
1,2:7,8 Dibenzanthracene	556	471[b]	18.0
2,3:6,7 Dibenzanthracene	639	544[b]	17.5
1,2:6,7 Dibenzophenanthrene	492	567[b]	13.2
2,3:6,7 Dibenzophenanthrene	556	530[b]	4.9
1,2:7,8 Dibenzophenanthrene	556	641[b]	13.3
Hexylbenzene	200	211[b]	5.2
p-Diisopropylbenzene	168	254[b]	33.9
Hexamethylbenzene	491	439[b]	11.8
2-Methyl-1-butene	117	135[b]	13.3
3-Methyl-1-butene	107	104[b]	2.9
1,4-Hexadiyne	196	<193[b]	>1.5
1,5-Hexadiyne	183	267[b]	31.5
2,4-Hexadiyne	351	342[b]	2.6
n-Octane	177	216[b]	18.1
2-Methylheptane	188	164[b]	14.6
2,2,3-Trimethylpentane	191	161[b]	18.6
2,2,3,3-Tetramethylbutane	296	374[b]	20.9
Bicyclo[2,2,2]-octane	370	443[b]	16.5

Values obtained from (a) the *Merck Index* and (b) the *CRC Handbook of Chemistry and Physics*.

A significant advantage of this method is that generally it can be used to predict the rank order of melting points for a series of isomers. Table 1.5 demonstrates this for several aromatic and aliphatic compounds.

The method does exceptionally well for most of the substituted benzenes in Table 1.5. 1,3,5-Trichlorobenzene is an exception. The high symmetry of this molecule leads to a low entropy of melting estimate and a relatively high predicted melting point. That this compound's melting point is overestimated by 25% is likely a function of a looser crystal structure than would be expected for a molecule as symmetrical as this. Crystal packing also is believed to be responsible for the differences between predicted and measured values for the conjugated ring systems above, as there appears no definite correlation between symmetry and melting point among the isomers.

Inaccuracies also arise in melting point estimations for the alkyl substituted aromatic compounds when these molecules display some degree of symmetry. If, for example, the isopropyl groups of p-diisopropylbenzene are considered freely rotating in the crystal, then a four-fold axis of symmetry is indicated. The entropy calculation based upon this symmetry classification results in a melting point error of 8%. Since it is flexible, however, this molecule is treated as unsymmetrical by the current method. The result is a high predicted entropy and a melting point that is underestimated by approximately 34%. Similarly, the melting point of n octane is underestimated because of the relatively stable crystal that results from the efficient packing of the symmetrical anti (all *trans*) form.

In spite of these particular limitations, this method is unique in both its ability to differentiate among many of the isomeric compounds mentioned above and in its level of accuracy. Using the data sets mentioned earlier, for example, Simamora et al. (1994) and Krzyzaniak et al. (1995) report the average error associated with melting point estimation for data ranging over several hundred K to be roughly 35 K. Since Equation (24) incorporates an improved method for the estimation of melting entropy, it is expected to have an even lower error.

1.5.2 An Alternate Method

For compounds which cannot be addressed with Equation (26), e.g., aliphatic hydrogen bonding compounds, we recommend the group contribution scheme of Joback and Reid (1987) (see Appendix 2). Joback and Reid use regression generated group contributions for the direct estimation of T_m and several other properties as part of a "unified approach" to physical property estimation. They were amongst the first to take this approach to melting point estimation and introduced not only an extremely practical method, but one with an unprecedented range of applicability. However, since Joback and Reid's method does not account for the effect of ΔS_m on the melting temperature, it is less accurate than the current method and should only be used to complement the latter when its range is exceeded.

1.6 Conclusion

The estimation method this chapter presents uses a group contribution scheme to calculate the enthalpy of melting and a non-group additive method to calculate the entropy of melting. The former addresses the effects of hydrogen bonding for aromatic compounds and the interactions arising from multiple electron withdrawing substituents. The method for estimating entropy uses a statistical equation to account for molecular symmetry and molecular flexibility. The melting point is ultimately calculated from the ratio of ΔH_m to ΔS_m as defined by the Gibbs free energy equation for equilibrium conditions.

This method can be used to estimate melting points for a very wide range of organic compounds with a relatively high degree of accuracy. It has an advantage over simple group contribution schemes in that it can distinguish among isomers and indicates a convergence temperature for both polymeric compounds and alkyl homologs. It is also practical because it is simple to use and requires only that a compound's structure be known.

1.7   Appendix 1: Application to Benchmark Chemicals

Anthracene

$\Delta H_m = \sum n_i m_i$
    $= 10(RYY\text{-}CH) + 4(RYY\text{-}C)$
    $= 10(1700) + 4(100)$
    $= 17400$ J/mol

$\Delta S_m = 50 - R\ln \sigma + R\ln \phi$
    $= 50 - 8.31\ln 4 + 8.31\ln 2.85^{(U)}$
    $= 50 - 12$
    $= 38$ J/K mol

$T_m = \Delta H_m / \Delta S_m$

$= \dfrac{17400 \ \ J \ / \ mol}{38 \ \ J \ / \ K \ \ mol}$

$= 458$ K   (491 K Merck Index)

Phenanthrene

$\Delta H_m = \sum n_i m_i$
    $= 10(RYY\text{-}CH) + 4(RYY\text{-}C)$
    $= 10(1700) + 4(100)$
    $= 17400$ J/mol

$\Delta S_m = 50 - R\ln \sigma + R\ln \phi$
    $= 50 - 8.31\ln 2 + 8.31\ln 2.85^{(U)}$
    $= 50 - 6$
    $= 44$ J/K mol

$T_m = \Delta H_m / \Delta S_m$

$= \dfrac{17400 \ \ J \ / \ mol}{44 \ \ J \ / \ K \ \ mol}$

$= 395$ K   (373 K Merck Index)

1α, 2α, 3β, 4α, 5α, 6β-hexachlorocyclohexane (Lindane)

$\Delta H_m = \sum n_i m_i$
    $= 6(X{>}CH\text{-}) + 6(X\text{-}Cl) + 6(RING6+) + 6(VIC)$
    $= 6(400) + 6(4000) + 6(-700) + 6(-200)$
    $= 21000$ J/mol

$\Delta S_m = 50 - R\ln \sigma + R\ln \phi$
    $= 50 - 8.31\ln 1 + 8.31\ln 2.85^{(U)}$
    $= 50$ J/K mol

$T_m = \Delta H_m / \Delta S_m$

$= \dfrac{21000 \ \ J \ / \ mol}{50 \ \ J \ / \ K \ \ mol}$

$= 420$ K   (385 K Merck Index)

Trichloroethylene

$\Delta H_m = \sum n_i m_i$
    $= 2(V{>}C{=}) + 3(V\text{-}Cl) + 1(GEM) + 2(VIC)$
    $= 2(1500) + 3(2200) + 1(-1100) + 2(-200)$
    $= 8100$ J/mol

$\Delta S_m = 50 - R\ln \sigma + R\ln \phi$
    $= 50 - 8.31\ln 1 + 8.31\ln 2.85^{(U)}$
    $= 50$ J/K mol

$T_m = \Delta H_m / \Delta S_m$

$= \dfrac{8100 \ \ J \ / \ mol}{50 \ \ J \ / \ K \ \ mol}$

$= 162$ K   (188 K Merck Index)

2, 6 - di (*tert*-butyl) phenol

$\Delta H_m = \sum n_i m_i$
    $= 6(X\text{-}CH_3) + 2(Y{>}C{<}) + 3(RYY\text{-}CH)$
       $+ 3(RYY\text{-}C) + 1(Y\text{-}OH)$
    $= 6(1700) + 2(-3100) + 3(1700)$
       $+ 3(100) + 1(6400)$
    $= 15800$ J/mol

$\Delta S_m = 50 - R\ln \sigma + R\ln \phi$
    $= 50 - 8.31\ln 1 + 8.31\ln 2.85^{(U)}$
    $= 50$ J/K mol

$T_m = \Delta H_m / \Delta S_m$

$= \dfrac{15800 \ \ J \ / \ mol}{50 \ \ J \ / \ K \ \ mol}$

$= 316$ K   (315 K CRC Handbook)

Adamantane

$\Delta H_m = \sum n_i m_i$
    $= 6(X\text{-}CH_2) + 4(X\text{-}CH) + 6(RING6+)$
       $+ 4(CHbridge)$
    $= 6(2200) + 4(100) + 6(-700)$
       $+ 4(1600)$
    $= 15800$ J/mol

$\Delta S_m = 50 - R\ln \sigma + R\ln \phi$
    $= 50 - 8.31\ln 12 + 8.31\ln 2.85^{(U)}$
    $= 29$ J/K mol

$T_m = \Delta H_m / \Delta S_m$

$= \dfrac{15800 \ \ J \ / \ mol}{29 \ \ J \ / \ K \ \ mol}$

$= 539$ K   (543 K CRC Handbook)

1.8   Appendix 2: Application of an Alternate Method to Benchmark Chemicals

Method of Joback and Reid (1987)

For the estimation of melting point (K), Joback and Reid give the following equation:

$$T_m = 122.5 + \sum n_i g_i \tag{27}$$

where n_i is the number of times a particular group appears in a molecule and g_i represents the group contribution. Table 1.6 presents the molecular fragments and their group values.

TABLE 1.6

Molecular fragments and Their Group Contribution Values for the Estimation of Melting Point (K).

Molecular Fragment	g_i	Molecular Fragment	g_i
Non-ring increments		*Oxygen increments*	
–CH$_3$	–5.10	–OH (alcohol)	44.45
–CH$_2$–	11.27	–OH (phenol)	82.83
>CH–	12.64	–O– (non ring)	22.23
>C<	46.43	–O– (ring)	23.05
=CH$_2$	–4.32	>C=O (non ring)	61.20
–CH=	8.73	>C=O (ring)	75.97
>C =	11.14	–CH=O (aldehyde)	36.90
=C=	17.78	–COOH (acid)	155.50
≡CH	–11.18	–COO– (ester)	53.60
–C≡	64.32	=O (except as above)	2.08
Ring increments		*Nitrogen increments*	
–CH$_2$–	7.75	–NH$_2$	66.89
>CH–	19.88	>NH (non ring)	52.66
>C<	60.15	>NH (ring)	101.51
–CH=	8.13	>N– (non ring)	48.84
>C =	37.02	N= (non ring)	
		N= (ring)	68.40
Halogen increments		=NH	68.91
–F	–15.78	–CN	59.89
–Cl	13.55	–NO$_2$	127.24
–Br	43.43		
–I	41.69		
Sulfur increments			
–SH–	20.09		
–S– (non ring)	34.40		
–S– (ring)	79.93		

References

Austin, J. 1930. A Relation Between the Molecular Weights and Melting Points of Organic Compounds. *J. Am. Chem. Soc.* 52, 1049.

Beacall, T. 1928. The Melting Points of Benzene Derivatives. *Recl. Trav. Chim. Pay-Bas* 47, 37–44.

Benko, J. 1959. New Relationships Between the Physical Constants of Organic Compounds. *Acta. Chim. Hung.* 21, 351.

Bondi, A. 1968. Physical Properties of Molecular Crystals, Liquids, and Glasses. John Wiley and Sons: New York, 139–187.

Budavari, S., Ed. 1996. *The Merck Index, 12th edition*. Merck and Co., Inc.: Rahway, N.J.

Carnelley, T. 1882. Chemical Symmetry, or the Influence of Atomic Arrangement on the Physical Properties of Compounds. *Philos. Mag.* 13, 112–130.

Chickos, J. and D. Hesse. 1990. Estimating Entropies and Enthalpies of fusion of Hydrocarbons. *J. Org. Chem.* 55, 12, 3833–3840.

Chickos, J. and D. Hesse. 1991. Estimating Entropies and Enthalpies of fusion of Organic Compounds. *J. Org. Chem.* 56, 3, 927–938.

Dannenfelser, R.M., N. Surendran, and S. Yalkowsky. 1993. Molecular Symmetry and Related Properties. *SAR and QSAR in Environmental Research*. 1, 273–292.

Dannenfelser, R.M. and S. Yalkowsky. 1996. Estimation of Entropy of Melting from Molecular Structure: A Non-Group Contribution Method. *Ind. Eng. Chem. Res.* 35, 1483–1487.

Dearden, J. and M. Rahman. 1988. QSAR Approach to the Prediction of Melting Points of Substituted Anilines. *Mathl. Comput. Modelling.* 11, 843.

Dearden, J. 1991. The QSAR Prediction of Melting Point, a Property of Environmental Relevance. *Sci. Total Environ.* 109/110, 59.

Dearden, J.C. 1999. The prediction of melting point. In *Advances in Quantitative Structure–Property Relationships*, vol. 2, Charton, M. and B.I. Charton, Eds. JAI Press, Stamford, CT. pp. 127–175.

Eaton, E. 1971. Correlation and Prediction of Melting Points. *Chem. Technol.* 362.

Flory, P. and A. Vrij. 1963. Melting Points of Linear-Chain Homologs. The Normal Paraffin Hydrocarbons. *J. Am. Chem. Soc.* 85, 3548.

Grigor'ev, S. and V. Pospelov. 1965. Melting Point Rules for Organic Compounds. *Sb. Nauchn. Tr., Ukr. Nauchn. – Issled. Uglekhim. Inst.* No. 16, 153.

Hanson, M. and D. Rouvray. 1987. The Use of Topological Indices to Estimate the Melting Points of Organic Molecules. In *Graph Theory and Topology in Chemistry*; King, R., Rouvray, D., Eds.; Elsevier, Amsterdam, 201.

Hirschfelder, J., D. Stevenson, and H. Eyring. 1937. A Theory of Liquid Structure. *J. Chem. Phys.* Vol 5, 896.

Holler, A. 1947. An Observation on the Relation between the Melting Points of the Disubstitution Isomers of Benzene and Their Chemical Constitution. *J. Org. Chem.* 13, 70–74.

Joback, K. and R. Reid. 1987. Estimation of Pure-Component Properties From Group-Contributions. *Chem. Eng. Comm.* Vol. 57, 233–243.

Keyes, R. 1959. High-Temperature Thermal Conductivity of Insulating Crystals: Relationship to the Melting Point. *Phys. Rev.* 115, 4, 564–567.

Krzyzaniak, J., P. Myrdal, P. Simamora, and S. Yalkowsky. 1995. Boiling Point and Melting Point Prediction for Aliphatic, Non-Hydrogen-Bonding Compounds. *Ind. Eng. Chem. Res.* 37, 7, 2530–2535.

Lennard-Jones, J., and A. Devonshire. 1939. Critical and Co-operative Phenomena: IV. A Theory of Disorder in Solids and Liquids and the Process of Melting. *Proc. R. Soc. London, Ser. A*, Vol. 170, 464.

Lindemann, F. 1910. On the Calculation of Molecular Eigen-frequencies. *Physik. Z.* 11, 609.

Longinescu, G. 1903. On the Polymerization of Organic Compounds in the Solid State. *J. Chim. Phys.* 1, 296.

Mackay, D., W. Shiu, A. Bobra, J. Billington, E. Chan, A. Yeun, C. Ng, and F. Szeto. 1982. Volatilization of Organic Pollutants from Water. *U.S. Environmental Agency Report PB 82-230939.* Athens, Georgia.

Mills, E. 1884. On Melting-Point and Boiling-Point as Related to Chemical Composition. *Phil. Mag.* 17, 173.

Myrdal, P. and S. Yalkowsky. 1994. A Simple Scheme for Calculating Aqueous Solubility, Vapor Pressure and Henry's Law Constant: Application to the Chlorobenzenes. *SAR and QSAR in Environmental Research.* 2, 17–28.

Prud'homme, M. 1920. Rule of Three Temperatures. *J. Chim. Phys.* 18, 359.

Przezdziecki, J. and T. Sridhar. 1985. Prediction of Liquid Viscosities. *AIChE Journal.* 31, 2, 333–335.

Simamora, P. and S. Yalkowsky. 1994. Group Contribution Methods for Prediction the Melting Points and Boiling Points of aromatic Compounds. *Ind. Eng. Chem. Res.* 37, 1404–1409.

Syunyaeva, R. 1981. Relationship Between Molecular Structure and Physicochemical Properties of n-Alkanes. *Chem. Technol. Fuels Oils.* 17, 161.

Temperley, H. 1956. Residual Entropy of Linear Polymers. *Journal of Research of the National Bureau of Standards.* 56, 2, 55–66.

Tsakinikas, P. and S. Yalkowsky. 1988. Estimation of Melting Point of Flexible Molecules: Aliphatic Hydrocarbons. *Toxicological and Environmental Chemistry.* 17, 19–33.

Ubbelohde, A. 1938. *The Molten State of Matter.* John Wiley and Sons, New York, 1978.

Yalkowsky, S. 1979. Estimation of Entropies of Fusion of Organic Compounds. *Ind. Eng. Chem. Fundam.* 18, 2, 108–111.

Yalkowsky, S., and S. Valvani. 1980. Solubility and Partitioning I: Solubility of Nonelectrolytes in water. *J. Pharm. Sci.* 69, 8, 4869–4873.

Wachalewski, T. 1970. A Relation Between the Melting Point of a Substance and its Chemical Composition and Configuration. *Postepy Fiz.* 21, 403.

Walters, A., P. Myrdal, and S. Yalkowsky. 1995. A Method for Estimating the Boiling Points of Organic Compounds From Their Melting Points. *Chemosphere.* 31, 4, 3001–3008.

Weast, R.C., Ed. 1972–73. *CRC Handbook of Chemistry and Physics, 53rd edition.* The Chemical Rubber Co., Cleveland, OH.

2

Boiling Point

Warren J. Lyman

CONTENTS

2.1 Introduction

The boiling point is defined as "the temperature at which a liquid's vapor pressure equals the pressure of the atmosphere on the liquid." If the pressure is exactly 1 atmosphere (101,325 Pa), the temperature is referred to as "the normal boiling point." Pure chemicals have a unique boiling point, and this fact can be used in some laboratory investigations to check on the identity and/or purity of a material. Mixtures of two or more compounds have a boiling point range. This chapter focuses on the estimation of boiling points for pure compounds only.

For organic compounds, boiling points range from –162°C for methane to over 700°C, but for most chemicals of interest the boiling points are in the range of 300–600°C. Many organic chemicals decompose at temperatures lower than their (hypothetical) boiling points, but estimating a boiling point for such thermally unstable chemicals remains important.

Having a value for a chemical's boiling point, whether measured or estimated, is significant first because it defines the uppermost temperature at which the chemical can exist as a liquid. Second, the boiling point itself serves as a rough indicator of volatility, with higher boiling points indicating lower volatility at ambient temperatures. Finally, the boiling point is a key input in equations that provide estimates of a chemical's vapor pressure as a function of temperature.

The boiling point is associated with a number of molecular properties and features. Most important is molecular weight; boiling points generally increase with this parameter. Next is the strength of the intermolecular bonding; boiling points increase with increasing bonding strength. This bonding, in turn, is associated with processes and properties such as hydrogen bonding, dipole moments, and acid/base behavior. A final set of associated parameters is related to molecular structure and descriptors that specifically describe molecular rigidity (or its opposite, flexibility), symmetry, and branching. The strong connection of boiling points to molecular weights and structural features has made it relatively easy to derive structure-based estimation methods.

2.2   Overview of Available Estimation Methods

With one exception, the estimation methods considered for this book are those that use only structural information as input, since that is all that is likely to be available; i.e., if no boiling point is available for a chemical, then it is unlikely that other property values will be available from which a boiling point could be estimated (via some property-property correlation). The one exception is for estimation methods using melting points as inputs, since a value for a melting point (but not a boiling point), especially for thermally unstable chemicals, is likely to be available in the literature.

Significant reviews of boiling point estimation methods have been provided by Reid and Sherwood (1966), Reid et al. (1977), Lyman et al. (1982), and Horvath (1992). The most recent work by Reid (Reid et al., 1987) does not contain a general review of boiling point estimation methods but does introduce one new method by Jobak (1984). Horvath's (1992) work is particularly valuable; it provides a historic summary of 36 estimation methods and covers publications up to about 1988, describing some methods in detail and briefly listing others. Horvath's work does provide a fairly detailed description of estimation methods based on topological indices such as the connectivity index. Other methods not covered by Horvath's review have been published by Cramer (1980a and 1980b), White (1986), Jobak and Reid (1987), Hansen and Jurs (1987), Kelly et al. (1988), Rordorf (1989), Screttas and Micha-Screttas (1991), Balaban et al. (1992), Stanton et al. (1991 and 1992), Stein and Brown (1994), and Yalkowsky et al. (1994 and 1995).

These reviews and publications show that:

- Many of the published estimation methods have been derived for specific homologous series, i.e., particular chemical classes such as n-paraffins, alcohols, substituted benzenes, etc. Table 2.1 provides a list of references for such methods. Within such a class, boiling point estimation can be fairly accurate (e.g., having average absolute errors under 10°C);

- There is significant interest in the use of topological indices for property prediction in general, including boiling point prediction;

- There is continuing interest in developing a consistent set of molecular fragments that will allow the prediction of a number of physicochemical properties for a wide variety of chemical classes. (See, for example, Cramer [1980a and 1980b], Jobak and Reid [1987], and Stein and Brown [1994]. The most ambitious of these is Cramer [1980a and 1980b], who developed a set of descriptors to predict 21 properties.)

TABLE 2.1

Availability of Boiling Point Estimation Methods for Specific Chemical Classes.[1]

Chemical Classes Covered	Reference
Alkanes.	Stiel and Thodos (1962)[2]
Alkanes, alkenes, alkynes, cycloalkanes, cycloalkenes.	Kinney (1938 and 1940)[2]
Normal alkanes.	Tsonopoulos (1987)[4]
Alkanes and N-heterocyclics.	Ogilvie and Abu-Elgheit (1981)[4]
Olefins.	Hansen and Jurs (1987)
Selected alkanes, halides (Cl, Br, I), ethers, alcohols, esters, aldelydes, carboxylic acids, sulfides, mercaptans, amines, nitrates.	Ogata and Tsuchida (1957)[2]
n-Alkyl compounds of the following: halides (F, Cl, Br, I), aldehydes, ketones, acids, amines (primary, secondary, and tertiary), benzenes, cychohexanes, cyclohex-1-enes, alkanes, alcohols, ethers, acetates, thiols. Also n-alk-1-enes and n-alk-1-ynes.	Somayajulu and Palit (1957)[2]
Saturated alcohols, primary and secondary amines, halides.	Bhatnagar et al. (1980)[4]
Substituted aromatics, including benzene, diphenyl and heterocyclic compounds.	Simamora and Yalkowsky (1994)[3]
Polycyclic aromatic hydrocarbons (PAH), including substituted PAH (all planar).	White (1986)
Halogenated dibenzo-p-dioxins and dibenzofurous.	Rordorf (1989)
Halo- and polyhaloalkanes with 1-4 carbon atoms (halogens = F, Cl, Br, I).	Balaban et al. (1992a)
Acyclic ethers, peroxides, acetyls, and their sulfur analogues.	Balaban et al. (1992b)
Pyrans and Pyrroles.	Stanton et al. (1992)
Compounds containing furan, tetrahydrofuran, and thiophene ring systems.	Stanton et al. (1991)
High-boiling compounds with normal alkyl chains. Classes include alkanes, halides (Cl, Br, I), alcohols, mercaptans, ethers, sulfides, amines (primary, secondary, and tertiary), acetates, and cyanides.	Screttas and Micha-Screttas (1991)
Perfluorinated saturated hydrocarbons.	Kelly et al. (1988)
Chlorofluoroethanes and chlorofluoropropanes.	Shigaki et al. (1986, 1988)[4]

[1] Selected list; see Horvath (1992) and Reid et al. (1966, 1977, and 1987) for other references.
[2] Method described in Lyman et al. (1982).
[3] Method included in this chapter.
[4] As cited by Horvath (1992).

- Estimation methods that are not limited to particular chemical classes usually are limited to molecules that contain only C, H, O, N, S, and halides (i.e., F, Cl, Br, and I);

- A focus on developing and expanding the number of fragment constants can improve the accuracy of an estimation method only to a limited extent. Some researchers (e.g., Lai et al., 1987) have found it necessary to develop sophisticated structural correction factors to achieve the highest accuracy. But a consequence of such development is added difficulty in use of the method, to the point where hand calculations become too tedious and prone to error, and only computerized calculations are reasonable. Lai et al. (1987) provide a good example of increased accuracy in estimation at the expense of simplicity and practicality. Their method achieves an average absolute percent error of only 1.3%, demonstrated with a 1,169-compound test set which included some with multi-functional groups. However, the method is complicated for hand calculations. (A computer package, called NBP, is available from the authors.)

- No one method can provide simple and accurate estimates of boiling points across all chemical classes. Most general methods (i.e., those not limited to a single chemical class) have average absolute errors of 10–30°C (about 2–5%)

when applied to independent test sets of chemicals with no more that one functional group. Estimation errors can be much higher for chemicals with two or more functional groups and/or complicated molecular structures.

- Along with the drive to obtain increased accuracy (with more sophisticated estimation equations and special rules) is a tendency to forget the simplest methods — admittedly less accurate — which are very easy to use and can be considered useful guides in an estimation process which may employ two or more methods. (Examples of such methods are those of Banks [1939] and Burnop [1938] which rely primarily on the molecular weight, and the equations using melting point developed by Yalkowsky et al. [1994] and Walters et al. [1995].)

2.3   Overview of Recommended Methods

Table 2.2 presents an overview of the seven methods for boiling point prediction that this chapter recommends and describes. The specific rationale for selecting these six methods was:

- To provide the reader with methods generally applicable to the widest possible classification of organic chemicals. With only one exception, this chapter omits methods limited to a single class of chemicals. (However, as noted below, such limited applicability may be useful for some hydrocarbons or monofunctional chemicals.)
- To provide the reader with methods which range from easy-to-use methods of limited accuracy to difficult-to-use but more accurate methods. Table 2.2 generally lists methods in this order, from less to more accuracy and difficulty.
- To present to the reader some "new" methods not previously highlighted in handbooks or textbooks. These methods may not be universally better than more well known methods, but we believe they are generally better in terms of range of applicability and accuracy. The "new" methods also avoid the need to calculate intermediate parameters such as molar refraction, parachor, or critical constants. (Such were required in the estimation methods of Meissner, Lydersen-Forman-Thodos, and Miller, recommended in the estimation methods handbook by Lyman et al. [1982].)

This chapter does not present the most accurate method recommended — number 7 in Table 2.2 (Lai et al., 1987) — in its entirety because it is too difficult and error prone for hand calculations. A brief description is provided, along with the basic equation that shows the method for estimating a boiling point while considering only two of the correction factors. Readers wishing to use the full method will need to obtain the original paper or, preferably, obtain the computer program that the researchers have made available (see below for details).

It is not possible to state which estimation method is the best, especially if ease of use (and time required) is a factor along with applicability and accuracy. The following suggestions may help in selecting one or more methods for a particular chemical:

1) If the chemical contains elements, functional groups, or even highly unusual structures not specifically covered by any method, use Method 1, 2 or 3. (But

TABLE 2.2

Overview of Boiling Point Estimation Methods Described in this Chapter

	Method	Applicability	Basis	Accuracy
1.	Banks (1939)	Generally applicable to all chemicals.	Simple correlation with molecular weight.	The author of this chapter found an average absolute error of 8.8% with a 70-compound test set containing a diversity of functional groups including 14 with mixed functional groups.
2.	Burnop's Rule (Gold and Ogle, 1969)	Generally applicable to all chemicals with C, H, O, N, and halides (F, Cl, Br, I).	Correlates boiling point with molecular weight and empirical structural factor.	Authors found average error of −1.46% for a 255-compound test set including many polar and hydrogen-bonded compounds.
3.	Melting Point Correlation (Yalkowsky et al., 1994; Walters et al. 1995)	Generally applicable to all organic chemicals, including hydrogen bonded compounds.	Correlates boiling point with melting point, a molecular symmetry number, a modified atom count, and the number of donor hydrogen atoms in a hydrogen-bonding molecule.	Authors found a root mean square error of 28.3°C with one derivation set of 979 compounds with no hydrogen bonding (Yalkowsky et al., 1994). A standard error of 28.1 K was reported for the 1,425-compound derivation set including hydrogen bonding compounds. A separate test set of 39 compounds yielded an average absolute error of 23.9 K (6.6%) (Walters et al., 1995).
4.	Simamora and Yalkowsky (1994)	Substituted aromatics, including heterocyclic compounds, with substituents (branches) based on groups containing C, N, O, S, and halides (F, Cl, Br, I).	Group contribution method covering 40 molecular fragments, groups, and structural features.	Using a test set of 444 compounds, authors found correlation between fragment constant sums and experimental boiling points had a standard deviation of 17.6 K.
5.	Cramer's BC(DEF) Parameters (Cramer, 1980)	Generally applicable to organic chemicals with selected structural features and substituent groups (containing O, N, S, and halides [F, Cl, Br, I]).	Group contribution method (5 parameters [B, C, D, E, F] per group) covering 34 specific groups or structural features.	The standard error of regression equation (derived from data on 114 compounds) is 3.3 K. Root mean square errors on three test sets of decreasing similarity to derivation set were 17.1, 38.7, and 100 K, respectively.
6.	Stein and Brown (1994)	Generally applicable to organic compounds with selected structural features and substituent groups (containing O, N, S, P, Be, Si, Sn, and halides [F, Cl, Br, I]).	Group contribution method covering 85 chemical groups.	For derivation set of 4,426 compounds, predicted boiling points had average absolute error of 15.5 K (3.2%). For an independent test set of 6,584 compounds, the average absolute error was 20.4 K (4.3%).
7.	Lai et al. (1987)	Generally applicable to organic compounds with selected structural features and substituent groups (containing O, N, S, and halides [F, Cl, Br, I]). Probably best for compounds with multifunctional groups.	Non-linear group contribution with extensive structural corrections which account for effects related to molecular shape and polar effects.	For 1,169 compounds (including 79 not included in derivation set), boiling points were predicted with an average absolute error of 1.29%.

you might try another method on a similar chemical with a known boiling point to see if it gives reasonable answers.)

2) Give preference to Method 3 over Method 1 or 2 if a measured melting point is available.

3) If the chemical is a hydrocarbon (i.e., contains only carbon and hydrogen) or contains only a single functional group on a relatively simple carbon structure, then look first to see if a published estimation method covers that chemical class and use it. (See Table 2.1 and the references cited therein to identify such methods.)

4) If the chemical is a substituted aromatic, or heterocyclic compound, give preference to Method 4.

5) If a reasonably accurate estimate is required, but the time available to obtain the estimate is limited, try Methods 5 or 6.

6) If the estimate must have the highest accuracy, use Method 7. (As noted above, this will require obtaining the original paper by Lai et al. [1987] or the computer program they have developed to carry out the calculations. Details appear below.)

7) If you are unclear about any method's applicability (including accuracy), try it out first with one or more chemicals, of similar structure and composition, that have known boiling points and compare the known points with the estimates. Some ad hoc adjustments may be appropriate.

2.4   Description of Recommended Methods

2.4.1   Banks' Molecular Weight Correlation

For a quick initial estimate — or for an estimate for a compound with unusual elements, functional groups, and/or structural features—consider using Banks' (1939) equation:

$$\log T_b = 2.98 - 4/\sqrt{M} \tag{1}$$

where:

T_b = boiling point (K)
M = molecular weight (g/mol)

In a test with 70 compounds (including 48 with one functional group and 14 with multi functional groups), the average absolute error was found to be 8.8% (this author's unpublished data).

Example: Estimate T_b for chlorpyrifos (molecular weight = 350.59 g/mol):

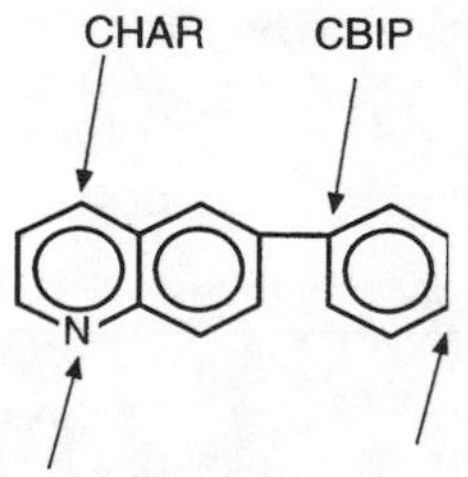

CHAR   CBIP

$$\log T_b \quad = \quad 2.98 - 4/\sqrt{350.59} = 2.766$$

$T_b \quad$ = 583 K (No measured value available, but the authors estimated a value of 630 K, based on a vapor pressure of 1.87×10^{-5} at 25°C.)

2.4.2 Burnop's Rule – A Modified Molecular Weight Correlation

For a quick initial estimate — if the chemical contains only C, H, O, N, or halides (F, Cl, Br and I), — consider Burnop's Rule, described initially by Burnop (1938) and recommended by Gold and Ogle (1969):

$$\log T_b = W/M - 8/\sqrt{M} \tag{2}$$

where:

$T_b \quad$ = boiling point (K)
$M \quad$ = molecular weight (g/mol)
$W \quad$ = sum of structural contributions given in Table 2.3

In a test of 255 compounds, Gold and Ogle (1969) found the method had an average error of –1.46%. Within this test set, the average error for various classes was as follows: –13.3% for hydrogen-bonded compounds (n = 49); +0.15% for polar hydrocarbons (n = 49); +1.72% for other polar hydrocarbons (n = 107); and +0.53% for nonpolar organics (n = 50).

TABLE 2.3

Structural Contribution for Burnop's Rule.

Atomic or Structural Group	Contribution to W
C	23.2
H	10.9
O	51.0
N	39.7
F	68.0
Cl	121.0
Br	255
I	398
Double Bond	16.1
Triple Bond	33.0
Six-membered ring	17.6

Source: Burnop, 1938.

Example: Estimate T_b for 2,6-di(tert-butyl)phenol (molecular weight = 206.33 g/mol):

Group	n	W	nW
C	14	23.2	324.8
H	22	10.9	239.8
O	1	51.0	51.0
6-memb. ring	1	17.6	17.6
Double bond	3	16.1	48.3
		Total =	681.5

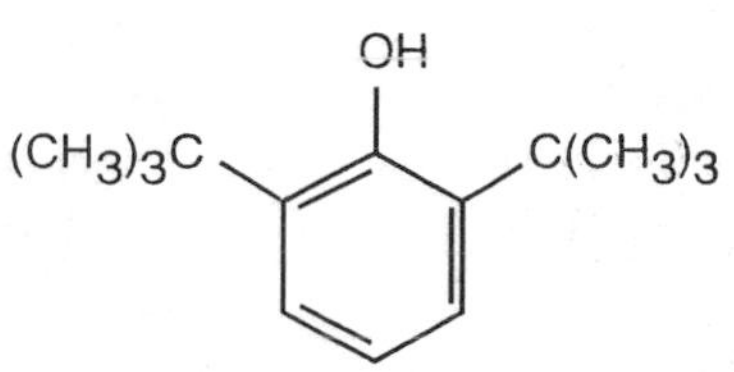

$$\log T_b = (681.5/206.33) - 8/\sqrt{206.33} = 2.746$$
$$T_b = 557 \text{ K (Measured value} = 509.6 \text{ K)}$$

2.4.3 Melting Point Correlation

The recommended melting point correlation comes from the work of Yalkowsky et al. (1994) and Walters et al. (1995). These authors show that, for certain idealized chemicals, one would expect the ratio of the melting point (T_m) and boiling point (T_b) to be constant. Their empirical correlation between T_m and T_b builds upon this basic precept and modifies it by accounting for non-ideality in the form of deviations from rigid spherical symmetry and hydrogen bonding. The deviations from spherical symmetry are represented by a molecular symmetry number (σ) and a modified count of the total number of atoms in the molecule (Z). Hydrogen bonding is accounted for in a single parameter (HB).

The primary features of this method are:

- It is easy to use.
- It uses one measured property of the chemical, the melting point, which should be readily available.
- It uses only 3 other parameters (σ, Z and HB), which are relatively easy to obtain from an inspection of the molecular formula and structure; correspondingly, a rigorous group contribution scheme is not required.
- It is reasonably accurate, having average absolute errors from various test sets of 24–28 K (about 7%).

Boiling points are estimated from the correlation equation (Walters et al. 1995):

$$T_b = 181 + 0.510 \, T_m - 47.6 \log \sigma + 15.8 \, Z + 12.5 \, HB \tag{3}$$

$$r^2 = 0.913 \quad s = 28.1 \text{ K} \quad n = 1419$$

where:

T_b = normal boiling point (K)
T_m = melting point (K)
σ = molecular symmetry number (described below)
Z = modified count of the number of atoms in the molecule (described below)
HB = number of donor hydrogen atoms in a hydrogen-bonding molecule (described below)

The correlation coefficient (r^2), standard deviation (s), and number of compounds (n), associated with Equation (3) are also provided above.

The molecular symmetry number (σ) is a measure of the rotational degeneracy of the molecule. It is defined as the number of indistinguishable positions that can be obtained by rigidly rotating the molecule about its center mass. Symmetry numbers for spherical, conical, and cylindrical molecules, shapes with infinite axes of rotation, have σ values of approximately 200, 20, and 20, respectively. Chemicals with no axes of symmetry have $\sigma = 1$.

The variable Z is determined by:

$$Z = \Sigma \, n_i z_i \tag{4}$$

where:

n_I = number of atoms of type i in molecule
z_i = defined atomic coefficient for each atom, as follows:
z = 0 for hydrogen and fluorine
z = 1 for carbon, nitrogen, oxygen and chlorine
z = 2 for bromine and sulfur (and other elements of similar polarizability)
z = 4 for iodine (and other elements that are more polarizable)

The parameter HB, as defined above, is the number of donor hydrogen atoms in a hydrogen-bonding molecule. Compounds subject to hydrogen bonding include alcohols, phenols, carboxylic acids, and primary and secondary amides.

Table 2.4 provides examples of the molecular descriptors Equation (3) requires and the resulting boiling point estimates.

TABLE 2.4

Examples of the Use of the Melting Point Correlation

Compound	Molecular Descriptors			Melting Point (K)	Estimated Boiling Point (K)[1]	Percent Error (%)[2]
	σ	Z	HB			
p-Bromophenol	2	9	1	339.5[3]	494	−3.3
m-Diiodobenzene	2	14	0	313.5[3]	548	−1.8
p-Aminobenzoic Acid	1	10	3	460[4]	611	−[6]
Benzene	12	6	0	278.7[5]	366.6	3.8
Thiophenol	2	8	0	258.3[3]	424.8	2.4
Tetrabromomethane	12	9	0	365[3]	458	−1.1
n-Butanol	1	5	1	183.9[5]	366.3	−6.3
Anthracene	4	14	0	489.8[5]	623	1.2
Lindane	12	12	0	385.6[6]	516	−13.5
2,6-Di(tert-butyl)phenol	2	15	0	329.6[3]	572	12.2
Chlorpyrifos	1	20	0	315[7]	658	(4.4)[7]
Trichloroethylene	1	5	0	186.75[5]	355	−1.4
				Avg. Absolute Error =		4.7%

1. Estimated from eq. 3.
2. % Error = [(Estimate − Literature Value)/(Literature Value)] × 100. Literature value taken from same reference as melting point, except where noted.
3. Lide (1993).
4. Verschueren (1983).
5. Reid et al. (1987).
6. No boiling point data found.
7. For comparison purposes, a boiling point was estimated from measured vapor pressure data (1.87 × 10^{-5} mm Hg @ 25°C), and this estimate was taken as the "true" boiling point for purposes of calculating a method error.

2.4.4 Simamora and Yalkowsky's Method for Aromatic Compounds

Simamora and Yalkowsky (1994) developed a structure-based estimation method just for aromatic compounds. This specifically includes substituted, heterocyclic, and polycyclic compounds. In total, the method employs a set of 40 molecular descriptors which consist of molecular fragments and structural correction factors. The 15 ring substituents covered include both non-hydrogen bonding and single hydrogen bonding groups.

The estimation method requires simply the addition of the appropriate number of molecular descriptors (b_i) based on their frequency of occurrence in the compound (n_i):

$$T_b = [\Sigma(n_i b_i)]/0.088 \tag{5}$$

$$n = 444 \quad r = 0.9994 \quad s = 17.62 \text{ K}$$

where:

T_b = normal boiling point (K)
n_i = frequency of occurrence of molecular descriptor i in molecule
b_i = value assigned to molecular i descriptor in Table 2.5

The number of compounds (n) the correlation analysis uses to derive Equation (5) appears above, along with the values for the correlation coefficient (r) and the standard deviation (s).

Table 2.5 defines the 40 molecular descriptors and provides their values. Figure 2.1 provides further definition of the different types of molecular fragments used while Figure 2.2 provides further definition of the hydrogen bonding and biphenyl ring corrections. Simamora and Yalkowsky (1994) consider the values in parentheses in Table 2.5 insignificant, based on the statistical analysis used to derive the molecular descriptor values.

The correlation parameters given above provide the only measures of method error; the standard error of 17.6 K (for the 444-compound test set) is indicative. By comparison, when the method of Jobak and Reid (1987) was used to estimate T_b for the same set of chemicals, the average error was 42.8 K (Simamora and Yalkowsky, 1994). This comparison did exclude compounds with structural groups not covered by the Jobak and Reid method, specifically: Y-SCN, Y-NCS, SO, and SO_2. This comparison shows that the Simamora and Yalkowsky method is clearly superior for aromatic compounds.

Example: Estimate T_b for anthracene:

Group	n	b	nb
CHAR	10	5.67	56.7
CRB	4	−0.704	−2.816
		Total =	53.884

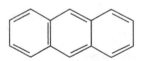

$$T_b = 53.884/0.088 = 612.3 \text{ K (Measured value is 613.1 K)}$$

2.4.5 Cramer's BC(DEF) Method

Cramer (1980a and 1980b) has developed an unusual structure-based estimation method that uses a common set of molecular fragments (and associated fragment constants) to

TABLE 2.5

Molecular Descriptors and the Associated Boiling Point Increments

Description	Designation	b_i
CAR	sp^2 carbon	−0.704
CBR	Bridgehead sp^2 carbon	−0.704
CHAR	sp^2 C–H	5.67
CBIP	Carbon involved in sp^2–sp^2 and sp^2–sp linkages	−4.98
NAR	sp^2 nitrogen	6.15
Y–F	Fluorine atom attached to an sp^2 carbon	5.89
Y–Cl	Chlorine atom attached to an sp^2 carbon	9.33
Y–Br	Bromine atom attached to an sp^2 carbon	10.96
Y–I	Iodine atom attached to an sp^2 carbon	13.63
Y–CH$_3$	Methyl group attached to an sp^2 carbon	8.04
Y-NO$_2$	Nitro group attached to an sp^2 carbon	13.11
Y–CN	Cyano group attached to an sp^2 carbon	11.61
Y–NCS	NCS group attached to an sp^2 carbon	15.10
Y–SCN	SCN group attached to an sp^2 carbon	16.86
Y–OH	Hydroxy group attached to an sp^2 carbon	11.61
Y–NH$_2$	Amino group attached to an sp^2 carbon	13.05
Y–CONH$_2$	Amide group attached to an sp^2 carbon	18.46
Y–COOH	Carboxyl group attached to an sp^2 carbon	17.07
Y–CHO	Aldehyde group attached to an sp^2 carbon	11.22
Y–SH	Mercapto group attached to an sp^2 carbon	11.54
YY–O	Oxygen atom attached to two sp^2 carbons	−9.64
YY–S	Sulfur atom attached to two sp^2 carbons	−5.35
YY–SO	Sulfoxide group attached to two sp^2 carbons	3.39
YY–SO$_2$	Sulfone group attached to two sp^2 carbons	(−1.38)
YY–CO	Keto group attached to two sp^2 carbons	(2.04)
YY–COO	Carboxyl group attached to two sp^2 carbons	−7.23
YY–NH	NH group attached to two sp^2 carbons	−8.77
YY–CH$_2$	Methylene group attached to two sp^2 carbons	(−0.80)
RYY–O	YY–O atom in a ring	6.11
RYY–S	YY–S atom in a ring	6.82
RYY–SO	YY–SO group in a ring	7.81
RYY–SO$_2$	YY–SO$_2$ group in a ring	0
RYY–CO	YY–CO group in a ring	0
RYY–COO	YY–COO group in a ring	9.98
RYY–NH	YY–NH group in a ring	6.13
RYY–CH$_2$	YY–CH$_2$ group in a ring	16.70
IHB-4	Four-membered ring intramolecular hydrogen bonding	4.81
IHB-5	Five-membered ring intramolecular hydrogen bonding	−1.42
IHB-6	Six-membered ring intramolecular hydrogen bonding	(−0.88)
IHB-7	Seven-membered ring intramolecular hydrogen bonding	(−2.36)
OBIP	Substituent present in biphenyl at the 2, 2′, 6, and/or 6′ positions	(−1.63)

Source: Simamora and Yalkowsky, 1994.
(Copyrighted by the American Chemical Society. Reproduced with permission.)

estimate 21 physicochemical properties, including the boiling point. The method has some general applicability to compounds with only C, H, O, N, S, and halides (F, Cl, Br, and I), with limitations on the types of functional groups containing O, N, and S. The method actually employs five fragment constants per molecular fragment, referred to as B, C, D, E, and F. For boiling point and several other properties, good correlation can be achieved with just the B and C parameters, but a slightly improved correlation is achieved with all five parameters. The B parameter is shown to represent molecular "bulk" (e.g., volume or mass) and the C parameter to represent molecular "cohesiveness" (i.e., intramolecular bonding strength).

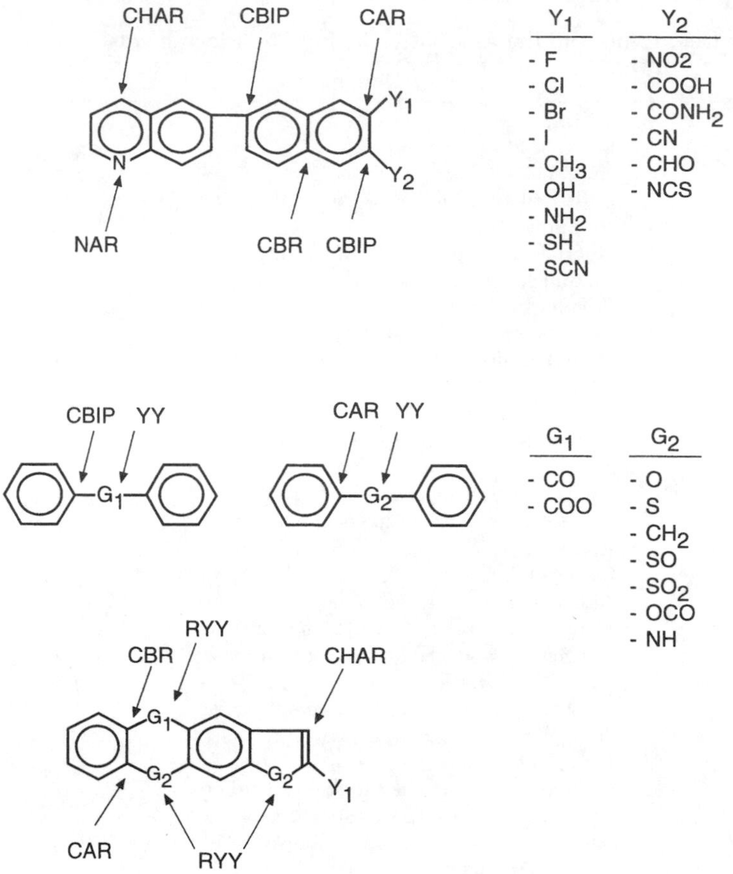

FIGURE 2.1
Designation of Molecular Fragments Used by Simamora and Yalkowsky. Source: Simamora and Yalkowsky, 1994. (Copyrighted by the American Chemical Society. Reproduced with permission.)

The correlation equation Cramer derived (1980a and 1980b) is:

$$T_b = 66.39 + 532.5B + 223.6C - 365.4D - 250.8E - 794.6F \tag{6}$$

$$n = 114 \quad r = 0.9996 \quad s = 3.31 \text{ K}$$

where:

$$T_b \qquad = \text{ normal boiling point (K)}$$
$$BCDEF = \text{ fragment constants to be summed for each molecular fragment. Table}$$
$$\qquad \qquad \text{ 2.6 provides values.}$$

The number of compounds (n) used to derive the regression equation, as well as the correlation coefficient (r) and standard error (s), appear above. Cramer (1980a) gives the 114 compounds used to derive Equation (6). This set appears too small to allow a broad range of chemical class applicability. Cramer himself notes that the compound set is not representative of chemistry as a whole, that most are liquids, none are ionic structures, and few are multifunctional or higher molecular weight substances. Nevertheless, the list does represent a fairly diverse group of chemical classes. Note that the value for "molecule," listed at

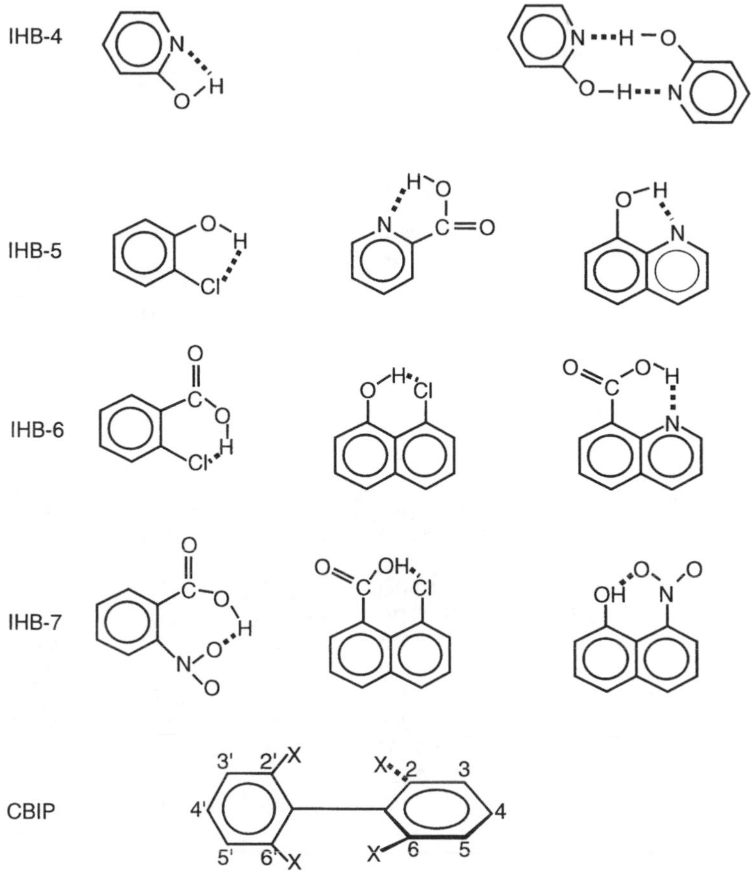

FIGURE 2.2
Correction Factors Considered by Simamora and Yalkowsky. Source: Simamora and Yalkowsky, 1994. (Copyrighted by the American Chemical Society. Reproduced with permission.)

the bottom of Table 2.6, is equivalent to the intercept of the correlation equation and is to be included in each calculation of Equation (6).

Cramer actually developed two schemes for applying the BC(DEF) values: a hierarchical additive-constitutive one and a linear additive-constitutive one. The two schemes are equivalent, but the latter is easier for hand calculations, and thus is the only one presented here. Table 2.6 provides the BC(DEF) fragment constants from the linear model.

The method error for Equation (6) is indicated by the correlation's standard error of 3.3 K. Cramer (1980b) also evaluated the equation with an independent test set of 138 compounds with the following results for the root mean square (rms) errors:

Set	Description	n	rms
I	Compounds that are either isomers or one- or two-carbon homologues of the original 114 compound test set	75	17.1
II	Compounds not in set I that contain no more that one structural fragment or grouping of structural fragments not found in the original 114	43	38.7
III	All other compounds	20	100.0
		138	

TABLE 2.6

Cramer's BC(DEF) Values for Individual Fragments

	B	C	D	E	F
–H	0.066	0.018	–0.027	–0.019	–0.019
–CH$_3$	0.142	–0.020	–0.016	–0.023	–0.015
–CH$_2$–	0.076	–0.038	0.011	–0.004	0.003
>CH–[c]	0.003	–0.058	0.053	0.018	0.015
>C<[c]	–0.075	–0.076	0.091	0.043	0.034
–CH=CH–	0.147	–0.043	0.028	0.010	0.003
–CH=CH$_2$	0.212	–0.025	0.000	–0.009	–0.015
>C=CH$_2$	0.147	–0.043	0.028	0.010	0.003
–C≡CH	0.171	0.074	0.027	0.002	–0.012
–C$_6$H$_5$	0.467	–0.007	0.012	0.007	–0.017
∂CH–(aromatic)	0.088	0.002	–0.007	0.001	–0.003
-naphthyl	0.766	0.018	–0.026	0.024	–0.028
-cyclohexyl	0.489	–0.148	0.004	–0.029	—0.009
–F[a,b]	0.078	0.088	0.009	–0.019	–0.020
–Cl[a,b]	0.165	0.087	–0.024	–0.012	–0.021
–Br[a,b]	0.213	0.095	–0.033	–0.008	–0.020
–I[a,b]	0.302	0.103	–0.056	–0.010	–0.031
–CF$_3$[b]	0.150	0.017	0.035	–0.037	–0.013
CCl$_3$[b]	0.410	0.015	0.009	–0.017	–0.017
–OH[a]	0.202	0.324	–0.012	–0.015	0.003
–O–[a]	0.044	0.155	0.061	0.019	–0.022
–C=O–[a]	0.135	0.246	0.061	0.023	–0.021
–CH=O–[a]	0.219	0.244	0.010	–0.014	–0.027
–COO–[a]	0.167	0.170	0.062	0.015	–0.027
–COOH[a]	0.323	0.342	–0.011	–0.017	0.008
–NH$_2$[a]	0.167	0.269	0.037	0.027	–0.014
–NH–[a]	0.082	0.251	0.095	0.056	–0.010
–N–[a]	–0.006	0.189	0.125	0.069	0.014
–CN[a]	0.241	0.269	–0.007	–0.023	–0.041
–N=[a](pyridine)	0.102	0.183	0.031	–0.011	–0.020
–NO$_2$[a]	0.238	0.241	–0.012	–0.027	–0.037
–CONH$_2$[a]	0.444	0.499	–0.019	–0.039	–0.012
–S–[a]	0.136	0.130	0.028	0.032	–0.020
–SH[a]	0.231	0.155	–0.026	–0.011	–0.013
No. of cycles (rings)	0.1045	0.0996	–0.1034	–0.0285	–0.042
No. of tertiary carbons[d]	–0.0074	–0.0012	0.0144	0.0030	–0.0067
No. of quaternary carbons[e]	–0.0199	–0.0016	0.0257	0.0095	–0.0061
No. of X–C–X tracings[f]	–0.0099	–0.0570	–0.0011	–0.0044	0.0019
No. of X$_3$C–CX$_3$ bonds[g]	–0.0077	–0.1762	–0.0063	–0.0142	–0.0051
No. of C=C–X tracings[h]	–0.0047	–0.0496	–0.0074	–0.0058	0.0031
No. of aromatic fusions	–0.0031	–0.0657	0.0181	–0.0140	0.00028
Molecule	–0.5065	0.056	0.007	0.031	0.028

[a] Value when attached to aliphatic system. Note the correction for "C=C–X" below to be applied once for this group when attached to an alkenyl carbon and twice when this group is attached to aromatic carbon. [b] Note correction for "X–C–X" below which must be applied when more than one halogen or other nonhydrogen, noncarbon atom is attached to the same carbon atom. The –CF$_3$ and –CCl$_3$ values already reflect the "X–C–X" correction. [c] Includes the "tertiary" or "quaternary" correction described below. [d] Count one for each sp^3 carbon having three nonhydrogen, nonhalogen attachments. [e] Count one for each sp^3 carbon having four nonhydrogen, nonhalogen attachments. [f] X = not hydrogen or carbon. Count one for *each* distinct path (i.e., for CF4, X–C–X = 3 + 2 + 1 = 6). [g] X = not hydrogen or carbon. Count one for each such C–C bond, and fractions for any bond having electronegative attachments (see text). [h] X = not hydrogen or carbon, unless carbon is C=X. One for *each* path (i.e., one for CH=CHCl but two for PhCl).

Source: Cramer, 1980b.

(Copyrighted by the American Chemical Society. Reproduced with permission.)

Example: Estimate T_b for Lindane:

Group	n	B	C	D	E	F
>CH	6	0.003	−0.058	0.053	0.018	0.015
Cl	6	0.165	0.087	−0.024	−0.012	-0.021
ring	1	0.1045	0.0996	−0.1034	−0.0285	−0.042
Molecule	1	−0.5065	−0.056	0.007	0.031	0.028
Weighted totals:		0.606	0.2176	0.0776	0.0385	−0.050

T_b=66.39 + 532.5(0.606) + 223.6(0.2176) − 365.4(0.0776) − 250.8(0.0385) − 794.6(−0.050)

T_b = 439.5 K (Measured value = 596.55 K)

2.4.6 Stein and Brown's Group Contribution Method

The group contribution method Stein and Brown (1994) developed is the most robust group contribution method that can be applied in a straightforward manner; i.e., it can cover a broader variety of chemicals than the Simamora and Yalkowsky and Cramer methods (described above) and is relatively easy to use compared to the method of Lai et al. (1987) (described below). The work is an extension of the work by Jobak and Reid (1987). The extension is primarily an increase in the number of fragment constants, from 41 to 85. However, many of the new groups are simply subdivisions of those Jobak and Reid used. The method assumes no interaction between fragments. A computerized version of the estimation method, called MPBPVP, is available from Syracuse Research Corporation (Syracuse, NY).

The fragment constants were derived by evaluating the boiling points of 4,426 compounds. This allowed the generation of fragment constants for several molecular groups containing C, N, O, S, and halides (F, Cl, Br, I), as well as four P-containing groups, three Si-containing groups, and one group each for B, Se, and Sn. Table 2.7 presents the constants. Note that some fragments contain up to 4 or even 5 atoms (e.g., C(O)OH, −N=NNH−, and −C(O)NH₂). Note also that the value of several fragment constants changes if the fragment is in (or on) a ring or is in some other specially defined structural position (e.g., on a secondary carbon). However, no separate structural correction factors need to be applied to the fragment constant sum.

The boiling point is estimated from the following three equations:

$$T_b = 198.2 + \Sigma n_i (\Delta T_b)_i \tag{7}$$

$$T_b(\text{corr}) = T_b - 94.84 + 0.5577 \, T_b - 0.0007705 \, T_b^2 \quad \text{for } T_b \leq 700 \text{ K} \tag{8}$$

$$T_b(\text{corr}) = T_b + 282.7 - 0.5209 \, T_b \quad \text{for } T_b > 700 \text{ K} \tag{9}$$

TABLE 2.7

Group Contributions for Stein and Brown's Method.

Structural Group	ΔT_b	Structural Group	ΔT_b
Carbon increments:		*Nitrogen increments: (cont'd.)*	
–CH$_3$	21.98	=NH	73.40
>CH$_2$	24.22	=N–	31.32
>C$_r$H$_2$[a]	26.44	=N$_r$–	43.54
>CH–	11.86	=N,N$_r$H–	179.43
>C$_r$H–	21.66	–N$_r$=C$_r$RN$_r$H–	284.16
>C<	4.50	–N=NNH–	257.29
>C$_r$<	11.12	–N=N–	90.87
=CH$_2$	16.44	–NO	30.91
=CH–	27.95	–NO$_2$	113.99
=C$_r$H–	28.03	–CN	119.16
=C<	23.58	ɸ–CN	95.43
=C$_r$<	28.19	*Halogen increments:*	
*aa*CH[b]	28.53	–F	0.13
*aa*C–	30.76	ɸ–F	–7.81
*aaa*C	45.46	–Cl	34.08
≡CH	21.71	1–Cl[c]	62.63
≡C–	32.99	2–Cl[c]	49.41
Oxygen increments:		3-Cl[c]	36.23
–OH	106.27	ɸ–Cl	36.79
1-OH[c]	88.46	–Br	76.28
2-OH[c]	80.63	ɸ–Br	61.85
3-OH[c]	69.32	–I	111.67
ɸ–OH[d]	70.48	ɸ–I	99.93
–O–	25.16	*Sulfur increments:*	
–O$_r$–	32.98	–SH	81.71
–OOH	72.92	ɸ–SH	77.49
Carboxyl increments:		–S–	69.42
–CHO	83.38	–S$_r$–	69.00
>CO	71.53	>SO	154.50
>C$_r$O	94.76	>SO$_2$	171.58
–C(O)O–	78.85	>CS	106.20
–C$_r$(O)O$_r$–	172.49	>C$_r$S	179.26
–C(O)OH	169.83	*Phosphorus increments:*	
–C(O)NH$_2$	230.39	–PH$_2$	59.11
–C(O)NH–	225.09	>PH	40.54
–C$_r$(O)N$_r$H–	246.13	>P–	43.75
–C(O)N<	142.77	>PO–	107.23
–C$_r$(O)N$_r$<	180.22	*Silicon increments:*	
Nitrogen increments:		>SiH–	27.15
–NH$_2$	61.98	>Si<	8.21
ɸ–NH$_2$	86.63	>Si$_r$<	–12.16
>NH	45.28	*Miscellaneous increments:*	
>N$_r$H	65.50	>B–	–27.27
>N–	25.78	–Sc–	92.06
>N$_r$–	32.77	>Sn<	62.89
>NOH	104.87		
>NNO	184.68		
*an*N	39.88		

[a] Atoms having the subscript *r* are in rings. [b] The symbol *a* denotes an aromatic bond. [c] Numbers 1, 2, and 3 denote attachments to primary, secondary, and tertiary carbons, respectively. [d] The symbol ɸ denotes an aromatic system.

Source: Stein and Brown, 1994.

(Copyrighted by the American Chemical Society. Reproduced with permission.)

where:

T_b = normal boiling point (K)
n_i = number of fragments of type i in molecule
$(\Delta T_b)_i$ = fragment constant for group i (from Table 2.7)
T_b (corr) = temperature-corrected value of normal boiling point

Equations (8) and (9) are used to correct for the fact that Equation (7) tends to overpredict the boiling point, especially above 500 K. Either Equation (8) orEquation (9) needs to be used to account for the temperature correction.

For the 4,426-compound data set used to develop this method, the predicted boiling points had an average absolute error of 15.5 K (3.2%). Stein and Brown (1994) also evaluated their method on an independent test set of 6,584 compounds and found the predicted boiling points had an average absolute error of 20.4 K (4.3%).

Example: Estimate T_b for 2,6-di(tert-butyl)phenol.

From Table 2.7:

Group	n	ΔT_b	$n(\Delta T_b)$
aaCH	3	28.53	85.59
aaC–	3	30.76	92.28
–CH$_3$	6	21.98	131.88
ϕ–OH	1	70.48	70.48
		Total =	380.23

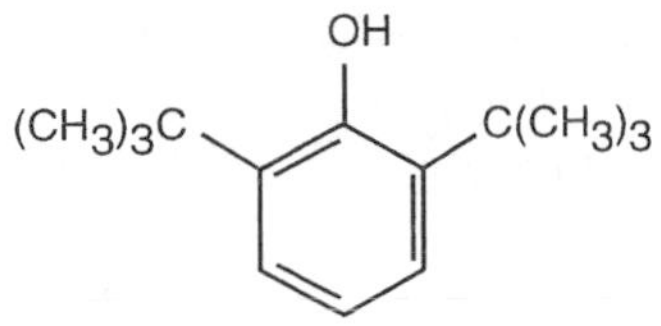

From Equation (7): $T_b = 198.2 + 380.23 = 573.03$ K

Using Equation (8): $T_b(\text{corr}) = 573.03 - 94.84 + 0.5577(573.03) - 0.0007705(573.03)^2$
$$= 544.8 \text{ K (Measured value} = 509.65 \text{ K.)}$$

Example: Estimate T_b for lindane.

From Table 2.7:

Group	n	ΔT_b	$n(\Delta T_b)$
>C,H	6	21.66	129.96
2-Cl	6	49.41	296.46
		Total =	426.42

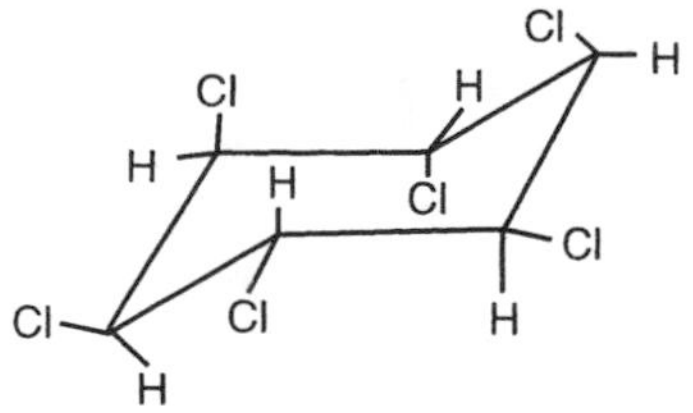

From Equation (7): $T_b = 198.2 + 426.42 = 624.62$ K

Using Equation (8): $T_b(\text{corr}) = 624.62 - 94.84 + 0.5577(624.62) - 0.0007705(624.62)^2$
$$= 577.5 \text{ K (Measured value} = 596.55 \text{ K.)}$$

2.4.7 Lai et al. Nonlinear Group Contribution Method

Lai et al. (1987) have developed what is likely the most accurate boiling point estimation method using a nonlinear group contribution method. The method is applicable to compounds with functional groups containing only C, N, O, S, and halides (F, Cl, Br, I). In this method, a basic group contribution scheme is supplemented by an extensive series of structural corrections that account for:

- Decreases in fragment constant values with increasing chain length (e.g., the decreasing value of the fragment constant for the $-CH_2-$ group as chain length increases);

- Interactions between two or more different functional groups on the molecule (e.g., two $-OH$ groups or a $-NH_2$ group and a $-Cl$ group);

- Hydrogen bonding (for alcohols, primary and secondary amines, carboxylic acids, and amides)

- Branching

- The position of a functional group along the main chain of a molecule

- cis vs. trans substitution

- Ring structures (aromatic, non-aromatic, heterocyclic, and fused)

- Unsaturation (i.e., double, triple, and conjugate double bonds)

The extensive series of structural corrections allows the method to achieve a significant increase in estimation accuracy over other available methods. In one major test with 1,169 compounds (including those with multi functional groups, hydrogen bonding, and complex structures) their method yielded an estimate with an average absolute error of 1.29% (Lai et al., 1987). Table 2.8 gives a summary of their test results. A smaller test set of 82 compounds was used to compare their method error with the errors associated with four other methods capable of handling chemicals with N, O, S, and halogens. Lai et al.'s method yielded an average absolute error of 1.12%, while the results for the other methods were as follows: Meissner, 4.1%; Lyderson et al., 8.83%; Miller, 6.83%; and Purarelli, 9.7%.

A significant consequence of the added sophistication of Lai et al.'s method is difficulty of use. The difficulty is such that the method is deemed unsuitable for hand calculations. To carry out a hand calculation for a complex molecule, one must use a series of up to 18 equations with a total of over 70 parameters, many of which themselves have multiple values for the different groups present in the molecule. Lai et al. (1987) provide a flow diagram in their paper to guide the user through the process, and two detailed examples. However, using this method accurately (without mistakes in application or calculation) requires much study and practice. We thus do not present this method here in full but only the main equation that embodies the basic fragment constants, correction for chain length, and interaction parameters (for compounds with two or more functional groups), to provide a preliminary view of the method. This main equation allows estimates of T_b without the other structural corrections mentioned above. For the full application of the method, see Lai et al. (1987) or the computer program available (see more below).

The basic equation — which does not incorporate all of their structural corrections — is:

$$T_b = \left(a + b_c \frac{1-r_c^n}{1-r_c} \right) + \sum_{i=1}^{1} \left(b_{fi} + b_{fic} \frac{1-r_c^n}{1-r_c} \right) \left(\frac{1-r_{fi}^{m_i}}{1-r_{fi}} \right) + \sum_{i=1}^{1} \sum_{j=2}^{j-i} b_{fifj} \left(\frac{1-r_{fi}^{m_i}}{1-r_{fi}} \right) \left(\frac{1-r_{fj}^{mj}}{1-r_{fj}} \right) \quad (10)$$

where:

$\quad\quad T_b \quad$ = normal boiling point (K)

TABLE 2.8

Method Errors in Boiling Point Estimates Using Lai et al.'s Method.[a]

Series	Data Pts	AAPE, $\lvert \bar{e} \rvert$, %	S.D., σ_e, %	Confidence Limit,[b] $\bar{e} \pm \sigma_e$ (1.96), %	MAPE $\lvert e \rvert_{max}$, %
Hydrocarbons	240	1.19	1.49	−0.01 ± 2.92	6.32
Alcohols	130	1.29	1.81	0.03 ± 3.54	7.87
Ethers	45	1.47	1.85	−0.92 ± 3.62	5.53
Ketones	42	1.35	1.69	−0.38 ± 3.30	5.39
Carboxylic acids	42	1.15	0.74	1.05 ± 1.45	2.70
Aldehydes	32	0.93	1.03	0.20 ± 2.02	2.22
Esters	78	0.75	0.96	0.23 ± 1.88	3.46
Amines					
Primary	65	1.11	1.61	−0.15 ± 3.16	6.02
Secondary	50	1.34	1.74	−0.21 ± 3.40	4.85
Tertiary	46	1.88	2.35	0.55 ± 4.60	5.45
Amides	19	0.71	1.00	0.02 ± 1.96	2.56
Nitros	38	1.04	1.58	−0.08 ± 3.10	5.34
Nitriles	30	1.07	1.54	0.43 ± 3.02	5.09
Thiols	45	1.23	1.58	−0.04 ± 3.10	3.86
Sulfides	35	0.84	0.77	0.59 ± 1.51	2.28
Fluorides	61	2.24	2.56	−0.74 ± 5.01	5.85
Chlorides	75	1.67	2.05	−0.87 ± 4.01	6.37
Bromides	60	1.68	2.04	−0.62 ± 4.00	5.94
Iodides	36	1.02	1.28	0.45 ± 2.51	4.33
Total	1169	1.29	1.56	−0.03 ± 3.06	7.87

[a] AAPE = average absolute percent error. MAPE = maximum absolute percent error. S.D. = standard deviation. [b] 95% reliability limit.

Source: Lai et al., 1987.

(Copyrighted by the American Chemical Society. Reproduced with permission.)

a = 103.59

b_c = 44.34

r_c = 0.94

n = number of carbon atoms

b_{fi} = characteristic constants of the functional group of the ith type (see Table 2.9)

r_{fi} = characteristic constants of the functional group of the ith type (see Table 2.9)

b_{fic} = characteristic constants of the functional group of the ith type (see Table 2.9)

m_i = number of functional groups of the ith type

l = number of functional group types

b_{fifj} = interaction parameter between functional groups of the ith type and the jth type ($i < j$) (see Table 2.10)

TABLE 2.9

Characteristic Constants of the Functional Groups for Use in Equation 10.[a]

Series	Functional Group	b_f	b_{fc}	r_f
Alcohols	OH	179.75	−15.35	0.88
Ethers	O	72.83	−6.98	0.93
Ketones	C(=O)	126.16	−10.17	(0.83)
Carboxylic acids	C(=O)OH	235.91	−16.64	(0.84)
Aldehydes	C(=O)H	119.29	−9.32	(0.84)
Esters	C(=O)O	124.51	−10.39	(0.83)
Amines				
Primary	NH_2	109.52	−6.81	0.94
Secondary	NH	86.20	−7.27	(0.94)
Tertiary	N	60.94	−4.51	0.55
Amides				
1	$C(=O)NH_2$	361.4	−35.80	(0.76)
2	C(=O)NHR	325.41	−30.32	(0.78)
3	C(=O)NRR′	276.30	−28.91	(0.63)
Nitriles	CN	171.31	−12.57	(0.77)
Nitros	NO_2	22.34	−16.08	0.70
Thiols	SH	139.04	−8.73	(0.81)
Sulfides	S	133.38	−7.47	(0.82)
Halides				
Fluorides	F	49.05	−3.42	0.83
Chlorides	Cl	108.14	−6.76	0.75
Bromides	Br	135.23	−8.61	0.83
Iodides	I	178.36	−9.79	0.77

[a] $a = 103.59$; $b_c = 44.34$; $r_c = 0.94$ in this method. An r_f value inside parentheses is predicted.
Source: Lai et al., 1987.
(Copyrighted by the American Chemical Society. Reproduced with permission.)

Example: Estimate T_b for 2,3-dibromo-1-propanol:

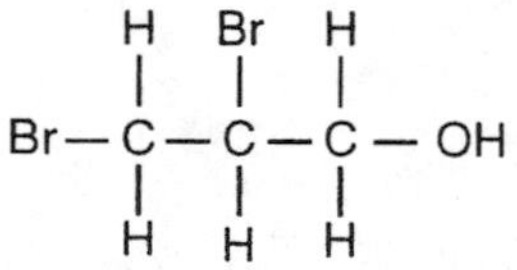

$$
\begin{array}{ccccccc}
 & H & & Br & & H & \\
 & | & & | & & | & \\
Br-&C&-&C&-&C&-OH \\
 & | & & | & & | & \\
 & H & & H & & H &
\end{array}
$$

From inspection of the structure: $n = 3$ and $l = 2$.
Also, for Br, set $m_1 = 2$ and use symbol j, and for OH set $m_2 = 1$ and use symbol i.

From Table 2.9:

Group	b_f	b_{fc}	r_f
Br	135.23	−8.61	0.83
OH	179.75	−15.35	0.88

From Table 2.10: $b_{12} = -39.75$ (interaction parameter for Br and OH)
Substituting in Equation (10):

$$T_b = 103.59 + 44.34\left[\frac{1-0.94^3}{1-0.94}\right] + \left(135.23 - 8.61\left[\frac{1-0.94^3}{1-0.94}\right]\right)\left[\frac{1-0.83^2}{1-0.83}\right]$$

$$+ 179.75 - 15.35\left[\frac{1-0.94^3}{1-0.94}\right] + 39.75\left[\frac{1-0.83^2}{1-0.83}\right]$$

$T_b = 495$ K (Chemical decomposes at 492 K [Lide, 1993])

One of the authors (D.H. Chen, Department of Chemical Engineering, Lamar University, Beaumont, Texas 77710), has written a computer program incorporating this complete estimation method and it is available for a nominal fee. This Fortran-based program is an executable program that will run out of DOS on an IBM-compatible PC. Structural input to the program is via a series of queries the user must respond to. Some familiarity with the estimation method is necessary to properly use the computer program.

TABLE 2.10

Values of the Interaction Parameter Between Different Functional Groups[a]

	Group A								
Group B	OH	C(=O)O H	O	NH$_2$	NH	CN	NO$_2$	Cl	Br
OH									
C(=O)OH	(−28.53)								
O	−14.88	(−14.08)							
NH$_2$	4.18	(−15.84)	(−6.02)						
NH	−4.26	(−13.97)	(−5.20)	(−5.87)					
CN	−21.98)	(−40.83)	(−14.37)	(−17.73)	(−25.05)				
NO$_2$	(−40.27)	(−53.17)	(−18.70)	(−23.10)	(−19.60)	(−51.26)			
Cl	−30.97	−27.26	−13.16	(−16.93)	(−14.02)	−9.86	−36.00		
Br	−39.75	−31.71	−16.26	(−13.37)	(−11.38)	(−27.61)	−32.58	−23.66	

[a] Values in parentheses are estimated.
Source: Lai et al., 1987.
(Copyrighted by the American Chemical Society. Reproduced with permission.)

References

Balaban, A.T., N. Joshi, L.B. Kier, and L.H. Hall. 1992a. Correlations between chemical structure and normal boiling points of halogenated alkanes $C_1 - C_4$. *J. Chem. Inf. Comput. Sci.* 32:233–37.

Balaban, A.T., L.B. Kier, and N. Joshi. 1992b. Correlations between chemical structure and normal boiling points of acyclic ethers, peroxides, acetals, and their sulfur analogues. *J. Chem. Inf. Comput. Sci.* 32:237–44.

Banks, W.H. 1939. Considerations of a vapor pressure-temperature equation, and their relationship to Burnop's boiling point function. *J. Chem. Soc.* 1939:292ff.

Burnop, V.C.E. 1938. Boiling point and chemical constitution. Part I. An additive function of molecular weight and boiling point. *J. Chem. Soc.* 1938: 826–29.

Cramer, R.D. 1980a. BC(DEF) parameters. 1. The intrinsic dimensionality of intermolecular interactions in the liquid state. *J. Am. Chem. Soc.* 102:1837–49.

Cramer, R.D. 1980b. BC(DEF) parameters. 2. An empirical structure-based scheme for the prediction of some physical properties. *J. Am. Chem. Soc.* 102:1849–59.

Jobak, K.G. 1984. A unified approach to physical property estimation using multivariate statistical techniques. SM Thesis, Chem. Eng., Mass. Inst. Technol., Cambridge, MA.

Hansen, P.J., and P.C. Jurs. 1987. Prediction of olefin boiling points from molecular structure. *Anal. Chem.* 59:2322–27.

Horvath, A.L. 1992. *Molecular Design – Chemical Structure Generation from the Properties of Pure Organic Compounds.* Elsevier, Amsterdam.

Howard, P.H. 1991. *Handbook of Environmental Fate and Exposure Data for Organic Chemicals.* Lewis Publishers, Chelsea, MI.

Jobak, K.G. and R.C. Reid. 1987. Estimation of pure-component properties from group contributions. *Chem. Eng. Comm.* 57:233–43.

Kelly, C.M., P.M. Mathias, and F.K. Schweighardt. 1988. Correlating vapor pressures of perfluorinated saturated hydrocarbons by a group contribution method. *Ind. Eng. Chem. Res.* 27:1732–36.

Kinney, C.R. 1938. A system correlating molecular structure of organic compounds with their boiling points. I: Aliphatic boiling point numbers. *J. Am. Chem. Soc.* 60:3032–39.

Kinney, C.R. 1940. Calculation of boiling points of aliphatic hydrocarbons. *Ind. Eng. Chem.* 32:559–62.

Lai, W.Y., D.H. Chen, and R.N. Maddox. 1987. Application of a nonlinear group-contribution model to the prediction of physical constants. 1. Predicting normal boiling points with molecular structure. *Ind. Eng. Chem. Res.* 26:1072–79.

Lide, D.R., Ed. 1993. *CRC Handbook of Chemistry and Physics.* CRC, Boca Raton, FL.

Lyman, W.J., D.H. Rosenblatt, and W.F. Reehl. 1982. *Handbook of Chemical Property Estimation Methods – The Environmental Behavior of Organic Chemicals.* McGraw-Hill, New York. (Now published under the same title by the American Chemical Society, Washington, DC.)

Ogata, Y., and M. Tsuchida. 1957. Linear boiling point relationships. *Ind. Eng. Chem.* 49:415–17.

Reid, R.C., J.M. Prausnitz, and B.E. Poling. 1987. *The Properties of Gases and Liquids, 4th ed.* McGraw-Hill, New York.

Reid, R.C., J.M. Prausnitz, and T.K. Sherwood. 1977. *The Properties of Gases and Liquids, 3rd ed.* McGraw-Hill, New York.

Reid, R.C., and T.K. Sherwood. 1966. *The Properties of Gases and Liquids – Their Estimation and Correlation, 2nd ed.* McGraw-Hill34
, New York.

Rordorf, B.F. 1989. Prediction of vapor pressures, boiling points and enthalpies of fusion for twenty-nine halogenated dibenzo-p-dioxins and fifty-five dibenzofurans by a vapor pressure correlation method. *Chemosphere* 18:783–88.

Screttas, C.G., and M. Micha-Screttas. 1991. Some properties and trends of enthalpies of vaporization and of Trouton's ratios of organic compounds. Correlation of enthalpies of vaporization and enthalpies of formation with normal boiling points. *J. Org. Chem.* 56:1615–22.

Simamora, P., and S.H. Yalkowsky. 1994. Group contribution methods for predicting the melting points and boiling points of aromatic compounds. *Ind. Eng. Chem. Res.* 33:1405–09.

Somayajulu, G.R. and S.R. Palit. 1957. Boiling points of homologous liquids. *J. Chem. Soc.* 1957:2540–44.

Stanton, D.T., L.M. Egolf, and P.C. Jurs. 1992. Computer-assisted prediction of normal boiling points of pyrans and pyrroles. *J. Chem. Inf. Comput. Sci.* 32:306–16.

Stanton, D.T., P.C. Jurs, and M.G. Hicks. 1991. Computer-assisted prediction of normal boiling points of furans, tetrahydrofurans, and thiophenes. *J. Chem. Inf. Comput. Sci.* 31:301–10.

Stein, S.E., and R.L. Brown. 1994. Estimation of normal boiling points from group contributions. *J. Chem. Inf. Sci.* 34:581–87.

Stiel, L.I., and G. Thodos. 1962. The normal boiling points and critical constants of saturated aliphatic hydrocarbons. *AIChE J.* 8:527–529.

Verschueren, K. 1983. *Handbook of Environmental Data on Organic Chemicals, 2nd edition.* Van Nostrand Reinhold, New York.

Walters, A.E., P.B. Myrdal, and S.H. Yalkowsky. 1995. A method for estimating the boiling points of organic compounds from their melting points. *Chemosphere* 31:3001–08.

White, C.M. 1986. Prediction of boiling point, heat of vaporization, and vapor pressure at various temperatures for polycyclic aromatic hydrocarbons. *J. Chem. Eng. Data* 31:198–203.

Yalkowsky, S.H., J.F. Krzyzaniak, and P.B. Myrdal. 1994. Relationships between melting point and boiling point of organic compounds. *Ind. Eng. Chem. Res.* 33:1872–77.

3

Vapor Pressure

Martin L. Sage and
Gloria W. Sage

CONTENTS

3.1   Introduction

The vapor pressure of a chemical is the pressure its vapor exerts in equilibrium with its liquid or solid phase. The vapor pressure's importance in environmental work results from its effects on the transport and partitioning of chemicals among the environmental compartments (air, water, and soil). The vapor pressure expresses and controls the chemical's volatility. The volatilization of a chemical from the water surface is determined by its Henry's law constant (see Chapter 4), which can be estimated from the ratio of a chemical's vapor pressure to its water solubility. The volatilization of a chemical from the soil surface is determined largely by its vapor pressure, although this is tempered by its sorption to soil solids and its Henry's law constant between soil, water, and air. A substance's vapor pres-

1-56670-456-1/00/$0.00+$.50
© 2000 by CRC Press LLC

sure determines whether it will occur as a free molecule in the atmosphere or will be associated with particulate matter (see Chapter 10).

This chapter provides a simple procedure for estimating the vapor pressure of a substance at normal environmental temperatures. For volatile substances that boil at or below 100°C, the vapor pressure is likely to be known, but, for many high-boiling substances with low vapor pressure, the value may be unknown or poorly known. An estimation procedure may be needed to help convert the known vapor pressure at the normal boiling point (i.e., 1 atmosphere) to the vapor pressure at the lower temperatures of environmental importance. For some of these high-boiling compounds, the actual boiling point may also be unknown, since the substance may decompose before it boils. In that case, the boiling point must be estimated using one of the procedures discussed in Chapter 2.

The vapor pressure of a chemical substance increases rapidly with temperature. Many equations have described this temperature dependence so that the vapor pressure can be calculated for a temperature of interest. The Antoine (1888) equation is most familiar. Riddick, Bunger, and Sakano (1996) provide an extensive discussion of many of the empirical equations used to describe the temperature dependence of the vapor pressure. To apply a particular equation to a chemical, the necessary parameters must be available from experimental data. If no such information exists, some method is needed to estimate the parameters.

Many compilations of vapor pressure are available, some for certain classes of chemicals and others for organic chemicals in general. In older compilations, the units of vapor pressure generally are given in mm Hg or torr, atmospheres (for more volatile substances), psi, or mbars. Most recent compilations report vapor pressures in Pa or kPa. Table 3.1 gives a list of conversion factors. The *CRC Handbook of Chemistry and Physics* (Lide 1995) contains data on over 1000 organic compounds; it reports the vapor pressures above 0.001 kPa, where available, at eight temperatures between 25°C and 150°C. The *Handbook of Physical Properties of Organic Chemicals* (Howard and Meylan 1997) contains vapor pressure data near 25°C on 12,800 chemicals, including estimated values when experimental values are unavailable. The *Handbook of Vapor Pressure* (Yaws 1994) contains the constants for a five-parameter expression for the vapor pressure of over 1000 compounds as does the *Physical and Thermodynamic Properties of Pure Chemicals* (Daubert et al. 1996). The *Illustrated Handbook of Physical-Chemical Properties and Environmental Fate for Organic Chemicals* (Mackay et al. 1999) includes vapor pressure data for a wide variety of molecules.

TABLE 3.1

Pressure Conversion Factors.[a]

	kPa	bar	torr	atm	psi
1 kPa	1.0000	1.0000×10^{-2}	7.501	9.869×10^{-3}	0.1450
1 bar	100.00	1.0000	750.1	0.9869	14.50
1 torr	0.1333	1.333×10^{-4}	1.0000	1.316×10^{-2}	1.934×10^{-2}
1 atm	101.33	1.0133	760.0	1.0000	14.70
1 psi	6.895	6.895×10^{-2}	51.71	6.805×10^{-2}	1.0000

[a] To use the table to convert from atm to kPa use the entry in the atm row and kPa column to find 1 atm = 101.33 kPa.

Grain (1982) presented several methods for estimating the vapor pressure of a wide variety of organic chemicals in an earlier handbook. He recommended the Antoine equation for liquids with vapor pressures above 10^{-4} kPa and the modified Watson method for solids and liquids with vapor pressures above 10^{-7} kPa. Section 3.2 discusses both methods in detail. Lyman (1984) recommended the Grain-Watson method and a method that Mackay and coworkers (1982) developed. Myrdal and Yalkowsky (1997) used a method similar to Mackay's except for a different procedure for determining the necessary constants. Jensen and coworkers (1982) used the UNIFAC group contribution method to estimate vapor pressures of a variety of organic compounds. Their applications were restricted to vapor pressures above 10 kPa. In addition, the needed contributions are available for only a limited number of groups. Recently, Site (1997) reviewed methods for estimating vapor pressure of environmentally significant organic compounds, but many of these methods were applied only to a few classes of molecules and required structural parameters not readily available.

In addition to these publications, software is available that allows the user to determine vapor pressures of a wide variety of compounds at room temperature. The Texas Research Center (TRC) (1996) distributes a PC DOS/Windows database that contains experimentally derived Antoine constants for approximately 6000 chemicals from which vapor pressures at user-selected temperatures can be calculated. Another Windows-based program, MPBPVP© by Meylan and Howard (1996), estimates the vapor pressure of organic compounds from their SMILES (Simplified Molecular Input Line Entry System) structure and their boiling points using the Antoine equation, the Grain-Watson method, and the Mackay method.

The fundamental relationship that allows the determination of the equilibrium vapor pressure, P, of a pure condensed phase as a function of temperature is the Clausius-Clapeyron equation

$$\frac{dP}{dT} = \frac{\Delta S_v}{\Delta V} = \frac{\Delta H_v}{T \Delta V} \tag{1}$$

where T is the absolute temperature, ΔS_v the molar entropy change, ΔH_v the molar enthalpy change (heat of vaporization), and ΔV the molar volume change on vaporization.

For pressures below one atmosphere, a number of simplifications can be made. The volume change on formation of vapor can be approximated reasonably by the volume of the vapor. The vapor is assumed to behave like an ideal gas.* In particular, $\Delta V = V_\omega = RT/P$. Substituting this value for ΔV into Equation (1) yields the Clausius-Clapeyron equation:

$$\frac{d \ln P}{dT} = \frac{\Delta H_v}{RT^2} \tag{2}$$

Integration of this equation yields

$$\ln \frac{P(T)}{P(T_0)} = \int_{T_0}^{T} \frac{\Delta H_v(T)}{RT^2} \, dT \tag{3}$$

which is immediately applicable to liquid-vapor equilibria. T_0 is some reference temperature, commonly taken to be the normal boiling point T_b, $\Delta H_v(T)$ the molar enthalpy of vaporization of liquid at temperature T, and $P(T)$ its equilibrium vapor pressure.

---

* The compressibility factor Z is defined to be PV/RT. Based on the detailed investigation of Miller (1964), Grain (1982) assumed the change of compressibility factor on vaporization, ΔZ for a number of chemicals at pressures below one atmosphere suggests that a value of 1 is more reasonable. Consequently, $\Delta Z = 1$ is used here.

For a relatively small temperature difference between the reference temperature, now assumed to be the normal boiling point, and the temperature of interest, $\Delta H_v(T)$ can be assumed to be a constant, ΔH_b.

$$\ln \frac{P_v(T)}{P_b} = \frac{\Delta H_b}{RT} \left(1 - T_\rho\right)$$ (4)

where $T_\rho = T/T_b$ and $P_b = 1$ atm.

However, when the reference temperature is much higher than the temperature of interest, as would be the case for estimating the room temperature vapor pressure of a relatively high boiling liquid, the variation of ΔH_v with temperature may introduce a significant error in the vapor pressure. Assuming that ΔH_v varies linearly over the range T to T_b

$$\Delta H_v = \Delta H_b + \Delta C_P \times (T - T_b)$$ (5)

where $C_p = dH_v/dT$ we obtain

$$\ln \frac{P(T)}{P_b} = \frac{\Delta H_b - T_b \Delta C_p}{RT} \left(1 - \frac{1}{T_\rho}\right) + \frac{\Delta C_p}{R} \ln T_\rho$$ (6)

For wider temperature ranges, $H_v(T)$ can be expressed as a polynomial or some other function of T. Integration of the Clausius-Clapeyron equation then leads to expressions given in the *Handbook of Vapor Pressure* (Yaws 1994) or in the *Physical and Thermodynamic Properties of Pure Chemicals* (Daubert et al. 1994).

So far we have considered only liquids. If the temperature for which the vapor pressure is to be estimated is below the melting point T_m, the liquid phase is not stable at that temperature. To find the vapor pressure of the equilibrium solid phase, we consequently must correct for the heat of melting. Sublimation is equivalent to melting of the supercooled liquid followed by vaporization, and therefore

$$\Delta H_s(T) = \Delta H_m(T) + \Delta H_v(T)$$ (7)

where ΔH_s, ΔH_m, and ΔH_v are respectively the enthalpies of sublimation, melting, and vaporization of the liquid. When the term $\Delta H_m(T)$ is ignored, the Clausius-Clapeyron equation yields the vapor pressure of the supercooled liquid. The ignored term contributes only below the melting point and yields the lowering in vapor pressure when the supercooled liquid freezes. This lowering may be found using the Clausius-Clapeyron equation and is

$$\ln \frac{P_s(T)}{P_l(T)} = \int_{T_m}^{T} \frac{\Delta H_m(T)}{RT^2} \, dT$$ (8)

where P_e and P_s are the vapor pressures of the liquid and solid, respectively.

3.2 Recommended Estimation Methods

The methods we recommend are the Antoine equation, which works best for liquids that boil below 200°C and have vapor pressures above 10^{-2} kPa at 25°C, and, for higher boiling, less volatile substances, the Grain-Watson method, which generally gives more reliable estimates. For solids, we recommend using the Grain-Watson method to find the vapor pressure lowering upon solidification of the supercooled liquid. Table 3.2 provides a summary of the information needed for these calculations and the accuracy of the estimated vapor pressures.

TABLE 3.2

Average Error in Estimated Vapor Pressure[a]

Pressure Range (kPa)	Antoine Method		Grain-Watson Method	
	Error in log P	Error (%)	Error in log P	Error (%)
1–100	0.05	7	0.02	4
10^{-3} – 1	0.16	37	0.18	43
10^{-7} – 10^{-3}	0.44	175	0.30	99

[a] Results calculated from Tables 3.5 and 3.6.

3.2.1 Antoine Equation Method for Liquids

The Antoine (1888) equation is an empirical equation

$$\ln \frac{P}{P_b} = B \left(\frac{1}{T-C} - \frac{1}{T_b - C} \right) \tag{9}$$

containing three parameters, T_b, B, and C. The best way of obtaining these parameters over a temperature range is to use a non-linear least-squares procedure to fit experimental vapor pressures. The Antoine equation is said to be the most reliable three-parameter equation for representing vapor pressure as a function of temperature from the melting point to 85% of the critical temperature,* although fitting experimental data in the range 1-100 kPa shows this is not generally the case. The Antoine equation (Equation (9)) using parameters chosen to give the best least-squares fit to experimental vapor pressures leads to larger deviations than did optimized parameters T_b, ΔH_b, and ΔC_p for the constant ΔC_p equation (Equation (6)). Thomson (1959) notes that the Antoine equation provides an "adequate fit" for pressures in the range 1-100 kPa, implying that caution should be used when considering high-boiling substances with vapor pressures lower than 1 kPa. Applying the Antoine equation requires knowledge of the vapor pressure at some reference temperature, typically the boiling point, and some method of determining the constants B and C. Stein and Brown's (1994) group correction method discussed by Lyman in the previous chapter may be used to estimate the normal boiling temperature when T_b has not been measured.

Taking the derivative of Equation (9) with respect to T at T_b and comparing the result with Equation (1) we obtain

---

* The normal boiling point is typically less than 70% of the critical temperature (Grain et al. 1982).

$$B = \frac{\Delta H_b (T_b - C)^2}{RT_b^2} \tag{10}$$

Therefore, B can be determined from C, ΔH_b, and T_b.

Thomson (1959) has determined that

$$C = -18 + 0.19\, T_b \tag{11}$$

based on vapor pressure measurements of 300 organic compounds of various types. The enthalpy of vaporization at the boiling point can be estimated using one of the extensions of Trouton's rule, although extensive compilations of this parameter now should be used when the value is known. Again, the *Handbook of Chemistry and Physics* (Lide 1995) contains the enthalpy of vaporization for over 1000 organic compounds. The method for estimating the enthalpy of vaporization suggested in this chapter is due to Fishtine (1963) and was used by Grain (1982). Fishtine considered the effects of van der Waals, dipole-dipole, and hydrogen bonding interactions and found that

$$\Delta S_b = \frac{\Delta H_b}{T_b} = K_F R \big(36.64 + \ln T_b\big) = K_F R \ln\big(82.06\, T_b\big) \tag{12}$$

where T_b is in Kelvin.* K_F (the Fishtine constant), is a dimensionless constant that can be evaluated using Tables 3.3 and 3.4. For benzene derivatives with substantial dipole moments that do not form hydrogen bonds,

$$K_F = 1 + 0.02\, \mu \tag{13}$$

where μ is the dipole moment in Debye (Fishtine 1962).

Table 3.5 includes results of applying this method to liquids. These results were obtained using the program MPBPVP© referred to above. The values for the supercooled liquid state that Table 3.6 presents in parentheses were found in the same manner.

3.2.2 Grain-Watson Method for Liquids

The problem with use of the Antoine equation is that its use can introduce unreasonable assumptions about the change in ΔH_v with temperature. This equation tends to overestimate the increase in enthalpy of vaporization with decreasing temperature. Grain (1982) used an approximation to the somewhat more realistic Watson[24] expression for this temperature dependence. To calculate the vapor pressure at temperature T_i lower than the boiling point, T_b, using the Clausius-Clapeyron equation, Watson suggested the function

$$\Delta H_v(T) = \Delta H_b \left(\frac{1 - T/T_c}{1 - T_b/T_c} \right)^{m_i} \tag{14}$$

for $T_i \le T_b$ where m_i is a constant determined by T_i/T_b.

---

* The usual expression for ΔS_b depends on the dimensions of the T_b and so cannot be dimensionally correct. However Kistiakowsky (1921) showed $\Delta S_b = R \ln (RT_b)$ gives reasonable values, provided RT_b is in cm^3 atm/mol.

TABLE 3.3

Fishtine Constants, K_F, for Aliphatic and Alicyclic[a] Organic Compounds.

Compound Type	Number of Carbon Atoms (N) in Compound, Including Carbon Atoms of Functional Group											
	1	2	3	4	5	6	7	8	9	10	11	12–20
Hydrocarbons												
n-Alkanes	0.97	1.00	1.00	1.00	1.00	1.00	1.00	1.00	1.00	1.00	1.00	1.00
Alkane isomers				0.99	0.99	0.99	0.99	0.99	0.99	0.99	0.99	0.99
Mono- and diolefins and isomers		1.01	1.01	1.01	1.01	1.01	1.01	1.01	1.01	1.01	1.01	1.01
Cyclic saturated hydrocarbons			1.00	1.00	1.00	1.00	1.00	1.00	1.00	1.00	1.00	1.00
Alkyl derivatives of cyclic saturated hydrocarbons				0.99	0.99	0.99	0.99	0.99	0.99	0.99	0.99	0.99
Halides (saturated or unsaturated)												
Monochlorides	1.05	1.04	1.03	1.03	1.03	1.03	1.03	1.03	1.02	1.02	1.02	1.01
Monobromides	1.04	1.03	1.03	1.03	1.03	1.03	1.02	1.02	1.02	1.01	1.01	1.01
Monoiodides	1.03	1.02	1.02	1.02	1.02	1.02	1.01	1.01	1.01	1.01	1.01	1.01
Polyhalides (not entirely halogenated)	1.05	1.05	1.05	1.04	1.04	1.04	1.03	1.03	1.03	1.02	1.02	1.01
Mixed halides (completely halogenated)	1.01	1.01	1.01	1.01	1.01	1.01	1.01	1.01	1.01	1.01	1.01	1.01
Perfluorocarbons	1.00	1.00	1.00	1.00	1.00	1.00	1.00	1.00	1.00	1.00	1.00	1.00
Compounds Containing the Keto Group												
Esters		1.14	1.09	1.08	1.07	1.06	1.05	1.04	1.04	1.03	1.02	1.01
Ketones			1.08	1.07	1.06	1.06	1.05	1.04	1.04	1.03	1.02	1.01
Aldehydes	—	1.09	1.08	1.08	1.07	1.06	1.05	1.04	1.04	1.03	1.02	1.01
Nitrogen Compounds												
Primary amines	1.16	1.13	1.12	1.11	1.10	1.10	1.09	1.09	1.08	1.07	1.06	1.05[b]
Secondary amines		1.09	1.08	1.08	1.07	1.07	1.06	1.05	1.05	1.04	1.04	1.03[b]
Tertiary amines			1.01	1.01	1.01	1.01	1.01	1.01	1.01	1.01	1.01	1.01
Nitriles	—	1.05	1.07	1.06	1.06	1.05	1.05	1.04	1.04	1.03	1.02	1.01
Nitro compounds	1.07	1.07	1.07	1.06	1.06	1.05	1.05	1.04	1.04	1.03	1.02	1.01
Sulfur Compounds												
Mercaptans	1.05	1.03	1.02	1.01	1.01	1.01	1.01	1.01	1.01	1.01	1.01	1.01
Sulfides		1.03	1.02	1.01	1.01	1.01	1.01	1.01	1.01	1.01	1.01	1.01
Alcohols												
Alcohols (single-OH group)	1.22	1.31	1.31	1.31	1.31	1.30	1.29	1.28	1.27	1.26	1.24	1.24[b]
Diols (glycols or condensed glycols)		1.33	1.33	1.33	1.33	1.33	1.33	1.33				
Triols (glycerol, etc.)			1.38	1.38	1.38							
Cyclohexanol, cyclohexyl methyl alcohol, etc.						1.20	1.20	1.21	1.24	1.26		
Miscellaneous Compounds												
Ethers (aliphatic only)		1.03	1.03	1.02	1.02	1.02	1.01	1.01	1.01	1.01	1.01	1.01
Oxides (cyclic ethers)		1.08	1.07	1.06	1.05	1.05	1.04	1.03	1.02	1.01	1.01	1.01

[a] Taken from Fishtine (1962).

a. Carbocyclic or heterocyclic compounds having aliphatic properties.

b. For N = 12 only; no prediction is made for K_F where N > 12.

Notes:

1. Consider any phenyl group as a single carbon atom.
2. K_F factors are the same for all aliphatic isomers of a given compound. For example, K_F = 1.31 for *n*-butyl alcohol, *i*-butyl alcohol, *t*-butyl alcohol, and *s*-butyl alcohol.
3. In organometallic compounds, consider any metallic atom as a carbon atom.
4. For compounds not included in this table, assume K_F = 1.06.

Source: Fishtine (1963) (*Reprinted with permission from the American Chemical Society.*)

$$m_i = 0.4133 - 0.2575\ T_i/T_b = 0.4133\backslash - 0.2575\ T_\rho \tag{15}$$

The expression for $H_v(T)$, Equation 14, cannot be strictly true, since the value of m_i depends on the temperature of interest T_i; different values of T_i will therefore lead to different temperature dependencies of $H_v(T)$. At best, Equation (14) gives an effective dependence of the heat of vaporization on temperature that is useful in calculating vapor pressure at a desired temperature. This is somewhat analogous to using an average ΔC_P over a temperature range. Grain notes that assuming $T_c/T_b \approx 1.5$ leads to at most a 5% error in Equation (14).

$$\Delta H_v(T) \approx \Delta H_b(3 - 2T_\rho)^m \tag{16}$$

Substituting this formula into the Clausius-Clapeyron equation and integrating by parts leads to

$$\ln\frac{P}{P_b} = \frac{\Delta H_b}{RT_b}\left[1 - \frac{\left(3 - 2T_\rho\right)^m}{T_\rho} - 2m\left(3 - 2T_\rho\right)^{m-1}\ln T_\rho\right] \tag{17}$$

3.2.3 Grain-Watson Correction for Solids

For solids, a factor must be included to correct the vapor pressure of the supercooled liquid phase to that of the solids using Equation (8). The two obvious ways to proceed are to either assume a constant value for $\Delta H_m(T)$ or to assume a Grain-Watson-like temperature dependence similar to that of $\Delta H_v(T)$ in Equation (16). In either case, we need to approximate ΔH_m, the enthalpy of melting at the melting point.

Several approximations are commonly used for $\Delta H_m = T_m\ \Delta S_m$. Grain in Lymon (1984) used

$$\Delta S_m = 0.6\ R\ \ln\ (82.06\ T_m) \tag{18}$$

which is similar to the Kistiakowsky form used for ΔS_b. Other expressions have been used to take molecular symmetry and flexibility into account, for example, Myrdal and Yalkowsky (1997). We will use the simple expression Equation (18).

TABLE 3.4

Fishtine Constants, K_F, for Hydrogen-Bonded Aromatic Organic Compounds.[a]

Aromatic Ring	Group	Substituents	
		K_F for Single	K_F for Multiple
Benzene	–OH	1.15	1.23
	–NH$_2$	1.09	1.14
	–NHR	1.06	1.06
Naphthalene	–OH	1.09	
	–NH$_2$	1.06	
	–NHR	1.03	

[a] Adapted from Fishtine (1963).
[b] Use the larger KF when several different groups are present.

TABLE 3.5

Estimated Vapor Pressure of Liquids at 25°C

| Compound[a] | T_b(°C) | $-\log (P/P_b)$ | | |
		Experiment[b]	Antoine[c]	Grain-Watson[c]
Acetone	56.1	0.52	0.49	0.50
Hexane	69.0	0.70	0.70	0.72
Trichloroethylene	86.7	0.98	1.00	1.03
Ethanol	78.3	1.11	1.08	1.11
1,4-Dioxane	101.3	1.30	1.25	1.29
2-Ethylbutanal	117.0	1.65	1.60	1.66
Acetic acid	117.9	1.68	1.62	1.67
2-Chloroethanol	129.0	2.02	2.21	2.29
Furfural	143.8	2.54	2.48	2.56
Aniline	184.0	3.19	2.94	3.01
Ethylene glycol	197.5	3.92	4.06	4.15
Glycerol	290	6.66	7.08	6.98

[a] Benchmark chemicals are in boldface.
[b] These values are from Howard and Meylan (1997).[6]
[c] These values were calculated using MPBPVB®. (Meylan and Howard (1996).

Assuming a constant ΔH_m, the correction becomes

$$\ln \frac{P_s(T)}{P_l(T)} = \frac{\Delta H_m}{RT}\left(1 - \frac{1}{T_{\rho m}}\right) \tag{19}$$

where $T_{\rho m} = T/T_m$. This correction is only reasonable for $T_{\rho m}$ close to one. If that is not the case the Grain-Watson temperature dependence will be used. With this assumed form, the correction factor is

$$\frac{P_s}{P_l} = 0.6\ln\left(82.06\,T_m\right)\left[1 - \left(\frac{3 - 2T_{\rho m}}{T_{\rho m}}\right)^{m'} - 2m'\left(3 - 2T_{\rho m}\right)^{m'-1}\ln T_{\rho m}\right] \tag{20}$$

where

$$m' = 0.4133 - 0.2575\,T_{\rho m} \tag{21}$$

Table 3.6 gives results of the calculations of vapor pressure for solids using the Grain-Watson expression.

TABLE 3.6

Estimated Vapor Pressure of Solids at 25°C.

Compound[a]	T_b(°C)	T_m(°C)	Experiment[b]	Antoine[c]	Grain-Watson[c]
				$-\log(P_s/P_b)$	
Phenol	170.0	40.9	3.34	3.29 (3.14)	3.37 (3.23)
2,6-Di(tert-butyl)phenol	253	37.0		5.00 (4.89)	5.02 (4.91)
Lindane	323	113	6.27	7.20 (6.32)	7.00 (6.12)
Aldrin	330[d]	105	6.80	7.29 (6.51)	7.07 (6.29)
Dicamba	329[d]	145	7.35	7.71 (6.49)	7.49 (6.27)
Chlorpyrifos	377[d]	42	7.65	8.07 (7.92)	7.57 (7.42)
Dieldrin	330	176	8.11	8.08 (6.52)	7.86 (6.28)
Anthracene	340	215	8.45[c]	8.82 (6.80)	8.55 (6.52)

[a]Benchmark chemicals are in boldface.
[b]These values are from Howard and Meylan (1997).
[c]These values were calculated using MPBPVB®. The values in parentheses are for the supercooled liquid.
[d]These values were estimated using the method of Stein and Brown (1994).
[e]This is an extrapolated value.

3.3   Step by Step Procedures

3.3.1   General Procedure

1. Determine the normal boiling point, T_b. If no measured value is available, use one of the estimation procedures described in the previous chapter.
2. Determine the Fishtine constant, K_F, from the tables.
3. Calculate ΔH_b using Equation (12).
4. Proceed as required by specific method 1 or 2 described below. Method 1 is more suited to substances with vapor pressures at 25°C above 0.1 kPa, while method 2 can be used for less volatile substances.
5. If the chemical is a solid, continue with method 3.

3.3.2   Specific Methods*

1. Antoine Method for Liquids
    i. Determine the constant C using Equation (11).
    ii. Determine the constant B using Equation (10).
    iii. Calculate $\log(P/P_b)$ at the desired temperature using Equation (9).
2. Grain-Watson Method for Liquids
    i. Determine the constant m using Equation (15).
    ii. Calculate $\log(P/P_b)$ at the desired temperature using Equation (17).
3. Grain-Watson Correction for Solids
    i. Find the melting point of the solid.

* For convenience, log x is calculated instead of lnx shown in equations (9), (17), and (20). lnx = 2.303 logx.

ii. Determine the constant m' using equation (21).

iii. Calculate log (P_s/P_l) using equation (20).

iv. Find P_s in terms of P_l.

3.4 Sample Calculations

3.4.1 General Procedure in Estimating the Vapor Pressure of Aniline at 25°C

1. The normal boiling point of aniline is 184.0°C or 184.0 + 273.2 = 457.2 K. T_ρ = 298.2/457.2 = 0.6522.

2. Table 3.4 gives the value of K_F = 1.09.

3. ΔH_b = 1.09 × 457.2 × 8.314 × ln (82.06 × 457.2 J/mol = 43.6 kJ/mol.

3.4.2 Estimating the Vapor Pressure of Aniline at 25°C Using the Antoine Method

1. C = − 18 + 0.19 × 457.2 = 69 K.

2. B = 43.6 × 10^3 × (457.2 − 69)²/(8.314 × 457.2²) K = 3.79 × 10^3 K.

3. Substituting into Equation (9), we obtain

$$\log \frac{P}{P_b} = \frac{3.79 \times 10^3}{2.303} \times \left[\frac{1}{457.2 - 69} - \frac{1}{298.2 - 69} \right] = -2.94$$

4. $P = 10^{-2.94}$ × 101.33 kPa = 0.12 kPa which compares with the observed value of 0.06 kPa.

3.4.3 Estimating the Vapor Pressure of Aniline at 25°C Using the Grain-Watson Method

1. m = 0.4133 − 0.2575 × 0.6522 = 0.2454.

2. Substituting into Equation (20) we obtain

$$\log \frac{P}{P_b} = \frac{43.6 \times 10^3}{2.303 \times 457.2 \times 8.314} \times$$

$$\left[1 - \frac{(3 - 2 \times 0.6522)^{0.2454}}{0.6522} - 2 \times 0.2454 \times (3 - 2 \times 0.6522)^{-0.7456} \times \ln(0.6522) \right] = -3.01$$

3. $P = 10^{-3.01}$ × 101.33 kPa = 0.10 kPa which compares with the observed value of 0.06 kPa.

3.4.4 Estimating the Correction Factor for Solid Anthracene at 25°C Using the Grain-Watson Method

1. T_m = 215°C or 215 + 273 = 488 K.
2. m′ = 0.4133 − 0.2575 × 0.6101 = 0.2560.
3. Substituting into Equation (20) we obtain

$$\log \frac{P_s}{P_l} = \frac{0.6 \times \ln(82.06 \times 488)}{2.303 \times 488 \times 8.314} \times$$

$$\left[1 - \frac{(3 - 2 \times 0.6101)^{0.2560}}{0.6101} - 2 \times 0.2560 \times (3 - 2 \times 0.6101)^{-0.7440} \ln(0.6101) \right] = -2.02$$

4. P_s = $10^{-2.02}$ × P_l = 9.6 × 10^{-3} × P_l. The Antoine and Grain-Watson methods determine log (P_l/P_b) of 6.80 and 6.52 respectively, leading to a value for P_s of 1.7 × 10^{-7} and 2.9 × 10^{-7} kPa. The extrapolated experimental value is 3.6 × 10^{-7} kPa.

3.5 Discussion

All the estimations of vapor pressure of liquids discussed above depend on the accuracy of the enthalpy of vaporization. Use of the Fishtine expression gives results that are not expected to be more accurate that 3% for ΔH_b. That implies that, even if the equation used to describe the temperature dependence of vapor pressure were exact, we still would expect no better than 3% accuracy in the estimated value of log (P_l/P_b); the lower the vapor pressure the greater the expected error. For a vapor pressure of 10 kPa, the expected error is 7%; for 100 Pa, 23%; for 1 Pa, 41%; and for 10 mPa, 62%.

For the corrections needed to convert from vapor pressure of supercooled liquid to vapor pressure of solids, the uncertainty is greater. The correction in this case may be accurate only to 20%. Since the ratio P_b/P_l is so much larger than the ratio P_e/P_s, this uncertainty is of less significance. If we consider a high-melting non-volatile substance like anthracene, the correction factor log (P_s/P_l) was calculated to be 2.02 at 25°C. A 20% uncertainty would lead to a 2.5-fold uncertainty in the prediction of P_s. However, when the vapor pressure is as low as 0.3 mPa, an uncertainty resulting in values of 0.1 mPa or 0.8 mPa would make little practical difference.

An easily implemented alternative for obtaining better vapor pressure estimates for high boiling chemicals would be to use an experimental value of the boiling point at reduced pressure rather than the normal boiling point. Then, using similar techniques to those described above, we would calculate a smaller change in pressure, so that the numerical uncertainty in vapor pressure at ambient temperature would be much less.

References

Antoine, C. 1888. Tensions des Vapeurs: Nouvelle Relation Entre les Tensions et les Tempé. *Compt. Rend.* 107:681–684.

Boublik, T., V. Fried, and E. Hala. 1984. *The Vapor Pressure of Pure Substances*, 2nd ed. Elsevier, Amsterdam.

Daubert, T.E., R.P. Danner, H.M. Sibul, and C.C. Stebbins. 1996. *Physical and Thermodynamic Properties of Pure Chemicals. Data Compilation.* Design Institute for Physical Property Data. American Institute of Chemical Engineers, New York.

Fishtine, S.H. 1963. Reliable Latent Heats of Vaporization. *Ind. Eng. Chem.* June 55:47–56.

Grain, C.F. In W.J. Lyman, W.F. Reehl, and D.H. Rosenblatt, Eds., *Handbook of Chemical Property Estimation Methods*. 1982. McGraw-Hill, New York, Chap. 14, pp. 1–20.

Howard, P.H., and W.H. Meylan, Eds. 1997. *Handbook of Physical Properties of Organic Chemicals.* Lewis, Boca Raton.

Jensen, T., Aa. Fredenslund, and P. Rasmussen. 1982. Pure-Component Vapor Pressures Using UNIFAC Group Contribution. *Ind. Eng. Chem. Fundam.* 20:239–246.

Kistiakowsky, V.A. 1921. Latent Heat of Vaporization. *J. Russ. Phys. Chem.* 53, I:256–264.

Lide, D.R., Ed. 1995. *CRC Handbook of Chemistry and Physics*, 76th ed. CRC Press, Boca Raton, pp. 6–77 to 6–108.

Lyman, W.J. 1984. Estimation of Physical Properties. In W.B. Neely and G.E. Blau, Eds., *Environmental Exposure from Chemicals*, Vol. 1. CRC Press, Boca Raton pp 13–47.

Mackay, D., A. Bobra, D.W. Chan, and W.Y. Shiu. 1982. Vapor Pressure Correlations for Low-Volatility Environmental Chemicals. *Environ. Sci. Technol.* 16:645–649.

Mackay, D., W. Y. Shiu, and K. C. Ma. 1999 *Illustrated Handbook of Physical-Chemical Properties and Environmental Fate for Organic Chemicals* (5 volumes). Also available as CDRom. CRC Press, Lewis Publ. Boca Raton, FL.

Meylan, W.M., and P.H. Howard. 1996. *MPBPVP©*. Syracuse Research Corporation, Syracuse. See also http://esc.syrres.com.

Miller, D.G. 1964. Estimating Vapor Pressures – a Comparison of Equations. *Ind. and Eng. Chem.* March 56:46–57.

Myrdal, P.B., and S.H. Yalkowsky. 1997. Estimating Pure Component Vapor Pressures of Complex Organic Molecules. *Ind. Eng. Chem. Res.* 36:2494–2499.

Ohe, S. 1976. Computer Aided Data Book of Vapor Pressure. Data Book Publishing Company, Tokyo.

Riddick, J.A., W.B. Bunger, and T.K. Sakano. 1996. *Organic Solvents*, 5th ed. Wiley-Interscience, New York.

Site, A.D. 1997. The Vapor Pressure of Environmentally Significant Organic Chemicals: A Review of Methods and Data at Ambient Temperatures. *J. Phys. Chem. Ref. Data.* 26:157–193.

Stein, S.E., and R.L. Brown. 1994. Estimation of normal boiling points from group contributions. *J. Chem. Inf. Sci* 34:581–587.

TRC Databases for Chemistry and Engineering: Vapor Pressure. 1996. Thermodynamics Research Center. The Texas A&M University System, College Station.

Thomson, G.W. 1959. In A. Weissberger, Ed., *Techniques of Organic Chemistry,* 3rd ed., Vol. I, Part 1. Interscience, New York, p 473.

Watson, K.M. 1943. Thermodynamics of the Liquid State – Generalized Predictions of Properties. *Ind. Eng. Chem.*, 35:398–406.

Yaws, C.E. 1994. *Handbook of Vapor Pressure*. Gulf, Houston.

Part II

Partitioning Properties

4

Henry's Law Constant

D. Mackay, W.Y. Shiu, and K.C. Ma

CONTENTS

4.1 Introduction

Along with the octanol-water and octanol-air partition coefficients, K_{AW} determines how a chemical substance will partition between the three primary media of accumulation in the environment, namely air, water, and organic matter present in soils, solids, and biota. "Volatile Organic Chemicals" such as chloroform with large values of K_{AW} evaporate appreciably from soils and water, and their fate and effects are controlled primarily by the rate of evaporation and the rate of subsequent atmospheric processes. For such chemicals, an accurate value of K_{AW} is essential. Even a very low value of K_{AW} of, say, 0.001 can be significant and must be known accurately, because the volume of the accessible atmosphere is much larger than that of water and soils by at least a factor of 1000; thus even a low atmospheric concentration can represent a significant quantity of chemical. Further, the rate of evaporation from soils and water is profoundly influenced by K_{AW}, because that process involves diffusion in water and air phases in series, or in parallel, and the relative concentrations which can be established in these phases control these diffusion rates.

Accurate values of K_{AW} are thus essential for any assessment of the behavior of existing chemicals or the prediction of the likely behavior of new chemicals.

4.2   Theoretical Background: Physical Chemical Fundamentals

Air-water partitioning can be viewed as the determination of the solubility of a gas in water as a function of pressure, as first studied by William Henry in 1803. A plot of concentration or solubility of a chemical in water expressed as mole fraction x, versus partial pressure of the chemical in the gaseous phase P, is usually linear at low partial pressures, at least for chemicals which are not subject to significant dissociation or association in either phase. This linearity is expressed as "Henry's Law." The slope of the P-x line is designated H′, the Henry's law constant (HLC) which in modern SI units has dimensions of Pa/(mol fraction). For environmental purposes, it is more convenient to use concentration units in water C_W of mol/m^3 yielding H with dimensions of Pa m^3/mol.

$$P(Pa) = H(Pa\ m^3/mol)\ C_W(mol/m^3) \tag{1}$$

The partial pressure can be converted into a concentration in the air phase C_A by invoking the ideal gas law

$$C_A = n/V = P/RT \tag{2}$$

where n is mols, V is volume (m^3), R is the gas constant (8.314 Pa m^3/mol K) and T is absolute temperature (K). It follows that

$$C_A = P/RT = (H/RT)C_W = K_{AW}C_W \tag{3}$$

The dimensionless air-water partition coefficient K_{AW} (which can be the ratio in units of mol/m^3 or g/m^3 or indeed any quantity/volume combination) is thus H/RT.

A plot of C_A versus C_W is thus usually linear with a slope of K_{AW} as Figure 4.1 illustrates. For organic chemicals which are sparingly soluble in water, these concentrations are limited on one axis by the water solubility and on the other by the maximum achievable concentration in the air phase which corresponds to the vapor pressure, as Figure 4.1 shows. To the right of or above the saturation limit, a separate organic phase is present. Strictly, this saturation vapor pressure is that of the organic phase saturated with water, not the pure organic phase. This consideration can be important for chemicals in which water is appreciably soluble, for example alcohols or esters, but for most chemicals this saturation vapor pressure can be assumed to be the pure compound vapor pressure. It is thus possible to estimate H or K_{AW} from the solubility C_W^S and vapor pressure P^S, the superscript S designating saturation.

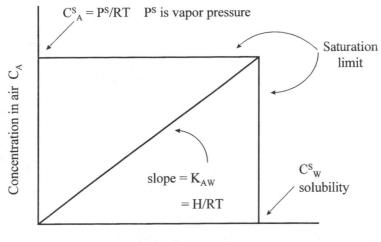

FIGURE 4.1
Plot of concentration in air versus concentration in water showing linearity with a slope K_{AW}. The line ends at saturation conditions in both phases. To the right and above the saturation limits, a pure solute phase is present.

$$K_{AW} = H/RT = P^S/C_W^S RT \tag{4}$$

This procedure is not recommended for chemicals soluble in water or vice versa to an extent exceeding a few percent. The vapor pressure and solubility can be of the chemical in a solid or liquid state, depending on the melting point. The same state must apply to both properties.

In more fundamental terms, the solubility of a chemical in water is determined by the activity coefficient in water γ_W which can be viewed as a "correction factor" to Raoults Law, i.e.,

$$P = \gamma_W x P_L^S \tag{5}$$

where x is mole fraction of chemical in solution and P_L^S is the reference vapor pressure or reference fugacity and is the partial pressure which the liquid chemical exerts when x is 1.0 (i.e., the chemical is pure) and (by definition) when γ_W is also 1.0. At saturation, when P equals P_L^S the group $\gamma_W x$ is unity and the mole fraction solubility is thus $1/\gamma_W$. Note that this "Raoult's Law" definition, in which γ becomes 1.0 when x is 1.0, differs from, and should not be confused with, the alternative "Henry's Law" in definition which γ'_W becomes 1.0 when x is zero and "infinite dilution" conditions apply, namely

$$P = \gamma'_W H x \tag{6}$$

The mole fraction concentration x is $C_W v$ where C has units of mol/m^3 and v is the molar volume of the solution (m^3/mol). Since the solutions are usually dilute, the molar volume is essentially that of water, i.e., approximately $18 \times 10^{-6}\ m^3/mol$. It follows that

$$P = \left(\gamma_W v P_L^S\right) C_W = H C_W \tag{7}$$

and

$$H = \gamma_W v P_L^S \tag{8}$$

This equation gives an additional method of estimating H, provided P_L^S and γ_W can be determined.

Note that P_L^S is the pressure which pure chemical will exert in the liquid state, at the system temperature. For liquids such as benzene, this can be assumed to be the measurable liquid phase vapor pressure. In two situations, P_L^S becomes experimentally inaccessible. When the pure chemical is a solid, for example naphthalene with a melting point of 80°C, the only measurable vapor pressure at 25°C is that of the solid P_S^S, which is lower than that of the hypothetical liquid state. The ratio of the solid to liquid state vapor pressures can be estimated from

$$\ln\left(P_S^S / P_L^S\right) = -\Delta S_F \left(\left(T_M / T\right) - 1\right) / R \tag{9}$$

where ΔS_F is the entropy of fusion (J/mol K), T_M is the melting point, and T is the system temperature (K).

Yalkowsky (1997) has shown, by applying Walden's Rule (and as discussed in Chapter 7 on solubility in water), that many organic compounds have entropies of fusion of approximately 56.5 J/mol K. Thus this relationship simplifies at a temperature T of 298 K to approximately

$$\ln\left(P_S^S / P_L^S\right) = -6.79 \left(\left(T_M / T\right) - 1\right) = -0.023 \left(T_M - 298\right) = \ln F \tag{10}$$

The ratio P_S^S / P_L^S is termed the fugacity "ratio, F."

Second, the chemical may be gaseous at the system temperature; i.e., the system temperature exceeds the critical pressure, as occurs with methane (critical temperature 191 K). In such cases, neither the hypothetical P_L^S or γ can be estimated accurately. The usual approach is to combine P_L^S and γ as a single parameter which can be determined experimentally as the mole fraction based Henry's law constant H′, such that P equals H′x.

It is possible to estimate or correlate γ for a series of chemicals by a number of approaches. Examples are the correlations of Yalkowsky and Valvani (1976) using total molecular surface area, Leinonen et al. (1971) using molar volume, Mackay, Shiu, and Ma (1992a,b), Tsonopolous and Prausnitz (1971), and Kabadi and Danner (1979) using carbon number. A satisfactory degree of correlation often is found between γ and such properties for a homologous series. Other approaches are the "group contribution" method developed by Derr and Deal (1968) and the more sophisticated UNIFAC and UNIQUAC methods, both of which Reid et al. (1987) and Yalkowsky and Banerjee (1992) review and Chapter 7 discusses more fully.

Such correlations are invaluable as a means of estimating γ for compounds for which no data exist and for checking the "reasonableness" of other data.

Values of γ are most easily obtained from measurements of solubility in water. For a liquid solute in equilibrium with its aqueous solution,

$$x_L \gamma_L P_L^S = x_W \gamma_W P_L^S \tag{11}$$

where subscripts L and W refer to the liquid chemical and the water solution phases, respectively. If the solubility of water in the liquid solute is negligible, then x_L and γ_L become unity; thus x_W the mole fraction solubility is simply the reciprocal of γ_W. At high dilutions (i.e., $x_W \ll 1$), γ_W can be assumed to be constant, since its logarithm normally varies approximately in proportion to $(1 - x_W)^2$, which is essentially constant at low values of x_W.

For solid solutes, if the water does not affect the solid phase vapor pressure,

$$P_S^S = x_W \gamma_W P_L^S$$

thus

$$\gamma_W = \left(P_S^S / P_L^S\right) / x_W = F / x_W \tag{12}$$

For solids, a knowledge of the fugacity ratio F or (P_S^S / P_L^S) is thus necessary to calculate γ_W from the solubility x_W.

Air-water partition coefficients and Henry's law constants are strongly temperature dependent because of the temperature dependencies of vapor pressure and of solubility. H is also slightly dependent on the temperature dependence of water density and, hence, molar volume. The constants may be concentration dependent because of variations in γ_W, although the effect is believed to be negligible at low concentrations of non-associating solutes. Noted that these simple relationships break down at high concentrations, i.e., at mole fractions in excess of approximately 0.01. For most environmental situations, the concentrations are (fortunately) usually much lower. For thermodynamic purposes, H' is usually preferred, whereas for environmental purposes, H is more convenient.

4.2.1 Methods of Estimating H or K_{AW}

Four approaches could be used. First is by direct measurement of C_A and C_W or P and C_W at the temperature of interest, and possibly using water containing electrolytes or other chemical species of environmental interest. In practice it may be preferable to measure the concentration in only one phase and infer the other from a mass balance.

Second, for substances which are sparingly soluble in water, and in which water is sparingly soluble, H can be estimated as P^S / C_W^S. (The estimation procedures for P^S and C_W^S are discussed in chapters on vapor pressure and solubility. This procedure also applies to solids, but the reported solubility and vapor pressure must both be of the solid, and strictly of the solid in the same crystalline form, and at the specified temperature. It is erroneous to use one liquid-state and one solid-state property. Occasionally, vapor pressure data are obtained by extrapolation from higher temperatures above the melting point; thus the estimated quantity is a subcooled liquid vapor pressure. If the chemical is liquid, the solubility of water in the liquid should be less than a few percent; otherwise, erroneous results will be obtained. Again both measurements must apply to the liquid. For gaseous chemicals, i.e., the boiling point is below the temperature of interest, C_W^S can be measured at either P_W^S (implying measurement at a higher than atmospheric pressure) or, if C_W^S is measured at atmospheric pressure, then that (lower) pressure should be used instead of P^S.

Third, K_{AW} can be estimated from the actual or hypothetical liquid-state vapor pressure and the water phase activity coefficient γ_W, using Equation 8. The activity coefficient γ_W can in turn be estimated by a variety of methods, including group contribution and UNIQUAC approaches. For sparingly soluble liquids, γ_W usually can be estimated with acceptable accuracy as $1/x^S$ or $1/C^S v_W$, where x^S is the mole fraction solubility and v_W is the molar volume of the solution. For solids, the fugacity ratio must be estimated, as discussed earlier, from the melting point and an estimated or measured entropy of fusion.

Finally, it is possible to estimate or correlate K_{AW} or H directly from molecular structure, avoiding the separate determination of vapor pressure and solubility.

As noted earlier, Henry's law constants are very temperature sensitive; thus a temperature correction may be necessary, the conventional method being to estimate the enthalpy change ΔH_S J/mol from gaseous to solution state. This dependence can be expressed.

$$H_1/H_2 = \exp(-\Delta H_S/R(1/T_1 - 1/T_2)) \tag{13}$$

where subscripts 1 and 2 refer to the two temperatures. Any factor that influences the activity coefficient of the chemical in water, and thus influences solubility, also will affect H. Notable are the presence of electrolytes, other dissolved organic chemicals, surface active material, colloids, and suspended matter.

4.3   Experimental Methods

When assessing the reliability of reported Henry's law constant data, an appreciation of the methods used to obtain experimental data is useful.

Mackay et al. (1993) recently reviewed these methods, which fall into six categories.

4.3.1   Types of Methods

4.3.1.1   Batch Stripping (Mackay et al., 1979)

In this, the most common method, air is bubbled at a known rate into a volume of water containing the solute, so that the exit air achieves equilibrium with the water. By measuring the decrease in water concentration, K_{AW} can be deduced from a mass balance. No air phase concentrations are measured. This method is ideal for fairly volatile chemicals, i.e., when K_{AW} exceeds 10^{-3} but can be applied down to about 10^{-4}. Yin and Hassett (1986) modified the method for less volatile chemicals; the air phase solute concentration is measured by trapping solute from the exit air stream. Hovorka and Dohnal (1997) have refined the test conditions to achieve greater accuracy.

4.3.1.2   EPICS Method (Lincoff and Gossett, 1984)

This method (Equilibrium Partitioning in Closed Systems) is suited particularly for highly volatile solutes, i.e., when K_{AW} exceeds 0.05. Known volumes of solute in water solution are equilibrated with air in sealed vials. Measurement is made of head space (air) concentrations. Only air concentrations are measured; thus the method has high precision, especially if the modified method described by Gossett (1987) is used. Ashworth et al. (1988) have obtained a large number of data using this method, including temperature dependencies, for hydrocarbons and halogenated hydrocarbons. Yurteri et al. (1987)

also have reported data obtained with the EPICS method and have studied the effect of solutes in the water. Munz and Roberts (1987) have described a multiple-equilibration approach and its application to systems at various temperatures and with cosolvents present.

4.3.1.3   Wetted-Wall Column Method (Fendinger and Glotfelty, 1988, 1990)

The solute is equilibrated between a thin flowing film of water and concurrent air flow in a vertical column and the air and water phase concentrations measured. The method has been applied to pesticides and other less volatile chemicals.

4.3.1.4   Headspace Analyses Techniques

In this approach, which takes various forms, accurate headspace analyses are done in systems in which known aqueous concentrations are established. The principle was first demonstrated quantitatively by Hussam and Carr (1985) and has been developed by Perlinger (1990) and Resendes et al. (1992) to probe partitioning in systems in which appreciable sorption occurs. Schoene and Steinhauses (1985) and Ettre et al. (1993) have described an automated system of this type.

4.3.1.5   Gas Chromatographic Retention Time

Tse et al. (1992) have reported a gas chromatographic retention time method which is simple and rapid for estimating of Henry's law constants.

4.3.1.6   Determining Concentrations in Air and Water

Leighton and Calo (1981) describe a system in which both phases are sampled and analysed, thus directly determining K_{AW}. Vejrosta et al. (1992) and Jönsson et al. (1992) describe a system in which water equilibrated with a gas stream of known composition is analyzed.

4.3.2   Structure-Property Relationships

When examining data for Henry's law constants, it is useful to compare values with data for structurally similar compounds. For a homologous series such as the chlorobenzenes, the increase in molar volume or area associated with substitution of chlorine for hydrogen causes a decrease in both solubility and vapor pressure; thus H may be fairly constant for such a series. The ideal situation is one in which reliable independent experimental data are available for P_L^S, C_W^S, and H which permit a consistency check of the three determinations.

4.3.3   Data Sources

For existing chemicals, several sources of data exist for H and K_{AW}, including Mackay and Shiu (1981), Mackay, Shiu, and Ma (1992a,b and 1993, 1995, 1997), and Eastcott et al. (1988). References to specific chemicals or classes appear in the reference section of this chapter. These sources can be an invaluable basis for estimation purposes because H and K_{AW} vary systematically with molecular structure. It is often possible to estimate H for a substance from the reported H of a homologous compound.

A comprehensive review by Staudinger and Roberts (1996) discusses both the experimental determination and estimation of Henry's law constant.

Mackay and Shiu (1981) and Shiu and Mackay (1986) reviewed the physical-chemical properties of a variety of organic chemicals, polychlorinated biphenyls, and recommended H from a selected value of vapor pressure and aqueous solubility. Eastcott et al. (1988) and Suntio et al. (1988) also gave H for hydrocarbons and pesticide chemicals, respectively. Mackay, Shiu, and Ma (1992a, 1992b, 1993, 1995, 1997) compiled physical-chemical properties, including Henry's law constants for many organic chemicals of environmental interest.

Other reports of data on Henry's law constants include: Volatile halogenated organic chemicals: Glew and Moelwyn-Hughes (1953) on methyl halides; Pearson and McConnell (1975) and McConnell et al.(1975) on chlorinated C_1 and C_2 hydrocarbons; Dilling et al. (1975) and Dilling (1977) on chlorinated VOCs; Smith and Bomberger (1978) on chlorinated hydrocarbons and PAHs; Chiou et al. (1980) on chlorinated VOCs, chlorophenols, PCBs, and some pesticides; Leighton and Calo (1981) on chlorinated hydrocarbons; Matter-Müller et al. (1981) on volatile solutes; Mackay and Yuen (1983) on VOCs, including some ketones and alcohols; Nicholson et al. (1984) on trihalomethanes; Lincoff and Gossett (1984) and Gossett (1987) on C_1 and C_2 hydrocarbons, by the EPICS method; Munz and Roberts (1987, 1989) on volatile organic solutes, by multiple equilibration; Tancrede and Yanagisawa (1990) on selected VOCs, by measuring concentration of both phases; Hansen et al. (1993) on volatile VOCs, by EPICS; Dewulf et al. (1995) and Moore et al. (1995) on chlorinated VOCs and monoaromatic hydrocarbons, by gas stripping.

Turner et al. (1996) review data and methods and give some data on temperature dependence. Hovorka and Dohnal (1997) report particularly accurate data for VOCs including temperature dependence.

PCBs

Atlas et al. (1982) report on PCB isomers in distilled water and seawater; Bopp (1983) on PCB isomer groups; Burkard et al. (1985) on all PCB congeners from the ratio of liquid vapor pressure and aqueous solubility; Oliver (1985) on chlorobenzenes, chlorotoluenes, and PCBs, by the gas purging method; Murphy et al. (1987) on individual PCB congeners in Aroclor mixtures; Dunnivant and Elzerman (1988) and Dunnivant et al. (1988, 1992) on PCB congeners, by the batch stripping technique; Sabljic and Güsten (1989) on calculated H of PCB congeners, from molecular connectivity indices; Brunner et al. (1990) on 58 PCB congeners in Clophen mixtures, by the wetted-wall column technique; Sandler and coworkers, Wright et al. (1992), Tse and Sandler (1992), and Tse et al. (1994) on halogenated hydrocarbons by measuring infinite dilution activity coefficients.

Hydrocarbons and aromatic hydrocarbons

Mackay and Wolkoff (1973), Mackay and Leinonen (1975) report on alkanes and pesticides; Jönsson et al. (1982) on *n*-pentane to *n*-nonane by direct measuring both phases and Arbuckle (1983), by estimating activity coefficients; Yurteri et al. (1987) on chlorinated hydrocarbons and aromatics by the EPICS method; Ashworth et al. (1988) on VOCs and aromatic hydrocarbons, by both EPICS and batch stripping, with an evaluation of the experimental data by the UNIFAC method; Fendinger and Glotfelty (1990) on PAHs, PCBs, and pesticides, by the wetted-wall column technique; Pankow and coworkers (1986, 1988, 1990) on a variety of VOCs; Hulscher et al. (1992) on selected chlorobenzenes, PCBs, and PAHs, by the gas stripping technique; Robbins et al. (1993) on monoaromatics, by a static headspace method; Friesen et al. (1993) and Santl et al. (1994) on chlorinated dibenzofuran and dibenzodioxins, by the gas stripping technique; Dewulf et al. (1995), on chlorinated VOCs and monoaromatic hydrocarbons by the gas stripping method; Alaee et al. (1996) on PAHs, by the gas stripping method, and Shiu and Mackay (1997) on PAHs and other compounds, by the gas stripping method.

Pesticides

Atkins et al. (1971) report on some organochlorines; Yin and Hassett (1986) on mirex, by the gas stripping technique; Mackay et al. (1986) on some organochlorine compounds including Aroclor mixtures and phthalates, estimated from vapor pressure and solubility; Warner et al. (1987) on halogenated VOCs, PAHs, and organochlorine compounds by the gas stripping method; Fendinger and Glotfelty (1989) and Fendinger et al. (1988) on selected pesticides, using a wetted-wall column; Jury et al. (1990) on VOCs and some pesticides; Kucklick et al. (1991) on hexachlorocyclohexanes, by the gas stripping method; Cotham and Bidleman (1991) on organochlorine compounds.

Others

Other reports include the studies by Buttery et al. (1969) on aldehydes, ketones, esters, and some alcohols; Buttery et al. (1971) on some volatile organic flavor compounds; Snider and Dawson (1985) on alcohols, carbonyls, and nitriles; Leuenberger et al. (1985) on phenols; Abd-El-Bary et al. (1986) and Dohnal and Fenclova (1995) on phenols; Betterton and Hoffmann (1988) on aldehydes, by gas stripping; Zhou and Mopper (1990) on carbonyl compounds, by gas stripping; Sagebiel et al. (1992) on phenols, by gas stripping; Jayasinghe et al. (1992) on methylanilines; Brimblecome et al. (1992) and Bowden et al. (1996) on acids, by measuring partial pressures over aqueous solutions; and Hamelink et al. (1996) on a silicone (octamethylcyclotetrasiloxane).

4.4 Estimation Methods

Lyman (1982) presented a comprehensive review of methods of estimating Henry's law constants available at that time, discussing the use of (i) vapor pressure/solubility ratio, (ii) the fragment constant method of Hine and Mookerjee, (iii) the Cramer fragment constant method, and (iv) the activity coefficient-vapor pressure method using the UNIFAC approach to estimate activity coefficients. He concluded that method (i) is accurate, provided the solubility of the chemical in water is less than 1 mol/L. Of the fragment constant methods, that of Hine and Mookerjee (1975) was generally preferred. For very soluble chemicals and at higher concentrations, method (iv) was preferred. Predictably, the fragment constant methods perform best on the types of chemical which were used to develop the correlations.

More recently, Brennan et al. (1997) compared five methods for estimating K_{AW}, namely the vapor pressure/solubility ratio, the group or bond contribution method, linear solvation energy methods, and molecular connectivity. The authors compared the methods by application to a common set of 150 chemicals and concluded that the Meylan and Howard (1991) bond contribution method and the molecular connectivity index method of Nirmalakhandan and Speece (1988) are comparably accurate, having standard deviations of, 0.29 and 0.34 log units, respectively.

4.4.1 Fragment Constant

Hine and Mookerjee (1975) correlated log K_{AW}, using bond and group contribution approaches with corrections for polar interactions. Data for 292 compounds were used, and the standard deviation was about 0.4 log units or a factor of 2.5 in K_{AW}. Greater accuracy was achieved for the better characterized homologous series, but reliable experimental data exist for these chemicals, thus estimation methods are rarely needed. From an inspection of the fit, it appears that, in most cases (i.e., 19 out of 20), accuracy is within a factor of 6. The method is now somewhat dated, since many new and better data have become available in the last 25 years.

Meylan and Howard (1991) essentially updated and revised this method. They list 59 bond contributions, derived from a data set for 345 chemicals. By using only the bond contributions a standard deviation of 0.45 log units was obtained, with a mean error of 0.30 log units. If correction factors are applied for certain classes, for example, for cyclic alkanes or epoxides, the standard deviation falls to 0.34 and the mean error to 0.21 units. When applied to a validation set not used to develop the correlation, the standard deviation was 0.46 log units and the mean error 0.31 log units. This method is simple and fairly accurate. It is also available as part of a suite of estimation methods (the EPIWIN Suite) from the Syracuse Research Corporation.

4.4.2 Connectivity Indices

Nirmalakhandan and Speece (1988) developed a correlation for $\log K_{AW}$ using connectivity indices χ (which have been developed, exploited, and discussed by Kier and Hall [1986]) and the polarizability Φ. The standard error was 0.445 log units.

A second improved correlation from data for 180 chemicals also was developed using only Φ and the first order connectivity $^1\chi^v$, but with addition of an indicator variable I which can adopt a value of 1 for hydrogen bonding compounds and zero for others. The correlation is

$$\log K_{AW} = 1.29 + 1.005 \ \Phi - 0.468 \ ^1\chi^v - 1.258 \ I \tag{14}$$

which has a standard error of 0.262 log units.

Nirmalakhandan et al. (1997) later applied this estimation method to 105 new chemicals, including oxygen, sulfur, and nitrogen, containing compounds which were not present in the original set. The average factor of error (AFE) was 3.57, with 80% of the estimations being less than an AFE of 3.0. They also developed an estimation method for K_{AW} as a function of absolute temperature in the form of the equation

$$\ln K_{AW} = A - B/T$$

where

$$A = 14.97 + 5.78°\chi - 11.79 \ ^1\chi^v$$

$$B = 4346 + 947°\chi - 1855 \ ^1\chi^v$$

The AFE for 170 data points obtained for 20 chemicals was 1.74.

This method is thus about equivalent in accuracy to the bond contribution method, but the connectivity index does contain information about molecular configuration or topology which is absent from the simple bond contribution method. It thus is inherently more likely to express differences between isomers. Its primary disadvantage is the need to deduce the indices, which can be difficult to the uninitiated. The indices lack physical meaning, which is worrisome to those who seek to understand fully the inherent nature and principles of the estimation method.

4.4.3 Other Methods

Suzuki et al. (1992) used principal components analysis to develop a combined connectivity, group contribution method for 229 monofunctional compounds from the Hine and Mookerjee data set. The average error was 0.14 log units.

Russell et al. (1992) developed a five parameter model for 63 compounds. The parameters include surface area and atomic charges selected from an initial list of 165 descriptors. The standard deviation was 0.375 units.

A linear solvation energy relationship (LSER) method was developed by Abraham et al. (1994) using five "solvatochronic" parameters for 408 chemicals. This method is related to the LSER method described in Chapter 7. Obtaining the parameter values can be demanding and difficult, but is potentially powerful.

Several groups have developed correlations specifically for PCBs. Sabljic and Gusten (1989) applied a similar approach to Nirmalakhandan and Speece, using two, fourth order connectivity indices and obtained a correlation with a standard error of 0.31 log units. Dunnivant et al. (1992) also have used connectivity and polarizability.

Hawker (1989) used total surface area, as did Burkhard et al. (1985). Shiu and Mackay (1986) used molar volume as a single descriptor.

Brunner et al. (1990) used the much simpler chlorine number and number of ortho chlorines to correlate K_{AW} for PCBs, as well as connectivity indices. These correlations are primarily of interest for revealing the molecular causes of variations in K_{AW} for this class of chemicals. They do not apply to other classes, but they do demonstrate the feasibility of producing fairly accurate correlations for a restricted class of compounds.

Correlations have also been reported for dioxins by Shiu et al. (1988), and for aromatic hydrocarbons by Eastcott et al. (1988).

4.5 Recommended Methods

In the absence of actual experimental data, we recommend method 1 below to estimate H or K_{AW}. We then recommend methods 2 and 3 as comparable in accuracy. It may be prudent to apply both methods. Finally (method 4), it may be possible to infer a value of K_{AW} from those reported for structurally similar compounds. Two of these methods are illustrated for four benchmark chemicals in the Appendix.

4.5.1 Method 1: Estimation from Vapor Pressure P^S and Solubility Data C_W^S

This method uses Equation 4 and is ideal when experimental data are available. It is also possible to use estimation methods for P^S and C_W^S, provided one recognizes that there is a compounding of estimation errors. The method should not be used when the solubility of the substance in water or water in the substance is appreciable, i.e., either exceeds 10%. In the range 1 to 10%, the result should be viewed with caution. Take care when the vapor pressure exceeds atmospheric pressure (i.e., the environmental temperature exceeds the normal boiling point) to ensure that the solubility was measured at high pressure.

TABLE 4.1

Bond Contributions as Suggested by Meylan and Howard (1991)[a]. Copyright by Pergamon Press. Reproduced with permission.

Bond[b]	Value	Bond[b]	Value
C–H	–0.1197	C_{ar}–OH	0.5967[c]
C–C	0.1163	C_{ar}–O	0.3473[c]
C–C_{ar}	0.1619	C_{ar}–N_{ar}	1.6282
C–C_d	0.0635	C_{ar}–S_{ar}	0.3739
C–C_t	0.5375	C_{ar}–O_{ar}	0.2419
C–CO	1.7057	C_{ar}–S	0.6345
C–N	1.3001	C_{ar}–N	0.7304
C–O	1.0855	C_{ar}–I	0.4806
C–S	1.1056	C_{ar}–F	–0.2214
C–Cl	0.3335	C_{ar}–C_d	0.4391
C–Br	0.8187	C_{ar}–CN	1.8606
C–F	–0.4184	C_{ar}–CO	1.2387
C–I	1.0074	C_{ar}–Br	0.2454
C–NO_2	3.1231	C_{ar}–NO_2	2.2496
C–CN	3.2624	CO–H	1.2102
C–P	0.7786	CO–O	0.0714
C=S	–0.0460	CO–N	2.4261
C_d–H	–0.1005	CO–CO	2.4000
C_d=C_d	0.0000[d]	O–H	3.2318
C_d–C_d	0.0997	O–P	0.3930
C_d–CO	1.9260	O–O	–0.4036
C_d–Cl	0.0426	O=P	1.6334
C_d–CN	2.5514	N–H	1.2835
C_d–O	0.2051	N–N	1.0956[e]
C_d–F	–0.3824	N=O	1.0956[e]
C_t–H	0.0040	N=N	0.1374
C_t≡C_t	0.0000[d]	S–H	0.2247
C_{ar}–H	–0.1543	S–S	–0.1891
C_{ar}–C_{ar}	0.2638[f]	S–P	0.6334
C_{ar}–C_{ar}	0.1490[g]	S=P	–1.0317
C_{ar}–Cl	–0.0241		

[a] At 25°C.

[b] C: single-bonded aliphatic carbon; C_d: olefinic carbon; C_t: triple-bonded carbon; C_{ar}: aromatic carbon; N_{ar}: aromatic nitrogen; S_{ar}: aromatic sulfur; O_{ar}: aromatic oxygen; CO: carbonyl (C=O); CN: cyano (C≡N). Note: the carbonyl, cyano, and nitro functions are treated as single atoms.

[c] Two separate types of aromatic carbon-to-oxygen bonds have been derived: (a) the oxygen is part of an –OH function, and (b) the oxygen is not connected to hydrogen.

[d] The C=C and C≡C bonds are assigned a value of zero by definition.

[e] Value is specific for nitrosamines.

[f] Intra-ring aromatic carbon to aromatic carbon.

[g] External aromatic carbon to aromatic carbon (e.g., biphenyl).

4.5.2   Method 2: Bond Contribution Method of Meylan and Howard (1991)

The Hine and Mookerjee method has lasted well but is now outdated, and the similar Meylan and Howard correlation supercedes it. This method has the significant advantage of transparency and simplicity; thus it is preferred by many over the more complex and less intuitively satisfying connectivity/polarizability methods. If time is available, and espe-

TABLE 4.2

Bond Contribution Correction Factors (Meylan and Howard, 1991). Copyright by Pergamon Press. Reproduced with permission.

Linear or branched alkane[a]	–0.75
Cyclic alkane[a]	–0.28
Monoolefin[a,b]	–0.20
Cyclic monoolefin[a,b]	+0.25
Linear or branched aliphatic alcohol[a]	–0.20
Adjacent aliphatic ether functions (–C–O–C–O–C–)	–0.70
Cyclic monoether	+0.90
Epoxide	+0.50
Each additional aliphatic alcohol function (–OH) above one	–3.00
Each additional aromatic nitrogen within a single ring above one	–2.50
A fluoroalkane with only one function	+0.95
A chloroalkane with only one chlorine	+0.50
A totally chlorinated chloroalkane	–1.35
A totally fluorinated fluoroalkane	–0.60
A totally halogenated halofluoroalkane	–0.90

[a] Can have no substituents except alkyl groups.
[b] Can have only one olefinic double bond.

cially if the estimator is familiar with the connectivity approach, both methods can be applied and the results compared and possibly averaged, or some weighting applied on the basis that the Meylan and Howard method is generally better for hydrocarbons, haloaromatics and haloalkanes, esters, alcohols, and phenols, whereas the Nirmalakhandan and Speece method performs better for haloalkanes and acids.

Table 4.1, reproduced from Meylan and Howard (1991), lists the 59 bond contributions, and Table 4.2 lists correction factors. These groups and factors when combined, yield the log (base 10) water to air partition coefficient. Table 4.3 compares the two methods and can be used to guide the method selection process. The original paper gives specimen calculations

4.5.3 Method 3: Molecular Connectivity Index Method of Nirmalakhandan and Speece (1988) and Nirmalakhandan et al. (1997)

For those familiar with connectivity indices, this is an excellent method, but for those who lack this familiarity, it will prove time-consuming and initially difficult.

4.5.4 Structure-Property Relationships with Similar Compounds

A fourth approach is to seek reported experimental data for substances similar in chemical class or structure to the substance of interest exploiting the data base listed earlier and develop a simple structure property relationship for this specific class, using molecular weight volume as a descriptor. This approach is likely to be fairly accurate, provided the original data are accurate.

TABLE 4.3

Comparison of Mean Class Errors for the Meylan and Howard Method and Nirmalakhandan and Speece Method. Copyright by Pergamon Press. Reproduced with permission.

	Mean Error (In Log Units)	
Chemical Class	Bond Contribution Method	Nirmalakhandan and Speece Method
Alkanes	0.062	0.096
Cycloalkanes	0.091	0.104
Alkenes	0.132	0.157
Alkynes	0.080	0.149
Alkylbenzenes	0.063	0.249
Halomethanes	0.272	0.258
Haloethanes	0.384	0.349
Halopropanes	0.310	0.249
Halobutanes	0.196	0.134
Halopentanes	0.067	0.113
Haloalkenes	0.425	0.560
Halobenzenes	0.085	0.249
Acids, aliphatic	0.330	0.140
Esters	0.121	0.152
Alcohols	0.061	0.106
Phenols	0.186	0.306

4.6 Appendix

In this appendix, H and K_{AW} are estimated for four of the benchmark chemicals using reported data on solubility and vapor pressure and the Meylan and Howard method.

Anthracene (MW 178.2)
From solubility and vapor pressure.
Solubility 0.045 g/m^3 = 2.53×10^{-4} mol/m^3
Vapor pressure 0.0010 Pa
H = 3.96 Pa m^3/mol
K_{AW} = H/RT = 3.96/(8.314 . 298) = 0.0016
measured values of H range from 1.96 to 6.59
Bond contribution method of Meylan and Howard
There are 16 intra-ring C – aromatic C bonds
and 10 aromatic C – hydrogen bonds
log K_{WA} = 16(0.2638) + 10(–0.1543) = 2.68
K_{WA} = 476, K_{AW} = 0.0021

Trichloroethylene (MW 131.4)
From solubility and vapor pressure
Solubility 1100 g/m^3 = 8.37 mol/m^3
Vapor pressure 9900 Pa
H = 1183 Pa m^3/mol
K_{WA} = 0.477

Reported values range from 0.3 to 0.5

Lindane (γ hexachlorocyclohexane) (MW 290.85)
Solubility = $7.3 \text{ g/m}^3 = 0.0251 \text{ mol/m}^3$
Vapor pressure = 0.00374 Pa
$H = 0.147 \text{ K}_{AW} = 6.0 \times 10^{-5}$
Reported values of H range from 0.1 to 0.3
There are six C–C bonds, six C–Cl bonds and six C–H bonds
$\log K_{WA} = 6(0.1163) + 6(0.3335) + 6(-0.1197) = 1.98$
$K_{WA} = 95, K_{AW} = 0.0105$

2,6 di(tert butyl) phenol
No solubilities, vapor pressures, or Henry's law constants are available.
There are 6 intra aromatic C-aromatic bonds, 2 C-aromatic C bonds
1 aromatic C-OH bond and 6 C-H bonds
$\log K_{WA} = 6(0.2638) + 2(0.1619) + (0.5967) + 6(-0.1197) = 1.785$
$K_{WA} = 61, K_{AW} = 0.0164$

References

Abd-El-Bary, M.F. and Hamoda, M.F. 1986. Henry's constants for phenol over its diluted aqueous solution. *J. Chem. Eng. Data* 31, 229–230.

Abraham, M.H., Andonian-Haftvan, J., Whiting, G.S., Leo, A., and Taft, S. 1994. Hydrogen bonding 34: the factors that influence the solubility of gases and vapors in water at 298 K, and a new method for its determination. *J. Chem. Soc. Perkin. Trans.* 2, 1777–1791.

Alaee, M., Whittal, R.M., and Strachan, W.M.J. 1996. The effect of water temperature and composition on Henry's law constant for various PAH's. *Chemosphere* 32, 1153–1164.

Arbuckle, W.B. 1983. Estimating acitivity coefficients for use in calculating environmental parameters. *Environ. Sci. Technol.* 17, 537–542.

Ashworth, R.A., Howe, G.B., Mullins, M.E., and Rogers, T.N. 1988. Air-water partitioning coefficients of organics in dilute aqueous solutions. *J. Hazard. Materials* 18, 25–36.

Atkins, D.H.F. and Eggleton, A.E.J. 1971. Studies of atmospheric wash-out and deposition of Δ-BHC, dieldrin and p,p'-DDT using radio-labelled pesticides. *Int. Atomic Energy Agency Symp., Vienna,* SM-142a/32, pp. 521–533.

Atlas, E., Foster, R., and Giam, C.S. 1982. Air-sea exchange of high molecular weight organic pollutants: Laboratory studies. *Environ. Sci. Technol.* 16, 283–286.

Betterton, E.A. and Hoffmann, M.R. 1988. Henry's law constants of some environmentally important aldehydes. *Environ. Sci. Technol.* 22, 1415–1418.

Brennan, R.A., Nirmalakhandon, N., and Speece, R.E. 1997. Comparison of predictive methods for Henry's law constants of organic chemicals. *Water Res.* (in press).

Brimblecombe, P., Clegg, S.L., and Khan, I. 1992. Thermodynamic properties of carboxylic acids relevant to their solubility in aqueous solutions. *J. Aerosol. Sci.* 23, S901–S904.

Bopp, R.F. 1983. Revised parameters for modeling the transport of PCB components across an air water interface. *J. Geophys. Res.* 88, 2521–2529.

Bowden, D.J., Clegg, S.L., and Brimblecombe, P. 1996. The Henry's law constant of trifluoroacetic acid and its partitioning into liquid water in the atmosphere. *Chemosphere* 32, 405–420.

Brunner, S., Hornung, E., Santl, H., Wolff, E., Piringer, O.G., Altschuh, J., and Brüggemann, R. 1990. Henry's law constants for polychlorinated biphenyls: Experimental determination and structure-property relationships. *Environ. Sci. Technol.* 24, 1751–1754.

Burkhard, L.P., Armstrong, D.E., and Andren, A.W. 1985. Henry's law constants for polychlorinated biphenyls. *Environ. Sci. Technol.* 19, 590–595.

Buttery, R.G., Ling, L.C., and Guadagni, D.G. 1969. Volatilities of aldehydes, ketones, and ester in dilute water solution. *J. Agr. Food Chem.* 17, 385–389.

Buttery, R.G., Bomben, J.L., Guadagni, D.G., and Ling, L.C. 1971. Some considerations of the volatilities of organic flavor compounds in foods. *J. Agr. Food Chem.* 19, 1045–1048.

Chiou, C.T., Freed, V.H., Peters, L.J. and Kohnert, R.L. 1980. Evaporation of solutes from water. *Environment International*, Vol. 3.

Cotham, Jr., W.E., and Bidleman, T.F. 1991. Estimating the atmospheric deposition of organochlorine contaminants to the arctic. *Chemosphere* 22, 165–188.

Derr, C.H. and Deal, E.L. 1968. Group contributions in mixtures. *Ind. Eng. Chem.* 60, 28–38.

Dewulf, J., Drijvers, D., and van Langenhove, H. 1995. Measurement of Henry's law constant as function of temperature and salinity for the low temperature range. *Atm. Environ.* 39, 323–331.

Dilling, W.L., Tefertiller, N.B., and Kallos, G.J. 1975. Evaporation rates and reactivities of methylene chloride, chloroform, 1,1,1-trichloroethane, trichloroethylene, tetrachloroethylene, and other chlorinated compounds in dilute aqueous solutions. *Environ. Sci. Technol.* 9, 833–838.

Dilling, W.L. 1977. Interphase transfer processed. II. Evaporation rates of chloromethanes, ethanes, ethylenes, propanes, and propylenes from dilute aqueous solutions. Comparisons with theorectical predictions. *Environ. Sci. Technol.* 11, 405–409.

Dohnal, V. and Fenclova, D. 1995. Air-water partitioning and aqueous solubility of phenols. *J. Chem. Eng. Data* 40, 478–483.

Dunnivant, F.M. and Elzerman, A.W. 1988. Aqueous solubility and Henry's law constant data for PCB congeners for evaluative structure-property relationships (QSPRs). *Chemosphere* 17, 525–541.

Dunnivant, F.M., Coates, J.T., and Elzerman, A.W. 1988. Experimentally determined Henry's law constants for 17 polychlorobiphenyl congeners. *Environ. Sci. Technol.* 22, 448–453.

Dunnivant, F.M., Eizerman, A.W., Jurs, P.C., and Hasan, M.N. 1992. Quantitative structure-property relationships for aqueous solubilities and Henry's law constants of polychlorinated biphenyls. *Environ. Sci. Technol.* 26, 1567–1573.

Eastcott, L., Shiu, W.Y., and Mackay, D. 1988. Environmentally relevant physical-chemical properties of hydrocarbons: A review of data and development of simple correlations. *Oil & Chem. Pollut.* 4, 191–216.

Ettre, L.S., Welter, C., and Kolb, B. 1993. Determination of gas-liquid partition coefficients by automatic equilibrium headspace-gas chromatography utilizing the phase ratio variation method. *Chromatographia* 35, 73–84.

Fendinger, N.J., and Glotfelty, D.E. 1988. A laboratory method for the experimental determination of air-water Henry's law constants for several pesticides. *Environ. Sci. Technol.* 22, 1289–1293.

Fendinger, N.J. and Goltfelty, D.E. 1990. Henry's law constants for selected pesticides PAHs and PCBs. *Environ. Toxicol. Chem.* 9, 731–735.

Fendinger, N.J., Goltfelty, D.E., and Freeman, H.P. 1989. Comparison of two experimental techniques for determining air/water Henry's law constants. *Environ. Sci. Technol.* 23, 1528–1531.

Friesen, K.J., Fairchild, W.L., and Lowen, M.D. 1993. Evidence for particle-mediated transport of 2,3,7,8-tetrachlorodibenzofuran during gas sparging of natural water. *Environ. Toxicol. Chem.* 12, 2307–2344.

Glew, D.N. and Moelwyn-Hughes, E.A. 1953. Chemical statics of the methyl halides in water. *Disc. Farad. Soc.* 15, 150–161.

Gossett, R. 1987. Measurement of Henry's law constants for C_1 and C_2 chlorinated hydrocarbons. *Environ. Sci. Technol.* 21, 202–208.

Hamelink, J.L., Simon, P.B., and Silberhorn, E.M. 1996. Henry's law constant volatilization rate, and aquatic half-life of Octamethylcyclotetrasiloxane. *Environ. Sci. Technol.* 30, 1946–1952.

Hansen, K.C., Zhou, Z., Yaw, C.L., and Aminabhavi, T.M. 1993. Determination of Henry's law constants of organic in dilute aqueous solutions. *J. Chem. Eng. Data* 38, 546–550.

Hawker, D.W. 1989. Vapor pressures and Henry's law constants of polychlorinated biphenyls. *Environ. Sci. Technol.* 23, 1250–1253.

Hine, J. and Mookerjee, P.K. 1975. The intrinsic hydrophilic character of organic compounds. Correlations in terms of structural contributions. *J. Org. Chem.* 40(3), 292–298.

Hovorka, S. and Dohnal, V. 1997. Determination of air-water partitioning of volatile halogenated hydrocarbons by the inert gas stripping method. *J. Chem. Eng. Data,* 42, 924–933.

Hussam, A. and Carr, P.W. 1985. Rapid and precise method for the measurement of vapor/liquid equilibria by headspace gas chromatography. *Anal. Chem.* 57, 793–801.

Jayasinghe, D.S., Brownawell, B.J., Chen, H., and Westall, J.C. 1992. Determination of Henry's constants of organic compounds of low volatility: methylanilines in methanol-water. *Environ. Sci. Technol.* 26, 2275–2281.

Jönsson, J.A., Vejrosta, J., and Novak, J. 1982. Air/water partition coefficients for normal alkanes (*n*-pentant to *n*-nonane). *Fluid Phase Equil.* 9, 279–286.

Jury, W.A., Russo, D., Streile, G., and El Abd, H. 1990. Evaluation of volatilization by organic chemicals residing below the soil surface. *Water Resour. Res.* 26, 13–20.

Kabadi, V.N. and Danner, R.P. 1979. Nomograph solves for solubilities of hydrocarbons in water. *Hydrocarbon Processing* 58, 245–246.

Khan, I. and Brimblecombe, P. 1992. Henry's law constants of low molecular weight (<130) organic acids. *J. Aerosol. Sci.* 23, S897–S990.

Kier, L.B. and Hall, L.H. 1986. *Molecular Connectivity in Structure-Activity Analysis.* Wiley, New York.

Kucklick, J.R., Hinckley, D.A., and Bidleman, T.F. 1991. Determination of Henry's law constants for hexachlorocyclohexanes in distilled water and artificial seawater as a function of temperature. *Marine Chem.* 34, 197–209.

Leighton, D.T. and Calo, J.M. 1981. Distribution coefficients of chlorinated hydrocarbons in dilute air-water systems for groundwater contamination applications. *J. Chem. Eng. Data* 26, 381–385.

Leinonen, P.J., Mackay, D., and Phillips, C.R. 1971. A correlation for the solubility of hydrocarbons in water. *Can. J. Chem. Eng.* 49, 288–290.

Leuenberger, C., Ligocki, M.P., and Pankow, J.F. 1985. Trace organic compounds in rain. 4. Identities, concentrations, and scavenging mechanisms for phenols in urban air and rain. *Environ. Sci. Technol.* 19, 1053–1058.

Lincoff, A.H. and Gossett, R.M. 1984. The determination of Henry's law constants for volatile organics by equilibrium partitioning in closed systems. In *Gas Transfer at Water Surfaces.* Brutsaert, W. and Jirka, G.H., Eds., pp. 17–26, D. Reidel Publishing Co., Dordrecht, Holland.

Lyman, W.J., Reehl, W.F., and Rosenblatt, D.H. 1982. *Handbook of Chemical Property Estimation Methods,* McGraw-Hill, New York.

Mackay, D., and Wolkoff, A.W. 1973. Rate of evaporation of low-solubility contaminants from water bodies to atmosphere. *Environ. Sci. Technol.* 7, 611–614.

Mackay, D. and Leinonen, P.J. 1975. Rate of evaporation of low-solubility contaminants from water bodies to atmosphere. *Environ. Sci. Technol.* 9. 1178–1180.

Mackay, D., Shiu, W.Y., and Sutherland, R.P. 1979. Determination of air-water Henry's law constants for hydrophobic pollutants. *Environ. Sci. Technol.* 13, 333–337.

Mackay, D. and Shiu, W.Y. 1981. A critical review of Henry's law constants for chemicals of environmental interest. *J. Phys. Chem. Ref. Data* 10, 1175–1199.

Mackay, D. and Yuen, A.T.K. 1983. Mass transfer coefficient correlations for volatilization of organic solutes from water. *Environ. Sci. Technol.* 17, 211–217.

Mackay, D., Paterson, S., and Schroeder, W.H., 1986. Model describing the rates of transfer processes of organic chemicals between atmosphere and water. *Environ. Sci. Technol.* 20, 810–816.

Mackay, D., Shiu, W.Y., and Ma, K.C. 1992a. *Illustrated Handbook of Physical-Chemical Properties and Environmental Fate for Organic Chemicals. Vol. I Monoaromatic Hydrocarbons, Chlorobenzenes and PCBs.* Lewis Publishers, CRC Press, Boca Raton, FL.

Mackay, D., Shiu, W.Y., and Ma, K.C. 1992b. *Illustrated Handbook of Physical-Chemical Properties and Environmental Fate for Organic Chemicals. Vol. II Polynuclear Aromatic Hydrocarbons, Polychlorinated Dioxins and Dibenzofurans.* Lewis Publishers, CRC Press, Boca Raton, FL.

Mackay, D., Shiu, W.Y., and Ma, K.C. 1993. *Illustrated Handbook of Physical-Chemical Properties and Environmental Fate for Organic Chemicals. Vol. III Volatile Organic Chemicals.* Lewis Publishers, CRC Press, Boca Raton, FL.

Mackay, D., Shiu, W.Y., and Ma, K.C. 1995. *Illustrated Handbook of Physical-Chemical Properties and Environmental Fate for Organic Chemicals. Vol. IV Oxygen, Nitrogen and Sulfur Compounds.* Lewis Publishers, CRC Press, Boca Raton, FL.

Mackay, D., Shiu, W.Y., and Ma, K.C. 1997. *Illustrated Handbook of Physical-Chemical Properties and Environmental Fate for Organic Chemicals. Vol. V Pesticide Chemicals.* Lewis Publishers, CRC Press, Boca Raton, FL.

Matter-Müller, C., Gujer, W., and Giger, W. 1981. Transfer of volatile substances from water to the atmosphere. *Water Res.* 15, 1271–1279.

McConnell, G., Ferguson, D.M., and Pearson, C.R. 1975. Chlorinated hydrocarbons and the environment. *Endeavour* 34, 13–18.

Meylan, W.M. and Howard, P.H. 1991. Bond contribution method for estimating Henry's Law Constants. *Environ. Toxicol. Chem.* 10, 1283–1293.

Moore, R.M., Geem, C.E., and Tait, V.K. 1995. Determination of Henry's law constants for a suite of naturally occurring halogenated methanes in seawater. *Chemosphere* 30, 1183–1191.

Munz, C. and Roberts, P.V. 1987. Air-water phase equilibria of volatile organic solutes. *J. Am. Water Work Assoc.* 79, 62–69.

Munz, C. and Roberts, P.V. 1989. Gas- and liquid-phase mass transfer resistances of organic compounds during mechanical surface aeration. *Water Res.* 23, 589–601.

Murphy, T.J., Mullin, M.D., and Meyer, J.A. 1987. Equilibration of polychlorinated biphenyls and toxaphene with air and water. *Environ. Sci. Technol.* 21, 155–162.

Nicholson, B.C., Maguire, B.P., and Bursill, D.B. 1984. Henry's law constants for the trihalomethanes: Effects of water composition and temperature. *Environ. Sci. Technol.* 18, 518–521.

Nirmalakhandan, N.N., and Speece, R.E. 1988. QSAR model for predicting Henry's law constant. *Environ. Sci. Technol.* 22, 1349–1357.

Nirmalakhandan, N.N., Brennan, R.A., and Speece, R.E. 1997. Predicting Henry's law constant and the effect of temperature on Henry's law constant. *Water Res.* 31, 1471–1481.

Oliver, B.G. 1985. Desorption of chlorinated hydrocarbons from spiked and anthropogenically contaminated sediments. *Chemosphere* 14, 1087–1106.

Pankow, J.M. 1986. Magnitude of artifacts caused by bubbles and headspace in the determination of volatile compounds in water. *Anal. Chem.* 58, 1822-1826.

Pankow, J.M. and Rosen, M.E. 1988. Determination of volatile compounds in water by purging directly to a capillary column with whole column cryotrapping. *Environ. Sci. Technol.* 22, 398–405.

Pankow, J.M. 1990. Minimization of volatilization losses during sampling and analysis of volatile organic compounds in water. In *Significance and Treatment of Volatile Organic Compounds in Water Supplies.* Ram, N.M., Christman, R.F., and Cantor, K.P., Eds., Chapter 5, pp. 73-86. Lewis Publishers, Inc., Chelsea, MI.

Pearson, C.R. and McConnell, G. 1975. Chlorinated C_1 and C_2 hydrocarbons in the marine environment. *Proc. R. Soc. Lond.* B. 189, 305–332.

Perlinger, J. 1990. M.Sc. Thesis, Department of Civil and Mineral Engineering, Institute of Technology, Unversity of Minnesota, Minnesota.

Reid, R.C., Prausnitz, J.M., and Poling, B.E. 1987. *The Properties of Gases and Liquids.* 3rd ed., McGraw-Hill, New York.

Resendes, J., Shiu, W.Y., and Mackay, D. 1992. Sensing the fugacity of hydrophobic organic chemicals in aqueous systems. *Environ. Sci. Technol.* 26, 2381–2387.

Robbins, G.A., Wang, S., and Stuart, J.D. 1993. Using the static headspace method to determine Henry's law constants. *Anal. Chem.* 65, 3113–3118.

Russell, C.J., Dixon, S., and Jrus, P. 1992. Computer-assisted study of the relationship between molecular structure and Henry's law constant. *Anal. Chem.* 64, 1350–1355.

Sabljic, A. and Güsten, H. 1989. Predicting Henry's law constants for polychlorinated biphenyls. *Chemosphere* 19, 1665–1676.

Sagebiel, J.C., Seiber, J.N., and Woodrow, J.E. 1992. Comparison of headspace and gas-stripping methods for determining the Henry's law constant H for organic compounds of low to intermediate H. *Chemosphere* 25, 1763–1768.

Santl, H., Brandsch, R., and Gruber, L. 1994. Experimental determination of Henry's law constant HLC for some lower chlorinated dibenzodioxins. *Chemosphere* 29, 2209–2214.

Schoene, K. and Steinhauses, J. 1985. Determination of Henry's law constant of automated head space-gas chromatography. *Fresenius Z Anal. Chem.* 321, 538–543.

Shiu, W.Y. and Mackay, D. 1986. A critical review of aqueous solubilities, vapor pressure, Henry's law constants, and octanol-water partition coefficients of the polychlorinated biphenyls. *J. Phys. Chem. Ref. Data* 15, 911–929.

Shiu, W.Y., Doucette, W., Gobas, F.A.P.C., Andren, A., and Mackay, D. 1988. Physical chemical properties of chlorinated dibenzo-p-dioxins. *Environ. Sci. Technol.* 22, 651–658.

Shiu, W.Y. and Mackay, D. 1997. Henry's law constants of selected aromatic hydrocarbons, alcohols, and ketones. *J. Chem. Eng. Data* 42, 27–30.

Smith, J.H. and Bomberger, D.C. 1978. Prediction of volatilization rates of chemicals in water. *AIChE Symposium Series* 75 (190) 375–381.

Snider, J.R. and Dawson, G.A. 1985. Tropospheric light alcohols, and actonitrile: concentrations in the southwestern United States and Henry's law data. *J. Geophys. Res.* 90, 3797–3805.

Staudinger, J. and Robert, P.V. 1996. A critical review of Henry's law constants for environmental applications. *Critical Rev. in Environ. Sci. & Technol.* 26, 205–297.

Suntio, L.R., Shiu, W.Y., Mackay, D., Seiber, J.N., and Glotfelty, D. 1988. Critical review of Henry's law contants for pesticides. *Rev. Environ. Contam. Toxicol.* 103, 1–59.

Suzuki, T., Ohtaguchi, K., and Koide, K. 1992. Application of principal components analysis to calculated Henry's constant from molecular structure. *Comput. Chem.* 16, 41–52.

Tancrede, M.V. and Yanagisawa, Y. 1990. An analytical method to determine Henry's law constants for selected volatile organic compounds at concentrations and temperatures corresponding to tap water use. *J. Air Waste Manage Assoc.* 40, 1658–1663.

Ten Hulscher, Th.E.M., Van der Velde, L.E., and Bruggeman, W.A. 1992. Temperature dependence of Henry's law constants for selected chlorobenzenes, polychlorinated biphenyls and polycyclic aromatic hydrocarbons. *Environ. Toxicol. Chem.* 11, 1595–1603.

Tse, G., Orbey, H., and Sandler, S.I. 1992. Infinite dilution activity coefficients and Henry's law coefficients of some priority water pollutants determined by a relative gas chromatographic method. *Environ. Sci. Technol.* 26, 2017–2022.

Tse, G. and Sandler, S.I. 1994. Determination of infinite dilutions activity coefficients and 1-octanol/water partition coefficients of volatile organic pollutants. *J. Chem. Eng. Data* 39, 354–357.

Tsonopoulos, C. and Prausnitz, J.M. 1971. Activity coefficients of aromatic solutes in dilute aqueous solutions. *Ind. Eng. Chem. Fundam.* 10, 593–600.

Turner, L.H., Chiew, Y.C., Ajlert, R.C., and Kosson, D.S. 1996. Measuring vapor-liquid equilibrium for aqueous-organic systems: review and a new technique. *AIChE Journal* 42, 1772–1788.

Vejrosta, J., Novak, J., and Jonsson, J.A. 1992. A method for measuring infinite-dilution partition coefficients of volatile compounds between the gas and liquid phases of aqueous systems. *Fluid Phase Equilibrium* 8, 25–35.

Warner, M.P., Cohon, J.M., and Ireland, J.C. 1987. Determination of Henry's law constants of selected priority pollutants. EPA/600/D-87/229, USEPA, Cincinnati, OH 45268.

Wright, D.A., Sandler, S.I., and DeVoll, D. 1992. Infinite dilution activity coefficients and solubilities of halogenated hydrocarbons in water at ambient temperatures. *Environ. Sci. Technol.* 26, 1828–1831.

Yalkowsky, S.H. and Valvani, S.C. 1976. Partition coefficients and surface areas of some alkylbenzenes. *J. Med. Chem.* 19, 727–728.

Yalkowsky, S.H. 1979. Estimation of entropies of fusion of organic compounds. *Ind. Eng. Chem. Fundam.* 18, 108–111.

Yalkowsky, S.H. and Banerjee, S. 1992. Aqueous solubility: Methods of estimation for organic compounds. Marcel Dekker Inc., NY.

Yin, C. and Hassett, J.P. 1986. Gas partitioning approach for laboratory and field studies of mirex fugacity in water, *Environ. Sci. Technol.* 20, 1213–1217.

Yurteri, C., Ryan, D.F., Callow, J.J., and Gurol, J.J. 1987. The effect of chemical composition of water on Henry's law constant. *J. WPCF* 59, 950–956.

Zhou, X. and Mopper, K. 1990. Apparent partition coefficients of 15 carbonyl compounds between air and seawater and between air and freshwater; implications for air-sea exchange. *Environ. Sci. Technol.* 24, 1864–1869.

5

Octanol/Water Partition Coefficients

Albert Leo

CONTENTS

5.1 Introduction

The usefulness of the ratio of the concentration of a solute between water and octanol as a model for its transport between phases in a physical or biological system has long been recognized (Leo 1971, 1981). It is expressed as: $P_{oct} = C_o/C_w = K_{ow}$ This ratio is essentially independent of concentration, and it usually is given in logarithmic terms (log P_{oct} or log K_{ow}) which are better suited for use as a free-energy based parameter in 'extra-thermodynamic' equations of the Hammett-Taft type (Hansch, 1995). The importance of bioconcentration in environmental hazard assessment (Kenaga, 1972) and the utility of the hydrophobic parameter in its prediction (Neely, 1974) led to an intense interest in the measurement of P_{oct} and also its prediction from structure. (Leo, 1993; Hansch, 1995).

Persistence is also a serious concern in the assessment of the environmental hazard of organic chemicals and a major pathway by which they are eliminated is through biodegradation. A great deal of biodegradation, from simple bacteria to whole animals, involves oxidation via the P-450 family of enzymes, the activity of which depends greatly upon hydrophobicity expressed as log P. (Hansch and Leo, 1995, Chapter 8). In 3,500 QSARs of biological activity of all sorts, 85% show a significant dependence upon log P. (Hansch, Hoekman, and Gao, 1996).

So many calculation methods have been published in the last five years that it is not possible to examine them all in detail in this chapter, which does discuss the strengths and weaknesses of the various approaches. For convenience and reproducibility, most users require computerized calculation, and so this chapter emphasizes these methods. Mannhold (1996) recently published a comparative study and van de Waterbeemed (1996) published a list of addresses for the programs. More information is available on the World Wide Web (http://www.pharma.ethz.ch/qsar/logp.html).

5.2   Experimental Methods

All calculation methods depend on the availability of a reliable 'test set' of measured values. Most structural features appear in a sufficient number of measured solutes that 'outliers' are readily detectable, but only one or two measurements evaluate some of the rarer features. Unfortunately, not enough experimental details accompany the published values for a considerable number of 'crucial' structures to allow judgment of their reliability.

Some researchers 'go overboard' with temperature control and with making sure equilibrium is reached between the phases; i.e., shaking for 24 hrs. in a thermostated bath. For all practical purposes, equilibrium is reached in the shake-flask procedure with two minutes of mild shaking, and log P_{oct} changes less than 0.01 units per degree (Leo, 1971). Especially when only one phase is analyzed, greater precision can be achieved if each phase after partitioning contains, as nearly as is possible, equal amounts of solute, and the amount in the non-analyzed phase is not the difference between two large numbers.

Many log P values have been reported in the 3.0 to 4.0 range when the octanol and water volumes were equal. Obviously the precision of such determinations would greatly increase the ratio of octanol to water were 1/500 or even 1/1000, making the amount of solute in each phase more nearly equal. However, when the solvent ratios are uneven, it is very important to maintain the stock solutions at saturation, and thermostating stock solutions is highly recommended. If the aqueous phase stock becomes cool (e.g., over the weekend in the winter), it will contain less octanol, even with a 'lens' of octanol present. When 500 ml of aqueous stock (now slightly unsaturated with octanol) is combined with 1 ml of octanol stock and the actual partitioning made in a warmer room, an appreciable amount of the 1 ml of octanol may be required to 're-saturate' the water phase, and the initial concentration of solute is higher than calculated.

Since a small amount of emulsion can greatly lower a log P value, it is inadvisable to shake the partitioning flask vigorously. Centrifuging the aqueous phase for 20 minutes is always recommended, but even this is insufficient to break some emulsions. For solutes which can act as mild surfactants or have log P values greater than 6.0, the two phases should not be shaken at all. For these solutes, one can use a Y-tube, which is gently rocked back and forth (Doluisio, 1964), or the slow-stir procedure (de Bruijn, 1989; Simpson, 1995).

Double-distilled or de-ionized water should be used and kept saturated with octanol in a thermostated bath. Octanol of sufficient purity to not interfere with analysis by UV or GC methods is commercially available.

Several newer, more convenient methods of determining log P_{oct} have come into widespread use, but not all actually measure the concentration of the completely solvated molecule in each phase at equilibrium. The filter probe method Tomlinson proposed (1986) circulates the aqueous phase through the UV detector after separating the octanol with a special filter. This method is well adapted to constructing a logP/pH profile. Sirius Analytical Instruments manufacturers a device especially designed for producing such profiles and this device is especially useful for simultaneously determining the pKa and log P_{oct} of amphoteric solutes with overlapping pKas (Takacs-Novak, 1995).

Centrifugal Partition Chromatography (CPC) has been shown to yield partition coefficients which compare quite favorably with those measured by traditional shake-flask (El Tayar, 1993). The procedure, which employs a coil planet type centrifuge in a horizontal flow-through system, has been described (El Tayar, 1991) The equipment can handle organic solvents with low viscosity (such as heptane) more readily than it handles octanol, but log octanol/water values in the range of –3.0 to +3.0 seem reliable. It has been used quite effectively in measuring the log Ps of amino acids (Tsai, 1991).

Microemulsion electrokinetic chromatography, MEEKC (or MECC), is a relatively new technique which holds some promise of delivering octanol/water partition coefficients much more conveniently than the shake-flask method (Gluck, 1996; Ishihama, 1994). MEEKC is claimed to have all the advantages of an HPLC method but it is not suitable for solutes with pKas much below 7.0 (Adlard, 1995). It has been used over a log P range of –1.0 to +4.0.

Although P_{oct} sometimes can be related to the more easily measured capacity factor in HPLC, the most reliable values still are obtained from traditional shake-flask methods or the 'slow-stir' technique (de Bruijn, 1990). Although slow and tedious, methods which equilibrate the solute directly between the mutually saturated phases can cover a log P range of –3.0 to +8.0.

Reversed phase HPLC methods have many supporters who insist that careful application of this technique can deliver log P_{oct} values very reliably (Klein, 1988). When the stationary support is octanol-saturated silica, the process most nearly imitates the completely solvated distribution between phases (Mirrlees, 1976), but great care must be taken to avoid "channeling" in the solid support, especially for hydrophobic solutes where column length is short.

The most common HPLC procedure uses a C-18 column with methanol/water as elutant. Many early workers recommended measuring the relative retention times at several methanol/water ratios and extrapolating to zero percent methanol. This is not only time-consuming but also prone to error, as it was shown that, depending upon solute structure, the relationship between the relative retention time, k' and % methanol is not a straight line at low methanol concentrations but could curve either up or down. Much of the recent work uses no more than three methanol concentrations and has shown a marked difference in the relationship of k', to log P_{oct}, depending on whether the solute was amphiphilic, a hydrogen bond acceptor only, or devoid of hydrogen bonding capability (Yamagami, 1993).

Remember that the partial desolvation which occurs in HPLC measurements may not be an appropriate model for the hydrophobic process under study. For example, epoxidation of a double bond in ethylene or propylene reduces log $P_{(oct)}$ by -1.43 log units. Epoxidation of aldrin to give dieldrin reduces log P by -1.30 log units when measured as equilibrium between two phases (slow-stir method; de Bruijn, 1989). However when using HPLC techniques, the epoxidation of the similar solute, heptachlor to heptachlor epoxide, is reported to reduce log P by only 0.04 log units (Veith, 1979).

It appears that the heptachlor epoxide is adsorbed onto the stationary phase so that it does not expose the oxygen lone pair electrons to the polar mobile phase. Incomplete desolvation by the stationary phase can result in an error of the opposite sign. This is probably the reason the log $P_{(oct)}$ of tricresyl phosphate by HPLC is reported as 3.42 (Veith, 1979), while it is calculated to be about 6.0 (CLOGP = 5.95; ACD = 6.23; KOWWIN = 6.34).

When a solute is calculated to be this hydrophobic, the preferred method of measurement is by slow-stir. A shake-flask measurement of tricresyl phosphate has been reported as 5.14 (Saeger, 1979), but this is very likely to be low for the reasons given above.

One should not lose sight of the fact that HPLC methods and shake-flask partitioning are both only *models* for a process one wants to parameterize. Transport of organic chemicals through soils depends greatly on their binding to the humic and fulvic acids present. The proper HPLC procedure is more likely to model this better than by partitioning between fully solvating liquid phases.

5.3   Databases of Measured Values

Volume 2 of "Exploring QSAR" by Hansch and Leo (1995) contains the most extensive hard-copy database of log P_{oct} values — about 18,000 values for over 16,500 structures. Sangster (1989) has published log P measurements and offers a comparable computer database on diskettes. Several of the computerized calculation programs access a similar log P_{oct} database. If values in other solvent/water systems are needed, the MASTERFILE database can accompany the CLOGP program on VAX and UNIX systems and contains about 40,000 measured log P values in over 300 solvent systems in addition to about 10,000 measured values of pKa.

Often the measure of hydrophobicity desired is an 'apparent log P' of an ionizable solute at a pH where it is partly ionized and thus more hydrophilic. This is most often termed a 'distribution coefficient' and reported as 'log D.' These can be calculated from a measured pKa and the neutral log P, if it assumed that the octanol phase contains a negligible portion of the ionic species. This is a good assumption if the pKa of the solute is no more than 3.0 log units on the ionized side of the pH of measurement.

For instance, the CLOGP calculation for the neutral form of the anti-arrhythmic drug, propafenone, is 3.2. Since it has a pKa of 9.6, the estimated log P value at the physiological pH of 7.4 would be 3.2 – (9.6 – 7.4) or 1.0. The log D of this drug was found to be 0.01 when measured at pH 5.0 (Mannhold, Dross, and Rekker, 1990). At this pH, the amount of ionic form in the octanol phase exceeds that of the neutral form. Since pKa – pH in this case is 4.6, it is *not* proper to correct for ionization simply by adding 4.6 to 0.01.

Software offered with version 4.1 of ProLogP and ProLogD attempts to calculate log D values from structure. This can be very difficult for the ionizable portions of many heterocycles which are common to pharmaceutical chemistry. For the more common acids and bases encountered in environmental research, the results may be satisfactory and, at times, very helpful. Note that the sources for these programs are in van de Waterbeemd (1996).

5.4 Solvent Forces that Determine Log P_{oct}

The clearest insights into the solvation forces which determine partition coefficients come from the work initiated by Kamlet, Taft, and their colleagues and referred to as the Solvatochromic Approach (Kamlet, 1988; Leahy, 1992). When considering octanol/water partitioning, the significant *solute* parameters are: size (V), polarity/polarizability (Π), and hydrogen bond acceptor strength (β). This is conveniently expressed as:

$$\log P_{oct} = aV + b\Pi + c\beta + d \tag{1}$$

5.4.1 Solute Size and Polarity

There is very little problem in calculating an acceptable measure of solute size. Simple calculations of either molecular volume or area based on either Bondi's (Bondi, 1964) or McGowan's (Abraham, 1987) methods work almost as well as those derived from molecular mechanics and quantum chemistry (Leo, 1993). When volume in cubic Ångstroms is used, V is normally scaled by 0.01 to produce a coefficient comparable to the others in the equation polarity/polarizability.

In spite of claims to the contrary, to date no completely satisfactory method exists to calculate the polarity/polarizability parameter, Π, as it applies to the equilibrium of solute between water and octanol. The 'excess molar refractivity' of the solute (compared to an alkane of equal size) can be estimated separately from polarizability/dipolarity (Abraham, 1994) and seems an attractive approach to this problem, but it needs further verification. The dipole moment of the entire molecule has been used as a polarity parameter (Bodor, 1992), but there are good reasons to believe it has marginal value at best. The square of the dipole moment also has been used (Leahy, 1992), and it, at least, has some theoretical basis (Kirkwood, 1934).

5.4.2 Hydrogen Bond Acceptor Strength

For all solutes except the simple hydrocarbons and their halogen derivatives, hydrogen bonding plays the most critical role in partitioning between water and an immiscible organic phase.

Both donor and acceptor strength of the solute are important for partitioning between water and most organic solvents. However, water-saturated octanol and water are about equally strong in H-bond *acceptor* strength, and so they do not distinguish solutes on the basis of their H-bond *donor* strength. This is the reason that Equation (1) above has no term in solute acceptor strength, α, while it is very important for the other solvent pairs in the 'critical quartet' (Leahy, 1992) as seen in Equation (2) and (3) for octanol/water and alkane/water respectively (Taft, 1996).

$$\log P_{oct} = 3.67\ V - 0.40\ (0.1\mu)^2 - 3.00\ \epsilon\beta + 0.24 \tag{2}$$

$$\log P_{alk} = 4.62\ V - 0.74\ (0.1\mu)^2 - 5.00\ \epsilon\beta - 3.20\ \epsilon\alpha + 0.02 \tag{3}$$

In both of the above equations, the symbol 'ε' denotes the 'effective sum of' and must be used when a solute contains more than one hydrogen bonding moiety in close proximity so that their effect is not additive (Taft 1996).

It is unlikely that Equation (1) will ever serve as a practical method of calculating log P_{oct}. In fact, its greatest utility comes in the reverse direction: calculating H-bond strengths from a series of log P measurements (Taft, 1996). In principle, one should be able to develop quantum chemical parameters which account for Π and β in Equation (1) and the α term needed for the other solvent pairs. It seems certain that both α and β are extremely sensitive to both electronic and steric effects, (Leahy, 1992; Taft, 1996), and it may be some time before M.O. calculations surpass the simple empirical approaches in evaluating them.

Although solute size is an important parameter in Equation (1) and appears in the quantum chemical methods as well (Bodor, 1992), one must keep in mind the fact that part of the energy 'cost' of creating the cavity in each solvent is 'paid back' when the solvent interacts favorably with parts of the solute surface. It is reasonable to conclude that this is taken care of by the negative signs on the coefficients 'b' and 'c' in Equation (1). But neither Equation (1) nor the quantum chemical calculations may be adequately taking into account the changes in the structure of the first layer of solvent water or the 'flickering clusters' (Frank, 1945) at a greater distance. Even for a hydrocarbon like propylbenzene, there may be a difference in the cavity 'wall' of water surrounding the aliphatic and the aromatic portions of the solute (Leo, 1976). Conceivably this could result from the weak H-bond acceptor strength residing in the π-cloud of the benzene ring (Taft, 1996, Figure 1b) and appears as a weak contribution to the β-term in Equation (1).

5.5   Methods for Estimating Log P_{oct} from Structure

Given the number of methods of calculating log P_{oct}, the ability to assign each to one of several categories would be helpful. Almost all the contributors to this field have attempted some sort of categorization, but so much overlap exists in the proposed classifications that much confusion remains.

5.5.1   Whole Molecule

Some researchers who use parameters derived from quantum chemical calculations on the **whole** solute **molecule** (Bodor, 1989) place their methods in a "theoretically-based" class. These are seen as preferable to methods which add the values assigned to structural parts, which then requires the assignment of 'interaction factors' or 'corrections,' depending on how the parts are attached to each other. Fragment methods then are said to be in a 'purely empirical' class and therefore 'obsolete.'

As it turns out, methods using quantum chemical parameters alone have been applied successfully only to very restricted structural types (Reddy, 1996). When applied to a larger variety of structures, non-quantum chemical parameters, such as an indicator for alkanes and a simple count of hetero atoms, are required (Bodor, 1992), and multivariate regression analysis of a database of measured values determines the contribution of each parameter. This certainly undermines any claim to a superior theoretical basis for quantum calculations.

5.5.2 Change of Parent Structure

Fujita proposed (1964) the first method of calculating log P_{oct}, based on the **change** in log P of a **parent** structure, resulting from the substitution of one or more of its hydrogen atoms. Of course, a measured value for the parent must be available, and it may contain most of the fragment interactions which are difficult to estimate. This method has not been computerized and will not be discussed further, even though it often produces the most dependable estimates when the parent structure has been measured.

5.5.3 Atom

Separating methods which use molecular 'parts' into two categories, 'atom-based' or 'fragment-based,' seems to make a distinction without a significant difference. In **'atom-based'** methods, each atom of the molecule is examined in respect to its connectivity index, and to the oxidation and hybridization state of all those atoms directly attached to it (Viswanadhan, 1989) or to some similar measure of 'atomic state' (Broto, 1984).

In the early method proposed by Broto et al., 222 of these atom-centered clusters were verified by a test set of 1868 measured log P_{oct} values of structures containing no intramolecular hydrogen bonds. The residual standard deviation by regression analysis was 0.43 log units. In a set of 500 solutes thought to contain intramol-H-bonds, the standard deviation doubled, which the authors attributed to variation in H-bond strength.

Using a more rigid definition of 'atomic state', Viswanadhan et al. (1989) were able to reduce the number of clusters to 120, which were derived from a test set of 893 solutes, and they obtained a standard deviation of 0.496. In a set of 47 nucleotides and nucleosides, which are notoriously difficult to predict due to long-range interactions of the polar atoms (see Table 5.3), their method resulted in a very respectable standard deviation of 0.51 (Viswanadhan, 1993). The difficulties encountered by these methods in dealing with long-range interactions are discussed in greater detail later, as well as the interpretation of the hydrophobicity of the constituent parts.

A method using the 'Atomistic' approach (Masuda, 1996) was published recently and claims an improved performance from consideration of solute SASA as well as proximity effects of substituent groups. Measured values for 500 solutes were taken as a test set, but just how substituent proximity was taken into account was not explained. For a set of 20 pharmaceuticals not in the original test set, five methods other than the SASA-scaled atomistic method were compared. Table 5.1 shows the statistical results for the best three methods.

TABLE 5.1

Statistical Results.

Method	r^2	s	F
SASA-scaled atomistic	0.839	0.638	94.2
Rekker	0.850	0.619	101.5
CLOGP	0.947	0.366	323.3

5.5.4   Fragment Values

Nys and Rekker (Nys, 1973) were the first to publish a method of calculating log P_{oct} by the addition of **fragment** values, and this procedure has been improved over time (Rekker, 1992). The designers determine what constitutes a valid fragment and verify the selection using a large database of measured values.

Possible alternative ways to 'assemble' a structure from fragments can give different values, but, the computer program decides this. For instance, in computing phenoxyacetic acid, the program finds the fragments ($-O-CH_2-CO_2H$) and C_6H_5, rather than adding a phenyl, ether oxygen, methylene, and carboxyl group with the appropriate proximity effect. Allowing for interaction corrections in complex solutes using multiples of a 'magic constant' is best left to the computer program. In chloramphenicol, for example, the magic constant is invoked seven times, which amounts to a correction of 1.97 log units (Rekker, 1992).

In the KOWWIN program (Meylan, 1995), the designers selected 144 '**atom/fragment contributions**' (AFCs) containing four atoms or less as justified by multiple regression analysis using a database of 8406 measured values. To account for interactions of the AFCs, ver. 1.5 of KOWWIN uses 248 'structural correction factors,' some of which, as the User Guide points out, "require a paragraph and a picture to completely describe it and/or resolve it from other factors." Some examples of this complexity appear later.

5.5.5   Reductionist

Perhaps the most useful distinction between calculation methodologies separate '**reductionist**' methods from 'constructionist.' In reductionist methods, the contribution of each structural feature (whether 'atom-cluster' or fragment) and each 'interaction factor' is evaluated by multiple regression analysis from a database of measured values. Each occurrence receives equal weight, regardless of whether it appears alone in the structure or is one of a dozen interacting fragments contributing to the total.

One naturally assumes that the larger the database used as a 'training set,' the more reliable will be the calculation for structures not represented therein. But databases reflect what chemists are interested in and what they measure. Thus chemists may 'over-represent' certain structural features (i.e., those which confer desirable properties), and evaluating these by regression analysis may give a 'skewed' perspective of what solvation forces are operating. Examples of this problem appear later.

5.5.6   Constructionist

The CLOGP program (Leo, 1993) is the only one in widespread use which is '**constructionist**' in nature; that is, the fragments and interaction factors upon which it depends are evaluated from the simplest examples in which they occur. Aliphatic carbons and hydrogens are evaluated from methane and ethane, which have been very carefully measured. Lengthening the chain, which introduces a possible effect of flexibility, and chain branching, which may expose less surface per hydrocarbon unit, both require a small but significant negative factor.

The constructionist approach requires the evaluation of the most common organic atom of all—hydrogen—by carefully measuring the partition coefficient of hydrogen gas. Its log P_{oct} was found to be +0.45, indicating that the hydrogen atom is significantly hydrophobic at about +0.2. It is interesting to note that reductionist methods derived using regression analysis, with the notable exception of the procedure of Rekker (1992), often find no significant contribution for hydrogen (Klopman, 1991).

In the constructionist method, aromatic carbons are evaluated from benzene, a solute which has been measured more times than any other. When the phenyl ring is fused to others, as in naphthalene or anthracene, or when it is joined to another, as in biphenyl, there is a change in the effective polarity of the pi electron cloud, and a slight but significant positive correction factor is introduced. Critics may disparagingly refer to these as 'fudge factors,' but these factors have been very helpful in distinguishing the solvation forces which operate in the two different phases (Taft, 1996).

5.5.7 Relative Advantages

The evaluation of the intrinsic hydrophobicity of the pyridine-type nitrogen provides a good example of one of the differences between 'constructionist' and 'reductionist' methods. For the constructionist approach used in CLOGP, this nitrogen was evaluated from a single solute, pyridine itself, using the careful measurements of Iwasa (1965). They determined a log P of 0.65 (since then, ten shake-flask measurements have been published, giving an average value of 0.651). However, when two such nitrogens are present in the same ring system (such as in pyrimidine or sym-triazine), or when a electron-rich substituent is attached, the pyridine nitrogen appears more hydrophobic (or less hydrophilic due to reduced $\varepsilon\beta$) and an appropriate correction factor is invoked.

It so happens that the log P databases where pyridine nitrogens are 'hydrophobically enhanced' contain many more solutes than those where they are 'isolated.' KOWWIN is one of the more advanced programs that derives fragment constants and structural corrections by multiple regression from a large database of measured values; i.e., it is one of the more efficient ones which use a 'reductionist' method. KOWWIN overpredicts the simple, unsubstituted pyridine at +0.80, even after applying a rather awkward "pyridine ring (non-fused) correction" of –0.16. Thus the 'reductionist' (statistically-derived) approach misses the *intrinsic* value for an aromatic nitrogen fragment by +0.31. Of course it then does not need to allow for (or call attention to) the real interaction of the pyridine nitrogen with a hydroxyl or methoxyl in 3-hydroxy or 3-methoxy-pyridine. Also note that the 'averaging' which occurs in evaluation using multiple regression results in an overprediction for pyrimidine of 0.34 log units, but an underprediction for phenazine of 0.55 log units.

The overprediction of the intrinsic fragment value of the aromatic nitrogen is also evident in thiazole, which KOWWIN predicts at 0.99, while the measured value is 0.44. In KOWWIN, the 47 analogs in a single paper (Naito, 1992) in which the measurements were made at a pH (2.0) where the nitrogen was largely protonated. Making the usual ionization corrections to log D (called log P' in this paper) is not recommended for protonated heteroaromatic bases, and the 'corrected' values were not considered as 'preferred' for CLOGP. They are probably higher than what would have been measured at a pH of 3.5 where the solute would be essentially neutral.

With only two exceptions (the –CF_3 fragment and the α-aminoacid fragment), CLOGP must dissect a structure into fragments according to a fixed rule. It does not allow its programmers to create a new 'composite fragment' to 'fit' a new data point. This restriction has some significant disadvantages, as might be expected, but it has often disclosed a type of electronic, steric, or conformational effect which was not appreciated previously. 'Polar' fragments as defined in CLOGP are those containing hetero atoms. They are what is left of an organic structure after all the 'Isolating Carbons' and their attached hydrogens are removed. Isolating Carbons are defined as those which are *not* doubly or triply bonded to a hetero atom, but they may be attached to a hetero atom with aromatic bonds. By this definition, the halogens are single atom polar fragments, a hydroxyl is a two atom polar fragment, a carboxyl group a four atom fragment, etc. These fragments exhibit their most negative value when attached to an sp^3 carbon which creates the strongest localized dipole and also maintains the greatest electron density on the hetero atom to act as a hydrogen bond acceptor. Conversely, a polar fragment has its highest value (most hydrophobic) when attached to an aromatic carbon. In both situations, the polar fragment values can be enhanced further by any electronegative groups nearby.

Hammett sigma constants (Hammett, 1970; Hansch, 1995) are useful in estimating the effect of this hydrophobic-enhancing interaction of polar fragments (Nakagawa, 1992), even though they originally were designed to predict the effect upon ionization (i.e. complete proton loss) rather than hydrogen donor/acceptor strength.* For instance, the CLOGP program makes a positive 'correction' to the calculation of nitrophenols, following a Hammett-like treatment using sigma/rho values of the NO_2 and OH substituents to characterize the electronic interaction between them which raises log P_{oct}. Klopman's CASE program (Klopman, 1991), for example, uses the string, NO_2–C=$CH_{(0)}$–$CH_{(0)}$=C–OH, to account for this interaction at a distance.

Just as in Hammett methodology, CLOGP treats the electron withdrawing power of a pyridine nitrogen *in* a ring in much the same way it treats a substituent *on* a ring, like the nitro group discussed above. For example, in 3-methoxypyridine, CLOGP informs the user that log P has been raised by +0.5 due to this electronic interaction. (Meas. = 0.99; Calc. = 1.06) ACD/LogP (see van de Waterbeemd 1996) sees this interaction needing +0.4338, but for some reasons uses a negative correction for the interaction with hydroxyl (–0.1494) and underpredicts 3-hydroxy-pyridine by 0.68 log units (meas. = 0.52; calc. = –0.16 ± 0.20).

5.6   Evaluation of Software Performance

5.6.1   Statistical Evaluation

Most suppliers of log P calculating software (see van de Waterbeemd, 1996) present a statistical evaluation of their program's performance. This usually is based on regression of calculated versus measured values and an examination of the regression coefficient and standard deviation.

While these statistical measures are important, a prospective user is well advised to examine the database that has been used as the 'test'. It should not be surprising that a 'test set' of a couple of dozen solutes can produce very misleading results, if only two or three measured values are in error (Leo, 1995b).

Even when the test set is large, one should see if the sub-structures of particular interest are calculated well. Out of the 9,000 selected log P values recently published (Hansch, 1995), it is easy to select three to four thousand monofunctional solutes which even the simplest program can calculate with a standard deviation less than 0.2 and a regression coefficient, r, greater than 0.99.** A recent comparison of computerized calculations (Mannhold, 1996) uses 138 measured solutes, 48 of which are complex drugs. While this might be an

---

* Note that the field effect, F, and Hammett sigma values can only be related to the loss of hydrogen bond acceptor strength, $\varepsilon\beta$. As noted by Taft (1996) the relationship between $\varepsilon\alpha$ and aqueous pKa is poor, but log P-octanol is unaffected by solute $\varepsilon\alpha$.
** CLOGP ver. 4.0, released since this manuscript was written, produced this equation for over 10,000 structures: Meas.LogP = 0.956 CLOGP + .084; r = 0.985; s = 0.278.

adequate test for many users, the relative performances may change considerably if a number of chlorinated hydrocarbons or phosphorous-containing pesticides are added or substituted. Nevertheless, because it represents such a great effort, it is unlikely that a more comprehensive study will be published in the near future.

TABLE 5.2

Log P of Benchmark Chemicals

Name	Meas.	KOWWIN	ACD/LogP	CLOGP
anthracene	4.45	4.35	4.68	4.49
lindane	3.72	4.26	3.94	3.75
2,6-di-t-butylphenol	4.92	4.48	4.86	5.13
trichloroethylene	2.61	2.47	2.26	2.63
chlorpyrifos	5.27*	4.66	4.77	4.68
	4.96			

* By 'slow-stir; usually preferred

Of the many listed by Mannhold, three programs, differing significantly in methodology, were chosen for comparison of performance: KOWWIN, ACD/LogP, and CLOGP. As Table 5.2 shows, all three perform well on the five 'benchmark' chemicals picked for this book.

5.6.2   Easy of Entry

Convenience of entering structures can be a matter of importance to the user, but this is often a matter of personal preference and should be checked out carefully in advance. Some users will require batch entry of large files of structures. Both KOWWIN and CLOGP use the SMILES notation (Weininger, 1988, 1989) as the primary entry medium, accepting SMILES files directly or providing conversion from other popular notations. Both accept CAS numbers. CLOGP also accepts compound names and synonyms; e.g., AZT or zidovudine.

5.6.3   Draw Structures

For users who prefer to draw in new structures, the later versions of Chem Draw* will deliver a SMILES notation which can be used via 'cut and paste' commands, and ISIS Draw** delivers a MOLfile which can be automatically converted to a SMILES file. ACD/LogP uses a proprietary drawing program, MolDraw, which is easy to learn and use. Entry by name in ACD is accomplished via a Dictionary with about 48,000 entries. Drawn structures can be saved to a file, but in version beta 0.9 of ACD/LogP there is no method of batch entry from either SMILES files or MDL MOLfiles.

5.6.4   Assignment of Probable Error

Assignment of probable error to the calculation of log P_{oct} is very difficult. Some of the programs (e.g., the three mentioned in the previous paragraph) retrieve a measured value when available, and this often is the most reliable measure of the error to be expected in solutes of similar structural complexity. KOWWIN offers no error estimate directly with

---

* Copyrighted by Cambridge Soft Corp., 875 Massachusettts Ave., Cambridge, MA 02139; Internet: support@cam-soft.com.

** Copyrighted by MDL Information Systems, Inc., 14600 Catalina St., San Leandro, CA 94577.

the calculation, but one can note the AFCs and Correction Factors used in the calculation of interest, and, referring to the tables in the User's Guide, see how many times each appeared in the training set. In its calculation details, CLOGP indicates whether each fragment was directly measured in the required environment, or whether it was estimated from other environments. For example, the program is allowed to estimate a 'benzyl-attached' fragment as +0.2 more hydrophobic than the measured 'aliphatic-attached' one. It is not readily apparent how ACD/LogP arrives at the error estimates it provides. Many appear reasonable, but in the ver.1.0 Tutorial (p. 29) hexazinone is calculated as −0.69 ± 0.67 but the measured value is +1.85; similarly, the value for chlorsulfuron is calculated as +0.18 ± 0.70 while the measured value is +2.14.

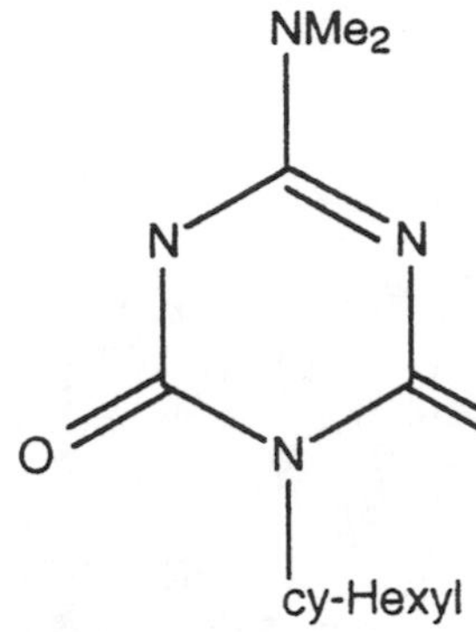

FIGURE 5.1
Hexazinone.

It is worthwhile to examine the herbicide hexazinone (see Figure 5.1) in some detail, as it illustrates an epistemological problem in log P calculation: How do you know what you know? The rules in CLOGP for 'fragmenting' a structure result in two polar fragments: the tertiary amine and 2,4-dioxo-1,3,5-triazo moiety. Furthermore, CLOGP uses the simple Hückel rule for ring aromaticity, and the heterocycle in hexazinone qualifies. This is important, for the lone pair electrons in such rings are much poorer hydrogen bond acceptors when the 4n + 2 condition is satisfied, even though, for ordinary organic chemistry, the ring would hardly be considered aromatic. The dioxo-triazo moiety is not common, and early versions of CLOGP failed to complete the calculation, returning the message, "Missing Fragment." When two or more polar fragments are fused together in an aliphatic environment, such as multiple amido or urea groups, the fragment value appears to level off at a low point of about -4.0. When incorporated into a ring which meets Hückel's rule for aromaticity, this is raised about one log unit, and thus the dioxo-triazo fragment could be estimated at -3.0. Until recently it was a policy *not* to include such estimations in the Fragment Database which drives CLOGP.

When the measured value for hexazinone was published (Yanase, 1992), the value for the dioxo-triazo fragment turned out to be −2.79 when allowance was made for its electronic interaction with the tertiary amino substituent. This is in good agreement with the prior estimate, and so the value appears in the latest fragment-database. The CLOGP calculation is, perforce, very good: 1.85 vs. the measured 1.85.

After dissecting out the amino, amide, urea and −N=C fragments (all considered as aliphatic attached), the KOWWIN program seeks out structural corrections to allow for the fact that part of the H-bond acceptor strength (εβ) is lost when they are attached together. As Figure 5.1 shows, the program appears to overcorrect in this instance.

LogP
Measured = +1.85
KOWWIN = +3.40
ACD/LogP = -0.69

In addition to the polar fragments, KOWWIN finds these strings which require structural corrections:

–NCO–N–CO	+0.6074
–CO–N–C(-amino)=N (aliphat. ring)	+1.2456
(N–C(=N)–N–CO–N–) amidino urea	+0.8615

Of course most calculations are not as complex, and the standard deviation for KOW-WIN for the entire 8406 in the training set has been reported to be a quite respectable 0.408.

The reason that the ACD/LogP program *under*estimates hexazinone by 2.54 log units is unexpected, considering the fact that it puts all the hetero atoms in a single fragment, leaving only the methyls and cyclohexyl to be evaluated separately. It errs in 'seeing' a guanidine moiety in the nine-atom fragment, evaluating it at –6.3898, and warning that a measurement would have to be made at a very high pH to assure a neutral solute molecule. It does label the fragment value as 'calculated', and so one wonders why it is carried out to four decimal places and how the 95% confidence limits are set at ±0.67. Of course the standard deviation for the training set of 3,600 values is much lower, and the warning is stated (in the Web site literature) that the 95% limits for some structures can be exceeded (presumably 5% of the time).

The recently described VLOGP program (Gombar, 1996) claims to approach the reliability problem in a more direct fashion. In deriving 363 pertinent 'electrotopological structural parameters' (E-states) from some 6675 measured log P values, it establishes an 'application domain' or Optimum Prediction Space (OPS). When a new structure is entered via SMILES, the program determines whether or not that structure falls within OPS. Initial tests reportedly indicate that the average deviation for those within OPS is 0.27 log units, but it is 1.35 log units for those falling 'outside.' Ninety-five percent confidence limits are provided in both cases. VLOGP was tested using hexazinone and nine other rather difficult structures, only one of which turned out to be outside OPS. Although hexazinone was seen to be within OPS, the deviation for the calculation was 0.74 log units (Calc. = +1.11). The average deviation for all nine lying within OPS was 0.46, which should be considered satisfactory in view of the level of difficulty involved. However, five of the ten calculated values fell outside the 95% confidence limits. Perhaps the biggest disadvantage of VLOGP is that the user can gain little insight into the solvation forces at work from the tabulation of 'Specific and Average E-state' values.

Perhaps the lesson for users of *all* programs that calculate log P_{oct} is simply this: It is very easy to think these programs 'know' more than they actually do. Although the partitioning process is extremely simple to carry out, the competing solvation forces involve more knowledge of physical chemistry than we can muster at the present time. But, realistically, very few readers of this book are involved in designing new chemicals and need, therefore, to ask *why* do certain components in a structure affect its hydrophobicity the way they do. The chemicals out in the environment are a 'given', and the question of greatest importance is *what* is the hydrophobicity of each.

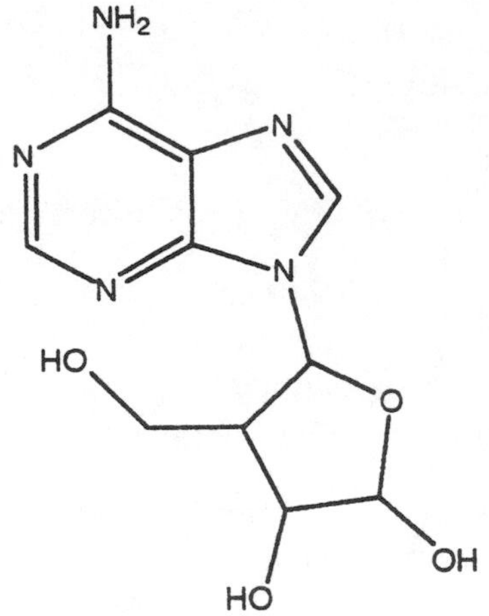

FIGURE 5.2
Adenosine (cis form).

5.6.5 Flexibility

Software that is flexible enough to allow programmers to insert values for a variety of 'atom strings,' as new measurements disclose the need for them, has some decided advantages. In most cases, the need for such 'special strings' arises from polar interactions taking place at a great topological distance because of 'unusual' conformations. The conformation of a solute in the octanol phase can be quite different from that in the aqueous phase. (This is also true of a solute's predominant tautomeric species, as is discussed later for keto/enol and thiol/thione pairs, as in Figure 5.7.) The *cis* form of adenosine (shown in Figure 5.2 with the sugar ring rotated close to the purine ring) is only about one Kcal less stable than the *trans* in an aqueous environment, but octanol would favor the *cis* since an internal hydrogen bond between N3 and 5'-OH would more than make up that difference.

Currently CLOGP lacks the ability to accept 'special strings,' and its poorer performance in calculating the ordinary nucleosides is evident from the values in Table 5.3. CLOGP allows for the interaction between the N3 of the base with the ring oxygen of the sugar because they are topologically close. It does not take into account any interaction of the nitrogen with the more distant hydroxyls, especially the 5'–OH. When those at 2' and 3' are removed in the di-deoxy analogs, the calculation improves.

5.6.5.1 *Performance with Nucleosides*

Although the performance of the ALOGP program with the nucleosides is commendable when judged from the final calculations, the user should not draw any conclusions regarding the 'atomic contributions' which make them up. For example, in adenosine (Table IV of Viswanadhan, 1993), all of the oxygen atoms are seen as hydrophobic, with values ranging from +0.14 to +0.17. It is the *hydrogens* attached to N6, O2', O3', and O5' which are seen as making negative contribution (−1.63) to the total.

But there is strong evidence, as the absence of an α-term in Equation (2) shows, that H-bond donor strength in the solute is unimportant in log P_{oct}, and that by its size, the hydrogen atom ought to make a positive contribution. The β-strength of the oxygen and nitrogen atoms must be responsible for the hydrophilicity of the nucleosides. When calculation *details* are examined (Leo, 1993), all of the 'atomic contribution' methods developed from

TABLE 5.3

Name	Meas.	CLOGP	KOWW	ACD	KLOGP	ALOGP	BLOGP
Ado	−1.23	−2.63	−1.38	−1.26	−2.13	−1.24	0.182
dAdo	−0.55	−2.05	−0.71	−0.53	−1.29	−0.63	0.356
ddAdo	−0.22	−0.63	−0.65	−0.49	−0.46	−0.19	0.406
ddeAdo	−0.36	−0.58	−0.86	−0.56	−0.42	0.04	0.379
Guo	−1.89	−3.62	−1.40	−1.89	−3.09	−1.63	−0.210
dGuo	−1.30	−3.03	−0.73	−1.28	−2.26	−1.01	−0.013
ddGuo	−1.01	−1.62	−0.67	−0.73	−1.43	−0.58	0.177
ddeGuo	−1.21	−1.57	−0.88	−0.80	−1.39	−0.35	0.115

Measured log Ps and CLOGP values are from BioByte software; KOWWIN values were calculated from version 1.5; ACD values calculated using ver. beta 0.9; ALOGP, KLOGP, and BLOGP values from Viswanadhan (1993). Ado is adenosine; Guo is guanosine; d = deoxy; dd = dideoxy; and dde = 2′, 3′-dideoxy, 2′,3′-dehydro.

the early work of Broto et al. (1984) seem to be misleading, even though the final results may be satisfactory.

5.6.5.2 *Performance with Chlorinated Hydrocarbons*

Chlorinated hydrocarbons are an important class of chemicals causing environmental concerns. A chlorine atom, even attached to an aliphatic carbon, is only a poor H-bond acceptor, i.e., it has a low β-value in Equation (2). Yet replacing one hydrogen in methane to yield methyl chloride *lowers* log P by 0.18. From Equation (1), we see that this must be the result of the localized C-Cl dipole (which enhances Π), overcoming the effect of greater solute size (V). Multiple chlorination appears to shield the C-Cl dipole, and the increased hydrophobicity *per chlorine atom* which results should be taken into account. KOWWIN neglects to do this, using an average value for the chlorines, and for the quaternary carbon in CCl_4 it increases the value from 0.27 (as in neopentane) to 0.97. It's calculation still falls 0.39 short of the measured value (see Table 5.4).

TABLE 5.4

Solute	Meas. log P	CLOGP	KOWWIN	ACD/LogP
CH3Cl	0.91	0.94	1.09	0.97
CH2Cl2	1.25	1.25	1.34	1.19
CHCl3	1.97	1.95	1.52	1.76
CCl4	2.83	2.88	2.44	2.86
Cl2C=CH2	2.13	2.37	2.12	1.77
ClC=CCl (C)	1.86	1.77	1.98	1.74
ClC=CCl(T)	2.09	1.77	1.98	1.74
Cl2C=CCl	2.61	2.63	2.47	2.26
Cl2C=CCl2	3.40	3.48	2.97	2.95

In multi-chlorinated, multi-fused alicyclics, chlorine atoms separated by a greater topological distance affect the shielding of C-Cl dipoles. Examples of such compounds are aldrin, dieldrin, and mirex. As Table 5.5 shows, CLOGP fails to account for this 'extra' proximity only 3-D shows, and it underestimates all three. The methodology ACD user appears to underestimate aldrin but greatly overestimates mirex where 81 Cl-Cl interactions are considered. KOWWIN also overestimates mirex but does well on aldrin. Difficulties in measuring these highly chlorinated solutes make it risky to reach definitive conclusions about program performance from these examples.

TABLE 5.5

Solute	Meas.	CLOGP	KOWWIN	ACD
Mirex	5.28[a]	4.65	7.35	9.91
Aldrin	6.50	5.41	6.75	4.84
Dieldrin	4.55[b]	3.63	5.45	4.79

[a] This shake-flask measurement may be slightly low. It was not made with the slow-stir technique. An HPLC value of 6.89 has been reported, but is suspect because no comparable 'spherical' solutes could be used as standards, and HPLC retention time has been shown to be shape dependent.

[b] This value was obtained by slow-stir technique, but a value of 5.20 was reported by other investigators using the same technique. Epoxidation of the double bond in propylene reduces log P by –1.74. Applying this as the value for aldrin we get 6.50 – 1.74 = 4.76, which supports the lower measured value.

As more chlorines are added in the polychlorinated biphenyl series (PCBs), the *measured* log P_{oct} values taper off asymptotically to a maximum value of 8.55 (de Bruijn, 1989). This is most probably due to the fact that the 10^{-3}M octanol which is present in the water phase is much greater than the amount of PCB, and it acts as a 'detergent,' sequestering the miniscule amount of any solute with log P>7.5. Of course, this may make *measured* octanol/water an improved model for many actual environmental systems. River water with a heavy load of sediment will adsorb hydrophobic solutes and appear to solvate them better than pure water. The only real problem in calculating the log P_{oct} of PCBs whose log P<7.0 is allowing for the decreased hydrophobicity of chlorines in the ortho position. Again, most programs handle this reasonably well, as the examples in Table 5.6 show.

TABLE 5.6

PCB	Meas. log P	CLOGP	KOWWIN	ACD/LogP
No Cl	4.01	4.03	3.76	3.98
2,2'–Cl2	4.97	4.96	5.05	4.93
2,6,2'6'–Cl4	5.94	5.88	6.34	5.91

5.6.5.3   *Performance with Halogens*

In an aliphatic environment, halogens can reduce the εβ of many polar groups through a field/inductive effect (Hansch, 1995) and a positive correction is required. Ideally, the Hammett methodology and the field/inductive parameters developed for that purpose would handle this. This presents considerable programming difficulties, however, and most software now uses some sort of pragmatic, empirical solution. The user must make some effort to decipher the calculation details in any of the current programs, as the following example shows using the pesticide, dipterex, shown in Figure 5.3, and the CLOGP details that follow:

To fully interpret these details, it is necessary to refer to the User's Guide. 'X' refers to a halogen other than fluorine; 'Y' refers to a polar fragment; 'Fragbranch' refers to the unique lowering of the hydrophobicity of chains radiating from phosphate, phosphonate, and tertiary amine fragments. The three X–C–X factors are the same as seen in chloroform; the

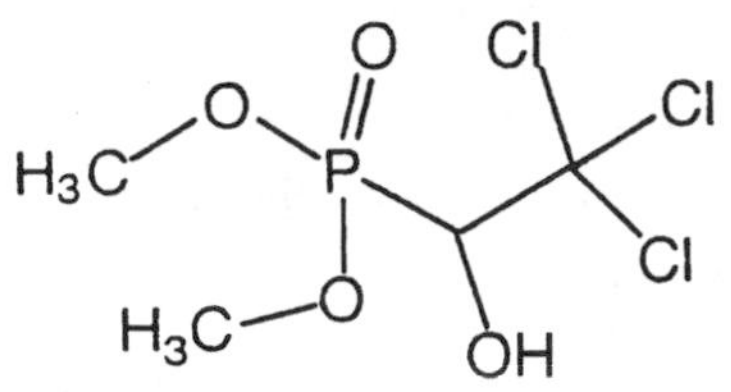

FIGURE 5.3
Dipterex Meas. Log P = 0.51

fragment	#1	phosphonate	meas.	−2.67
fragment	#2	hydroxy	meas.	−1.64
fragment	#3	chloride	meas.	0.06
fragment	#4	chloride	meas.	0.06
fragment	#5	chloride	meas.	0.06
Isolating	carbon	4 aliphatic Isol. carbons		0.78
Exfragment	branch	1 non-halogen, polar group	group	−0.22
Exfragment	hydrogen	7 hydrogens on Isol. C.		1.589
Exfragment	bonds	7 chain & 0 alicyclic		−0.84
Fragbranch	Frag. 1	5 net bonds (out of 8)		−0.95
proximity	X–C–X	3 interacting fragments		1.59
proximity	X–CC–Y	0 Fluorine & 3-non Fluorine		1.05
proximity	Y–C–Y	Frag 1 & 2: −0.42(−2.67–1.64)		1.81
RESULT	v.2.10	All fragments meas.		0.679
			meas	0.51

X–CC–Y is the electronic interaction of the chlorines reducing the εβ of the phosphonate and hydroxyl fragments; Y–C–Y shows that 42% of the hydrophilic contribution of the phosphonate and hydroxyl fragments have been lost because of their proximity to one another; i.e., log P is raised by +1.81.

The ACD/LogP 'explanation' shows that it finds the same fragments as CLOGP but gives them slightly different values because it does not directly count bonds and branches. It also accounts for the interaction of the three chlorines with each other in the same way (+1.61 instead of +1.59 above). However, it considers the interaction of the three chlorines with the hydroxyl and the phosphonate fragments as additive, totaling +3.04, in contrast to the +1.05 found in CLOGP. Although it computes the interaction between hydroxyl and phosphonate lower (+1.21 vs. +1.81), it still arrives at a final result of +2.02 ± 0.66, which is 1.51 log units too high.

As previously noted, the KOWWIN program does not consider any interaction of geminal chlorines with each other. It also neglects any of their interaction with the nearby H-bond acceptors. Thus, as Table 5.7 shows, it starts out with a much higher value for each chlorine (0.3102 vs. 0.18 in ACD and 0.06 in CLOGP), but it then underestimates the total when three chlorines are on one carbon. As a result KOWWIN underpredicts dipterex by 0.79 log units and trichloroacetonitrile, Cl_3CCN, by 0.88 log units (Measured = 2.09; KOWWIN = 1.21).

The 'constructionist' approach used in CLOGP makes it difficult to incorporate even a rather commonly observed 'anomaly' if it cannot be expressed in simple structural terms. For instance, the decreased hydrophilicity of a hydroxyl group in the 11 position of a ste-

TABLE 5.7

KOWWIN Calc. of Dipterex.

Type	Num.	Fragment Description	Coeffic.	Value
Frag	2	–CH3 (aliphatic carbon)	0.5473	1.0946
Frag	1	–CH (aliphatic carbon)	0.3614	0.3614
Frag	1	C (aliph. carbon; no H; not tertiary)	0.9723	0.9723
Frag	1	OH (hydroxy, aliphatic attached)	–1.4086	–1.4086
Frag	3	–Cl (chlorine, aliphatic attached)	0.3102	0.9306
Frag	2	–O–P (aliphatic attached)	–0.0162	–0.0324
Frag	1	O=P	–2.4239	–2.4239
Const		equation constant		+0.2290
			log Kow	–0.2770

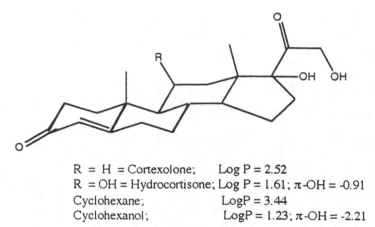

R = H = Cortexolone;   Log P = 2.52
R = OH = Hydrocortisone; Log P = 1.61; π-OH = -0.91
Cyclohexane;   LogP = 3.44
Cyclohexanol;   LogP = 1.23; π-OH = -2.21

phenol: Meas. = 1.46
Calc. = 1.47

2,6-di-s-butyl phenol: Meas. = 4.36
without branch ortho Calc. = 5.39
with branch ortho Calc. = 4.36

FIGURE 5.4

roid nucleus is well documented (Leo, 1993). One can easily appreciate this by noting (see Figure 5.4) that adding the 11β-OH to cortexolone (2.52) to produce hydrocortisone (1.61) changes log P by only –0.91 log units, while adding OH to cyclohexane changes its log P by –2.21 log units.

It is difficult to state a generalized rule which recognizes this feature and does not apply it where it is not warranted (Leo, 1995-A). It is tempting to ascribe this decrease in the hydrophilicity of the OH to the fact that it is surrounded by hydrocarbons which block its access to the solvent, and to correlate its less negative fragment value to the reduced solvent-accessible surface area, SASA (this is often referred to as 'hydrophobic shielding'). But this approach does not consider the fact that log P is not simply a measure of solubility but is a competition between water and octanol, and water can gain access to a highly hindered site that may exclude octanol. This would have the effect of *lowering* the expected log P.

In fact, a similar reduction in the solvent-accessible-surface-area of the hydroxyl group occurs in 2,6-di-sec-butyl phenol, which has a partition coefficient one log unit more hydrophilic than expected at 4.36 (Leo, 1995-A). This has been allowed for in the latest version of CLOGP, which gives a value of 4.39 for this solute. KOWWIN uses a special ortho factor to obtain a reasonable result (4.56), but ACD/LogP calculates it at 5.22. A tentative hypothesis has been given to explain the seemingly opposite effect in a steroid (Leo, 1995-A), but a *generalized* way to program this effect, if it is appropriate for other solutes, has not been devised. Later versions of CLOGP have been equipped with a correction that deals specifically with steroids. Ver 4.0 of CLOGP replaces the early 'patch' with a more general and informative solution.

Another example that supports the postulate that a decrease in log P_{oct} results when water 'wins' the competition to a hindered β-site comes from the DDT analogs shown in Table 5.8.

TABLE 5.8

Solute	Meas.	CLOGP	KOWWIN	ACD
DDT	6.91	6.76	6.79	5.92
DDE	6.96	6.94	6.00	6.51
Dicofol*	4.28	6.06	5.81	5.74
CH3CCl3	2.49	2.48	2.68	2.10
HOCH2CCl3	1.42	1.54	1.21	1.38

* Dicofol is an analog of DDT with an OH on the carbon joining the two rings.

Dicofol

With the exception of Dicofol, the three programs tested give reasonable predictions for these and other DDT analogs. All three give satisfactory values for 1,1,1-trichloroethane and its hydroxy analog, which has the same proximity of hydroxyl to trichloro groups as Dicofol shows. The VLOGP program (Gombar, 1996) calculates Dicofol at 4.90, which is outside the 95% confidence limits but still much better than the three results Table 5.8 shows. The reason for the better performance is not apparent from the calculation details.

One important shortcoming in earlier versions of CLOGP involved phenyl ether analogs and applies mostly to substituted diphenyl ether herbicides and to the chloro-dibenzo-dioxins. The most frequently encountered phenyl ethers are the methoxy analogs, or other alkyl-phenyl ethers. In this type of structure, the oxygen still has an appreciable H-bond acceptor strength, β, and is sensitive to electron withdrawing power of substituents on the ring. Therefore it has been assigned a moderately high rho value. The most common of these substituents, methoxyl, is much smaller than the phenoxy and is given a class which rates a smaller negative ortho correction (less twisting out of plane). In nitrofluorofen (see Figure 5.5), the overprediction of the positive electronic and the underprediction of the negative ortho effect work in concert, and the result was an overprediction of 1.26 log units. The newer version takes into account the increased ortho effect, and the deviation has been reduced to 0.46. With diclofop-methyl, the electronic correction is relatively unimportant, and CLOGP's deviation is 0.03. ACD/LogP needs to adjust for the special ortho effect, as it calculates nitrofluorofen at 6.12. KOWWIN uses an appropriate ortho correction and calculates both very well.

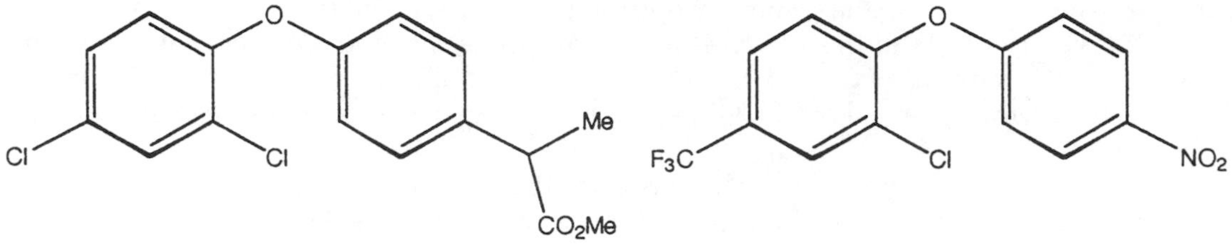

FIGURE 5.5

Diclofop-Methyl			Nitrofluorofen		
Meas.	=	4.80	Meas.	=	4.54
CLOGP	=	4.83; (old = 5.63)	CLOGP	=	5.00; (old = 5.80)
ACD/logP	=	4.65	ACD/logP	=	6.12
KOWWIN	=	4.54	KOWWIN	=	4.63

Bodor claimed (1991) that the "whole molecule" approach using AM1 calculations can estimate the difference in log P of diastereoisomers. The supporting examples are betamethasone and dexamethasone, where the latter was calculated as 0.06 log units lower (1.90 vs. 1.96). Eight measured values from the literature (Hansch, 1995) show the same range in the measurements of each isomer, namely, 1.84 to 2.01. (The latest version of CLOGP estimates both at 2.00.) When the formula used to calculate log P via MO has 17 variables (plus I-alkane = 0) and results in a difference less than half of that shown in the measured data, one hardly has accounted for a subtle difference.

Based on this 'proof' of effectiveness, the authors calculate the log P for each diastereoisomer of two peptide analogs (Cpd. III and IV of Figure 5.6.), each having two chiral centers. This calculation also requires 18 terms. The calculated values for the isomers of Cpd. III span a range of 2.48 to 2.63, while those for Cpd. IV span a range of 2.73 to 2.78. The authors did not report measured log Ps of these (perhaps the compounds were not made). This is regrettable, since four other methods of calculation predict negative log Ps for these peptides—almost 3.0 log units lower than the average of the values Bodor calculated (1991).

5.6.6   Ancillary Features

The ACD/LogP program volunteers some potentially useful information, other than log P, after a structure is processed. In the beta 0.9 version tested, this feature needed considerable improvement. As noted in the hexazinone example discussed above (See Figure 5.1.), the program appeared to detect a 'guanidine' moiety in the heterocyclic ring and warned that, to obtain the neutral solute, partitioning would need to be performed in a basic buffer. Obviously the tertiary amine substituent cannot be a part of a guanidine tautomer, and the two carbonyls would reduce the basicity even if –C(=NH)- were present. As quantum chemical programs become more efficient and available on desktop computers, it should be possible to deliver this sort of information interactively, but even then one should be aware that not all the calculations are appropriate to log P estimation.

Another feature given prominence in ACD promotions is the ability to present the user with various tautomeric forms of the structure entered and to calculate the log P_{oct} for each.

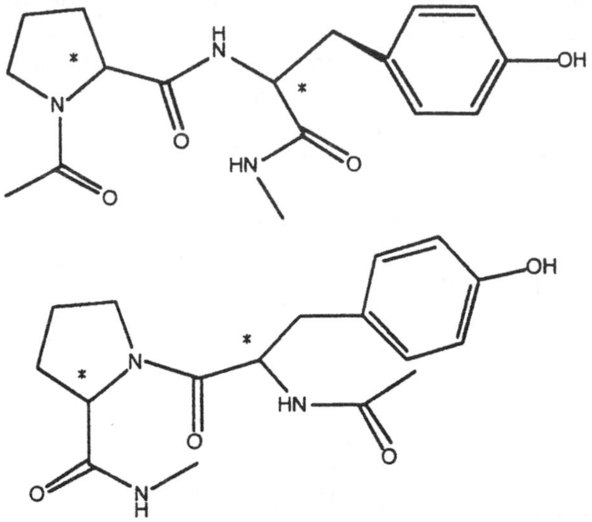

FIGURE 5.6
Diastereisomers Cpd. III

Again this is a feature which needs improvement (beyond that in ver. beta 0.9) to be reliable, as a number of equilibria in the two phases must be considered (Abraham, 1995). Two of the examples given in the ACD Tutorial to illustrate this feature are 6-thiopurine (CAS 50-44-2; Figure 5.7) and thioguanine (CAS 154-42-7).

FIGURE 5.7
6-Purinethiol. Left, OL-Tautomer; Right, ONE-Tautomer.

Good theoretical and experimental evidence (Masuda, 1984; Emsley, 1987) indicates that the "-OL" tautomer is much better accommodated by octanol than is the "-ONE". When 6-thiopurine is selected from the ACD dictionary, it is depicted in the -OL form, with the notation that this is the 'supposedly minor form'—presumably as it exists in the aqueous phase. (Merck displays it in the -OL form, while the '97 USP Dictionary shows the -ONE, again presumably for the aqueous phase.) ACD calculates the -OL form at +0.84 and the -ONE form at –0.55, while the measured value is intermediate, -0.18. The difference is in the right direction (higher value for the -OL) and the spread is large, as expected (but not as great as shown in CLOGP). When a 2-amino group is added to give thioguanine and that name is selected from the ACD dictionary, four tautomeric forms are suggested, none of which are the -OL. When the -OL form is drawn in, it again tags it as 'supposedly minor form' and it suggests the four other tautomeric forms which differ only in which nitrogen holds the hydrogen. However, in this case, the -OL form calculates slightly more *negative* than the -ONE (-0.12 vs -0.08), both of which are remarkably close to the measured value,

−0.07. Like the purine analog, CLOGP calculates the tautomers widely different (+0.72 vs −1.54), and the calculated values bracket the measured ones.

The KOWWIN program calculates the -ONE and -OL forms very nearly the same and close to the measured value for both the thiopurine and thioguanosine. The almost unlimited freedom to incorporate atom strings to fit measured values certainly makes for improved statistics, but it might possibly be misleading when the measurement that is 'fit' is actually for an equilibrium mixture of structures having quite different hydrophobic properties.

Other kinds of tautomerism are not so obvious, and the ability of the ACD/LogP program to call attention to them may make it quite valuable. Camptothecin, as Figure 5.8 shows, provides such an example.

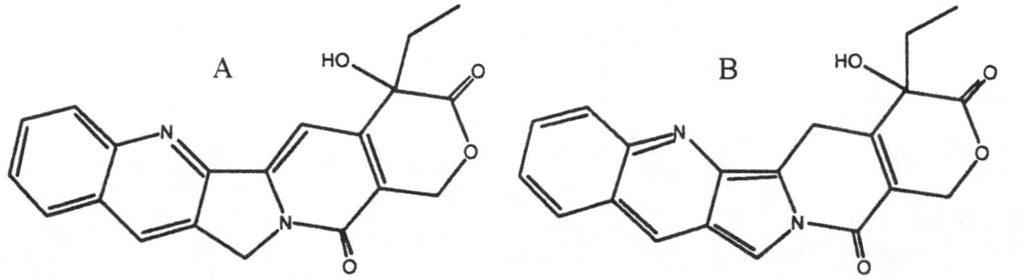

FIGURE 5.8
Camptothecin Tautomers.

Structure A is the one depicted in Merck and the one stored in the ACD dictionary but labeled as 'minor form' when a log P calculation is made. The ACD/LogP calculations for A and B are +1.55 and +2.04, respectively. Two measured values have been reported, 1.74 and 1.85. KOWWIN and CLOGP calculate structure B very nearly the same (1.29 and 1.22), but both see structure A as considerably more hydrophilic (−0.15 and +0.60).

5.7   Summary

Accuracy is certainly a major objective of any program to calculate log P_{oct}. Statistics do not always unambiguously determine this, since there is no 'standard test set' of measured values. An effort should be made to produce such a set, but there can be honest disagreement, for a significant number of solutes, as to which of several measurements is most accurate; e.g., whether a shake-flask, a 'slow-stir', or an HPLC determination is most reliable.

Aside from this concern, a prospective user should ascertain what precision level he/she needs and what structural features can be included while retaining that precision. In some research, the estimation of the relative hydrophobicity of a set of analogs is crucial, but the absolute values need not be precise. For instance, if one wants to know if hydrophobicity contributes to the activity of a certain toxiphore, regression analysis using a calculated log P can provide an answer even if the parent is mispredicted, as long as all of the analogs deviate by this same amount. Of course, the equation intercept, which is a measure of *intrinsic* activity, is then in error.

Occasionally a study in toxicology can reverse the direction of information flow. In a set of anilines which seemed to be more toxic as their log P was increased, there was one outlier, the 2,6-di-isopropyl aniline shown in Figure 5.9. It was much less toxic than predicted by the hydrophobicity an early version of CLOGP provided (Veith, 1988). A partition coefficient measurement showed that shielding of the amino group did, indeed, *lower* the hydrophobicity, and this was programmed as a negative ortho effect into the later versions, as noted previously for phenols.*

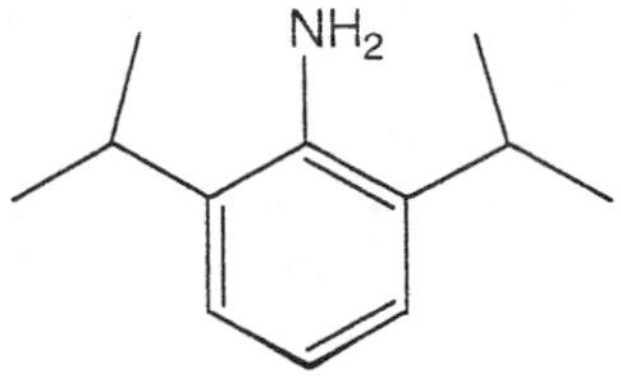

Meas. CLOGP old CLOGP KOWW ACD

3.18   3.17   3.67   3.99   3.61

FIGURE 5.9
2,6-di-isopropyl-aniline.

Often environmental work can tolerate an estimation error of 0.5 log units in log P, but in this instance it was crucial for the correct mechanistic conclusion. This is the sort of information, then, which can be valuable to both the pesticide designer and the environmental toxicologist.

A computer calculation often can detect errors in published values that may not be immediately apparent to the average researcher. For instance, a measured log $P_{(oct)}$ value for toluene diamine was reported as 3.16 (Veith, 1979) but three computer programs show it to be three log units lower: ACD = -0.36; KOWWIN = 0.16; CLOGP = 0.14. One then wonders if a typographical error occurred, and the measured value was 0.16, not 3.16.

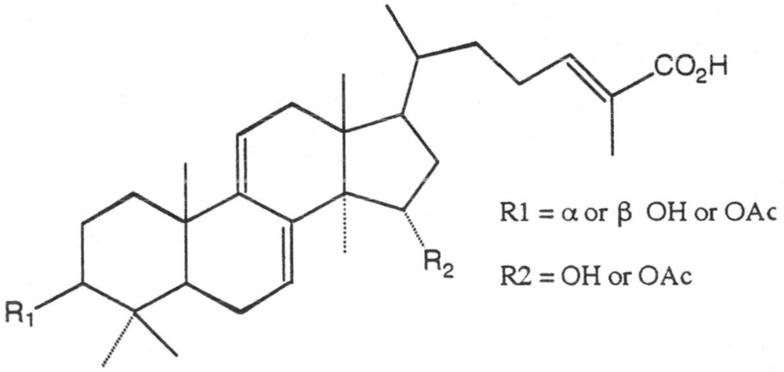

R1 = α or β OH or OAc

R2 = OH or OAc

FIGURE 5.10
Steroid Analogs.

In a recent paper dealing with steroid analogs, the hydrophobicity was determined by both HPLC and shake-flask procedures (Lin, 1995). Figure 5.10 shows the structures.

The relative retention times, k′, show an interesting difference in retention of amphiphilic (OH) and OAc (H-acceptor only) as the alpha or beta epimers at position 3 (R1), but no attempt was made to convert these values to log $P_{(oct)}$. To convert the acids to the neutral form prior to measurement, Lin et al. added 0.5% acetic acid. The shake-flask values they obtained for the 3,15-dihydroxy analogs were about 2.68. The CLOGP value of 6.55 would have made it apparent that they were measuring the largely ionic form and needed a *strong* acid to protonate these analogs. The program also would have led them to

---

* Note that the ortho correction KOWWIN used to achieve a satisfactory prediction for 2,6-di-sec-phenol was not written to apply to anilines.

expect a log unit increase upon acetylation of each hydroxyl, and when their shake-flask measurements showed only an increase of +0.04 for the diacetyl, they would have realized that their experimental procedure was faulty. Most programs can be of some value in this respect; namely, in giving some indication when measurements may be in error.

Of course price and convenience can be a very important factor in the choice of a program to calculate log P_{oct}. Availability on one's computer — e.g. UNIX workstation, VAX station, or the popular desktops, Macintosh or Windows for P.C. — can be the deciding factor. The ability to import structural files in a format that can be handled in batch mode may be a critical need for some. Entry by either name or CAS number might be necessary to others. Providing pKas or tautomeric forms may be very attractive to still others. The manner in which structures are displayed may be a factor.

The 'bottom line' is that a prospective user should spend sufficient time with an adequate demonstration version of any computerized log P calculator, including the many not directly mentioned in this chapter, to make sure it will perform to meet his/her specific needs. Perhaps the best way a prospect can 'shop' for vendors supplying such programs is through the contacts van de Waterbeemd lists (1996) or http://www.pharma.ethz.ch/qsar/logp.html.

References

Abraham, M., and J. McGowan. 1987. The use of characteristic volumes to measure cavity terms in Reverse Phase Liquid Chromatography. *Chromatographia* 23, 243–246.

Abraham, M.H., and A.J. Leo. 1995. Partition between phases of a solute that exists as two interconverting species. *J. Chem. Soc. Perkin Trans.* 2, 1839–1842.

Abraham, M., H. Chadha, and R. Mitchell. 1994. Hydrogen Bonding. 33. Factors that influence the distribution of solutes between blood and brain. *J. Pharm. Sci.* 83, 1257–1268.

Adlard, M., G. Okafo, E. Meenan, and P. Camilleri. 1995. *J. Chem. Soc., Chem. Comun.* 2241–2243.

Bodor, N., and M.-J. Huang. 1991. Predicting Partition Coefficients for Isomeric Diasterioisomers of Some Tripeptide Analogs. *J. Comput. Chem.* 12, 1182–1186.

Bodor, N., and M.-J. Huang. 1992. An Extended Version of a Novel Method for the Estimation of Partition Coefficients. *J. Pharm. Sci.* 81, 272–281.

Bodor, N., A. Gabanyi, and C.-J. Wong. 1989. A New Method for the Estimation of Partition Coefficients. *J. Am. Chem. Soc.* 111, 3783–3786.

Bondi, A. 1964. van der Waals Volume and Radii. *J. Phys. Chem.* 68, 441–451.

Broto, P., G. Moreau, and C. Vandycke. 1984. Molecular Structures: Perception, Autocorrelation Descriptors and SAR Studies. *Eur. J. Med. Chem.* 19, 71–78.

de Bruijn, J., and J. Hermens. 1990. Relation Between Octanol/Water Partition Coefficients and Total Molecular Surface Area and Total Molecular Volume of Hydrophobic Organic Chemicals. *Quant. Struct.-Act. Relat.* 9, 11–21.

de Bruijn, J., F. Busser, W. Seinen, and J. Hermens. 1989. Determination of Octanol/Water Partition Coefficients for Hydrophobic Organic Chemicals with the "Slow-Stirring" Method. *Environ. Toxicol. Chem.* 8, 499–512.

Doluisio, J., and J. Swintosky. 1964. Drug Partitioning II: In Vitro Model for Drug Absorption. *J. Pharm. Sci.* 53, 597–601.

El Tayar, N., A. Mark, P. Vallat, R. Brunne, B. Testa, and W. van Gunsteren. 1993. Solvent Dependent Conformation of Cyclosporin-A: Evidence from Partition Coefficients and Molecular Dynamics Simulations. *J. Med. Chem.* 36, 3757–3764.

El Tayar, N., R.-S. Tsai, P. Vallat, C. Altomare, and B. Testa. 1991. Measurement of Partition Coefficients by Various Centrifuge Partition Chromatographic Techniques: A Comparative Evaluation. *J. Chromatog.* 556 181–194.

Emsley, J., N.J. Freeman, R. Parker, R. Kuroda, and R. Overill. 1987. β-diketone interactions, Part 3. *J. Mol. Struct.* 159, 173–182.

Frank, H., and M. Evans. 1945. Structure and Thermodynamics in Aqueous Electrolytes; Free Volume and Entropy in Condensed Systems. III. *J. Chem. Phys.* 13, 507–532.

Fujita, T., J. Iwasa, and C. Hansch. 1964. A New Substituent Constant, pi, Derived from Partition Coefficients. *J. Am. Chem. Soc.* 86 5175–5179.

Gluck, S.J., M.H. Benkoe, R.K. Hallberg, and K.P. Steele. 1996. Indirect determination of octanol/water partition coefficients by microemulsion electrokinetic chromatography. *J. Chromatog. A.* 744, 141–146.

Gombar, V., and K. Enslein. 1996. Assessment of n-Octanol/Water Partition Coefficient: When Is the Assessment Reliable? *J. Chem. Inf. Comput. Sci.* 36, 1127–1134.

Hammett, L.P. 1970. *Physical Organic Chemistry,* 2nd Ed., McGraw-Hill, New York.

Hansch, C., and A. Leo. 1995. Electronic Effects on Organic Reactions, In *Exploring QSAR,* S. Heller, Ed., Vol. 1., Chpts. 1, 4, and 5, and Vol. 2, Parameter Table, ACS Reference Books, Washington, D.C.

Hansch, C., D. Hoekman, and H. Gao. 1996. Comparative QSAR: Toward a Deeper Understanding of Chemicobiological Interactions. *Chem. Rev.* 96, 1045–1075.

Ishihama, Y., Y. Oda, K. Uchikawa, and N. Asakawa. 1994. Correlation of Octanol/Water Partition Coefficients with Capacity Factors by Micellar Electrokinetic Chromatography. *Chem. Pharm. Bull.* 42, 1525–1527.

Iwasa, J., T. Fujita, and C. Hansch. 1965. Substituent Constants for Aliphatic Functions Obtained from Partition Coefficients. *J. Med. Chem.* 8, 150–153.

Kamlet, M., R.M. Doherty, M.H. Abraham, Y. Marcus, and R.W. Taft. 1988. Linear Solvation Energy Relationships. 46. An Improved Equation for Correlation and Prediction of Octanol/Water Partition Coefficients of Organic Nonelectrolytes. (Including Strong Hydrogen Bond Donor Solutes). *J. Phys. Chem.* 92, 5244–5255.

Kenaga, E. 1972. Determination of Bioconcentration Potential. *Residue Review.* 44, 73–113.

Kirkwood, J. 1934. Theory of Solutions of Molecules Containing Widely Separated Charges, with special application to zwitterions. *J. Chem. Phys.* 351–361.

Klein, W., W. Kordel, M. Weis
, and J.J. Poremski. 1988. Updating of the OECD Test Guideline 107 "Partition Coefficient N-Octanol/Water:" OECD Laboratory Intercomparison Test on the HPLC Method. *Chemosphere.* 17, 361–386.

Klopman, G., J.-Y. Li, S. Wang, and M. Dimayuga. 1994. Computer Automated log P Calculations Based on an Extended Group Contribution Approach. *J. Chem. Inf. Comput. Sci.* 34, 752–781.

Klopman, G., and S. Wang. 1991. A Computer Automated Structure Evaluation (CASE) Approach to Calculation of Partition Coefficients. *J. Comput. Chem.* 8 1025–1034.

Leahy, D., J.J. Morris, and P.J. Taylor. 1992. Model Solvent Systems for QSAR. Part 3. An LSER Analysis of the "Critical Quartet:" New Light on Hydrogen Bond Strength and Directionality. *J. Chem. Soc. Perkin Trans.* 2, 705–731.

Leo, A.J. 1995a. The Future of log P Calculation. In *Lipophilicity in Drug Action and Toxicology,* V. Pliska, B. Testa, and H. van de Waterbeemd, Eds., Chpt. 9. 157–172, VCH, Weinheim.

Leo, A.J. 1995b. Critique of Recent Comparison of Log P Calculation Methods. *Chem. Pharm. Bull.* 43, 512–513.

Leo, A., C. Hansch, and P. Jow. 1976. Dependence of the Hydrophobicity of Apolar Molecules on their Molecular Volume. *J. Med. Chem.* 19, 611–615.

Leo, Albert J. 1981. Hydrophobicity, the Underlying Property in Most Biochemical Events. In *Environmental Health Chemistry,* J. McKinney, Ed., Chpt. 16, 323–336, Ann Arbor Science, Ann Arbor MI.

Leo, Albert J. 1993. Calculating log P_{oct} From Structures. *Chem. Rev.* 93, 1281–1306.

Leo, Albert, C. Hansch, and D. Elkins. 1971. Partition Coefficients and their Uses. *Chem. Rev.* 71: 525–616.

Mannhold, R., and K. Dross. 1996. Calculation Procedures for Molecular Lipophilicity: a Comparative Study. *Quant. Struct.-Act. Relat.* 15, 403–409.

Masuda, S., M. Tanaka, and T. Ota. 1984. Solvent effects on the co-polymerization of 5-hexene-2,4-dione and styrene. *Chem. Lett.* 1327–1330.

Masuda, T., T. Jikhara, K. Nakamura, A. Kimura, T. Takagi, and H. Fujiwara. 1997. Introduction of Solvent-Accessible Surface Area in the Calculation of the Hydrophobicity Parameter log P from an Atomistic Approach. *J. Pharm. Sci.* 86, 57–63.

Meylan, W.M., and P.H. Howard. 1995. Atom/Fragment Contribution Method for Estimating Octanol-Water Partition Coefficients. *J. Pharm. Sci.* 84, 83–92.

Mirrlees, M., S. Moulton, C. Murphy, and P. Taylor. 1976. Direct Measurement of Octanol/Water Partition Coefficients by High Performance Liquid Chromatography. *J. Med. Chem.* 19, 615–626.

Naito, Y., Y. Yamaura, Y. Inoue, C. Fukaya, K. Yokoyama, Y. Nakagawa, and T. Fujita. 1992. Quantitative structure-activity relationships of 2-[4-(thiazol-2-yl)phenyl]propionic acids derivatives inhibiting cyclooxygenase. *Eur. J. Med. Chem.* 27, 645-654. (Abnormality of back calculation from log P' of protonated base confirmed by T. Fujita, private communication.)

Nakagawa, Y., K. Izumi, N. Oikawa, T. Sotomatsu, M. Shigemura, and T. Fujita. 1992. Analysis and Prediction of Hydrophobicity Parameters of Substituted Acetanilides, Benzamides and Related Aromatic Compounds. *Environ. Toxicol. Chem.* 11, 901–916.

Neely, W.B., D.R. Branson, and G.E. Blau. 1974. Partition Coefficients to Measure Bioaccumulation Potential of Organic Chemicals in Fish. *Environ. Sci. Technol.* 8, 1113–1115.

Nys, G., and R. Rekker. 1974. Concept of hydrophobic fragment constant (f-value) II. *Eur. J. Med. Chem.* 9, 361–375.

Reddy, K., and M. Locke. 1996. Molecular Properties as Descriptors of Octanol-Water Partition Coefficients of Herbicides. *Water, Air and Soil Pollution.* 86, 389–405.

Rekker, R., and R. Mannhold. 1992. *Calculation of Drug Lipophilicity.* VCH, Weinheim.

Saeger, V., O. Hicks, R. Kaley, P. Michael, J. Mieure, and E.S. Tucker. 1979. Environmental Fate of Selected Phosphate Esters. *Environ. Sci. Technol.* 13, 840–844.

Sangster, J. 1989. *J. Phys. Chem. Ref. Data.* 18, 1111.

Simpson, C.D., R.J. Wilcock, T.J. Smith, A.L. Wilkins, and A.G. Langdon. 1995. Determination of Octanol/Water Partition Coefficients for Major Components of Technical Chlordane. *Bull. Environ. Contam. Toxicol.* 55, 149–153.

Taft, R., M. Berthelot, C. Laurence, and A. Leo. 1996. Hydrogen Bonds and Molecular Structure. *CHEMTECH.* 20, 20–29.

Takacs-Novak, K., M. Jozan, and G. Szasz. 1995. Lipophilicity of Amphoteric Molecules Expressed by the True Partition Coefficient. *Intern. J. Pharmaceut.* 113, 47–55.

Tomlinson, E., and T. Hafkensheid. 1986. The Filter Probe Extractor: A Versatile Tool for the Rapid Determination of Solute Oil/Water Distribution Behavior. In *Partition Coefficient Determination and Estimation,* W.J. Dunn III, J.H. Block, and R.S. Pearlman, Eds., 83–99, Pergamon Press, N.Y.

Tsai, R.-S., B. Testa, N. El Tayar, and P.-A. Carrupt. 1991. Structure-Lipophilicity Relationships of Zwitterionic Amino Acids. *J. Chem. Soc. Perkin Trans. 2,* 1797–1802.

van de Waterbeemd, H., and R. Mannhold. 1996. Programs and Methods for Calculation of log P-values. *Quant. Struct.-Act. Relat.* 15, 410–412.

Veith, G. 1988. Environmental Research Laboratory, Duluth, MN, Personal Communication.

Veith, G., D. Defoe, and B. Bergstead. 1979. *J. Fish Res. Board Can.* 36, 1040–1047.

Viswanadhan, V., A. Ghose, G. Revankar, and R. Robins. 1989. Atomic Physicochemical Parameters for Three-dimentional Structure-directed QSAR. *J. Chem. Inf. Comput. Sci.* 29, 163–172.

Viswanadhan, V., M. Reddy, R. Bacquet, and M. Erion. 1993. Assessment of Methods Used for Prediction of Lipophilicity: Applied to Nucleosides and Nucleotide Bases. *J. Comput. Chem.* 14, 1019–1026.

Weininger, D. 1988. SMILES, a Chemical Language and Information System. 1. Introduction to Methodology and Encoding Rules. *J. Chem. Inf. Comput. Sci.* 28, 31–36.

Weininger, D., A. Weininger, and J. Weininger. 1989. SMILES. 2. Algortihm for Generation of Unique SMILES Notation. *J. Chem. Inf. Comput. Sci.* 29, 97–101.

Yamagami, C., and N. Takao. 1993. Hydrophobicity Parameter Determination by Reversed Phase Liquid Chromatography VII; Hydrogen-bond effect in prediction of Log P values for benzyl-N,N-dimethylcarbamate. *Chem. Pharm. Bull.* 41, 694–698.

Yanase, D., and A. Andoh. 1992. Translocation of Photosynthesis-Inhibiting Herbicides in Wheat Leaves Measured by Phytofluorography, the Chlorophyll Fluorescence Imaging. *Pestic. Biochem. Physiol.* 44, 60–67.

6

Vegetation-Air Partition Coefficient

Michael S. McLachlan

CONTENTS

6.1 Introduction

The vegetation-air or plant-air partitioning coefficient K_{PA} describes the equilibrium partitioning of a chemical between the gas phase and aerial vegetation. The below-ground part of the plant is not included in the vegetation-air partition coefficient. This separation of the plant into two parts makes sense from an environmental fate perspective, since the aerial and below-ground plant parts reside in distinctly different media.

Vegetation plays an important role in the fate of many chemicals. Plants are the first link in terrestrial food chains, and hence the accumulation of chemicals in plants is a crucial step determining exposure of higher terrestrial organisms, including humans, to environmental chemicals. Plants affect the atmospheric transport of chemicals by scavenging them from the air, and they also serve as a medium for transfering chemicals between the soil and the atmosphere. Plant tissues have potential as biomonitors of air contamination, provided that the partitioning relationship between air and plant concentrations can be quantified. Finally, the accumulation of environmental chemicals in plants can have adverse impacts on the plants themselves.

The plant-air partition coefficient is often the parameter determining the relationship between the concentration of a chemical in the gas phase of the atmosphere and its concentration in vegetation, and as such it is a parameter of great significance. However, remember that the quotient of the vegetation and gaseous concentrations only equals K_{PA} if a state of equilibrium exists between the plant and the gas phase. Changes in plant properties, temperature, or concentration in the gas phase can disturb an existing state of equilibrium, resulting in plant concentrations that may be higher or lower than the equilibrium value. For some compounds, K_{PA} is so large that the vegetation can never approach a partitioning equilibrium, either because the plant dies before equilibrium or because of rapid growth dilution (McLachlan et al., 1995). Also, particle bound deposition — either wet or dry — can induce a disequilibrium, potentially increasing the plant concentration above that expected from equilibrium partitioning. Refer to McLachlan (1999) for a framework for determining whether equilibrium partitioning, kinetically limited gaseous deposition, or particle-bound deposition is responsible for the chemical concentration in a given situation. One must always keep the equilibrium prerequisite in mind when using K_{PA}.

The properties of the two phases, the properties of the chemical, and the temperature, control the partitioning process. For trace concentrations of chemicals in the ambient atmosphere, the air compartment can be treated as a perfect gas. Methods to predict K_{PA} thus must consider three factors: the properties of the plant, the properties of the chemical, and temperature. The temperature dependence of K_{PA} is a function of the chemical and the plant; the influence of chemical properties on K_{PA} is a function of temperature and the plant. Understanding and describing these interdependencies quantitatively is a considerable challenge which has been addressed only recently; although much progress has been made, much remains to be done.

6.2   Measurement Methods

6.2.1   Chamber Experiment

Several methods have been employed for measuring K_{PA}. The most common is the chamber experiment, in which plants are exposed to a constant gaseous concentration in a contamination chamber. This technique has been used for volatile compounds for which partitioning equilibrium is rapidly achieved and the quotient of the plant and gaseous concentrations gives K_{PA} (Frank and Frank, 1989). It also has been applied to less volatile compounds where the partition coefficient was determined by extrapolation on the basis of the uptake and clearance kinetics (Bacci and Gaggi, 1987; Reischl et al., 1989; Hauk et al., 1994). Batch experiments in which chemical is injected into a chamber containing plant foliage also have been used for volatile compounds (Figge, 1990).

6.2.2   Fugacity Meter

The solid phase fugacity meter is a more recent development which allows the direct measurement of partition coefficients for less volatile chemicals (Horstmann and McLachlan, 1992; Tolls and McLachlan, 1994). It involves a brief exposure of the vegetation to a contaminated atmosphere, a subsequent equilibration phase, and then measurement of the air concentration in a column under flow conditions which allow the establishment of an equilibrium between the air and the surface of the vegetation.

6.2.3 Plant-Water Partition Coefficients

Measurements of plant-water partition coefficients (K_{PW}) also have been an important source of data for several investigators, who divided these by the air-water partition coefficients (K_{AW}) to obtain estimates of K_{PA}. This approach presumes that immersion in water does not influence the partitioning properties of the plant.

Plant-water partition coefficients have been determined in batch experiments, in which the plants were immersed in an aqueous solution and K_{PW} was calculated from the concentrations in the plant and in water after equilibrium was reached (Wolf et al., 1991). They also have been measured in flow-through systems where the plants were exposed to constant aqueous concentrations (Gobas et al., 1991).

6.2.4 Cuticle-Water Partition Coefficients

A particularly large dataset has come from the study of cuticle-water partition coefficients determined using batch experiments with isolated cuticles (Kerler and Schönherr, 1988). The cuticle, a lipophilic membrane covering all leaves, is an important storage compartment for lipophilic organic compounds in vegetation.

6.2.5 Octanol-Air Coefficients

The octanol-air partition coefficient (K_{OA}) has played an important role in the study of plant-air partitioning. At the end of the 1980s, when interest in this subject started to grow, octanol was already well established as a model for the partitioning properties of organic carbon and lipids in aquatic systems. Several researchers borrowed from this experience and postulated that the partitioning properties of the hydrophobic or lipid-like portions of plants also could be modelled using octanol (Schramm et al., 1987; Paterson and Mackay, 1989; McLachlan et al., 1989). Since the partner phase was air, the octanol-water partition coefficient could not be used and it became necessary to define an octanol-air partition coefficient.

The octanol-air partition coefficient was calculated initially as the quotient of the octanol-water coefficient (K_{OW}) and air-water partition coefficient (K_{AW}) (Paterson et al., 1991a). Recently, direct measurements obtained by passing purified air over an octanol solution of the chemical have become available (Harner and Mackay, 1995; Harner and Bidleman, 1996; Harner and Bidleman, 1997; Kömp and McLachlan, 1997a), and an interpolation method based on GC retention times has been proposed (Zhang et al., 1999). The measurements indicate that the calculated values may underestimate the true K_{OA} values. It has been postulated that this is because the octanol–water partition coefficient does not represent partitioning between pure octanol and pure water, but rather between octanol saturated with water and water saturated with octanol (Harner and Mackay, 1995).

However, before more comprehensive data sets of measured values become available, the calculation of K_{OA} from K_{OW} and K_{AW} is the most viable approach. Indeed, most of the investigations of plant-air partitioning to date have been done with calculated K_{OA} values, and hence the derived correlations are valid only with calculated values. In order to avoid confusion, this chapter will refer to calculated K_{OA} values as K_{OW}/K_{AW}.

6.3   Estimation Methods

Current estimation methods for K_{PA} are based on the assumption that the plant can be modelled as a mixture of air, water, and one or several more lipophilic phases. The general equation for K_{PA} at a reference temperature (25°C) is:

$$K_{PA}(25°C) = v_A + v_W/K_{AW} + v_{L1}K_{L1A} + v_{L2}K_{L2A} + ... + v_{Ln}K_{LnA} \qquad (1)$$

where v is the volume fraction of a compartment in the plant, K is an equilibrium partition coefficient at 25°C, and the subscript A refers to air, W to water, and L1 through Ln to the 1st through nth lipophilic plant compartments. Note that K_{PA} is defined here and throughout this chapter on a volume/volume basis, i.e., concentration per unit plant volume divided by concentration per unit gas volume. The air compartment generally can be neglected since it is of interest only for $K_{PA} < 1$. The estimation methods differ in their treatment of the lipophilic compartments, each author having chosen a particular dataset with which to derive correlations to predict K_{L1A} through K_{LnA}.

The several early efforts to use K_{OW} or the Henrys law constant H as descriptors for the influence of physical-chemical properties on K_{PA} (Bacci and Gaggi, 1987; Travis and Hattemer-Frey, 1988; Reischl et al., 1989) were based on very limited data sets. As mentioned above, at the end of the 1980s an apparent consensus emeged that the quotient of K_{OW} and K_{AW} (or K_{OA}) was a more suitable descriptor for partitioning into the lipid-like compartments. The first and simplist model assumed that there were one or several lipid-like compartments whose partition coefficients were all equal to K_{OW}/K_{AW} (Schramm et al., 1987; Paterson and Mackay, 1989; Paterson et al., 1991b). Equation (1) then becomes

$$K_{PA}(25°C) = v_A + v_W/K_{AW} + v_{L1}K_{OW}/K_{AW} + v_{L2}K_{OW}/K_{AW} + ... + v_{Ln}K_{OW}/K_{AW} \qquad (2)$$

which can be reduced to

$$K_{PA}(25°C) = v_A + v_W/K_{AW} + v_L K_{OW}/K_{AW} \qquad (3)$$

6.3.1   Linear Method

This will be referred to as the linear K_{OA} method.

Although originally derived from theoretical considerations, support for this method was found in the data from the chamber experiments of Bacci and co-workers. They found a linear relationship between K_{PA} and K_{OW}/K_{AW} for a wide range of compounds in azalea leaves (Bacci et al., 1990), while v_L was obtained by fitting Equation (3) to the data (Paterson et al., 1991a). Fugacity meter measurements of K_{PA} in ryegrass also yielded a linear relationship between K_{PA} and K_{OA} for a range of lipophilic compounds, and in this case an independent estimate of v_L based on cuticle volume and extractable lipids agreed well with the v_L value obtained by fitting Equation (3) to the data (Tolls and McLachlan, 1994).

6.3.2   Non-Linear Variation — Trapp Method

A variation of this approach has been proposed by Trapp and coworkers. Although these authors originally used the linear K_{OA} method (Trapp et al., 1990), they later proposed an equation in which K_{PA} is not linearly proportional to K_{OW}/K_{AW}, but rather to $K_{OW}^{0.95}/K_{AW}$ (Trapp et al., 1994; Trapp and Matthies, 1995). Correcting a small inconsistency in their papers (the mass fraction of lipid is transformed to a volume fraction using the density of water instead of the density of lipid), their equation can be expressed as:

$$K_{PA}(25°C) = v_W/K_{AW} + v_L K_{OW}^{0.95}/K_{AW} \tag{4}$$

Note that these authors do not include an air compartment. The basis for the exponent of 0.95 on the K_{OW} term was the experimental work of Briggs et al. (1983), who measured the partitioning of two series of organic compounds between barley shoots and water.

6.3.3   Two-Compartment Approach – Riederer Method

Riederer (1990) published a more complex method based on two lipid-like compartments, an acylglycerol lipid compartment and a cuticle compartment. The acylglycerol-air partition coefficient was assumed to equal K_{OW}/K_{AW}, while measured values of the cuticle-water partition coefficient were employed for the cuticle compartment. Riederer (1995) later modified this model to include a predictive equation for the cuticle-water partition coefficient, based on Kerler and Schönherr's measurements (1988) of eight chemicals with log K_{OW} values ranging from 1.92 to 7.86. They used isolated citrus and rubber plant leaf cuticles as well as tomato and green pepper fruit cuticles. The resulting equation is

$$K_{PA}(25°C) = v_A + (v_W + v_G K_{OW} + 1.14\, v_C K_{OW}^{0.97})/K_{AW} \tag{5}$$

where the subscript G refers to acylglycerol lipids and C to cuticle. An alternative expression based on a molecular connectivity index (Sabljic et al., 1990) also is given for those cases where reliable K_{OW} values are not available.

6.3.4   Müller Method

Müller et al. (1994) adapted and extended the two-compartment approach. They defined four plant compartments in addition to water: cellular lipids, cuticle, structural carbohydrates, and proteins. Their predictive equation is:

$$K_{PA}(25°C) = (v_W + v_{CL} K_{OW} + 1.11\, v_C K_{OW}^{0.97} + 0.0372\, v_{CA} K_{OW}^{0.95} + v_P(86.2 K_{OW}^{-1} + 3.70))/K_{AW} \tag{6}$$

where the subscripts are defined as follows: CL = cellular lipids; C = cuticle; CA = structural carbohydrates; P = proteins. The expression for partitioning to the cuticle was also taken from Kerler and Schönherr's measurements of cuticle-water partitioning (1988). Measurements of cellulose-water partitioning coefficients served to define partitioning to the structural carbohydrates, and measurements of protein-water partitioning were used for the protein compartment. The authors compared the predictions of their method with 25 K_{PA} values measured in spruce needles, azalea leaves, and grass, estimating the volume fractions using a complex series of assumptions. They reported "relatively good agreement," but the deviation between measured and predicted values exceeded an order of magnitude for about one quarter of the data points.

6.3.5   Agreement of Methods

To summarize, there are four current methods for predicting $K_{PA}(25°C)$: the linear K_{OA} method (Equation (3)), the Trapp method (Equation (4)), the Riederer method (Equation (5)) and the Müller method (Equation (6)). In Table 6.1 these methods are employed to calculate K_{PA} (25°C) of lindane, anthracene and 2,2′,5,5′-tetrachlorobiphenyl in grass using the volume fractions for the different compartments given in Müller et al. (1994): $v_W = 0.65$, $v_{CL} = 0.003$, $v_C = 0.004$, $v_{CA} = 0.078$ and $v_P = 0.039$. v_A was set to 0.1; v_G was set equal to v_{CL} for the Riederer method; and v_L for the linear K_{OA} and Trapp methods was assumed to be equal to the sum of the cellular lipid and cuticle compartments, namely 0.007.

TABLE 6.1

Comparison of the Four Methods for Prediction of K_{PA}.

	Lindane	Anthracene	2,2′,5,5′-Tetrachlorobiphenyl
log K_{OW}	3.8[a]	4.54[b]	6.1[c]
log K_{AW}	−4.28[a]	−2.80[b]	−1.72[c]
log K_{OW}/K_{AW}	8.08	7.34	7.82
log K_{PA}			
Linear K_{OA} Method	5.93	5.19	5.67
Trapp Method	5.74	4.96	5.36
Riederer Method	5.90	5.14	5.60
Müller Method	6.01	5.24	5.69

[a]Suntio et al., 1988.
[b]Mackay et al., 1992b.
[c]Mackay et al., 1992a.

6.3.5.1   *General Agreement*

The results indicated good agreement among the four methods. This is not surprising, as the methods are very similar. Since the protein compartment in the Müller method can be neglected for compounds with log $K_{OW} > 2$, this method as well as the Trapp and Riederer methods express K_{PA} as a near-linear function of K_{OW}/K_{AW}, and hence are close to the linear K_{OA} method.

6.3.5.2   *Temperature Dependence*

While these methods allow prediction of K_{PA} at 25°C, it is often necessary to determine K_{PA} for other temperatures. It has been suggested that K_{PA} should be exponentially proportional to reciprocal temperature, and that the temperature dependence can be expressed using a van't Hoff-type equation (McLachlan et al., 1995)

$$K_{PA}(T) = K_{PA}(25°C)\exp((1/T - 1/298.15)_{\Delta}H_{PA}/R) \tag{7}$$

where $_{\Delta}H_{PA}$ is the enthalpy of phase change from plant to air. In the absence of measured values it was proposed that $_{\Delta}H_{PA}$ can be approximated by the enthalpy of vaporization from the sub-cooled liquid $_{\Delta}H_{vap}$.

     Recently, the temperature dependence of K_{PA} was measured for polychlorinated biphenyls (PCBs) in ryegrass (Kömp and McLachlan, 1997b). It was found that the temperature dependence indeed could be described using Equation (6). However, $_{\Delta}H_{PA}$ was not equal to $_{\Delta}H_{vap}$. Rather, it was described by the equation

$$_{\Delta}H_{PA} = 2.67 - _{\Delta}H_{vap} - 130 \quad kJ/mol \tag{8}$$

6.3.5.3   Difficulty of Defining Volume Fractions

Combining equations (7) and (8) with one of equations (3), (4), (5), or (6) provides a complete method to estimate K_{PA} at any environmental temperature. However, the application of these methods is hampered by the difficulty defining the volume fractions of the different compartments.

The estimates of Müller et al. (1994) for just three plant species yielded a range of a factor of 12 for the volume fraction of the cellular lipids v_{CL} and a factor of 7 for the cuticular membrane v_C, which would indicate that these parameters are variable and that they must be estimated for each plant species or canopy of interest. However, only in one case has an independent measure of the volume fractions been successfully used to fit measured K_{PA} data (Tolls and McLachlan, 1994); in most other cases the volume fractions were deduced from the regression of measured K_{PA} against K_{OW}/K_{AW}.

Our current understanding is insufficient about the identity of the different lipid-like compartments and how they can be estimated or measured for different plant species. The only endeavor in this direction has been that of Müller et al. (1994), and their estimation methods are not amenable to generalization.

6.3.5.4   Limits of Data

Current plant-air partitioning data are limited to a very few species, namely azalea leaves and rye grass and a few data for spruce needles. The plant-water partitioning data used to develop methods also are limited to isolated cuticles from just four plants and macerated barley leaves. We thus cannot expect that currently available methods adequately account for the interspecies variability in partitioning.

Indeed, recent measurements of K_{PA} for PCBs in four herbal species common to central European grasslands have produced strong evidence that K_{PA} is not always linearly proportional to K_{OA} (Kömp and McLachlan, 1997c). Using an equation of the form $K_{PA} = a\,K_{OA}^n$ (the aqueous compartment could be neglected for these compounds), the exponent n was found to vary between 0.57 and 1.15 for these four species and ryegrass.

This calls all four methods presented in this paper into question, since they have n approximately equal to 1. Furthermore, the enthalpy of phase change $_\Delta H_{PA}$ for a given compound varied by up to 25 kJ/mol between the five species. The linear K_{OA} model and equations (6) and (7) with an estimated v_L of 0.01 were used to predict K_{PA} at 10°C for 2,3-dichlorobiphenyl in these five plant species. The predicted values differed from the measured values by up to a factor of 13 (see Table 6.2).

TABLE 6.2

Comparison of Measured Values of K_{PA} at 10°C for 2,3-Dichlorobiphenyl in Five Different Grass and Herb Species with the Predicted Value from the Linear K_{OA} Model.[a]

	log K_{PA}
Ryegrass (*Lolium multiflorum*)	6.01
Clover (*Trifolium repens*)	6.59
Plantain (*Plantago lanceolata*)	6.49
Hawk's Beard (*Crepis biennis*)	6.78
Yarrow (*Achillea millefolium*)	7.25
Linear K_{OA} Model	6.14

[a] Calculated using measured values of log K_{OA} (6.98) and $_\Delta H_{vap}$ (72.2 kJ/mol) (Kömp and McLachlan, 1997a), and assuming $v_L = 0.01$.

6.4 Summary

In summary, current methods of predicting K_{PA} agree well with each other, and we recommend using all four methods to gain some appreciation of error. It is difficult to assign error, or to recommend a specific method, because of the insufficient understanding of specific plant properties that determine the size of the lipid-like compartments. Furthermore, the influence of both the physical-chemical properties and temperature on K_{PA} are also a function of plant properties, which adds further complexity to the problem of predicting K_{PA}. Much more research is required for an understanding of these interactions.

In the meantime, the methods presented here can at best give only an approximate value of K_{PA}. The potential error associated with these methods is more than an order of magnitude but, due to the limited database of measured K_{PA} values, it is not possible to estimate the maximum error or the error frequency distribution associated with extrapolating these methods to plant species and chemical classes other than those used in the original experiments.

References

Bacci, E., M.J. Cerejeira, C. Gaggi, G. Chemello, D. Calamari, and M. Vighi. 1990. Bioconcentration of Organic Chemical Vapours in Plant Leaves: The Azalea Model. *Chemosphere* 21, 525–535.

Bacci, E. and C. Gaggi. 1987. Chlorinated Hydrocarbon Vapours and Plant Foliage: Kinetics and Applications. *Chemosphere* 16, 2515–2522.

Briggs, G.G., R.H. Bromilow, A.A. Evans, and M. Williams. 1983. Relationships between lipophilicity and the distribution of nonionized chemicals in barley shoots following uptake by the roots. *Pestic. Sci.* 14, 492–500.

Figge, K. 1990. Luftgetragene, organische Stoffe in Blattorganen. UWSF - Z. Umweltchem. Ökotox. 2, 200–207.

Frank, H. and W. Frank. 1989. Uptake of Airborne Tetrachloroethene by Spruce Needles. *Environ. Sci. Technol.* 23, 365–367.

Gobas, F.A.P.C., E.J. McNeil, L. Lovett-Doust, and G.D. Haffner. 1991. Bioconcentration of Chlorinated Aromatic Hydrocarbons in Aquatic Macrophytes. *Environ. Sci. Technol.* 25, 924–929.

Harner, T. and T.F. Bidleman. 1996. Measurements of octanol-air partition coefficients for polychlorinated biphenyls. *J. Chem. Eng. Data* 41, 895–899.

Harner, T. and T.F. Bidleman. 1998. Measurements of Octanol-Air Partition Coefficients for Polycyclic Aromatic Hydrocarbons (PAHs) and Polychlorinated Naphthalenes (PCNs). *J. Chem. Eng. Data*, 43, 40–46.

Harner, T. and D. Mackay. 1995. Measurement of Octanol-Air Partition Coefficients for Chlorobenzenes, PCBs, and DDT. *Environ. Sci. Technol.* 29, 1599–1606.

Hauk, H., G. Umlauf, and M.S. McLachlan. 1994. Uptake of Gaseous DDE in Spruce Needles. *Environ. Sci. Technol.* 28, 2372–2379.

Horstmann, M. and M.S. McLachlan. 1992. Initial Development of a Solid-Phase Fugacity Meter for Semivolatile Organic Compounds. *Environ. Sci. Technol.* 26, 1643–1649.

Kerler, F. and J. Schönherr. 1988. Accumulation of Lipophilic Chemicals in Plant Cuticles: Prediction From Octanol/Water Partition Coefficients. *Arch. Environ. Contam. Toxicol.* 17, 1–6.

Kömp, P. and M.S. McLachlan. 1997a. Octanol/Air Partitioning of Polychlorinated Biphenyls. *Environ. Toxicol. Chem.* 16, 2433–2437.

Kömp, P. and M.S. McLachlan. 1997b. The Influence of Temperature on the Plant/Air Partitioning of Semivolatile Organic Compounds. *Environ. Sci. Technol.* 31, 886–890.

Kömp, P. and M.S. McLachlan. 1997c. Interspecies Variability of the Plant/Air Partitioning of Polychlorinated Biphenyls. *Environ. Sci. Technol, 31*, 2944–2948.

Mackay D., W.Y. Shiu, and K.C. Ma. 1992a. *Illustrated Handbook of Physical-Chemical Properties and Environmental Fate for Organic Chemicals, Vol. 1.* Lewis, Boca Raton.

McLachlan, M.S. 1999. Framework for the interpretation of measurements of SOCs in plants. *Environ. Sci. Technol. 33*, 1799–1804.

Mackay D., W.Y. Shiu, and K.C. Ma. 1992b. *Illustrated Handbook of Physical-Chemical Properties and Environmental Fate for Organic Chemicals, Vol. 2.* Lewis, Boca Raton.

McLachlan, M.S., A. Reischl, M. Reissinger, and O. Hutzinger. 1989. Some Thoughts on Plant Accumulation of Airborne Organic Pollutants. In D. Mackay and S. Paterson, Eds., *Human Exposure to Chemicals*, Institute of Environmental Studies, University of Toronto.

McLachlan, M.S., K. Welsch-Pausch, and J. Tolls. 1995. Field Validation of a Model of the Uptake of Gaseous SOC in *Lolium multiflorum* (Rye Grass). *Environ. Sci. Technol. 29*, 1998–2004.

Müller, J.F., D.W. Hawker, and D.W. Connell. 1994. Calculation of Bioconcentration Factors of Persistent Hydrophobic Compounds in the Air/Vegetation System. *Chemosphere 29*, 623–640.

Paterson, S., D. Mackay, E. Bacci, and D. Calamari. 1991a. Correlation of the Equilibrium and Kinetics of Leaf-Air Exchange of Hydrophobic Organic Chemicals. *Environ. Sci. Technol. 25*, 866–871.

Paterson, S., D. Mackay, and A. Gladman. 1991b. A Fugacity Model of Chemical Uptake by Plants from Soil and Air. *Chemosphere 23*, 539–565.

Paterson, S. and D. Mackay, 1989. Modeling the Uptake and Distribution of Organic Chemicals in Plants. In D.T. Allen, Y. Cohen, and I.R. Kaplan, Eds., *Intermedia Pollutant Transport*, Plenum Publishing Corporation, 269–282.

Reischl, A., M. Reissinger, H. Thoma, and O. Hutzinger. 1989. Uptake and Accumulation of PCDD/F in Terrestrial Plants: Basic Considerations. *Chemosphere 19*, 467–474.

Riederer, M. 1990. Estimating Partitioning and Transport of Organic Chemicals in the Foliage/Atmosphere System: Discussion of a Fugacity-Based Model. *Environ. Sci. Technol. 24*, 829–837.

Riederer, M. 1995. Partitioning and Transport of Organic Chemicals between the Atmospheric Environment and Leaves. In S. Trapp and C. McFarlane, Eds., *Plant Contamination - Modeling and Simulation of Organic Chemical Processes*, Lewis, Boca Raton, 153–190.

Sabljic, A., H. Güsten, J. Schönherr, and M. Riederer. 1990. Modeling Plant Uptake of Airborne Organic Chemicals. 1. Plant Cuticle/Water Partitioning and Molecular Connectivity. *Environ. Sci. Technol. 24*, 1321–1326.

Schramm, K.-W., A. Reischl, and O. Hutzinger. 1987. UNITTree – A Multimedia Compartment Model to Estimate the Fate of Lipophilic Compounds in Plants. *Chemosphere 16*, 2653–2663.

Suntio, L.R., W.Y. Shiu, D. Mackay, J.N. Seiber, and D.E. Glotfelty. 1988. Critical Review of Henry's Law Constants for Pesticides. *Rev. Environ. Contam. Toxicol. 103*, 1–59.

Tolls, J. and M.S. McLachlan. 1994. Partitioning of Semivolatile Organic Compounds between Air and *Lolium multiflorum* (Welsh Ray Grass). *Environ. Sci. Technol. 28*, 159-166.

Trapp, S., M. Matthies, I. Scheunert, and E.M. To 1990. Modeling the Bioconcentration of Organic Chemicals in Plants. *Environ. Sci. Technol. 24*, 1246-1252.

Trapp, S., C. McFarlane, and M. Matthies. 1994. Model for Uptake of Xenobiotics into Plants: Validation with Bromacil Experiments. *Environ. Toxicol. Chem. 13*, 413-422.

Trapp, S. and M. Matthies. 1995. Generic One-Compartment Model for Uptake of Organic Chemicals by Foliar Vegetation. *Environ. Sci. Technol. 29*, 2333-2338.

Travis, C.C. and H.A. Hattemer-Frey. 1988. Uptake of Organics by Aerial Plant Parts: A Call for Research. *Chemosphere 17*, 277-283.

Wolf, S.D., R.R. Lassiter, and S.E. Wooten. 1991. Predicting chemical accumulation in shoots of aquatic plants. *Environ. Toxicol. Chem. 10*, 665-680.

Zhang, X., K.-W. Shramm, B. Henkelmann, C. Klimm, A. Kaune, A. Dettrup, and P. Lu. 1999. A method to estimate the octanol–air partition coefficient of semivolatile organic compounds. *Anal. Chem. 71*, 3834–3838.

7

Solubility in Water

D. Mackay

CONTENTS

7.1   Introduction

Solubility in water is one of the most important physical chemical properties of a substance, having numerous applications to the prediction of its fate and its effects in the environment. It is a direct measurement of hydrophobicity, i.e., the tendency of water to "exclude" the substance from solution. It can be viewed as the maximum concentration which an aqueous solution will tolerate before the onset of phase separation.

Substances which are readily soluble in water, such as lower molecular weight alcohols, will dissolve freely in water if accidentally spilled and will tend to remain in aqueous solution until degraded. On the contrary, sparingly soluble substances dissolve more slowly and, when in solution, have a stronger tendency to partition out of aqueous solution into other phases. They tend to have larger air-water partition coefficients or Henry's law

1-56670-456-1/00/$0.00+$.50
© 2000 by CRC Press LLC

constants, and they tend to partition more into solid and biotic phases such as soils, sediments, and fish. As a result, it is common to correlate partition coefficients from water to these media with solubility in water, as other chapters discuss.

7.2   Theoretical Foundation

Solubility normally is measured by bringing an excess amount of a pure chemical phase into contact with water at a specified temperature, so that equilibrium is achieved and the aqueous phase concentration reaches a maximum value. It follows that the fugacities or partial pressures exerted by the chemical in these phases are equal. Assuming that the pure chemical phase properties are unaffected by water, the pure phase will exert its vapor pressure P^S (Pa) corresponding to the prevailing temperature. The superscript S denotes saturation. In the aqueous phase, the fugacity can be expressed using Raoult's Law with an activity coefficient γ:

$$P^S = x\gamma P^S_L \tag{1}$$

where x is the mole fraction of the chemical in aqueous solution, and P^S_L is the vapor pressure of pure liquid phase chemical, again at the specified temperature.

7.2.1   Equilibrium Situations

Four possible equilibrium situations may exist, depending on the nature of the chemical phase, each of which requires separate theoretical treatment and leads to different equations for expressing solubility. These equations form the basis of the correlations discussed later.

7.2.1.1   Pure Chemical is an Immiscible Liquid (e.g., Benzene)

In this case, P^S is also P^S_L. Thus the product $x\gamma$ is 1.0 and x is $1/\gamma$. Sparingly soluble substances are such because the value of γ is large.

   For example, at 25°C benzene has a solubility in water of 1780 g/m^3 or 22.8 mol/m^3. Since 1m^3 of solution contains approximately $10^6/18$ mol water (1m^3 is 10^6 g and 18 g/mol is the molecular mass of water), the mole fraction x is $22.8/(10^6/18)$ or 0.00041. The activity coefficient γ is thus 2440; i.e., a benzene molecule in aqueous solution behaves as if its concentration were 2440 times higher.

   Substances such as PCBs can have activity coefficients exceeding 1 million. Hydrophobicity thus is essentially an indication of the magnitude of γ. Some predictive methods focus on estimating γ, from which solubility can be deduced.

   As is discussed later, a molar activity coefficient γ_m may be defined such that $S\gamma_m$ is 1.0 where S is the molar solubility; i.e., in units of mol/litre. Since S is 55.5 x, γ_m is $\gamma/55.5$, 55.5 being the number of moles of water in 1 litre of solution.

7.2.1.2   Pure Chemical is an Immiscible Solid (e.g., Naphthalene)

In this case, P^S is the vapor pressure of the solid, P^S_S, e.g., naphthalene, while P^S_L is of liquid naphthalene. The fugacity ratio F or P^S_S/P^S_L, i.e., the solid to liquid ratio, can be estimated as:

$$P_S^S / P_L^S = \exp\left(-\Delta S_F \left(T_M/T - 1\right)/R\right) = F \tag{2}$$

since the entropy of fusion, ΔS_F, is typically 56 J/mol at the melting point, as Walden's Rule (Walden 1908) suggests:

$$P_S^S / P_L^S = \exp\left(-6.79\left(T_M / T - 1\right)\right) \tag{3}$$

The fugacity ratio F is 1.0 at the melting point and lower at lower temperatures. When T exceeds T_M, the substance is liquid and the fugacity ratio is 1.0.

For example, naphthalene with a melting point of 80°C has a fugacity ratio of 0.286 at 25°C; thus $x\gamma$ is 0.286. The solubility of solid naphthalene (molecular weight 128) is 31.7 g/m³ or 0.25 mol/m³, x is 4.5×10^{-6}, and γ is 64000. If naphthalene could exist as a liquid at 25°C, its solubility would be 31.7/0.286 or 111 g/m³. F is thus the ratio of both vapor pressures and solubilities.

This fugacity ratio correction is critically important and can have an enormous effect on solubility. Since chemicals in solution are in the liquid state they behave as if, in the limit, they were pure liquids, and the reference fugacity or vapor pressure must be of the liquid. Correlations of partition coefficients such as octanol-water with solubility must include this "correction."

7.2.1.3 Pure Chemical is a Gas or Vapor (e.g., Methane)

If the substance's boiling point is less than the environmental temperature, it can not exist as a stable liquid or solid at atmospheric pressure. Thus the pure chemical phase is a gas, usually at atmospheric pressure, and P^S is 1 atm or 101325 Pa. Solubilities may, however, be reported at the higher vapor pressure. For more dilute gases at partial pressures less than 1 atm, the solubility is then proportional to the partial pressure as expressed by Henry's Law. The reference pressure P_L^S may be known from data obtained at higher pressures, or it may be unknown if the environmental temperature exceeds the critical temperature.

It is usual to lump γ and P_L^S together as one parameter H′, a Henry's law constant, i.e.,

$$P = x\gamma P_L^S = xH' \tag{4}$$

H′ thus contains information about both the volatility P_L^S and the hydrophobicity γ. The units of H′ depend on the units of P and concentration. The Henry's law constant often is expressed with pressure in Pa and concentration in mol/m³, thus giving H units of Pa m³/mol.

Relatively few environmental contaminants are gases, and they are well characterized. The most common gaseous examples are vinyl chloride and freons.

7.2.1.4 Pure Chemical is a Miscible Substance (e.g., Ethanol)

If the activity coefficient is relatively small, i.e. <20, it is likely that the liquid is miscible with water and no solubility can be measured. The relevant descriptor of hydrophobicity in such cases is the activity coefficient. Correlations of other environmental partitioning properties with solubility are then impossible.

7.2.2   Complicating Factors

Solubility is a function of temperature because both P_L^S and γ are temperature dependent. Usually γ falls with increasing temperature; thus solubility increases. This implies that the process of dissolution is endothermic. Exceptions are frequent and in some cases, such as benzene, there may be a solubility minimum as a function of the temperature at which the enthalpy of dissolution is zero.

The presence of electrolytes in the aqueous phase (as occurs most commonly in seawater) generally causes a solubility reduction or "salting out" effect. This usually is expressed by the Setschenow constant K, namely

$$\log_{10}(S_O/S) = KC \tag{5}$$

where S_O and S are the solubility in pure water and in the electrolyte solution and C is the concentration (molarity) of the electrolyte. Xie et al. (1997) reviews such data. This effect is important when assessing the fate of organic chemicals in salt water.

In partially miscible systems with γ typically in the range 20 to 200, appreciable dissolution of water in the chemical phase may occur, thus the vapor pressure P^S is not that of the pure chemical but of the somewhat diluted chemical phase.

Surface active materials may not display a solubility because they form micelles in solution, which are essentially soluble aggregations of the pure substance.

In environmental situations, dissolved organic matter such as fulvic acids frequently increase the apparent solubility. This is the result of sorption of the chemical to organic matter which is sufficiently low in molecular mass to be retained permanently in solution. The "true" solubility or concentration in the pure aqueous phase probably is not increased. The apparent solubility is the sum of the "true" or dissolved concentration and the quantity which is sorbed.

The solubility of substances such as carboxylic acids, which dissociate or form ions in solution, is also a function of pH, a common environmental example being pentachlorophenol. Data must thus be at a specified pH. Alternatively, the solubility of the parent (non-ionic) form may be given, and pKa or pKb given, to permit the ratio of ionic to non-ionic forms to be calculated as

$$\text{ionic/non-ionic} = 10^{\,(\text{pH-pKa})} \tag{6}$$

The total solubility is then that of the parent and ionic forms.

7.3   Experimental Methods

The conventional approach is to equilibrate an excess amount of the pure chemical with an aqueous solution at controlled temperature and measure the concentration of the solution. Common analytical methods include gravimetric or volumetric techniques, spectroscopy, gas chromatography, and scintillation counting. Difficulties may be encountered with the presence of micro-particles of pure chemical phase which may give an apparently high solubility.

The use of "generator columns," in which water is flowed slowly through a tube packed with a solid support coated with the chemical, reduces this risk (Wasik et al., 1993). Hydrophobic substances tend to sorb appreciably to solid surfaces such as glassware, and they may evaporate from solution; thus extreme care is required when handling these solutions.

Solubilities are generally reported in units such as mg/L or the equivalent g/m³, but other unit systems are used, especially molarity and molality. Solutions exceeding 100 mg/L are relatively easy to handle and analyze. In the range 1 to 100 mg/L, more care is required and accuracy is reduced. In the range 0.001 to 1 mg/L, extreme care is required and reported data are often suspect. Below 0.001 mg/L, solutions are so dilute that it is very difficult to obtain reliable data. Mackay et al. (1997) and Yalkowsky and Banerjee (1992) review methods of measuring solubility and other environmentally relevant properties.

7.4   Data Sources

The most critically reviewed solubility data are those in the IUPAC Solubility Data series (1984, 1985, 1989a, 1989b). Other sources include the Handbooks by Howard et al. (1989, 1990, 1991, 1993), Mackay et al. (1997), and Verschueren (1996) and the papers and texts by Horvath (1973, 1975, 1982) and Yalkowsky and Bannerjee (1992). Commercial databases are also available from a variety of sources, e.g., the AQUASOL database of Yalkowsky and Dannenfelser (1991).

7.5   Estimation Methods

Andren et al. (1987) have reviewed a variety of estimation methods.

Three general classes of estimation method are in frequent use. The first exploits the considerable data on octanol water partition coefficient K_{OW} and the well developed estimation methods for that quantity by seeking a relationship between solubility and K_{OW}. The second attempts to relate molecular structure to solubility by a variety of techniques involving counting bonds or groups or calculating molecular volumes, areas, or other topological indices such as connectivity.

The third uses another measured property of the chemical which can be related to solubility. Most notable is HPLC retention time, which is a measure of the partitioning of the chemical between an aqueous phase (or more usually a water-cosolvent mixture) and a stationary organic phase. Other properties exploited include boiling point, parachor, and solubility parameters.

We recommended two of these methods for general use, estimation from octanol-water partition coefficient and a group contribution method named AQUAFAC. Three other methods are also valuable under certain circumstances, the connectivity, UNIFAC, and solvatochromic approaches.

7.5.1   Estimation from Octanol-Water Partition Coefficient

Hansch et al. (1968) showed that solubility and K_{OW} are well correlated for liquid solutes. This correlation is expected on theoretical grounds because solubility is inversely proportional to activity coefficient γ in the aqueous phase, and K_{OW} can be shown to be proportional to γ.

K_{OW} is $\gamma_W v_W / \gamma_O v_O$ where v is molar volume and subscripts w and o refer to the water and octanol phases (Mackay, 1991). A plot of log K_{OW} versus log solubility is thus expected to have a slope of approximately –1, and this is observed. Numerous further studies have explored and refined this relationship (e.g., Miller et al., 1985; Chiou et al., 1977, 1982; Valvani et al., 1981; Yalkowsky and Valvani, 1979; Yalkowsky et al., 1983a, 1983b; Banerjee et al., 1980).

For solids, it is essential to correlate K_{OW} with the liquid solubility, not the solid solubility; thus the fugacity ratio expression must be included in any correlation. This is readily done, since melting point $T_M(K)$ is usually available and Walden's Rule can be applied. Following Yalkowsky (1979), the fugacity ratio F can be estimated at 25°C (T = 298 K) as

$$\log F = 6.79(1 - T_M/T)/2.303$$

$$= -0.01(T_M - 298) \tag{7}$$

The solid molar solubility S_S thus can be calculated from the liquid molar solubility S_L and F as

$$S_S = FS_L \tag{8}$$

or

$$\log S_S = \log S_L + \log F = \log S_L - 0.01\ (T_M - 298)$$

The usual correlation approach is then to determine the parameters A and B in the equation

$$\log S_L = A - B \log K_{OW} \tag{9}$$

or

$$\log S_S = A - B \log K_{OW} - 0.01\ (T_M - 298) \tag{10}$$

A considerable number of such correlations have been developed with the constant 0.01 being allowed to vary, and in some cases it is (wrongly) ignored, or because of inherent error in the data it appears to be unimportant (e.g., Isnard and Lambert, 1989). As was discussed earlier, there are compelling theoretical reasons to include the fugacity ratio term.

For rigid molecules, Yalkowsky et al. (1983) obtained a simple correlation for molar solubility with B equal to 1.0, namely:

$$\log S_S = 0.8 - \log K_{OW} - 0.01(T_M - 298) \tag{11}$$

This equation also can be written in terms of the liquid solubility S_L as

$$S_L = 6.31/K_{OW} \tag{11a}$$

This implies that the pseudo solubility of these substances in octanol is 6.31 mol/L. This equation is ideal for screening purposes or when a fast check of the consistency of reported values of S_L and K_{OW} is needed.

Note that, if the melting point is expressed in K, and T_M is less than 298 K, the log F term containing $(T_M - 298)$ should be set to zero, i.e., $(T_M - 298)$ can never be negative. If entropy of fusion data are available, it is preferable to include these data rather than rely on Walden's Rule. In that case, the constant 0.01, which is $56.4/(2.303 \times 8.314 \times 298)$ can be rewritten as $0.000175 \Delta S$ where ΔS has units of J/mol K. If ΔS is in cal/mol K, the constant changes to 0.00073.

For long chain flexible molecules, Yalkowsky and Valvani (1980) expressed ΔS in terms of chain length as

$$\Delta S = 56.5 + 10.5 \ (n\text{-}5) \ \text{J/mol K}$$

or

$$13.5 + 2.5 \ (n\text{-}5) \ \text{cal/mol K} \tag{12}$$

They obtained a correlation for 167 substances of

$$\log S_L = -0.000813 \ \Delta S(T_M - 298) - \log K_{OW} + 0.54 \tag{13}$$

Many other correlations, for example those compiled by Lyman et al. (1982), have coefficients B in equation 9 of 1.1 to 1.3. This implies that, as K_{OW} increases, the activity coefficient of the chemical in octanol γ_0 also increases. For example, if γ_W increases by a factor of 100, and γ_0 increases by a factor of about 2.5; then K_{OW} increases by a factor of only 40. The result is that solubility is proportional to K_{OW} raised to a power of -1.25, since $40^{-1.25}$ is 0.01. This effect of an increase in γ_0 or a decreasing solubility in octanol is very important for hydrophobic substances. Examination of these correlations, especially those of Andren et al. (1987) and Isnard and Lambert (1989), suggests that a coefficient B of 1.25 is more generally applicable. This causes a change in the coefficient A to about 1.10, giving the correlation

$$\log S = -0.01(T_M - 298) - 1.25 \log K_{OW} + 1.10 \tag{14}$$

This correlation is recommended for more hydrophobic substances. Equations (11) and (14) give the same solubility when $\log K_{OW}$ is 1.2.

As expected, restricting the class of chemicals to structurally similar compounds such as alcohols or PCBs permits an improved correlation, Table 7.1, adapted from Lyman et al. (1982) and Yalkowsky and Banerjee (1992), lists a number of correlations for specific chemical classes.

In summary, for general use, apply Equation (14), which implies a knowledge of K_{OW} and melting point. If necessary, both quantities can be estimated (see Chapters 1 and 5).

If entropy of fusion data are also available, the constant 0.01 can be "corrected" to $0.000175\Delta S$ with ΔS in units of J/mol K, or to $0.00073 \ \Delta S$ with units of cal/mol K.

The standard deviation is expected to be about 0.3 log units, corresponding to a factor of 2. For a narrower class of substances, using the equations in Table 7.1, the standard deviation is expected to fall to about 0.2 log units or a factor of 1.6. The Syracuse Research Corporation offers an estimation method WSKOWWIN as part of its Estimation Program Interface for Windows (EPI/WIN); details are available at the web site http://syrres.com/interskow/epi.htm.

7.5.2 Estimation from Molecular Structure

Irmann (1965) pioneered the correlation of solubilities of hydrocarbons and halogenated hydrocarbons with molecular structure using a scheme with the form

TABLE 7.1

Regression Equations Relating Solubility S (mol/L) to K_{OW}. The Solubility is of the Liquid or Subcooled Liquid State. For Solids, a Fugacity Ratio Term Must be Included

Chemical Class	logS = A − B log K_{OW} (Equation (9))		
	A	B	Reference
Alcohols	0.926	1.113	Hansch et al. (1968)
Alcohols	0.338	0.971	Tewari et al. (1982)
Ketones	0.720	1.229	Hansch et al. (1968)
Ketones	0.431	0.927	Tewari et al. (1982)
Esters	0.520	1.013	Tewari et al. (1982)
Esters	0.306	1.073	Tewari et al. (1982)
Alkylholides	0.832	1.221	Hansch et al. (1968)
Alkynes	1.043	1.294	Hansch et al. (1968)
Alkenes	0.248	1.294	Hansch et al. (1968)
Alkenes	0.275	1.101	Tewari et al. (1982)
Mono Aromatics	0.339	0.996	Hansch et al. (1968)
Mono Aromatics	0.727	0.947	Tewari et al.(1982)
Alkanes	−0.248	1.237	Hansch et al. (1968)
Halogenated HCs	0.356	1.103	Tewari et al. (1982)
Halogenated HCs	1.50	0.962	Chiou & Freed (1977)
Acids, bases, neutrals	0.845	1.163	Hafkenscheid & Tomlinson (1983)
Dyes	0.453	0.820	Hou & Baughman, quoted in Yalkowsky & Banerjee (1992)
PAHs	0.262	0.880	Yalkowsky & Valvani (1979)

$$-\log S = x + \Sigma y_i n_i + \Sigma z_j n_j \tag{15}$$

The constant x depends on the class of compound (e.g., aromatic vs. aliphatic), n_i is the number of specific atoms (e.g., C or Cl), y_i is a constant specific to that atom and how it is bonded, n_j is the number of certain bonds or structural types (e.g., double or triple bonds or branching), and z_i is a constant specific to these bonds or structural elements. Lyman et al. (1982) gives a full account of this approach and recommend it as the preferred method when no partition coefficient data are available. However, this method is dated and does not include functional groups containing oxygen, nitrogen, and sulfur atoms.

Yalkowsky, Myrdal, and co-workers have developed the AQUAFAC group contribution method which predicts the molar activity coefficient γ_m from which the molar solubility S can be deduced, using supplemental information on the fugacity ratio F (Myrdal et al., 1992, 1993, 1995). The method exploits the AQUASOL database described by Dannenfelser and Yalkowsky (1991) and Yalkowsky and Dannenfelser (1991). The fugacity ratio (which is termed the "ideal solubility" in their publication) is expressed as

$$\log F = -(13.5 - 4.6 \log \sigma)(T_M - 298)/1364 \tag{16}$$

where σ is a symmetry number, i.e., the number of indistinguishable positions in which a compound may be oriented. Anthracene has a σ of 4, while phenanthrene has a σ of 2. The units in this equation are cal/mol K, and 1364 is $1.987 \times 2.303 \times 298$ where 1.987 is the gas constant R in units of cal/mol K. With SI units of Joules the equation is

$$\log F = -(56.5 - 19.2 \log \sigma)(T_M - 298)/5706 \tag{17}$$

The prediction equations for molar solubility are thus:

$$\text{liquids, } S = 1/\gamma_m \quad \text{or} \quad \log S = -\log\gamma_m \tag{18}$$

$$\text{solids, } S = F/\gamma_m \quad \text{or} \quad \log S = \log F - \log \gamma_m \tag{19}$$

The molar activity coefficient is given by:

$$\log \gamma_m = \Sigma n_i q_i \tag{20}$$

where n_i is the number of times a group appears and q_i is the contribution of that group. Table 7.2 lists values of q_{ui}.

TABLE 7.2

AQUAFAC Group Contribution q-Values. (supplied by S. Yalkowsky)

X_a-G	q-Values	X_nY-G	q-Values	X_nY_2-G	q-Values
X_2–CH$_2$	0.545	XY–CH$_2$	0.03	Y$_2$–CH$_2$	0.149
X_3–CH	0.305	X$_2$Y–CH	0.085	XY$_2$–CH	−0.127
X_4–C	0.019	X$_3$Y–C	−0.308	X$_2$Y$_2$–C	−0.52
X_2–C=	0.583	XY–C	0.525	Y$_2$–C	0.319
X_2–O	−1.51	XY–O	−0.664	Y$_2$–O	−0.017
X_2–BARB	−2.689	XY–BARB	−2.593	Y$_2$–BARB	−2.667
X_3–N	−3.428	X$_2$Y–N	0.379	XY$_2$–N	0.320
X_2–C=O	−0.968	XY–C=O	−0.722	Y$_2$–C=O	−0.41
X_2–NH	−2.233	XY–NH	−0.110	Y$_2$–NH	N/D[a]
X–CH$_3$	0.706	Y–CH$_3$	0.204	N/E[b]	
X–CH=	0.636	Y–CH=	0.321	N/E[b]	
X–F	0.251	Y–F	−0.141	N/E[b]	
X–Cl	0.389	Y–Cl	0.409	N/E[b]	
X–Br	0.379	Y–Br	0.645	N/E[b]	
X–I	0.49	Y–I	0.887	N/E[b]	
X–CHO	−1.111	Y–CHO	−0.772	N/E[b]	
X–NH$_2$	−1.911	Y–NH$_2$	−1.193	N/E[b]	
X–N=	−0.668	Y–N=	−0.969	N/E[b]	
X–N=C=S	1.203	Y–N=C=S	1.266	N/E[b]	
X–CONH	−1.509	Y–CONH	−0.847	N/E[b]	
X–CONH$_2$	−2.126	Y–CONH$_2$	−0.508	N/E[b]	
X–CON	−1.601	Y–CON	N/D[a]	N/E[b]	
X–C≡N	−0.619	Y–C≡N	−0.427	N/E[b]	
X–NO$_2$	−0.127	Y–NO$_2$	0.082	N/E[b]	
X–NHCON(CH$_3$)$_2$	−2.190	Y–NHCON(CH$_3$)$_2$	−1.229	N/E[b]	
X–(COO)	−1.117	Y–(COO)	−0.796	N/E[b]	
X–C≡CH	0.438	Y–C≡CH	N/D[a]	N/E[b]	
X–OOCH	−1.283	Y–OOCH	N/D[a]	N/E[b]	
X–COOH	N/D[a]	Y–COOH	−1.419	N/E[b]	
X–OH	−2.285	Y–OH	−1.810	N/E[b]	
X=CH$_2$	0.579	Y=CH$_2$	N/E[b]	N/E[b]	
EPOXIDE	−0.301	EPOXIDE	N/E[b]	N/E[b]	
X_2–S	N/D[a]	XY–S	N/D(a)	Y$_2$–S	−0.310
C$_{AR}$				0.525	
C$_{BRIDGE\ HEAD}$				0.319	
C$_{HAR}$				0.321	
N$_{AR}$	−0.969				
C$_{RING}$				−0.063	
ORTHOBIPHENYL				−0.123	

[a] N/D, group not yet defined
[b] N/E, non-existent group

The "groups" include atoms of H, F, Cl, Br, etc., with different values for carbon and oxygen atoms, depending on their bonding environment. An X or sp^3 neighboring atom is a hydrogen, aliphatic carbon, halogen, amine nitrogen, or ether oxygen. A Y or sp^2 neighboring atom includes aromatic carbon and nitrogen, including nitro groups. A YY neighbor is a bridgehead aromatic C, as in naphthalene, or a group which is joined to two aromatic carbons, such as the carbonyl in diphenyl ketone. Two correction factors also are included, "C_{RING}" for alicyclic rings and "ORTHOBIPHENYL" for the number of halogens in ortho positions in a biphenyl ring.

Table 7.3 gives the results of an analysis of the predictive capabilities of the AQUAFAC and partition coefficient approaches for 97 compounds.

TABLE 7.3

Average Absolute Errors in Log Solubility as Predicted by AQUAFAC and from Log K_{OW} (Myrdal et al., 1995).

Compound Type	Number of Compounds	AQUAFAC	Log K_{OW}
All compounds	97	0.41	0.61
Liquids	49	0.36	0.40
Solids	48	0.45	0.82
Non hydrogen bonding solids & liquids	78	0.42	0.63
Hydrogen bonding solids & liquids	19	0.34	0.53

7.5.3 Molecular Connectivity

Kier and Hall (1976) and Hall et al. (1975) have pioneered the use of the connectivity index as a descriptor of molecular structure. It is an expression of the sum of the degrees of connectedness of each atom in a molecule. Indices can be calculated to various degrees or orders, thus encoding increasing information about the structure. Although the index has been used with success in a number of applications, it is not entirely clear on theoretical grounds why this is so. It appears that the index generally expresses molar volume or area.

Nirmalakhandan and Speece (1988a, 1988b, 1989) have used this index in conjunction with polarizability to correlate solubilities of some 470 compounds, with a standard error of 0.332 in log solubility. The correlation include three parameters, a modified polarizability parameter and two connectivity indices. For those familiar with connectivity indices, this method is very convenient, but for occasional use by those unfamiliar with these concepts, the method requires a considerable learning period. It is discussed more fully in Chapter 8.

7.5.4 Solvatochromic or Linear Solvation Energy Methods

Kamlet, Taft, and co-workers have developed and advocated the use of the "solvatochromic" method, in which solubility is predicted from molar volume melting point and two parameters which express dipolarity/polarizability and hydrogen bond basicity (Kamlet et al., 1986, 1988; Taft et al., 1985). The method has been applied to a large number of properties, including partitioning and toxicities. Yalkowsky et al. (1988) have commented critically on the method, suggesting that the K_{OW} approach is more generally applicable. The approach has merit in that it exploits fundamental information on molecular interactions, but apparently it has not been widely used. It could form the basis of improved estimation methods in the future. It is not recommended for general use at this time.

7.5.5 UNIFAC

The UNIFAC (UNIQUAC functional group activity coefficient) method is an extension of the UNIQUAC (Universal quasi chemical) method, which has been used widely in chemical process engineering to describe partitioning in organic systems as occur in petroleum and chemical processing (Fredenslund et al., 1975, 1977). It has been applied less frequently to aqueous systems. It expresses the activity coefficient as the sum of a "combinational" component, which quantifies the nature of the area "seen" by the solute molecule, and a "residual" component, which is deduced from group contributions. Arbuckle (1983, 1986), Banerjee (1985), Banerjee and Howard (1988), and Campbell and Luthy (1985) have tested the applicability of the method to water solubility.

The method involves some computation which, while not difficult, requires extensive accurate coding and thus discourages use for occasional purposes. Since it was designed to treat complex mixtures, it is likely to find favor for estimating solubilities in water-cosolvent systems. Compilations of group parameters are available (e.g., Magnussen et al., 1981; Gmehling, 1983; Gmehling et al., 1982; and Tiegs et al., 1987) but the focus is more on chemicals of commerce than on environmental contaminants.

This method is well established and undoubtedly will be refined and extended; thus it is likely to play an increasing role in environmental situations. It is not yet as fully tested as the two recommended methods; thus it is not recommended at this time except for those who already use the technique and have available the required computer programs and some experience in their use.

7.6 Summary

Two methods are recommended for routine use for estimating solubility; the correlation with octanol water partition coefficient method and the AQUAFAC group contribution method. When possible, use both approaches and compare the results. Insights can also be obtained if solubility data are available for structurally similar compounds. Both methods are illustrated below.

Three other methods are applicable in certain circumstances, particularly if there is prior experience in their use. These are the connectivity, solvatochromic, and UNIFAC methods.

The estimation of solubility is an area of current research. Thus, existing methods are being refined and expanded, and new methods are evolving. The reader should keep abreast of these developments by consulting the current literature.

7.7 Appendix

Illustrations of Estimation Methods at 25°C

Anthracene and phenanthrene

These two isomeric three-ring polycyclic aromatic hydrocarbons have similar values of K_{OW} but different melting points, illustrating the importance of the fugacity ratio term. Both have a molecular formula $C_{14}H_{10}$ and a molecular mass of 178.2.

Anthracene has an mp of 216°C (489K), and log K_{OW} is 4.54.
Phenanthrene has an mp of 101°C (374 K), and log K_{OW} is also 4.54.

Estimation from K_{OW}
Equation (14) logS = –0.01 (T_M–298) –1.25 log K_{OW} + 1.10
Anthracene: logS = –0.01(191) –1.25 (4.54) + 1.10 = -6.49
S = 3.27×10^{-7} mol/L = 0.058 g/m³ (expl value 0.045 g/m³)
Phenanthrene logS = –0.01(76) –1.25 (4.54) + 1.10 = –5.34
S = 4.62×10^{-6} mol/L = 0.82 g/m³ (expl value 1.1 g/m³)
Equation (11) gives higher values of 0.40 g/m³ and 5.6 g/m³.

Ignoring the fugacity ratio term would introduce a considerable error.

AQUASOL method
Anthracene
$\Sigma n_i q_i$ = 10(Y–CH=) + 4(YY>C=) = 10(0.321) + 4(0.319) = 4.486
logS = –0.01(MP – 25)-$\Sigma n_i q_i$= –0.01(218–25) –4.486 = –6.586
S = 2.59×10^{-7} mol/L = 0.046 g/m³
For phenanthrene, only the fugacity ratio term changes, and S is 1.01 g/m³.

Lindane
Molecular mass is 291, melting point is 113°C, and log K_{OW} is reported to be 3.0.
Applying Equation (14) as above gives:
logS = –3.53, S = 2.95×10^{-4} mol/L = 86 g/m³
6(X–CH) + 6(XCl) + 6(C Ring) = 6(0.305) + 6(0.389) + 6(–0.062)
$\Sigma n_i q_i$ = 6*CH+6*XC1+6*C_{ring} = 3.79
logS = –0.01(MP – 25)-$\Sigma n_i q_i$ = -0.01(112–25)–3.79 = –4.656
S = 2.21×10^{-5} mol/L = 6.4 g/m³
The reported value is 7.3 g/m³.

Trichloroethylene
Molecular mass is 131.4 g/mol, melting point is –73°C (200 K), and log K_{OW} is reported to be 2.53.
Applying equation 14 as above with no fugacity ratio term:
logS = –2.06, S = 8.7×10^{-3} mol/L = 1144 g/m³
The reported solubility is 1100 g/m³.
$\Sigma n_i q_i$ = (XCH=) +X_2C = +3(XC1) = 2.446
Since this compound is a liquid, the melting point term is set to zero. Thus, the predicted solubility is equal to the sum of AQUAFAC q-values.
logS = –0.01(MP – 25) – $\Sigma n_i q_i$ = –2.446
S = 3.58×10^{-3} mol/L = 471 g/m³

References

Andren, A.W., Doucette, W.J., and Dickhut, R.M. 1987. Methods for estimating solubilities of hydrophobic organic compounds: Environmental modeling efforts. In *Sources and Fates of Aquatic Pollutants*. R.A. Hites and S.J. Eisenreich, Eds., pp 3-26, Advances in Chemistry Series 216. American Chemical Society, Washington, D.C.

Arbuckle, W.B. 1983. Estimating Activity Coefficients for Use in Calculating Environmental Parameters. *Environ. Sci. Technol.* 17: 537–542.

Arbuckle, W.B. 1986. Using UNIFAC to Calculate Aqueous Solubilities. *Environ. Sci. Technol.* 20: 1060–1064.

Banerjee, S., S.H. Yalkowsky, and S.C. Valvani. 1980. Water Solubility and Octanol/Water Partition Coefficients of Organics. Limitations of the Solubility-Partition Coefficient Correlation. *Environ. Sci. Technol.* 14: 1427–1429.

Banerjee, S. 1985. Calculation of Water Solubility of Organic Compounds with UNIFAC-Derived Parameters. *Environ. Sci. Technol.* 19: 369–370.

Banerjee, S. and P.H. Howard. 1988. Improved Estimation of Solubility and Partitioning through Correction of UNIFAC-Derived Activity Coefficients. *Environ. Sci. Technol.* 22: 839–848.

Campbell, J.R. and R.G. Luthy. 1985. Prediction of Aromatic Solute Partition Coefficients Using the UNIFAC Group Contribution Model. *Environ. Sci. Technol.* 19: 980–985.

Chiou, C.T., V.H. Freed, D.W. Schmedding, and R.L. Kohnert. 1977. Partition Coefficients and Bioaccumulation of Selected Organic Compounds. *Environ. Sci. Technol.* 11: 475–478.

Chiou, C.T., D.W. Schmedding,and M. Manes. 1982. Partitioning of Organic Compounds in Octanol-Water Systems. *Environ. Sci. Technol.* 16: 4–10.

Dannenfelser, R.-M. and S.H. Yalkowsky. 1991. Database of aqueous solubility for organic non-electrolytes. *Sci. Total Environ.* 109/110, 625–628.

Fredenslund, A., R.L. Jones, and J.M. Prausnitz. 1975. Group-Contribution Estimation of Activity Coefficients in Nonideal Liquid Mixtures. *AICHE J.* 21:1086–1099.

Fredenslund, A., J. Gmehling, and P. Rasmussen. 1977. *Vapor-Liquid Equilibria Using UNIFAC.* Elsevier, Amsterdam.

Gmehling, J., P. Rasmussen, and A. Fredenslund. 1982. Vapor-Liquid Equilibria by UNIFAC Group Contribution. Revision and Extension. 2. *Ind. Eng. Chem. Process. Des. Dev.* 21: 118–127.

Gmehling, J. 1983. Vapor-Liquid Equilibria by UNIFAC Group Contribution. Revision and Extension. 3. *Ind. Eng.Chem. Process Des. Dev.* 22:676–678.

Hafkenscheid, T.L. and E. Tomlinson. 1983. Isocratic Chromatographic Retention Data for Estimating Aqueous Solubilities of Acid in Basic and Neutral Drugs. *Int. J. Pharm.* 16, 1–20.

Hall, L.H., L.B. Kier, and W.J. Murray. 1975. Molecular Connectivity II: Relationship to Water Solubility and Boiling Point. *J. Pharm. Sci.* 64: 1974–1977.

Hansch, C., J.E. Quinlan, and G.L. Lawrence. 1968. The Linear Free Energy Relationships Between Partition Coefficients and the Aqueous Solubility of Organic Liquids. *J. Org. Chem.* 33: 347–350.

Horvath, A.L. 1973. Solubilities of Members of Homologous Series. Correlation and Prediction. *Chem. Engin. Sci.* 28: 299–304.

Horvath, A.L. 1975. Similarity-Symmetry-Analog Principle. Solubility of Halogenated Hydrocarbons in Water. *Chem. Ingin. Tech.* 47: 815.

Horvath, A.L. 1982. *Halogenated Hydrocarbons: Solubility-Miscibility with Water.* Marcel Dekker, New York.

Howard, P.H., Ed. 1989. *Handbook of Fate and Exposure Data for Organic Chemicals. Vol. I.* Large Production and Priority Pollutants. Lewis Publishers, Chelsea, Michigan.

Howard, P.H., Ed. 1990. *Handbook of Fate and Exposure Data for Organic Chemicals. Vol. II – Solvents.* Lewis Publishers, Inc., Chelsea, Michigan.

Howard, P.H., Ed. 1991. *Handbook of Fate and Exposure Data for Organic Chemicals. Vol. III Pesticides.* Lewis Publishers, Inc. Chelsea, Michigan.

Howard, P.H. 1993. *Handbook of Fate and Exposure Data for Organic Chemicals. Vol. IV. Solvent 2.* Lewis Publishers, Inc., Chelsea, Michigan.

Irmann, F. 1965. Eine einfache Korrelation zwischen Wasserloslichkeit und Struktur von Kohlenwasserstoffen und Halogenkohlenwasserstoffen. *Chemie. Ing. Techn.* 37: 789–798.

Isnard, P. and S. Lambert. 1989. Aqueous solubility/n-octanol-water partition coefficient correlations. *Chemosphere* 18, 1837–1853.

IUPAC Solubility Data Series. 1984. *Vol. 15: Alcohols with Water.* A.F.M. Barton, Ed., Pergamon Press, Oxford, England.

IUPAC Solubility Data Series. 1985. *Vol. 20: Halogenated Benzenes, Toluenes and Phenols with Water.* A.L. Horvath and F.W. Getzen, Eds., Pergamon Press, Oxford, England.

IUPAC Solubility Data Series. 1989a. *Vol. 37: Hydrocarbons (C_5-C_7). with Water and Seawater.* D.G. Shaw, Ed., Pergamon Press, Oxford, England.

IUPAC Solubility Data Series. 1989b. *Vol. 38: Hydrocarbons (C_8-C_{36}). with Water and Seawater.* D.G. Shaw, Ed., Pergamon Press, Oxford, England.

Kamlet, M.J., R.M. Doherty, J.-L.M. Abboud, M.H. Abraham, and Taft, R.W. 1986. Solubility: A New Look. Chemtech: 566–576.

Kamlet, M.J., R.M. Doherty, P.W. Carr, D. Mackay, M.H. Abraham, and R.W. Taft. 1988. Linear Solvation Energy Relationships. 44. Parameter Estimation Rules which Allow Accurate Prediction of Octanol/Water Partition Coefficients and other Solubility and Toxicity Properties of Polychlorinated Biphenyls and Polycyclic Aromatic Hydrocarbons. *Environ. Sci. Technol.* 22:503–509.

Kier, L.B. and L.H. Hall. 1976. *Molecular Connectivity in Chemistry and Drug Research.* Academic, New York.

Lyman, W.J., W.F. Reehl, and D.H. Rosenblatt. 1982. *Chemical Property Estimation Methods.* McGraw-Hill, New York.

Mackay, D. 1991. *Multimedia Environmental Models: The Fugacity Approach.* Lewis /CRC Press Boca Raton, FL.

Mackay, D., W.Y. Shiu, and K.C. Ma. 1997. *Illustrated Handbook of Physical Chemical Properties and Environmental Fate for Organic Compounds. Vol. 5 Pesticide Chemicals, and Vol. 1 and 4.* CRC Press, Boca Raton, FL.

Magnussen, T., P. Rasmussen, and A. Fredenslund. 1981. UNIFAC Parameter Table for Prediction of Liquid-Liquid Equilibria. *Ind. Eng. Chem. Process Des. Dev.* 20: 331–339.

Miller, M.M., S.P. Wasik, G.-L. Huang, W.-Y. Shiu, and D. Mackay. 1985. Relationships between Octanol-Water Partition Coefficients and Aqueous Solubility. *Environ. Sci. Technol.* 19: 522–529.

Myrdal, P.B., G.H. Ward, R.-M. Dannenfelser, D. Mishra, and S.H. Yalkowsky. 1992. AQUAFAC 1: Aqueous Functional Group Activity Coefficients; Application to Hydrocarbons. *Chemosphere* 24, 1047–1061.

Myrdal, P.B., G.H. Ward, P. Simamora, and S.H. Yalkowsky. 1993. AQUAFAC: Aqueous Functional Group Activity Coefficients. *SAR and QSAR in Environmental Research*, Vol. 1: 53–61.

Myrdal, P., A.M. Manka, and S.H. Yalkowsky. 1995. Aquafac 3: Aqueous Functional Group Activity Coefficients; Application to the Estimation of Aqueous Solubility. *Chemosphere.* 30, No. 9: 1619–1637.

Nirmalakhandan, N.N. and R.E. Speece. 1988a. Predicition of Aqueous Solubility of Organic Chemicals Based on Molecular Structure. *Environ. Sci. Technol.* 22:325–338.

Nirmalakhandan, N.N. and R.E. Speece. 1988b. QSAR Model for Predicting Henry's Law Constant. *Environ. Sci. Technol.* 22: 1349–1350.

Nirmalakhandan, N.N. and Speece, R.E. 1989. Prediction of Aqueous Solubility of Organic Chemicals based on Molecular Structure. 2. Applications to PNA's, PCB's, PCDD's, etc. *Environ. Sci. Technol.* 23: 708–713.

Taft, R.W., J.-L.M. Abboud, M.J. Kamlet, and M.H. Abraham. 1985. Linear Solvation Energy Relations. *J. Solution. Chem.* 14: 153–186.

Tewari, Y.B., M.M. Miller, S.P. Wasik, and D.E. Martire, 1982. Aqueous solubility and octanol water partition coefficients of organic compounds at 25°C. *J. Solution Chem.* 11, 435–445.

Tiegs, D., J. Gmehling, P. Rasmussen, and A. Fredenslund. 1987. Vapor-Liquid Equilibria by UNIFAC Group Contribution. 4. Revision and Extension. *Ind. Eng. Chem. Process Des. Dev.* 261: 159–161.

Valvani, S.C., S.H. Yalkowsky, and T.J. Roseman. 1981. Solubility and Partitioning IV: Aqueous Solubility and Octanol-Water Partition Coefficients of Liquid Nonelectrolytes. *J. Pharm. Sci.* 70: 502–507.

Verschueren, K. 1996. *Handbook of Environmental Data on Organic Chemicals, 3rd Edition.* Van Nostrand Reinhold, New York, N.Y.

Walden, P. 1908. Uber die Schmelzwrme, Spezifische Kohsion und Molekulargrsse bei der Schmelztemperatur. *Z. Elektrochem.* 14: 713–728.

Wasik, S.P., M.M. Miller, Y.B. Tewari, W.E. May, W.J. Sonnefeld, H. DeVoe, and W.H. Zoller. 1993. Determination of the vapor pressure, aqueous solubility, and octanol/water partition coefficient of hydrophobic substances by coupled generator column/liquid chromatographic methods. *Res. Rev.* 85, 29–42.

Xie, W.H., W.Y. Shiu, and D. Mackay. 1997. A review of the effect of salts on the solubility of organic compounds in seawater. *Marine Environ. Res.* 44, 429–444.

Yalkowsky, S.H. 1979. Estimation of Entropies of Fusion of Organic Compounds. *I&EC Fundam.* 18: 108–111.

Yalkowsky, S.H. and S.C. Valvani. 1979. Solubility and Partitioning 2: Relationships Between Aqueous Solubility, Partition Coefficients, and Molecular Surface Areas of Rigid Aromatic Hydrocarbons. *J. Chem. Eng. Data.* 24: 127–129.

Yalkowsky, S.H. and S.C. Valvani. 1980. Solubility and Partitioning I: Solubility of Non-elctrolytes in Water. *J. Pharm. Sci.* 69: 912–922.

Yalkowsky, S.H., S.C. Valvani, and D. Mackay. 1983a. Estimation of the Aqueous Solubility of Some Aromatic Compounds. *Residue Reviews.* 85: 43–55.

Yalkowsky, S.H., S.C. Valvani, and T.J. Roseman. 1983b. Solubility and Partitioning VI: Octanol Solubility and Octanol-Water Partition Coefficient. *J. Pharm. Sci.* 72: 866–870.

Yalkowsky, S.H., R. Pinal, and S. Banerjee. 1988. Water Solubility: A Critique of the Solvatochromic Approach. *J. Pharm. Sci.* 77: 74–77.

Yalkowsky, S.H. and R.-M. Dannenfelser. 1991. *AQUASOL DATABASE of Aqueous Solubility, 5th edition*. University of Arizona, Tucson, AZ.

Yalkowsky, S.H. and S. Banerjee. 1992. *Aqueous solubility: Methods of estimation for organic compounds.* Marcel Dekker Inc., NY.

8

Soil and Sediment Sorption Coefficients

William J. Doucette

CONTENTS

1-56670-456-1/00/$0.00+$.50
© 2000 by CRC Press LLC

8.1   Introduction

8.1.1   Use of Sorption Coefficients in Fate Assessments

Sorption processes play a major role in determining the environmental fate and impact of organic chemicals. Sorption affects a variety of specific fate processes, including volatilization (Chiou and Shoup, 1985), bioavailability (Alexander, 1999; Ditoro et al., 1991), biodegradability (Scow and Johnson, 1997; Miller et al., 1987; Mihelcic and Luthy, 1991; Shelton and Parkin, 1991; Smith, 1991; Webb et al., 1991; Falatko and Novak, 1992; Scow and Alexander, 1992; Weissenfels et al., 1992), photolysis (Tao et al., 1999; Miller et al., 1989), and hydrolysis (Deeley et al., 1991; Somasundaram et al., 1991). Sorption on solids is one of the fundamental processes controlling the removal of toxic organic compounds in municipal wastewater treatment plants, affecting both the efficiency of the treatment system and the management of wastewater solids (Dobbs et al., 1989).

Sorption coefficients quantitatively describe the extent to which an organic chemical is distributed at equilibrium between an environmental solid (i.e., soil, sediment, suspended sediment, wastewater solids) and the aqueous phase it is in contact with. Sorption coefficients depend on (1) the variety of interactions occurring between the solute and the solid and aqueous phases and (2) the effects of environmental and/or experimental variables such as organic matter quantity and type, clay mineral content and type, clay to organic matter ratio, particle size distribution and surface area of the sorbent, pH, ionic strength, suspended particulates or colloidal material, temperature, dissolved organic matter (DOM) concentration, solute and solid concentrations, and phase separation technique.

Models use sorption coefficients to predict concentrations of organic solutes in water leaching through the soil profile or in runoff from land surfaces into lakes or streams. Along with other properties such as Henry's law constants and bioconcentration factors, sorption coefficients are essential in modeling the overall distribution of organic chemicals in the environment.

8.1.2 Mechanisms of Sorption

Adsorption, absorption, and *sorption* are terms used to describe the uptake of a solute by another phase. *Adsorption* describes the concentration of a solute at the interface of two phases, while *absorption* describes the process when a solute is transferred from the bulk state of one phase into the bulk state of the other phase (Hassett and Banwart, 1989). The term *sorption* is used frequently in environmental situations to denote the uptake of a solute by a solid (soil or sediment or component of soil) without reference to a specific mechanism, or when the mechanism is uncertain.

Sorption occurs when the free energy of the interaction between an environmental solid and an organic chemical sorbate is negative. The sorption process can be either enthalpy or entropy driven, depending on the properties of the solid sorbent and chemical solute (Hamaker and Thompson, 1972). Enthalpy-related forces include van der Waals interactions, electrostatic interactions, hydrogen bonding, charge transfer, ligand exchange, direct and induced dipole-dipole interactions, and chemisorption, while hydrophobic bonding or partitioning is considered the primary entropy driven force (Bailey, 1970; Hamaker and Thompson, 1972; Burchill et al., 1981; Stevenson, 1982; Sposito, 1984; Koskinen and Harper, 1990; and Von Oepen et al., 1991).

The complex and heterogeneous nature of environmental solids makes it difficult, if not impossible, to identify specific sorption mechanisms for most solid–chemical combinations (Green and Karickhoff, 1990) and, in most situations, several mechanisms operate simultaneously. In most soils, and under most conditions, organic chemicals are sorbed on both organic and inorganic constituents. The relative importance of organic versus inorganic constituents depends on the amount, distribution, and properties of those constituents and the properties of the organic chemical. As the polarity, number of functional groups, and ionic nature of the organic chemical increase, so too does the number of potential sorption mechanisms (Koskinen and Harper, 1990). Fortunately, for many solid–organic chemical interactions, one or two mechanisms dominate the sorption process and generalizations regarding sorption behavior can be made.

For instance, the sorption of most neutral, hydrophobic organic chemicals by environmental solids correlates highly with the organic matter content of the solid (Hamaker and Thompson, 1972; Chiou et al., 1979; Karickhoff, 1984). The extent to which clay minerals contribute to sorption depends on both the ratio of clay mineral to organic carbon fractions of the soil or sediment and on the nature of the organic sorbate. Green and Karickhoff (1990) suggested a ratio of 40 as the cutoff for organic carbon dominated sorption. Of the various inorganic soil constituents, smectites have the greatest potential for sorption of organic chemicals, due to their large surface area and abundance in agricultural soils (Laird et al., 1992).

Another generalization is that cationic organics and weak bases, protonated at low pH, are highly sorbed on negatively charged soils (Hamaker and Thompson, 1972; Podoll and Irwin, 1988), while the sorption of weak acids on soils is generally greater at low pH when the molecular form of the acid dominates.

Many detailed reviews of sorption mechanisms have been written, including Hamaker and Thompson, 1972; Guenzi, 1974; Theng, 1974; Greenland and Hayes, 1978; Hassett et al., 1981; Cheng, 1990; Koskinen and Harper, 1990; Von Oepen et al., 1991; Weber et al., 1991; and Pignatello and Xing, 1996.

8.2   Theoretical Background

8.2.1   Characterizing Environmental Solids as Sorbents

8.2.1.1   *Soils*

Soil is a dynamic and life-sustaining system composed of solids, liquid, and gas, with solids typically accounting for about one-half to two thirds by volume. Living organisms are also very important parts of the soil and contribute greatly to its general properties and behavior. The solid phase of soil is comprised of fragmented mineral matter, derived from the weathering of hard rock at the earth's surface, and from organic matter, consisting of a mixture of plant and animal residues in various stages of decomposition, of substances synthesized microbiologically, and/or chemicals formed from the breakdown products, and of the bodies of live and dead microorganisms and small animals and their decomposing remains (Schnitzer and Khan, 1972; Sposito, 1984).

Soil water acts both as a solvent for the organic chemical and as a solute with which the organic chemical has to compete for sorption sites on the solid surface. Typically, soil water is a solution comprised mainly of Ca^{+2}, Mg^{+2}, Na^+, K^+, SO_4^{-2}, CO_3^{-2}, and HCO_3^-. Ionic strengths are typically 0.5 mol/L or higher; pH values of 5–8.5 are common. The characteristics of the solution phase determine the reaction chemistry of the solid and the direction of dissolution/precipitation reactions, and they influence ion activity, ion pairing, and speciation. All these potentially can influence a chemical's sorptive behavior.

8.2.1.2   *Sediments and Suspended Sediments*

The natural organic matter suspended in the oceans at sites far from land consists of altered and linked biomolecules such as amino acids, sugars, and triglycerides that have been linked together. In large lakes and estuaries, the natural organic material in sediments and suspended sediments is derived from a mixture of the remains of terrestrial and planktonic organisms.

Generally, soils and sediments differ in the amount and type of organic matter they contain. Typically, soils contain higher percentages of cellulose and hemicellulose, while sediments contain higher percentages of lipid-like material. For neutral organic compounds, sorption is generally greater in sediments than in soils, even where normalized to organic carbon content (Gerstl and Kliger, 1990; Kile et al., 1995; Kile et al., 1999).

8.2.2   Soil Sorption Coefficients – Definitions

Sorption coefficients quantitatively describe the extent to which an organic chemical distributes itself between an environmental solid (i.e., soil, sediment, suspended sediment, wastewater solids, etc.) and the aqueous phase that it is in contact with at equilibrium.

Sorption coefficients generally are determined from an isotherm, a diagram that depicts the distribution of the test chemical between a solid sorbent and the solution in equilibrium with it over a range of concentrations at constant temperature. These isotherms can be linear or nonlinear, depending on the properties of the test chemical and solid and on the aqueous phase concentration of the chemical. In many cases, sorption isotherms are linear at low concentration but tend to become nonlinear (sorption tends to decrease) as the concentration of chemical in the aqueous phase increases, especially for polar or ionizable chemicals or soils that are low in organic carbon and high in clay. Linear sorption isotherms often are observed if the equilibrium aqueous phase organic compound concentrations are below 10^{-5} M or one half the aqueous phase solubility (whichever is lower) and the organic content of the solid is greater than 0.1% (Karickhoff, 1981; Karickhoff, 1984).

If the sorption isotherm is linear, the concentration of chemical sorbed by solids is directly proportional to the concentration of the chemical in water, and the slope of the isotherm is referred to as the linear sorption coefficient (K_d):

$$K_d = C_s/C_w \qquad (1)$$

where C_s and C_w are the concentrations of the organic chemical sorbed by the solid phase (mg/Kg) and dissolved in aqueous phase (mg/L), respectively. Units of K_d typically are given as L/kg, mL/g, or cm^3/g.

For nonlinear isotherms, the Freundlich model most often is used to describe the relationship between the sorbed (C_s) and the solution phase concentrations (C_w):

$$C_s = K_f C_w^N \qquad (2)$$

where K_f is the Freundlich sorption coefficient and N, (values of N are less than one and typically range between 0.75 and 0.95) generally is a constant. (Hamaker and Thompson, 1972; Rao and Davidson, 1980). However, in some cases, N has been observed to exceed one. When N equals one, a linear equation results, and K_f and K_d are equivalent.

The Langmuir and Brumnauer, Emmett, and Teller (BET) models also have been used to describe nonlinear sorption behavior for environmental solids, particularly for mineral dominated sorption (Ruthven, 1984; Weber et al., 1992). The Langmuir model assumes that maximum adsorption corresponds to a saturated monolayer of solute molecule on the absorbent surface, that there is no migration of the solute on the surface phase, and that the energy of adsorption is constant. The BET model is an extension of the Langmuir model that postulates multilayer sorption. It assumes that the first layer is attracted most strongly to the surface, while the second and subsequent layers are more weakly held.

8.2.3   Experimental Methods

Batch methods (i.e., ASTM, 1987a) have been used most often to determine sorption coefficients and examine the sorption kinetics for organic chemicals. Batch techniques involve agitating a small amount of soil with a solution containing the chemical of interest, separating the phases, and measuring the chemical concentration in one or both phases at selected times. Batch techniques are conceptually simple but are subject to a variety of operational difficulties and experimental artifacts related to the separation of phases, agitation speed, insufficient time for equilibration, and exposure of new sorptive surfaces during agitation. In addition, obtaining kinetic information in batch systems can be difficult when sorption is rapid.

Miscible displacement or packed column techniques also are widely used (i.e., Brusseau et al., 1990; Lee et al., 1991). In this approach, the chemical of interest is pumped through a soil column and the concentration of the chemical in the column effluent is monitored over time. The resulting concentration versus time curve is subtracted from that of a conservative (non-sorbed) tracer, pumped through the column at the same time or in a different experiment, to calculate the extent of sorption.

Column studies more closely mimic the real soil environment and more readily allow the determination of kinetic information than batch methods. However, equipment costs are higher, and column studies are subject to a variety of experimental artifacts, including failure to attain local equilibrium, loss of sorbent particles through column end retainers, wall and end effects, and column flow channeling (MacIntyre et al., 1991). In addition, columns are difficult to pack and maintain, and column experiments can be very time consuming for soils with large K_d values.

General agreement between batch and column results have been found, although significant differences also have been reported. For example, sorption coefficients determined from column experiments run at velocities of 10^{-3} cm s^{-1} were quite similar to those determined in 18-hour batch equilibrium experiments, while column experiments run at 10^{-2} cm s^{-1} showed the effect of slow sorption kinetics (Schwarzenbach and Westall, 1981; Maraga et al., 1998; Benker et al., 1998; Bayard et al., 1998).

Sorption coefficients also have been determined from the motion of an unconfined sorbate plume in a simulated aquifer box model. MacIntyre et al. (1991) compared the sorption coefficients for naphthalene on low soil organic carbon (less than 0.1%) aquifer materials determined by the batch, column, and box methods, and found good agreement among the three methods.

Continuous stirred tank reactors (CSTRs) have been used to study the sorption behavior of inorganic chemicals and of both organic and inorganic chemicals to activated carbon, but they have not been used widely to study the sorption of organic chemicals to soil and sediments. As in the batch method, the soil is agitated, usually with a magnetic stir bar, to provide more complete mixing of the soil and aqueous solution. However, the sorption and desorption reactions are followed by monitoring of the leachate concentration, similarly to the packed column method. Although CSTR methods for determining sorption coefficients for organics have been described have been described (Green et al., 1980; Su, 1994; deJonge et al., 1999), application of this method to soils has been limited primarily to inorganic chemicals (Carski and Sparks, 1985; Miller et al., 1989; Eick et al., 1990).

8.2.4   Factors Affecting Sorption Coefficients

Many factors potentially can affect the distribution of an organic chemical between an aqueous and solid phase. These include environmental variables, such as temperature, ionic strength, dissolved organic matter concentration, and the presence of colloidal material, and surfactants and cosolvents. In addition, factors related specifically to the experimental determination of sorption coefficients, such as sorbent and solid concentrations, equilibration time, and phase separation technique, can also be important. A brief discussion of several of the more important factors affecting sorption coefficients follows.

8.2.4.1   Temperature

The effect of temperature on sorption equilibrium is a direct indication of the strength of the sorption process. The weaker the interaction between sorbent and sorbate, the less the effect of temperature (Hamaker and Thompson, 1972). While temperature can influence sorption, the strength and direction of the effect depends on the properties of the sorbent and sorbate and on the sorption mechanism. Adsorption processes are generally exothermic, so the higher the temperature, the less the adsorption (Hamaker and Thompson, 1972). Hydrophobic sorption, however, has been shown to be relatively independent of temperature (Chiou et al., 1979).

ten Hulscher and Cornelissen (1996) reviewed the influence of temperature on equilibrium sorption and found that, in most cases, equilibrium sorption decreases with increasing temperature. However, a few examples of increasing equilibrium sorption with increasing temperature and of no effect of temperature also were observed. For example,

Talbert and Fletchall (1965) found that increasing temperature resulted in decreased sorption of simazine and atrazine. Similarly, Santana-Casiano and Gonzalez-Davila (1992) found that the sorption of lindane to chitin decreased as temperature increased from 5 to 45°C. In addition, sorption-desorption hysteresis, observed at lower temperatures, was reduced dramatically at the highest temperature. However, Chiou et al. (1979) observed decreased aqueous solubility and increased sorption at higher temperature for 1,1,1-trichloroethane.

Using literature data, ten Hulscher and Cornelissen (1996) calculated sorption enthalpies with the following formula:

$$d\ln K_d / d(1/T) = -\Delta H/R$$

where K_d is a linear sorption coefficient, ΔH is the change in sorption enthalpy (J/mol), T is absolute temperature (K), and R is the universal gas constant. Slightly negative enthalpy changes were observed when hydrophobic interactions were dominant. When electrostatic interactions were likely to occur, solution enthalpies were more exothermic.

8.2.4.2   pH

For neutral chemicals, sorption coefficients usually are unaffected by pH. However, for ionizable organic chemicals, sorption coefficients can be affected greatly, since pH affects not only the speciation but also the surface characteristics of natural sorbents.

Typically, for weak acids the free acid form (HA) is more strongly sorbed than the anionic form (A-). For example, Shimizu et al. (1992) showed that pentachlorophenol (PCP) sorption decreased with increasing pH over the entire pH range tested (2 to 12). For weak bases, the cationic form dominates at low pHs and is more highly sorbed than the free base. Green and Karickhoff (1990) suggested that the soil pH can have a significant effect on the sorption of a basic compound if the pH of the soil-suspension minus 3 units is less than the pKa of the compound. In addition, polar compounds such as diuron may not actually form cationic or anionic forms, but are capable of forming hydrogen bonds and will show differing sorption with pH (Bouchard et al., 1989).

8.2.4.3   Salinity or Ionic Strength

Salts can affect sorption of organic compounds by displacing cations from the soil ion exchange matrix, by changing the activity of the sorbate in solution, and by changing the charge density associated with the soil sorption surface (Hamaker and Thompson, 1972). Salt effects are most important for basic sorbates in the cation state, where an increase in salinity can significantly lower the sorption coefficient. Salt effects are least important for neutral compounds, which may show either increases or decreases in sorption as salinity increases.

8.2.4.4   Effect of Dissolved or Colloidal Organic Matter (Suspended Particulates or "Third Phase")

The presence of dissolved or colloidal organic matter has been shown to influence sorption depending on the nature of the chemical and the organic matter (Enfield et al., 1985; Bouchard et al., 1989). Santos-Buelga et al. (1992), using an equilibrium dialysis method, found that ethofumesate was associated extensively with the dissolved organic matter, and its sorption by several sorbents (two soils, montmorillonite and humic acid) decreased significantly in the presence of dissolved organic matter (DOM). Herbert et al. (1993) characterized several size fractions of water soluble organic carbon (WSOC) and found that the effect of dissolved organic matter of the sorption of pyrene may be limited, but the presence of colloidal organic matter suspended in the soil solution may have significant impact on the sorption of pyrene.

8.2.4.5   Cosolvents

Rao et al. (1990) investigated the effect of nonpolar cosolutes (trichloroethylene, toluene p-xylene), polar cosolutes (1-octanol, chlorobenzene, nitrobenzene, o-cresol) and polar cosolvents (methanol and dimethyl sulfoxide) on sorption of several polycyclic aromatic hydrocarbons (PAHs). The nonpolar cosolutes did not significantly influence PAH sorption, while the polar cosolutes (nitrobenzene, o-cresol), having sufficiently high aqueous solubilities, caused a significant decrease in PAH sorption.

Miscible organic solvents, such as methanol and ethanol, have been shown to increase solubility of hydrophobic organics and to decrease sorption (Nkedi-Kizza, 1985; Fu and Luthy, 1986; Nkedi-Kizza et al., 1987; Walters and Guiseppi-Elle, 1988; Wood et al., 1990; Lee et al., 1991). This is presumably the result of (i) reducing the activity coefficient of the sorbate chemical in the aqueous phase, and (ii) competition for sorbing sites.

8.2.4.6   Competitive Sorption

At concentrations normally encountered in environmental situations, sorption often has been observed to be relatively non-competitive. For example, Chiou et al. (1983, 1985) found no competition in the sorption of binary solutes m-dichlorobenzene and 1,2,4-trichlorobenzene and between parathion and lindane. MacIntyre and deFur (1985) measured the sorption of methyl and dimethyl naphthalene, individually and as components of JP-8 and synthetic jet fuel mixtures on two sediments and montomorillonite clay in water. The sorption coefficients of the naphthalenes generally varied by less than a factor of two.

However, more recently competitive sorption has been reported and is thought to be the result of site-specific sorption occurring in soil organic matter (Xing et al., 1996; Xing and Pignatello, 1998).

8.2.4.7   Soil to Solution Ratio/Solids Effect (Concentration of Sorbing Solid Phase and "Third Phase" Effects)

The soil to solution ratio used in the experimental determination of sorption coefficients is typically one or two to ten (ATSM, 1987), but it often depends on the analytical considerations associated with the sorbent and sorbate being studied. For example, in soils with low sorption capacity, the soil to solution ratio should be large enough to precisely measure the smallest possible change in solution concentration during sorption.

Several investigators have reported a particle concentration effect associated with the sorption of hydrophobic solutes. The decrease in the amount sorbed with increasing particle concentration is most pronounced for compounds characterized by large partitioning ($K_d > 1000$ mL/g). O'Conner and Connolly (1980) reported observations that linear partition coefficients are inversely dependent upon the concentration of the solids in the system. Voice et al. (1983) also observed a significant increase in sorption as the solids concentration decreased and attributed it to the presence of microparticles contributed by the solids and not removed from suspension in the separation procedure. This suggests that sorption coefficients produced in studies using different techniques are not necessarily comparable.

While several explanations for the particle concentration effect have been presented, VanHoof and Andren (1990) suggested a kinetic explanation rather than a thermodynamic one. Equilibrium sorption is independent of particle size, but kinetics is not, with small particles reaching equilibrium faster. A solids concentration effect on 4-chlorobiphenyl was observed in an aqueous suspension of polystyrene microspheres. The apparent inverse relationship of 4-chlorobiphenyl partitioning with particle concentration diminished with time, demonstrating that nonattainment of equilibrium results in the observed phenomenon.

However, more recently, Perlinger et al. (1993) observed a solids concentration effect with benzene on aluminum oxide that could not be explained by incomplete phase separation, nonattainment of sorption equilibrium, or aggregation of particles. Perlinger et al. (1993) suggested that the observed effect resulted from changes in the activities of water and/or benzene sorbed at the mineral surface. This was thought to occur when the structure of water near the mineral surface increased due to particle interactions.

8.2.4.8 Loss of Compound

Loss of compound by sorption onto the walls of the equilibration vessels, volatilization, and chemical or biological degradation also can affect the experimental determination of sorption coefficients. These potential loss mechanisms must be eliminated or accounted for if accurate sorption coefficients are to be determined. It is preferable to measure the concentration of the chemical in both phases and determine the mass balance to quantify potential loss mechanisms. Singh et al., (1990) found that K_{oc} values obtained by measuring only the concentration in the solution phase were consistently higher than with those generated using a mass balance approach when the concentrations in both phases are measured.

8.2.4.9 Organic Matter Type and Origin

While the constancy of K_{oc} values suggests an uniformity of organic matter with regard to sorption behavior, it is becoming increasingly apparent that organic matter type can be an important sorption variable for some sorbent/sorbate combinations (Garbarini and Lion, 1986; Gauthier et al., 1987; Grathwohl, 1990; Rutherford et al., 1992; Kile et al., 1995; Kile et al., 1999). For example, Gerstl and Kliger (1990) found that the sorption of naproamide, a nonionic herbicide, was greater in the sediment than in the soils, even on an organic carbon basis. The increased sorption in sediment was attributed to the fact that soils contained a higher percentage of cellulose and hemicellulose material, whereas the sediments contain a higher lipid-like fraction.

8.2.5 Kinetic Considerations — The Implications of Non-Equilibrium Behavior

Sorption generally is regarded as a rapid process and, in many laboratory sorption experiments, equilibrium often is observed within several minutes or hours. An equilibration time of 24 hours often is used for convenience. True sorption equilibrium, however, may require weeks to months to achieve (Karickhoff, 1981; Karickhoff, 1984), depending on the chemical and environmental solid of interest. In many instances, an early period of rapid and extensive sorption, followed by a long slow period, is observed (Karickhoff, 1980; Wu and Gschwend, 1986; Brusseau and Rao, 1989; Brusseau et al., 1990; Ball and Roberts, 1991).

Talbert and Fletchall (1965) observed that sorption slowly increased with time for several s-triazines. Experimental determination of sorption coefficients requires preliminary kinetic experiments to determine the time to reach equilibrium.

Two processes govern rate-limited or nonequilibrium sorption: transport of the substance to the sorption sites and the sorption process itself (Hamaker and Thompson, 1972; Brusseau et al., 1991a; Brusseau et al., 1991b; Brusseau and Rao, 1991a,b). Transport-related nonequilibrium typically results from the existence of a heterogeneous flow domain. Sorption-related nonequilibrium, caused by rate-limited interactions between the sorbate and sorbent, may be the result of chemical nonequilibrium (i.e., chemisorption) or diffusive mass transfer limitations (i.e., diffusion of solute within pores of microporous particles or molecular diffusion into macromolecular organic matter) (Brusseau and Rao, 1991a,b). Sorption kinetics are likely to be environmentally important in short contact situations such as sediment resuspension, soil erosion, and infiltrating ground water (Schwarzenbach et al., 1993).

In general, adsorption processes tend to be rapid, nearly instantaneous, whereas non-surface sorption tends to be slower. For neutral organic chemicals, the more hydrophobic the compound, the larger the sorption coefficient, and the longer it takes to reach equilibrium between the solid and aqueous phases. This is because the sorbent must remove chemical from a larger volume of water.

Generally, sorption estimates are based on equilibrium conditions only; however, incorporation of kinetic considerations into sorption estimation techniques is likely to be an important area of future work. For example, the assumption of equilibrium sorption in dynamic field systems may result in calculating too much pesticide in the sorbed state.

Sorption kinetics have been discussed in detail by Pignatello (1989), Brusseau (1991b), Weber (1991), Pignatello and Xing (1996), and Schwarzenbach et al. (1993).

8.2.6 Hysteresis

Hysteresis, in which sorption and desorption isotherms differ in extent and/or time, also is observed frequently in laboratory sorption experiments. However, in many cases, experimental artifacts, such as the nonattainment of equilibrium during the sorption step and the loss of compound due to degradation or volatilization, have been suggested as reasons for the observed hysteresis (Pignatello, 1989). While reversibility has been demonstrated experimentally for nonpolar compounds like PAHs (Karickhoff et al., 1979), PCB congeners (Gschwend and Wu, 1985), and solvents like tetrachloroethylene and toluene (Garbarini and Lion, 1985), desorption coefficients generally are observed to be greater than sorption coefficients.

8.2.7 Physical and Chemical Properties Affecting the Sorption of Organic Compounds to Environmental Solids

For neutral organic compounds, in soils having a low clay/organic carbon ratio, sorption coefficients tend to increase as the hydrophobicity of the compound increases. Aqueous solubility or octanol/water partition coefficients often are used as indicators of a compound's hydrophobicity. An increase in polarity, number of functional groups, and ionic nature of the chemical will increase the number of potential sorption mechanisms for a given chemical (Garbarini and Lion, 1985). For ionizable compounds, pK_a is of particular importance because it determines the dominant form of a chemical at the specific environmental pH.

8.2.8   Relationships Between Sorption and Soil Organic Matter: The "K_{oc} Approach"

For a given organic chemical, sorption coefficients (K_d or K_f) vary considerably from soil to soil or sediment to sediment, depending on the properties of the sorbent. However, for many organic chemicals, and in particular neutral hydrophobic organics, sorption is directly proportional to the quantity of organic matter associated with the solid (Hamaker and Thompson, 1972; Chiou et al., 1979; Kenaga and Goring, 1980; Rao and Davidson, 1980; Briggs, 1981; Karickhoff, 1981). Thus, as Equation (3) shows, normalizing soil or sediment specific sorption coefficients to the organic carbon content of the sorbent yields a new coefficient, K_{oc}, that is considered a unique property or "constant" of the organic chemical being sorbed:

$$K_{oc} = K_d/oc \text{ or } K_f/oc \tag{3}$$

where K_{oc} is the organic carbon normalized sorption coefficient, K_d and K_f are the linear and Freundlich sorption coefficients specific to a particular sorbent and chemical combination, and oc is the organic carbon content of that sorbent in units of g oc/g dry soil. For K_{oc} to be a true constant, K_d should be used, since it is independent of concentration. However, K_f also can be used if N is known or can be predicted.

Sorption coefficients also have been expressed on an organic matter basis (K_{om}) by assuming that the organic matter content of a soil or sediment equals some factor, usually between 1.7 to 1.9, times its organic carbon content on a mass basis (Hamaker and Thompson, 1972; Lyman et al., 1982). Often 1.724 is used as this factor, implying that the carbon content of organic matter is 1/1.724 or 60%. However, K_{oc} is considered a more definite and less ambiguous measure than K_{om} (Hamaker and Thompson, 1972).

Assumptions inherent in the use of a K_{oc} (or K_{om}) are that: sorption is exclusively to the organic component of the soil; all soil organic carbon has same sorption capacity per unit mass; equilibrium is observed in the sorption-desorption process; and the sorption and desorption isotherms are identical (Green and Karickhoff, 1990). Both K_{oc} and K_d have units of L/kg or cm^3/g.

This interaction of neutral organic solutes with soil or sediment organic matter (SOM) has been referred to as hydrophobic sorption or partitioning. Hassett and Banwart (1989) describe hydrophobic sorption as an entropy driven process resulting from the removal of the solute from solution. The entropy change is largely due to the destruction of the highly structured water shell surrounding the solvated organic. Chiou (1989) used the term *partitioning* to denote an uptake in which the sorbed organic chemical permeates into the network of an organic medium by forces common to the solution, analogous to the extraction of an organic compound from water with an organic liquid. By either description, hydrophobic sorption or partitioning should increase as compounds become less water soluble or more hydrophobic.

Additional characteristics typically associated with hydrophobic sorption or partitioning include sorption isotherms that are linear over a relatively wide range of concentrations, sorption coefficients that are not strongly temperature dependent, and a lack of competition between sorbates (Chiou, 1989).

8.3   Methods for Estimating K_{oc}

Organic carbon normalized sorption coefficients (K_{oc} or K_{om}) have been correlated with a variety of physical/chemical properties and/or structural descriptors often related to the hydrophobicity of the chemical. Many of the reported methods for estimating K_{oc} are based on the relationship between K_{oc} and the octanol/water partition coefficient (K_{ow}) or aqueous solubility (S) values as determined by regression analysis (i.e., Kenaga and Goring, 1980; Briggs, 1981; Karickhoff, 1981; Chiou et al., 1983). These regression models usually are expressed by relating log K_{oc} to log K_{ow} or log S.

Numerous estimation methods based on correlations with structurally derived parameters such as molecular connectivity indices (MCIs) (Sabljic, 1984; Sabljic, 1987; Bahnick and Doucette, 1988; Meylan et al., 1992), parachor (Briggs, 1981), molecular weight (Kanazawa, 1989), molecular surface area (Fugate, 1989; Holt, 1992, Brusseau, 1993), the characteristic root index (CRI) (Sacan and Balcioglu, 1996), fragment constants (Tao et al., 1999 and linear solvation energy relationships (LSERs) (Hong et al., 1996; Baker et al., 1997; Xu et al., 1999) also have been reported. These correlations are especially valuable when experimental K_{ow} and S values are unavailable. MCI based methods are probably the most widely used of the structurally based methods. A method using both MCIs and correction factors based on specific functional groups also has been reported (Meylan et al., 1992) that appears to extend the applicability of MCI-based estimation methods to a wider variety of compounds.

Experimental retention times or capacity factors generated by reverse phase high performance liquid chromatography (RP-HPLC) (Vowles and Mantoura, 1987; Hodson and Williams, 1988; Pussemier et al., 1990; Szabo et al., 1990a,b; Hong et al., 1996) also have been correlated with K_{oc}.

Selecting and applying the most appropriate method for estimating K_{oc} depends on several factors, including the availability of required input, the appropriateness of the model to the chemical of interest, and the methodology for calculating the necessary topological or structural information.

While the so-called "K_{oc} approach" for estimating sorption coefficients is most appropriate for neutral, hydrophobic organic chemicals on environmental solids containing a significant amount of organic matter, this approach has been applied to a wide variety of chemical and soil types. The main reasons for the wide acceptance of this approach is that it works reasonably well for a large number of organic chemicals, and the organic carbon content of the environmental solid is usually available. In addition, an estimate of hydrophobic sorption or partitioning based on K_{oc} often represents a minimum value or conservative estimate for the sorption of a particular hydrophobic organic compound. If specific sorbate-sorbent interactions also are involved, additional sorption may occur (Khan et al., 1979).

Generally, other sorbent properties, such as the type and amount of clay, soil pH, cation exchange capacity (CEC), and hydrous oxide content have less effect on the sorption process except in situations where the organic carbon content of the sorbent is low or when the clay content is high (Hassett and Banwart, 1989). In these situations, correlations between sorbent specific sorption coefficients (K_d and K_f) and CEC (Hicken, 1993), clay content (Hassett and Banwart, 1989; Hassett et al., 1981), and surface area (Pionke and DeAngelis, 1980) have been reported. However, these correlations typically are valid only for specific soil/sorbent combinations and are considered far less general in their application. In addition, soil properties such as surface area and CEC often are related directly to the amount of organic carbon (Tate, 1987; Green and Karickhoff, 1990).

The remainder of this chapter will focus on the use, applicability, and limitations of methods used to estimate K_{oc}. Specific examples using recommended methods for K_{oc} will be presented. The reader is encouraged to examine several previous reviews of sorption estimation techniques (Lyman et al., 1982; Gerstl, 1990; Green and Karickhoff, 1990; Sabljic et al., 1995; Baker et al., 1997; Gawlik et al., 1997).

TABLE 8.1

Representative Examples of Regression Models Used to Estimate Log K_{oc} from Log K_{ow}.

Eq. No	Equation	n	r2	Chemical Classes	References
	General				
	$\log K_{oc} = 0.903 \log K_{ow} + 0.094$	72	0.91	Wide variety	Baker et al., 1997
	$\log K_{oc} = 0.679 \log K_{ow} + 0.663$	419	0.831	Wide variety	Gerstl, 1990
	$\log K_{oc} = 0.544 \log K_{ow} + 1.377$	45	0.74	Variety, mostly pesticides	Kenaga and Goring, 1980
	$\log K_{oc} = 0.81 \log K_{ow} + 0.10$	81	0.887	Hydrophobics	Sabljic et al., 1995
	$\log K_{oc} = 0.52 \log K_{ow} + 1.02$	390	0.631	Nonhydrophobics	Sabljic et al., 1995
	Class Specific				
	$\log K_{oc} = 0.63 \log K_{ow} + 0.90$	54	0.865	Subst. phenols, anilines, nitrobenzenes, and chlorinated benzonitriles	Sabljic et al., 1995
	$\log K_{oc} = 0.47 \log K_{ow} + 1.09$	216	0.681	"Agricultural" chemicals: acetamilides, carbamates, esters, phenylureas, phosphates, triazines, triazoles, and uracils	Sabljic et al., 1995
	$\log K_{oc} = 0.545 \log K_{ow} + 0.943$	57	0.713	ureas	Gerstl, 1990
	$\log K_{oc} = 0.433 \log K_{ow} + 0.919$	39	0.863	carbamates	Gerstl, 1990
	$\log K_{oc} = 0.402 \log K_{ow} + 1.071$	15	0.69	Pesticides	Kanazawa, 1989
	$\log K_{oc} = 0.904 \log K_{ow} - 0.539$	12	0.99	PAHs, chlorinated hydrocarbons	Chiou et al., 1983
	$\log K_{oc} = 0.937 \log K_{ow} - 0.006$	9	0.95	Triazines	Brown and Flagg, 1981
	$\log K_{oc} = 1.029 \log K_{ow} - 0.18$	0.13	0.94	Chlorinated hydrocarbons, pesticides	Rao and Davidson, 1980
	$\log K_{oc} = 1.00 \log K_{ow} - 0.21$	10	1.00	PAHs, aromatics	Karickhoff et al., 1979

8.3.1 Correlation with Physical and Chemical Properties (K_{ow}) — The Octanol/Water Partition Coefficient and Aqueous Solubility (S)

Probably the most widely used and accepted approach for estimating K_{oc} is based on correlations with physical/chemical properties such as K_{ow} or S. The literature presents many such relationships and Tables 8.1 and 8.2 list several representative examples. Also see several previous compilations of K_{oc}-property correlations by Lyman et al. (1982) and Gawlik et al. (1997).

Selecting and applying the most appropriate regression model depends mainly on the quality and extent of the database used to develop the model and the structural similarity between the chemical of interest and the chemicals used to develop the model. For example, many of these correlations were developed for specific classes of compounds and are less generally applicable than those developed using a wide variety of chemical types. However, the use of an appropriate "class-specific" relationship, if available for the compound of interest, should provide the best estimate (i.e., the estimate associated with the least amount of uncertainty). An example illustrating the hierarchy of K_{ow}–K_{oc} relationships potentially available for polynuclear aromatic hydrocarbons (PAHs) appear below: one developed using only PAHs, one using aromatics including PAHs, and one using a wide variety of compound types (Doucette and Holt, 1992). These expressions have been incorporated into a computer program designed for estimating a variety of physical/chemical properties in addition to K_{oc} (Holt, 1992).

TABLE 8.2

Representative Examples of Regression Models Used to Estimate Log K_{oc} from Log S.

Eq. No	Equation	n	r2	Chemical Classes	Reference
	General				
1	$\log K_{oc} = (-0.508 \times \log S + 0.953) - Fc*$	419	0.757	Wide variety	Gerstl, 1990
1	$\log K_{oc} = -0.55 \log S + 3.64$ (S in mg/L)	106	0.71	Variety, mostly pesticides	Kenaga and Goring, 1980
	$\log K_{oc} = 0.51 [\log S + (0.01 MP - 0.25)] + 0.8$	38	0.77	Variety, mostly pesticides	Briggs, 1981
4	$\log K_{oc} = -0.83 \log S - 0.01 (MP - 25) - 0.93$ (S in mole fraction)	47	0.93	PAHs, chlorinated hydrocarbons, pesticides	Karickhoff, 1984
	Class Specific				
	$\log K_{oc} = -0.381 \log S + 1.177$ (S is in moles/L)	57	0.616	ureas	Gerstl, 1990
	$\log K_{o}c = -0..410 \log S + 0.97$ (S is in moles/L)8	39	0.601	carbamates	Gerstl, 1990
	$\log K_{oc} = -0.356 \log S + 3.01$ (S in ppm)	15	0.79	pesticides	Kanazawa, 1989
	$\log K_{om} = -0.729 \log S + 0.001$ (S is in moles/L for supercooled liquid)	12	0.99	PAHs, chlorinated hydrocarbons	Chiou et al., 1983
	$\log K_{oc} = -0.68 \log S + 4.273$ (S in μmol/L)	23	0.93	PAHs, hetroPAHs, aromatic amines, chlorinated hydrocarbons	Hassett et al., 1980
	$\log K_{oc} = -0.58 \log S + 4.24$ (S in umol/L)	15	0.41	apolar hydrocarbons	Mingelgrin et al., 1983
	$\log K_{oc} = -0.557 \log S + 4.277$ (S in μ moles/L)	15	0.99	chlorinated hydrocarbons	Chiou et al., 1979
3	$\log K_{oc} = \log S + 0.44$	10	0.94	PAHs, aromatics	Karickhoff et al., 1979
	$\log K_{oc} = 0.921 \log S - 0.00953 (MP-25) - 1.405$	5	0.954	PAHs	Karickhoff, 1981

* Fc = semi empirical polarity correction factor derived for several classes of compounds (i.e., acetanilides = 0.31; amides = 0.66; halogenated aromatic hydrocarbons = -0.20; non-halogenated aromatic hydrocarbons = –0.27; carbamates = 0.27; dinitroanilines = 0.19; organophosphorous pesticides = 0.01; PAHs = –0.93, triazines = 0.24, triazoles = –0.22; ureas = 0.19).

$\log K_{oc} = 0.823 \log K_{ow} + 0.727$, $r^2 = 0.828$, $n = 10$ (PAHs)

$\log K_{oc} = 0.529 \log K_{ow} - 0.916$, $= 0.664$, $n = 38$ (non-halogenated aromatics, (Gerstl, 1990)

$\log K_{oc} = 0.588 \log K_{ow} + 1.001$, $r^2 = 0.686$, $n = 82$ (Universal, variety of compounds)

Using a similar hierarchical approach, Sabljic et al., (1995) reported several general (i.e., hydrophobic and nonhydrophobic), subgeneral (phenol type, agricultural, alcohols, and organic acids), and class-specific (acetanilides, alcohols, amides, anilines, carbamates, dintiroanilines, esters, nitrobenzenes, organic acids, phenols and benzonitriles, phenylureas, phosphates, triazines, and triazoles) log K_{oc}-log K_{ow}. Table 8.1 also lists several of these.

Gao et al. (1996) recently developed a nonlinear model for estimating K_{oc} from both K_{ow} and S, using an artificial neural network approach. The authors concluded that this nonlinear equation outperformed linear models in fitting K_{oc} values for the training set and predicting them for the test set. Although a training set of 119 nonpolar and polar compounds organics was used, 66 of the 119 compounds in the training set were PCBs. Because of the relative simplicity of the neutral net, an analytical equation, relating K_{oc} to K_{ow} and S, was developed. By starting with a neural network, converging the bias and weight values using the available values of S, K_{ow} and K_{oc}, and then combining the equations for each node in the final neural network, the following equation was reported:

$$\log K_{oc} = \frac{6.786}{\exp\left[1.573 + 0.282 \log S - 0.301 \log K_{ow}\right]} + 0.954$$

After selecting the most appropriate regression model, class-specific or "universal", an accurate K_{ow} or S value must be input. In addition, several of the S-K_{oc} regression models Table 8.2 lists require the additional input of MP for solids in order to correct for fugacity ratio (Briggs, 1981; Karickhoff, 1981). Unfortunately, for many compounds, K_{ow} or S values are unavailable, because these parameters, like K_{oc} itself, are also difficult to determine experimentally. While the use of estimated K_{ow} or S values for estimating K_{oc} has been reported, this tends to increase the unreliability of the estimate. Generally, the log-log relationships listed in Tables 8.1 and 8.2 allow the prediction of a compounds K_{oc} value within a factor of 2 to 10 for soils and sediments with an organic carbon content greater than 0.1%, if accurate values of S or K_{ow} are available.

The recommended general approach for estimating K_{oc} from K_{ow} or S is to:

1) Obtain an accurate experimental value of K_{ow} or S (and melting point) for the chemical of interest.

2) Select the most appropriate equation from Tables 8.1 or 8.2, based on the quality and extent of the database used to develop the model and the structural similarity between the chemical of interest and the chemicals used to develop the model.

3) Calculate K_{oc}, using the selected regression equation.

4) Estimate the site-specific sorption coefficient (K_d or K_f) from the estimated K_{oc} and the organic carbon content of the soil or sediment, using Equation (3).

At the end of this chapter, several examples are given illustrating the use of this approach.

8.3.2 Molecular Connectivity Indices (MCIs)

Another widely used approach for estimating K_{oc}, especially when experimental values of K_{ow} or S are unavailable, is to employ correlations between MCIs and K_{oc} (Koch, 1983; Sabljic, 1984; Gerstl and Helling, 1987; Sabljic, 1987; Bahnick and Doucette, 1988). Molecular connectivity is a method of bond counting from which topological indices can be derived from chemical structures. For a given molecular structure, several types and orders of MCIs can be calculated. Information on the molecular size, branching, cyclization, unsaturation, and heteroatom content of a molecule is encoded in these various indices, (Kier and Hall, 1976). One significant advantage of using MCI-K_{oc} regression models over property-K_{oc} regression models is that, once the model has been developed, only the structure of the chemical of interest is required as input, and no additional experimental parameters are needed. Kier (1980) and Kier and Hall (1976, 1986) present a detailed discussion of molecular connectivity and its application.

8.3.2.1   Choosing Indices

MCIs have been shown to be related closely to molecular size and hence to the hydrophobicity of a chemical. First order simple ($^1\chi$) and valence ($^1\chi^v$) indices typically have been used in developing MCI-K_{oc} relationships. For example, Koch (1983) found a strong relationship between sorption coefficients and the first order valence MCI ($^1\chi^v$) for 18 neutral organic compounds.

Similarly, Sabljic (1984) showed that the simple first order MCI ($^1\chi$) highly correlated to K_{om} for several classes of non-polar organic compounds, including polycyclic aromatic hydrocarbons (PAHs), PCBs, and chlorinated benzenes. Polar organic compounds did not fit the same K_{oc}-$^1\chi$ MCI regression line developed for non-polar hydrophobic compounds but instead systematically deviated below the line (Sabljic, 1987). A series of semi-empirical polarity correction factors are derived from the difference between the general regression line obtained for nonpolar compounds and the almost parallel regression lines obtained for the several specific classes of polar compounds such as anilines, nitrobenzenes, and phenylureas.

Gerstl and Helling (1987) also found that, while good predictions of K_{oc} were possible for specific groups of compounds, the ability of any one equation to predict log K_{oc} based upon one or two MCIs was rather limited when a diverse group of compounds was considered.

Thus, while individual indices such as $^1\chi$ or $^1\chi^v$ have been successful in modeling the sorption of non-polar hydrophobic compounds, they are not by themselves good indicators of any non-hydrophobic sorption component, since, for a given class of compounds, all MCIs show a general increase with molecular size.

In an attempt to develop a single expression that could be used to predict K_{oc} for both non-polar and polar organic compounds using only structurally-derived MCIs as input, Bahnick and Doucette (1988) proposed a new index to describe the non-hydrophobic contribution to K_{oc}. By subtracting the first order valence MCI ($^1\chi^v$) from the first order valence MCI of a hypothetical molecule, formed by replacing any highly electronegative atoms in the molecule of interest, such as oxygen or nitrogen with carbon atoms ($^1\chi^v_{np}$), a new index ($\Delta^1\chi^v$), with the molecular size contribution approximately removed, was obtained, as Equation (4) shows.

$$\Delta^1\chi^v = (^1\chi^v)_{np} - {}^1\chi^v \tag{4}$$

Using a set of 56 compounds (24 nonpolar and 32 polar), they developed a regression model that used $^1\chi$ to describe the hydrophobic sorption component and $^1\chi^v$ to describe the non-hydrophobic component of sorption. They subsequently tested this model on a set of 40 (16 nonpolar and 24 polar) organic chemicals not used in developing the original regression model. Log K_{oc} values could be estimated for a large variety of nonpolar and polar organic compounds, most within the experimental uncertainties in their measurement.

Meylan et al. (1992) described another attempt to extend MCI-K_{oc} relationships to polar compounds. This method uses the first order molecular connectivity index ($^1\chi$) and a series of statistically derived fragment contribution factors for polar compounds. To develop the model, they performed two separate regression analyses. The first related log K_{oc} to $^1\chi$ for a set of 64 nonpolar compounds, including a variety of halogenated and nonhalogenated aromatics, PAHs, halogenated aliphatics, and phenols. The second regression, using 125 polar compounds, was performed to generate the polarity correction factors. The polar correction factors assume that a single numerical value represents the contribution of each polar functional group (i.e., a specified atom, a group of atoms bonded together, or structural factor) to log K_{oc} and that the contributions made by each group are independent of each other. By summing the values of the various polarity correction factors and adding the resulting value to the $^1\chi$, K_{oc} can be calculated directly. The combination of the regression model developed for the nonpolar compounds with the correction factors yields the following equation for directly estimating log K_{oc}:

$$\log K_{oc} = 0.53\ ^1\chi + 0.63 + \Sigma\ P_fN \tag{5}$$

where $^1\chi$ is the first order MCI, and $\Sigma\ P_fN$ is the summation of the products of all applicable correction factors multiplied by the number of times (N) that the fragment occurs in the structure with the exception of organophosphorous, nitrogen to nonfused aromatic ring, organic acids and nitriles, and cyanide fragments, that are counted only once no matter how many times they occur in a structure.

The method was tested on a validation set consisting of 205 compounds (41 nonpolar and 164 polar chemicals). In addition, the method was compared to regression methods using S and K_{ow} as the input variables. Results show that this model outperforms and covers a wider range of chemical structures than models based on K_{ow} aqueous solubility, or single MCIs. Meylan et al. (1992) concluded that the MCI-fragment contribution method was clearly superior in the prediction of K_{oc}, although in many cases experimental values of S and K_{ow} were not available and estimated values were used. They presented no direct comparison with other MCI-based estimation methods.

The method Meylan et al. (1992) described also has been encoded in a computer program, PCKOCWIN. After the user enters the structure of the chemical of interest represented in SMILES notation (Anderson et al., 1987; Weininger, 1988), the program automatically calculates $^1\chi$ and determines the appropriate fragments and correction factors.

8.3.2.2   Choosing the Most Appropriate MCI-K_{oc} Method

Table 8.3 lists several examples of MCI-K_{oc} regression models that have appeared in the literature. A more comprehensive compilation of MCI-K_{oc} regression models appear in Gawlik et al. (1997). As with the property-K_{oc} relationships previously described, if an appropriate class-specific relationship can be found, it is likely to provide the best estimate. However, if an appropriate class-specific relationship is not available or if the compound of interest has multiple functional groups that make it difficult to classify, one of the more general expressions should be used.

For non-polar organics, the regression models several investigators reported using a relatively wide range of compound types and K_{oc} values are essentially equivalent. For polar organics, the use of one of the "non-polar" or size-based regression models will result in estimates that are too low. While several models have been developed to correct for the polarity of the molecule, the method by Meylan et al. (1992) is probably the best defined and, therefore, is recommended.

8.3.2.3   Calculating MCIs

Regardless of the regression model used to estimate K_{oc} using a MCI-K_{oc} relationship, the first requirement is to calculate the MCI(s). This can be done by hand or by using one of

TABLE 8.3

Representative Examples of K_{oc}-MCI Relationships.

Eqn #	Equation	n	r2	Chemical Classes	Reference
	General				
	$\log K_{oc} = 0.53\ {}^1\chi + 0.63 + \Sigma\ P_fN$	189	0.96	Wide variety	Meylan et al., 1992
	$\log K_{oc} = 0.53\ (\pm0.04){}^1\chi - 2.09\ (+0.22)\ \Delta^1\chi^v + 0.64$	56	0.94	Wide variety	Bahnick and Doucette, 1988
	$\log K_{oc} = 0.52\ {}^1\chi + 0.70$	81	0.96	Hydrophobics	Sabljic et al., 1995
	$\log K_{oc} = 1.94 - 0.492\ {}^0\chi + 0.776\ {}^5\chi + 0.440\ {}^0\chi^c$	419	0.535	Wide variety	Gerstl, 1990
	Class Specific				
	$\log K_{om} = 0.55\ {}^1\chi\ 0.54$	72	0.95	PAHs, halogenated hydrocarbons	Sabljic, 1987
	$\log K_{oc} = 0.567\ {}^1\chi^v - 0.29$	12	0.86	Alcohols	Gerstl and Helling, 1987
	$\log K_{oc} = 1.146\ {}^3\chi^v + 0.54$	14	0.83	PAHs, heterocyclic PAHs	Gerstl and Helling, 1987
	$\log K_{om} = 0.55\ {}^1\chi\ 0.45$	37	0.94	PAHs, halogenated hydrocarbons, chlorophenols	Sabljic, 1984
	$\log K_{oc} = 0.673\ {}^1\chi + 0.455$	18	0.95	Halogenated aromatics	Koch, 1983

several available computer programs, including Moleconn-X (Hall, 1992), PEP (Doucette and Holt, 1992), and Graph III (Sabljic and Horvatic, 1993).

First order MCIs, most commonly used in estimating K_{oc}, can be hand calculated, although for large molecules the use of one of the computer programs previously mentioned greatly simplifies the task. A brief description of the procedure to calculate MCIs appears here. For a more detailed discussion of the procedure for calculating MCIs, see Kier and Hall (1986).

MCIs are calculated from the hydrogen suppressed skeleton of a molecule. First, each non-hydrogen atom is assigned a delta value (δ). For simple indices, δ is equal to the number of atoms to which it is bonded; for valence indices, δ values are based upon the number of valence electrons not involved in bonds to hydrogen atoms. Simple and valence indices of different orders and types can be calculated for a given molecule.

The *order* refers to the number of bonds in the skeletal substructure or fragment used in computing the index: zero order defines individual atoms, first order uses individual bond lengths, second order uses two adjacent bond combinations, and so on.

The *type* refers to the structural fragment (path, cluster, path/cluster or chain) used in computing the index. Only path indices are possible for orders less than 3.

The symbol ${}^2\chi$ represents a simple second order simple index, whereas the symbol ${}^1\chi^v$ represents a first order valence index.

Each index is computed by an algorithm introduced by Randic (1975), which sums the reciprocal square roots of the assigned δ values over all molecular fragments, as illustrated below, for zero (${}^0\chi$), first (${}^1\chi$), and second order (${}^2\chi$) MCIs:

$$^0x = \sum \frac{1}{\sqrt{\delta_i}} \quad \text{for all i atoms}$$

$$^1x = \sum \frac{1}{\sqrt{\delta_i \times \delta_j}} \quad \text{for all bonded pairs of atoms}$$

$$^2x = \sum \frac{1}{\sqrt{\delta_i \times \delta_j \times \delta_k}} \quad \text{for all bonded trios of atoms}$$

The general procedure for estimating K_{oc} from MCI-K_{oc} relationships is to:

1. Draw out the hydrogen suppressed structure for the chemical of interest.
2. Calculate the first order MCIs (by hand or using a computer program).
3. For non-polar hydrophobic compounds, select the most appropriate equation from Table 8.3 based on the quality and extent of the database used to develop the model and the structural similarity between the chemical of interest and the chemicals used to develop the model.
4. For polar compounds, use the method reported by Meylan et al. (1992), and pick the appropriate functional groups from Table 8.4.
5. Calculate K_{oc} using the selected regression equation.
6. Estimate the site-specific sorption coefficient (K_d or K_f) from the estimated K_{oc} and the organic carbon content of the sorbent of interest, using Equation (1).

Several examples illustrating the use of this approach appear at the end of this chapter.

TABLE 8.4

Fragment Correction Values for Method of Meylan et al. (1992).

Fragment	Coeff or P_f	Fragment	Coeff or P_f
Amine, aromatic (non-fused ring)	−0.7770[a]	Thiocarbonyl (C=S)	−1.1002
Ether, aromatic (–C–O–C–)	−0.6431[b]	OrganoPhosphorus [P=S]	−1.2634[a]
Nitro (–NO2)	−0.6317	OrganoPhosphorus [P=O], aliphatic	−1.6980[a,d]
N–CO–C (acetamide-type)	−0.8112	N–CO–O–Phenyl Carbamate	−2.0022
Urea (N–CO–N)	−0.9222	Ether, aliphatic (–C–O–C–)	−1.2643
Nitrogen to Carbon (aliphatic) (–N–C)	−0.1242[c]	Ester (–C–CO–O–C–) or (HCO–O–C)	−1.3089
Carbamate (N–CO–O) or (N–CO–S)	−0.0249	Sulfone (–C–SO2–C–)	−0.9945
Triazine ring	−0.7521	Azo (–N=N–)	−1.0277
Nitrogen-to-Cycloalkane (aliphatic)	−0.822	N–CO–O–N Carbamate	−1.9200
Uracil (–N–CO–N–CO–C=C– Ring)	−1.8060	Aromatic ring with 2 nitrogens	−0.9650
Organic Acid (–CO–OH)	−1.7512[a]	OrganoPhosphorus [P=O], aromatic	−2.8781[a]
Ketone (–C–CO–C–)	−1.2477	Miscellaneous Carbonyl (C=O) Group	−1.0000[e]
Aliphatic Alcohol (–C–OH)	−1.5193	Pyridine ring (NO other fragments)	−0.7001[f]
Nitrile/Cyanide (–C=N)	−0.7223[a]		

[a] Counted only once per structure, regardless of number of occurrences.
[b] Either one or both carbons are aromatic; if both carbons are aromatic, cannot be cyclic.
[c] Any nitrogen attached to a double bond is not counted; also, carbonyl and thiocarbonyl are not counted as carbons.
[d] This is the only fragment counted, even if other fragments occur.
[e] Not included in regression derivation; estimated from other carbonyl fragments.
[f] A pyridine ring is counted only when no other fragments in this list are present.

8.3.3　Correlations with Capacity Factors Obtained from High Performance Liquid Chromatography (RP-HPLC)

Experimentally determined retention times or capacity factors (k) generated by reverse phase, usually octadecylsilane (ODS), high performance liquid chromatography (RP-HPLC) have been used widely to estimate K_{ow} values (McDuffie, 1981; Haky and Young, 1984; Sarna, 1984; Doucette and Andren, 1988). More recently, this approach has been used to directly estimate K_{oc} (Vowles and Mantoura, 1987; Hodson and Williams, 1988; Szabo et al., 1990; Kordel et al., 1993; Kordel et al., 1995; Hong et al., 1996). This is not strictly an estimation method because it relies on the acquisition of experimental retention times.

Briefly, the method involves determining the capacity factors (retention time corrected for an unretained substance) for a suitable set of reference substances (having known K_{oc} values) using RP-HPLC. The relationship between the capacity factors and K_{oc} for the reference or calibration compounds is determined from regression analysis of a log-log plot of the two properties. The capacity factors of compounds having unknown K_{oc} values then are determined using the identical experimental conditions, and K_{oc} values then are calculated from the regression expression.

Vowles and Mantoura (1987) investigated the relationship between RP-HPLC capacity factors obtained using ODS and alkylcyano columns and K_{oc} for 14 alkylbenzene and PAHs. They found that the correlations obtained with the capacity factors measured with the alkylcyano column were slightly better and more general (less class specific) than those obtained from the octadecylsilane (ODS) column. A significant correlation between K_{oc} and K_{ow} also was reported.

Similarly, for 22 compounds, Hodson and Williams (1988) also found that log K_{oc} correlated more highly with HPLC capacity factors measured on a cyanopropyl column than an ODS column and that the correlation improves with increased mobile phase water content. The authors also reported that the correlation between log K_{oc} and log k was better than that between log K_{oc} and log K_{ow}.

Pussemier et al. (1989) found that log K_{oc} for 12 aryl N-methylcarbamates correlated significantly to a linear combination of two parameters, π hydrophobic effect (calculated from RP-HPLC measurements) and δ Hildebrand solubility parameter. The authors also used data for 16 phenylureas and 13 anilides from Briggs (1981) and obtained similar results.

In a series of papers, Szabo et al. (1990 a,b) used a variety of stationary phases, including octadecylsilica (ODS), cyanopropyl, ethylsilica, and immobilized humic acid, to investigate the relationship between K_{oc} and RP-HPLC capacity factors for 11 aromatic hydrocarbons. While capacity factors generated with all the stationary phases showed significant correlations with log K_{oc}, the authors concluded that the immobilized humic acid column capacity factors, obtained by extrapolation of retention data from binary elements to 100% water, gave the best correlation.

Kordel et al. (1993) evaluated ODS, cyanopropyl, and trimethylammoniumpropyl columns for estimating K_{oc} and found that ODS columns were suitable only for a limited range of compounds and did not work well for very polar or very hydrophobic compounds. Cyanopropyl and trimethylammoniumpropyl columns were found to work better than ODS, with cyanopropyl being recommended because it was better for dissociating substance that required buffered elements. Kordel et al. (1995) later described an eleven laboratory ring test used to validate the HPLC-screening method for estimating K_{oc}. Each laboratory used a cyanopropyl column, isocratic elution, either a methanol/water (55/54 % v/v) or methanol/0.01 M citrate buffer pH 6.0 (55/45 % v/v) mobile phase and the following seven test substances: n-phenylacetamide (log K_{oc} = 1.61), nonuron (log K_{oc} = 1.99), atrazine (log K_{oc} = 1.81), fenthion (log K_{oc} = 3.31), linuron (log K_{oc} = 2.59), triapenthenol (log K_{oc} = 2.37), and trifluralin (log K_{oc} = 3.94). The authors concluded that the method was relatively insensitive to the column manufacturer and the method used to determine dead time and variation in flow rate. Deviations obtained in the ring test were smaller than those obtained by direct experimental determination using a standard batch method. However, the test compounds evaluated varied only by about two orders of magnitude in K_{oc} (log K_{oc} 1.61 to 3.94)

As the examples in Table 8.5 illustrated, a variety of investigators have reported significant correlations between log K_{oc} and capacity factors generated by RP-HPLC. Many of the reported correlations have been developed using relatively small data sets with a limited range of K_{oc} values. Highly polar (log $K_{oc} < 1$) and highly hydrophobic compounds (log $K_{oc} > 4$) have not been tested adequately. It also has been observed that the column type and mobile phase water content influence the strength of the correlations. However, by standardizing column type, mobile phase composition, and the calibration standards, the approach has been shown to be fast, precise, and considerably less expensive than direct experimental determination. Nevertheless, structurally based methods of estimation do not require experimentally generated retention times, have been evaluated over a wider range of compound types, and are still faster and less expensive. Gawlik et al. (1997) provide a more complete compilation of log K_{oc}-capacity factors (k) relationships.

TABLE 8.5

Examples of Log K_{oc} – Log Capacity Factors (k') Relationships Generated by RP-HPLC.

Equation[a]	R2	n	Column[b]	Compound Types	Reference
log K_{oc} = 1.8 log k' + 2.4	0.98	48	CN	mainly pesticides,	Kordel et al.,
log K_{oc} = 1.4 log k' + 2.5	0.87	30	ODS	amides, triazines	1993
log K_{oc} = 2.0 log k' + 3.0	0.91	48	TMAP		
log K_{oc} = 0.893 log k + 1.803	0.971	11	humic acid	aromatics	Pussemier et al., 1990
log K_{oc} = 0.95 log kw' + 1.781	0.986	10	humic acid	aromatics	Szabo et al., 1990
log K_{oc} = 1.441 log kw' + 1.488	0.947	10	ethylsilica	aromatics	Szabo et al., 1990
log K_{oc} = 2.70 log k' + 2.04	0.992	7	CN	aromatic and aliphatic hydrocarbons	Hodson and Williams, 1988

[a] Equations: k' = Capacity Factor; K_w = Capacity Factor Extrapolated a 100% Water Mobile Phase

[b] Column: ODS = Octadecylsilane Column; CN = Cyanopropylic Column; TMAP = Trimethylammonium-propylic Column

8.3.4 Miscellaneous Methods for Estimating K_{oc}

In addition to MCIs, several other structurally derived parameters have been correlated with K_{oc}, including parachor (Briggs, 1981), molecular weight (Kanazawa, 1989), molecular surface area (Fugate 1989; Holt 1992), the characteristic root index (CRI) (Sacan and Balcioglu, 1996), and several quantum mechanical parameters. Linear solvation energy relationships (LSERs) also have been reported (Hong et al., 1996; Baker et al., 1997). In general, these methods have not been as rigorously validated or as widely used as the methods previously described.

8.3.4.1 Parachor

A strong correlation between the parachor (P), a structurally derived property previously used by McGowan (1954) to calculate aqueous solubilities and partition coefficients, and K_{om} was first observed by Lambert (1967) for two classes of compounds, substituted phenylurea herbicides and homologs of dinitroaniline herbicides. The rationale for using parachlor came from it being an approximate measure of the molar volume (Lambert, 1967).

Direct calculation of P requires the surface tension and density of the liquid and the density of the vapor. However, tables of measured P values are available, and a fragment constant approach also can be used to estimate P (Quayle, 1953).

To extend the application of K_{om}-P correlations, Hance (1969) proposed an additional term for compounds capable of hydrogen bonding. The resulting factor (parachlor-45N), where N is the number of sites in a molecule that can participate in the formation of a hydrogen bond (such as primary, secondary, and tertiary amino, carbonyl heterocyclic nitrogen, and ether oxygen) was shown to correlate with log K_f for 29 aromatic herbicides. Internal H-bonding or steric effects were ignored.

Briggs (1981) developed a similar relationship for 38 chemicals:

$$\log K_{om} = 0.0062 \ (P{-}100n) + 0.58 \quad r = 0.92$$

where n is taken as 1 for each O atom not bonded or conjugated to an aromatic ring, 1 for each singly bonded N atom, 1 for each heterocyclic aromatic ring (no matter how many heteroatoms are in the ring), and 0.25 for each halogen attached to a saturated carbon atom.

Lyman et al. (1982) reviewed the use of P values to estimate K_{oc} and also provided several examples illustrating the use of P in estimating K_{oc} or K_{om}.

8.3.4.2 Relationship with Molecular Surface Area

Fugate (1989) reported the following relationship between total molecular surface area (TSA) and log K_{oc} for 36 nonpolar organic compounds, including chlorobenzenes, PAHs, and PCBs ranging from 1.44 to 5.95 in log K_{oc}:

$$\log K_{oc} = 0.02 \ TSA - 0.29, \quad r^2 = 0.83$$

where TSA is the total molecular surface area (Å^2) calculated by the Pearlman (1980) method.

Holt (1992) examined the relationship between log K_{oc} and surface area for 167 nonpolar and polar compounds. Surface area contributions from the polar portions of the molecule, determined by calculating separate surface areas for nitrogen (N-TSA), oxygen (O-TSA), phosphorous (P-TSA), and sulfur (S-TSA) and aromatic nitrogens (ARN-TSA), were calculated in addition to the total surface area. The multilinear relationship obtained using the partial TSAs was significantly better than the regression model obtained by using just TSA alone.

$$\log K_{oc} = 0.22 + 0.015 \ TSA{-}0.041 \ (N{-}TSA) - 0.25 \ (O{-}TSA) - 0.045 \ (ARN{-}TSA), \quad r^2 = 0.49$$

8.3.4.3 Group Contribution Method

Karickhoff (1983) described a simple fragment constant method for unsubstituted condensed ring aromatics. Also, since K_{oc} and K_{ow} are approximately linearly related, Karickhoff suggested that fragment constants for sorption should approximate those for the comparable octanol/water systems, at least for relatively simple mono- and di-substituted compounds.

8.3.4.4 Corrections Between K_{oc} and Characteristic Root Index (CRI)

Sacan and Balcioglu (1996) reported the following correlation between K_{oc} and the Characteristic Root Index (CRI) for 36 chlorinated biphenyls, phenols, and benzenes:

$$K_{oc} = (\pm 0.034)\ 1.034\ CRI + (\pm 0.113)\ 0.441,\ r = 0.982,\ s = 0.22,\ F = 912$$

The CRI, like MCIs, is a structural parameter calculated from the hydrogen suppressed skeleton of a molecule. Sacan and Inel (1995) provide a discussion of the procedure used to calculate the CRI. The authors state that the CRI comprises more structural features than the first order MCI because CRI includes all possible orders of MCIs except zero order, and CRI is more sensitive to the chlorine substitutions' pattern and branching.

8.3.4.5 Corrections Between K_{oc} and Quantum Mechanical Parameters

For 50 carboxylic acids, esters, amines, amides, and three different soils varying in organic carbon content, pH, and clay content, Von Oepen et al. (1991) examined the relationship between K_{oc} and a variety of electronic, geometric, and topological parameters, including highest occupied and lowest unoccupied molecular orbitals (HOMO and LUMO), ionization potential, electronegativity, dipole moment, charge distribution, self-polarizability, total and average charge, probability of nuclephilic/electrophilic attack, molar refraction, molecular volume, molecular connectivity indices, and calculated octanol/water partition coefficient.

Overall, a large variability in K_{oc} values was observed, especially for the more polar compounds. For relatively nonpolar esters, K_{oc} was considered a useful model for sorption to soil with high organic carbon and low clay content, with variation by a factor of 3-5. Significant correlations between K_{oc} and molecular size, self polarizability, and MCIs were observed. For polar acids and amines, K_{oc} values varied up to two orders of magnitude, and poor correlations were obtained for most of the parameters examined.

8.3.4.6 Estimation of K_{oc} Using Aqueous Phase and Soil Organic Matter (SOM) Phase Activity Coefficients Calculated from UNIFAC and ELBRO-FV

Ames and Grulke (1995) used aqueous phase and soil organic matter (SOM) phase activity coefficients, calculated with the group contribution approaches UNIFAC (Gmehling et al., 1982) and ELBRO-FV (Kontogeorgis et al., 1993), to estimate K_{oc} values for 18 nonionic organic compounds. A model humic acid molecule represented SOM as a polymeric phase. Predictions were typically within an order of magnitude, and the use of an experimental activity coefficient, when available, significantly improved the predictions. The relatively large predictive errors and the small data set used to develop and test this method make it difficult to evaluate its potential applicability.

8.4 Limitations of the K_{oc} Approach

Although widely used for many chemical/sorbent combinations, the K_{oc} approach is most appropriate for predicting the sorption of neutral hydrophobic compounds onto sorbents having an organic carbon content greater than about 0.1%. In situations where sorbents have low organic carbon high clay contents, and for chemicals that have highly polar and/or ionizable functional groups that may significantly interact with polar or charged sites on sorbent surfaces, the K_{oc} approach may not be suitable. However, even in these situations, the K_{oc} approach still may be used because it is often the only available method for estimating sorption behavior. In addition, the estimate of hydrophobic sorption or partitioning based on K_{oc} often represents a minimum value or conservative estimate for the sorption of a particular hydrophobic organic compound. If specific sorbate-sorbent interactions also are involved, additional sorption may occur (Khan et al., 1979).

Finally, it is important to note that the K_{oc} model is not appropriate for describing the sorption of organics onto dry and subsaturated soils (Chiou, 1989). In these situations, sorption is primarily a function of the mineral type, mineral content, and humidity, or water content.

8.4.1   Low Organic Carbon/High Clay Content Solids

In soils having low organic carbon content (>0.1%), especially in those with high clay content, the potential contribution of soil minerals to the sorption process may be equally important or even dominate the overall sorption process (Huang et al., 1984; Hassett et al., 1981; Khan et al. 1979). Khan et al., (1979) found that, for acetophenone, the amount of clay appeared to be important in cases where the organic carbon was low and did not mask the effect of the clay minerals. Green and Karickhoff (1990) suggested that the use of the K_{oc} approach may be inappropriate when soil has clay/oc > 40. Soil properties such as particle size distribution, pH, the type and amount of clay minerals, cation-exchange capacity (CEC), and the hydroxide content can have significant effects on the sorption process in low organic carbon/high clay situations.

However, even in cases where a strong interaction between the solute and clay minerals is involved, a positive linear correlation with SOM (Hassett and Banwart, 1989) may still be observed because, in many soils, SOM and clay content also are correlated (Mingelgrin and Gerstl, 1983).

8.4.2   Polar and Ionizable Organic Compounds

The sorption of polar organics may not follow the K_{oc} model even in soils having sufficient organic carbon content. As previously mentioned, K_{oc} was considered a useful model for the sorption of relatively nonpolar esters to soils with high organic carbon and low clay content, with variation by a factor of 3-5. Significant correlations between K_{oc} and molecular size, self polarizability, and MCIs were observed. However, for polar acids and amines, K_{oc} values varied up to two orders of magnitude, and poor correlations were obtained for most of the parameters examined (Von Oepen et al., 1991).

The sorption of highly polar or ionizable organics may correlate more readily with other properties such as CEC, clay content, and total surface area of sorbing surface. For example, Podoll et al. (1987) reported that the sorption of poly(ethylene glycol) and poly(ethyleneamines) correlated with CEC and sediment clay content and not with sediment organic content. Brownawell et al. (1990) reported the sorption of organic cations to be more strongly correlated with CEC than organic carbon content. Davis (1993) showed that 1,2 ethanediamine and N(2-aminoethyl)-1,2-ethanediamine sorption correlated closely with CEC and organic content for six soils. While a strong relationship between the foc and CEC of the six soils was observed, the relationship was not strictly linear.

Most weakly acidic organic chemicals are in a negatively charged form at the pH of most natural soils, while most weakly basic chemicals are in their non-ionized or molecular form. The K_{oc} approach may still be applied to organic acids and bases, but it is necessary to consider the sorption of both ionic and neutral forms. For example, Fontaine et al. (1991)

investigated the sorption of the neutral and anionic forms of a sulfonamide herbicide, flumetsulam and was able to calculate K_{oc} values for the neutral (640) and anionic (12) forms of the molecule. Describing the soil sorption as a combination of anionic and neutral forms provided an adequate description of the measured sorption. Similarly, the sorption of the weak acids tetrachlorophenol and pentachlorophenol onto two soils showed no correlation with SOM content, until the K_{om} values were corrected for the pH/pKa effects. The pH-corrected log K_{om} values correlated with log K_{ow} (van Gestel et al., 1991).

Using log K_{ow}, pK$_a$, soil pH, and fraction organic carbon content (foc), Bintein and Devillers (1994) developed a model to directly estimate soil/water distribution coefficients (K_p) applicable to both organic acids and bases, in addition to non-ionized organic chemicals. The proposed model was developed using 229 K_p values recorded for 53 chemicals and tested on 87 additional chemicals having 500 K_p values. Only K_p values recorded in soils having a %OC greater than 0.1 were included in the development and testing of the model. The final model is listed below:

$$\log K_p = 0.93 \log K_{ow} + 1.09 \log foc + 0.32 C_{fa} - 0.55 C_{fb'} + 0.25, S = 0.433, r = 0.966$$

where C_{fa} is the relationship between the anionic species concentration and pH values, and $C_{fb'}$ is the relationship between the protonated species concentration and pH values.

For acids, $C_{fa} = \log \dfrac{1}{1 + 10^{pH-pKa}}$ and C_{fa} is 0 for bases and neutral compounds.

Similarly for bases, $C_{fb'} = \log \dfrac{1}{1 + 10^{pKa-(pH-2)}}$ with C_{fb} is equal to zero for acids and other neutral compounds.

8.4.3 Availability of K_{oc} Values

Numerous compilations of experimental K_{oc} values have appeared in the literature over the past 20 years (for example, Bahnick and Doucette, 1988; Meylan et al., 1992; Sabljic et al., 1995; Baker et al., 1997). Table 8.6 provides an abbreviated list of K_{oc} values, taken from two compilations (Bahnick and Doucette, 1988; Baker et al., 1997). Computerized databases of environmentally relevant physical/chemical property values, including K_{oc} values, are also commercially available (for example, Syracuse Research Corporation, 1994). One potential problem in using K_{oc} values from these compilations is that, in many cases, the original reference is not listed and specific details describing the experimental determination of the K_{oc} values are unavailable (i.e., was K_{oc} calculated from a single concentration or range of concentrations? was the isotherm linear or nonlinear? were multiple soils containing various amounts of organic carbon evaluated?).

Considerable variation in K_{oc} values appear in the literature. For example, Mackay et al. (1992) provided 24 values of K_{oc} for benzene, ranging from 0.11 to 2.08 L/kg. Variation is likely the result of differences in the sorption characteristics of SOM, variation in the methods used to determine K_{oc} (separation of phases, mass balance, single point, or isotherm, kinetics), impact of other soil properties, and the properties of chemicals being sorbed. The variability in K_{oc} values is generally greater for the more polar compounds.

The ASTM (1991) describes a standard method for determining K_{oc} that requires the use of several soils of differing SOM content, verification of equilibrium, sterilization of the soil, analysis of degradation byproducts, or complete mass balance determination (ASTM, 1987b; Seth et al., 1999). However, in most cases, K_{oc} values reported in the literature were

TABLE 8.6

List of Experimental Log K_{oc} Values Compiled by Bahnick and Doucette (1988) and Baker et al., (1997).

Compound	log K_{oc}	Ref.*
1,1,1-trichloroethane	2.26	10
1,1,2,2-tetrachloroethane	1.9	10
1,1,2-trichloroethane	1.89	5
1,2,3,4-tetrachlorobenzene	3.84	3
1,2,3,4-tetrachlorobenzene	4.28	1a
1,2,3,5-tetrachlorobenzene	3.20	9
1,2,3,5-tetrachlorobenzene	4.25	1a
1,2,3-trichlorobenzene	3.91	1a
1,2,3-trimethylbenzene	2.8	3
1,2,4,5-tetrachlorobenzene	4.27	2a
1,2,4-trichlorobenzene	2.94	2
1,2,4-trichlorobenzene	4.02	1a
1,2,5,6-dibenzanthracene	6.31	8
1,2,7,8-dibenzocarbazole	6.11	7
1,2-dibromo-3-chloropropane	2.11	10
1,2-dibromoethane	1.64	4
1,2-dichlorobenzene	2.5	2
1,2,5,6-dibenzanthracene	6.33	3a
1,3,5-trichlorobenzene	2.85	5
1,3,5-trichlorobenzene	4.13	1a
1,3,5-trimethylbenzene	2.82	3
1,3-dichlorobenzene	2.47	2
1,4-dichlorobenzene	2.44	2
1,4-dimethylbenzene	2.52	3
1-naphthol	2.64	6
1-naphthol	2.72	3a
1-naphthylamine	3.26	4a
13H-dibenzo[a,i]carbazole	6.02	3a
2,2′-biquinololine	4.02	3a
2,3,4,2′,3′,4′-hexachlorobiphenyl	5.05	9
2,3,4,2′,5′-pentachlorobiphenyl	4.54	9
2,3,4,5,6,2′,5′-heptachlorobiphenyl	5.95	9
2,3,4,5-tetrachlorophenol	4.12	5a
2,3,4,6-tetrachlorophenol	2.66	5
2,3,4,6-tetrachlorophenol	3.79	5a
2,3,7,8-TCDD	6.66	6a
2,3-dichlorophenol	2.39	5a
2,4,5,2′,5′-pentachlorobiphenyl	4.63	4
2,4,5-T	1.72	4
2,4,5-trichlorophenol	3.26	5a
2,4,5-trichlorophenoxyacetic acid	2.76	4a
2,4,6-trichlorophenol	2.52	5
2,4,6-trichlorophenol	2.94	5a
2,4-D	1.30	4
2,4-D	2.26	4, 12a
2,4-dichlorophenol	2.75	5
2,4-dichlorophenol	2.47	5a
2,5,2′5′-tetrachlorobiphenyl	4.91	10
2-aminoanthracene	4.45	3a
2-chloroacetanilide	1.58	1
2-methylnaphthalene	3.93	4
2-methylnaphthalene	3.93	7a
3,4,5-trichlorophenol	3.48	5a
3,4-dichloroacetanilide	2.34	1
3,4-dichloronitrobenzene	2.53	1

TABLE 8.6 (CONTINUED)

List of Experimental Log K_{oc} Values Compiled by Bahnick and Doucette (1988) and Baker et al., (1997).

Compound	log K_{oc}	Ref.*
3,5-dinitrobenzamide	2.20[a]	4a
3-(3,4-dichlorophenyl)-1-methylurea	2.46	1
3-(3-chlorophenyl)-1,1-dimethylurea	1.79	1
3-(3-chlorophenyl)-1-methylurea	1.93	1
3-(4-bromophenyl)-1-methyl-1-methoxyurea	2.02	1
3-(4-chlorophenyl)-1-methyl-1-methoxyurea	1.84	1
3-(4-methylphenyl)-1,1-dimethylurea	1.51	1
3-(trifluoromethyl)aniline	2.36	1
3-chloro-4-methoxyaniline	1.93	1
3-chloroacetanilide	1.86	1
3-methyl-4-bromoaniline	2.26	1
3-methylacetanilide	1.45	1
3-methylaniline	1.65	1
3-methylcholanthrene	6.25	8
3-methylcholanthrene	6.1	3a
3-nitrobenzamide	2.10[a]	4a
3-phenyl-1-cyclohexylurea	2.07	1
3-phenyl-1-cyclopentylurea	1.93	1
4-aminobenzoic acid	2.50[a]	4a
4-aminonitrobenzene	1.88	1
4-bromonitrobenzene	2.42	1
4-bromophenol	2.41	1
4-hydroxybenzoic acid	2[a]	4a
4-methoxyacetanilide	1.40	1
4-methylaniline	2.85[a]	4a
4-methylbenzoic acid	2.28	4a
4-methylbenzoic acid ethyl ester	2.84	4a
4-nitroaniline	2.77[a]	4a
4-nitrobenzamide	2.10[a]	4a
4-nitrobenzoic acid	2.07	4a
4-nitrobenzoic acid ethyl ester	2.67	4a
6-aminochrysene	5.21	8
6-aminochrysene	5.16	3a
7,12-dimethylbenzanthracene	5.37	8
7,12-dimethylbenz[a]anthracene	5.35	3a
9-methylanthracene	4.81	7a
acetanilide	1.86[a]	4a
acetophenone	1.63	3
acetophenone	1.54[a]	3a
acridine	4.11	7
aniline	2.14[a]	4a
anisole	1.54	2
anthracene	4.42	4
anthracene-9-carboxylic acid	2.71	8
anthracene-9-carboxylic acid	2.62	3a
atrazin	2.17	4
atrazine	2.33	8a
benzamide	1.87[a]	4a
benzene	1.92	7a
benzoic acid	1.99	4a
benzoic acid ethyl ester	2.5	4a
benzoic acid methyl ester	2.32	4a
benzoic acid phenyl ester	3.53	4a
bromacil	1.86	4
butralin	3.91	4

TABLE 8.6 (CONTINUED)

List of Experimental Log K_{oc} Values Compiled by Bahnick and Doucette (1988) and Baker et al., (1997).

Compound	log K_{oc}	Ref.*
butyl benzyl phthalate	4.23	9a
carbaryl	2.02	1
chlorobenzene	2.41	3
cyanazine	2.30	4
cyanizine	2.26	8a
DDT	5.38	4
di-(2ethylhexyl) phthalate	4.94	9a
diallate	3.28	4
dibenzothiophene	4	6
dibenzothiophene	4.05	3a
dibutyl phthalate	3.14	9a
dichlobenil	2.37	4
dichloromethane	1.44	9
diethyl acetamide	1.15[a]	4a
diethyl phthalate	2.58	4a
dinoseb	2.09	4
diphenylamine	2.78	1
fluchloralin	3.56	4
fluorene	3.68	10a
hexachlorbiphenyl	6.08a	7a
hexachlorobenzene	3.59	4
hexanoic acid	1.38	4a
isocil	2.11	4
lindane	2.96	4
m-xylene	2.22	11a
methoxychlor	4.9	7a
methyl N-phenylcarbamate	1.73	1
metribuzin	1.98	4
N,N-dimethylaniline	2.63	4a
N,N-dimethylbenzamide	1.60[a]	4a
n-butyl N-phenylcarbamate	2.26	1
n-butylbenzene	3.39	3
N-methylaniline	2.79[a]	4a
N-methylbenzamide	1.76[a]	4a
n-propyl N-phenylcarbamate	2.06	1
naphthalene	3.11	4
nitrobenzene	1.94	1
o-xylene	2.11	11a
p-xylene	2.31	11a
PCB-138	5.93[a]	1a
PCB-149	5.79[a]	1a
PCB-153	5.86[a]	1a
PCB-52	5.41	1a
PCB-87	5.73	1a
PCB-95	5.55	1a
PCB-97	5.69	1a
pebulate	2.8	4
pentachlorobenzene	3.50	9
pentachlorobenzene	4.61	2a
pentachlorophenol	2.95	4
pentachlorophenol	4.59	5a
phenanthrene	4.36	4
phenol	1.43	4
phenylacetic acid	1.51[a]	4a
phenylacetic acid ethyl ester	2.11	4a

TABLE 8.6 (CONTINUED)

List of Experimental Log K_{oc} Values Compiled by Bahnick
and Doucette (1988) and Baker et al., (1997).

Compound	log K_{oc}	Ref.*
phenylurea	1.35	1
phthalic acid	1.63[a]	4a
picloram	1.23	4
propazine	2.56	8a
pyrene	4.92	4
pyrene	4.8	3a
quintahclorobenzene	4.49	1a
simazine	2.33	8a
tetracene	5.81	4
tetrachloroethylene	2.42	11a
tetrachloroguaiacol	2.85	5
tetrachloromethane	1.85	9
toluene	2.06	11a
trichloroacetamide	0.58[a]	4a
trichloroethene	2.00	5
trichloroethylene	1.81	11a
trichloromethane	1.65	9
trietazine	2.78	4
trietazine	2.74	8a
trifluralin	4.49	8a

* References for log K_{oc} values (Bahnick and Doucette, 1988): 1.
  Briggs (1981); 2. Chiou et al., (1983); 3. Schwarzenbach and
  Westfall (1981); 4. Kenaga and Goring (1980); 5. Seip et al.,
  (1986); 6. Hassett et al., (1981); 7. Banwart et al., (1982); 8. Means
  et al., (1980); 9. Koch (1983); 10. Chiou et al., (1979)
  References for log Koc values (Baker et al., 1997): 1a. Paya-Perez
  et al., (1991); 2a. Barber et al., (1992); 3a. Hasset et al., (1980);
  4a. von Oepen et al., (1991); 5a. Schellenberg et al., (1984); 6a.
  Walters et al., (1989); 7a. Karickhoff et al., (1979); 8a. Brown and
  Flagg (1981); 9a. Russell and McDuffie (1986); 10a. Abdul et al.,
  (1986); 11a. Abdul et al., (1987); and 12a. Rippen et al., (1982)

not obtained using this protocol. Often, K_{oc} values are calculated from a single soil and, in
some cases, from a single concentration.

An interlaboratory test program was used to evaluate the ASTM standard method for
measuring K_{oc} (ASTM, 1987b). Conducted at four laboratories using trifluralin and 13
soils types, it was found that the K_{oc} values varied from 2400 to 14990. The coefficient of
variation was about 50%. In general, one can expect a variation of 20 to 50% in K_{oc},
(ASTM, 1987b).

8.5   Summary and Conclusions

Despite its imperfections, the K_{oc} approach is currently the most widely used and generally
applicable method for predicting the sorption of organic compounds to soils and sediments. It is most appropriate for predicting the sorption of non-polar organic compounds

onto soils or sediments having greater than 0.1% organic matter content. However, even in situations where the organic carbon content of the sorbent is lower than 0.1% and/or polar organics are being considered, the K_{oc} approach generally provides a conservative estimate of sorption. For example, compounds having specific interactions with soil components other than SOM generally sorb to a greater extent than the K_{oc} approach predicted.

For organic acids and bases, the K_{oc} approach still can be applied if the neutral form of the compound dominates at the pH of the soil solution. If both neutral and ionized forms of the chemical are present in significant quantities, then the extent of sorption will depend on the fraction of each form present (Bintein and Devillers, 1994).

As previously discussed, the two approaches considered most generally appropriate for estimating K_{oc} are: 1) correlations with log K_{ow} and/or S, and 2) correlations with MCIs. Other approaches may work well for specific classes of compounds but are not as generally applicable. The addition of group contribution factors, as Meylan et al. (1992) documented, generally improve MCI/K_{oc} correlations for chemicals containing polar functional groups and should be used. For organic acids and bases, the method outlined by Bintein and Devillers (1994) is the most well defined. The remainder of this chapter provides the reader with several detailed examples outlining the use of these approaches for a small set of benchmark chemicals which Table 8.7 lists.

TABLE 8.7

Benchmark Chemicals and Information Needed to Estimate K_{oc} and Experimental Log K_{oc}.

Name	CAS #	Formula	MW	MP °C	log K_{ow}	log S (M)	$^1\chi$	log Koc
anthracene	120-12-7	$C_{14}H_{10}$	178.2	218	4.54	−6.39	6.93	4.42
chlorpyrifos	2921-88-2	$C_9H_{11}Cl_3NO_3PS$	350.6	41	5.27	−6.07	8.71	3.70
lindane	58-89-9	$C_6H_6Cl_6$	290.8	112.5	3.72	−6.07	5.46	2.96
2,6-di (t-butyl) phenol	128-39-2	$C_{14}H_{22}O$	206.3	35-38	5.43*	−3.92	6.64	3.41
trichloroethylene	79-01-6	$ClCH=CCl_2$	131.4	−84.8	2.53	−1.98	2.27	2.00
pentachlorophenol (pKa = 4.75)	87-86-5	C_6HCl_5	266.4	190	5.01	−4.28	5.46	2.95
quinoline (pKa = 4.94)	91-22-5	C_9H_7N	129.2	−15	2.0	−0.333	4.97	3.10

* estimated log K_{ow}

8.6   List of Symbols

K_f = Freundlich sorption coefficient
N = Freundlich constant, typically range between 0.75 and 0.95
C_s = concentration of the organic chemical sorbed by the solid phase (mg/Kg)
C_w = concentration of the organic chemical dissolved in aqueous phase (mg/L)
K_d = linear sorption coefficient; units of K_d are typically given as L/kg, mL/g or cm^3/g
K_{oc} = organic normalized sorption coefficients
R^2 = correlation coefficient
S = aqueous solubility (moles/L)
K_{ow} = octanol/water partition coefficient
MCIs = molecular connectivity indices
RP-HPLC = reverse phase high performance liquid chromatography

8.7 Appendix: Examples for Estimating K_{oc} – Benchmark Chemicals

Table 8.7 lists the five benchmark chemicals used throughout this book, along with one organic acid and one base. These compounds will be used to illustrate the estimation of K_{oc} and subsequent calculation of the linear sorption coefficient K_d.

Example 1(a) Estimate K_{oc} and K_d for anthracene in a soil containing 2% organic carbon using the K_{ow} value of 4.42 listed in Table 8.7.

Anthracene is a three-ring polyaromatic hydrocarbon (PAH) with the following structure.

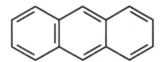

Table 8.1 lists a variety of general (larger, more hetrogeneous data sets) and class-specific equations (smaller, more homogeneous data sets) for estimating K_{oc} from K_{ow}. Class specific relationships generally give more accurate estimates, but only if the compound of interest is truly structurally representative of the compounds used to develop the equation. Unfortunately, this is not always easy to determine, especially for compounds containing multiple functional groups. Also, for many compounds, appropriate class specific relationships have not been developed.

If there is uncertainty that a particular class specific relationship is suitable for the compound or an appropriate class specific relationship can not be found, use one of the general expressions published by Gerstl (1990), Sabljic et al. (1995), and Baker et al. (1997). These expressions are recommended because they were developed with well documented and/or large data sets of K_{oc} values. For anthracene, the use of several class-specific equations developed specifically for PAHs are demonstrated here, along with several more general equations.

General equations

$\log K_{oc} = 0.679 \log K_{ow} + 0.663$ (Gerstl, 1990)
$\log K_{oc} = (0.679 \times 4.54) + 0.663 = 3.75$ L/kg
$K_{oc} = 5570$ L/kg
$K_d = K_{oc} * oc = 5570 \times 0.02 = 111$ L/kg

The calculation of K_d shown above assumes a linear relationship between the concentration sorbed and the concentration of the chemical in the aqueous phase (i.e., a linear isotherm). If sorption was known to be non-linear, the Freundlich constant, N, would need to be known or an estimate of it would need to be made. Generally values of N range from 0.7 to 1.2.

$\log K_{oc} = 0.81 \log K_{ow} + 0.10$ (Sabljic et al., 1995)
$\log K_{oc} = (0.81 \times 4.54) + 0.10 = 3.78$ L/kg
$K_{oc} = 5990$ L/kg
$K_d = 5990 * 0.02 = 120$ L/kg

log K_{oc} = 0.903 log K_{ow} + 0.094 (Baker et al., 1997)
log K_{oc} = (0.903 × 4.54) + 0.094 = 4.19 L/kg
K_{oc} = 1.56 × 10^4 L/kg
K_d = 1.56 × 10^4 × 0.02 = 312 L/kg

The average value of log K_{oc} estimated using the three general equations is 3.91. This is slightly lower than the experimental value of 4.42 Kenaga and Goring (1980) reported.

Class specific equation for PAHs

log K_{oc} = 1.00 log K_{ow} − 0.21 (Karickhoff et al., 1979)
log K_{oc} = (1.00 x 4.54) − 0.21 = 4.33
K_{oc} = 2.14 × 10^4 L/kg
K_d × foc = (2.14 × 104 × 0.02) = 428 L/kg

In this example, the class-specific equation yielded an estimated value closer to the experimental value than the more general expressions.

As shown in the example below, a similar approach can be used if a suitable value for the aqueous solubility of anthracene is available. Again, Table 8.2 gives both class specific and more general equations for estimating K_{oc} from S.

Example 1b) Estimate K_{oc} and K_d for anthracene in a soil containing 2% organic carbon using the log S value of −6.39 m listed in Table 8.7.

Table 8.2 lists both a general and three class specific equations for estimating log K_{oc} from log S.

General equation

log K_{oc} = −0.508 log S + 0.953 Fc (Gerstl, 1990)
log K_{oc} -0.508 (−6.390) + 0.953 + 0 = 4.20 L/kg
K_{oc} = 158 × 10^4 L/kg
K_d = 158 × 10^4 L/kg × 0.02 = 1050 L/kg

Class specific equations for PAHs

log K_{oc} = -0.439 log S + 1.910 (Gerstl, 1990)
log K_{oc} = −0.439 (−6.390) + 1.910 = 4.72
K_{oc} = 52500 L/kg
K_d = 5250 L/kg × 0.02 L/kg

log K_{oc} = −0.68 log S + 4.273 (S in μmol/L), (Hassett et al., 1980)
log K_{oc} = (−0.68 × −0.390) + 4.273 = 4.54 L/kg
K_{oc} = 3.45 × 10^4 L/kg
Kd = 372 × 10^4 L/kg

The average value of log K_{oc} estimated using the two class specific equations is 4.37. This is within 0.05 log units of the experimental value of 4.42 Kenaga and Goring (1980) reported.

Example 1c) Assuming no experimental values of K_{ow} or S are available for anthracene, estimate K_{oc} for anthracene from MCIs using the approach of Meylan et al. (1992). This approach is based on the following relationship:

$$\log K_{oc} = 0.53\ ^1\chi + 0.63 + \Sigma\ P_f N \tag{6}$$

where $^1\chi$ is the first order MCI and $\Sigma P_f N$ is the summation of the products of all applicable correction factors multiplied by the number of times (N) that the fragment occurs in the structure (with the exception noted at the bottom of Table 8.4).

The solution of this expression involves the following steps:

1. Calculate the first order molecular connectivity index for anthracene.
   a. To calculate the value of $^1\chi$ for anthracene by hand, first draw out its hydrogen suppressed structure and assign delta values to each of its atoms, as shown below. For a first order MCI (1χ), the delta value for each atom is equal to the number of atoms it is bonded to.

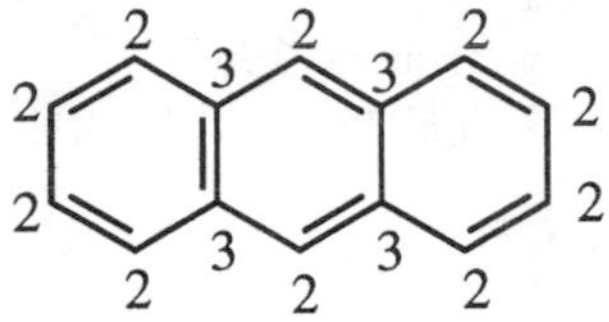

   b. Calculate 1χ using Equation (2).

$$^1\chi = \sum \left(6\times\left(1/\sqrt{2\times2}\right)\right)+\left(8\times\left(1/\sqrt{3\times2}\right)\right)+\left(2\times\left(1/\sqrt{3\times3}\right)\right) = 6.933$$

2. Determine if any polar correction fragments are necessary by looking in Table 8.2. No appropriate fragments appear in Table 8.2 for anthracene.
   c. Calculate $\log K_{oc}$ using Equation (3)

$$\log K_{oc} = 0.53\ ^1\chi + 0.63 + \Sigma\ P_f N$$
$$\log K_{oc} = (0.53)\ (6.93) + 0.62 + 0 = 4.29$$
$$K_{oc} = 1.97 \times 10^{-4}$$
$$K_d = 1.97 \times 10^{-4} \times 0.02 = 394$$

This estimated value is within 0.13 log units of the $\log K_{OC}$ value of 4.42 listed in Table 8.7.

Example 2a) Estimate K_{oc} and K_d for chlorpyrifos in a soil containing 4% organic carbon using the K_{ow} value of 5.27 listed in Table 8.7.

Chlorpyrifos is a organophosphate pesticide with the structure listed below.
Table 8.1 lists both general and class specific equations (organophosphates) for estimating K_{oc} from K_{ow}.

General equations
$$\log K_{oc} = 0.679 \times \log K_{ow} + 0.663 \text{ (Gerstl, 1990)}$$

$\log K_{oc} = (0.679 \times 5.27) + 0.663 = 4.24 \text{ L/kg}$
$K_{oc} = 1.74 \times 10^4 \text{ L/kg}$
$K_d = 1.74 \times 10^4 \text{ L/kg} \times 0.04 = 697 \text{ L/kg}$

Note: K_d is assumed to be a linear sorption coefficient. A non-linear sorption coefficient (K_f) could be calculated if a value for N is known or could be estimated.

$\log K_{oc} = 0.81 \log K_{ow} + 0.10$ (Sabljic et al., 1995)
$\log K_{oc} = (0.81 \times 5.27) + 0.10 = 4.37 \text{ L/kg}$
$K_{oc} = 23400 \text{ L/kg}$
$K_d = 23400 \text{ L/kg} \times 0.04) = 938 \text{ L/kg}$
$\log K_{oc} = 0.903 \log K_{ow} + 0.094$ (Baker et al., 1997)
$\log K_{oc} = (0.903 \times 5.27) + 0.094 = 4.85 \text{ L/kg}$
$K_{oc} = 70800 \text{ L/kg}$
$Kd = (70800 \text{ L/kg} \times 0.04) = 2830 \text{ L/kg}$

The average value of $\log K_{oc}$ estimated using the three general equations is 4.49. This is 0.79 log units higher than the experimental value of 3.70 Kenaga and Goring (1980) reported.

Class-specific expressions for organophosphates

$\log K_{oc} = 1.17 + 0.49 \log K_{ow}$, phosphates (Sabljic et al., 1995)
$\log K_{oc} = 1.17 + 0.49 (5.25) = 3.75 \text{ L/kg}$
$K_{oc} = 5650 \text{ L/kg}$
$K_d = 56520 \text{ L/kg} \times 0.04 = 226 \text{ L/kg}$

$\log K_{om} = 0.58 \times \log K_{ow} + 0.32$ (Briggs, 1981)
$\log K_{om} = 0.58 \times 5.7 + 0.32 = 3.38 \text{ L/kg}$
$\log K_{oc} = \log K_{om} - 0.237 = 3.14$
$K_{oc} = 1380 \text{ L/kg}$
$K_d = 1380 \text{ L/kg} \times 0.04 = 55 \text{ L/kg}$

For chlorpyrifos the class-specific expressions gave values that were closer to the log K_{OC} value of 3.70 listed in Table 8.5 than the general equations.

Example 2b) Estimate K_{oc} and K_d for chlorpyrifos in a soil containing 4% organic carbon using the aqueous solubility value provided in Table 8.7 (log S = −6.07).
Table 8.2 lists both a general and a class specific equation for estimating log K_{oc} from log S.

General equation

$\log K_{oc} -0.508 \log S + 0.953$ Fc(Gerstl, 1990)
$\log K_{oc} -0.508 \times -6.07 + 0.953 - 0.01 = 4.03 \text{ L/kg}$

$K_{oc} = 1.06 \times 10^4 \text{ L/kg}$
$K_d = 1.06 \times 10^4 \text{ L/kg} \times 0.04 = 425 \text{ L/kg}$

Using the class-specific equation for organophosphates:

$\log K_{oc} = -0.609 \times \log S \text{ (moles/L)} + 0.564$, (Gerstl, 1990)
$\log K_{oc} = (0.609 \times -6.07) + 0.564 = 4.26 \text{ L/kg}$
$K_{oc} = 18200 \text{ L/kg}$
$K_d = 18200 \text{ L/kg} \times 0.04 = 728 \text{ L/kg}$

Both estimates are within 0.56 log units of the log K_{oc} (3.70) value reported in Table 8.7.

Example 2c) Assuming no experimental values of K_{ow} or S are available, estimate K_{oc} for chlorpyrifos (SMILES String: Cl-c(nc(c1Cl)OP(=S)(OCC)OCC)c(Cl)c1) from MCIs using the following expression (Meylan et al., 1992):

$$\log K_{oc} = 0.53 \, {}^1\chi + 0.63 + \Sigma \, P_f N \qquad (7)$$

First, draw out the hydrogen suppressed structure for chlorpyrifos and assign delta values to each of its atoms as shown below. For ${}^1\chi$, the delta value for each atom is equal to the number of atoms it is bonded to.

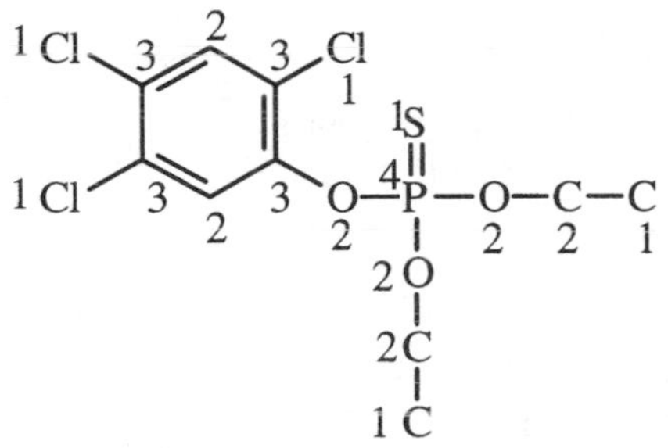

Calculate the ${}^1\chi$ using Equation 2, as follows:

$$
{}^1\chi = \sum \left(\left(3/\sqrt{1\times3}\right) + \left(5/\sqrt{3\times2}\right) \right) + \left(2/\sqrt{3\times3}\right) + \left(2/\sqrt{2\times2}\right)
$$
$$
+ \left(2/\sqrt{2\times2}\right) + \left(1/\sqrt{4\times1}\right) + \left(3/\sqrt{4\times2}\right) \Big)
$$

$${}^1\chi = 8.41$$

Check the table of fragment correction values (Table 8.2) Meylan et al. (1992) provided to see if any appropriate fragments are listed for chlorpyrifos. Since chlorpyrifos is an organohosphorous compound, the [P=S] fragment, having a value of −1.263, should be used.

Plug the values for ${}^1\chi$ and the [P=S] fragment constant into Equation (3) and solve for log K_{oc} as shown below:

$\log K_{oc} = 0.53 \times {}^1\chi + 0.62 + \text{sum (fragments} \times \text{\# of occurrences)}$

$\log K_{oc} = (0.53 \times 8.41) + 0.62 + (1 \times -1.263) = 3.81 \text{ L/kg}$
$K_{oc} = 6456 \text{ L/kg}$

The estimate is within 0.11 log units of the log K_{oc} (3.70) value listed in Table 8.7.

Example 3a) Estimate K_{oc} and K_d for lindane in a soil containing 4% organic carbon using the log K_{ow} value of 3.72 listed in Table 8.7.
Lindane is a chlorinated pesticide with the following structure:

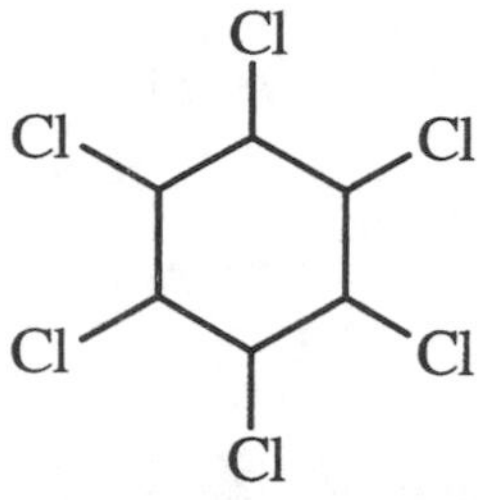

Both general and class-specific equations (chlorinated hydrocarbon) are available for estimating K_{oc} from K_{ow} and appear in Table 8.1.

General equations

$\log K_{oc} = 0.679 \times \log K_{ow} + 0.663$ (Gerstl, 1990)
$\log K_{oc} = (0.679 \times 3.72) + 0.663 = 3.19 \text{ L/kg}$
$K_{oc} = 1.54 \times 10^3 \text{ L/kg}$
$K_d = 1.54 \times 10^3 \times 0.04 = 61.8 \text{ L/kg}$

$\log K_{oc} = 0.81 \times \log K_{ow} + 0.10$ (Sabljic et al., 1995)
$\log K_{oc} = (0.81 \times 3.72) + 0.10 = 3.11$
$K_{oc} = 1.30 \times 10^3$
$K_d = 1.30 \times 10 \text{ L/kg}^3 \times 0.04 = 51.9 \text{ L/kg}$

$\log K_{oc} = 0.903 \log K_{ow} + 0.094$ (Baker et al., 1997)
$\log K_{oc} = 0.903 \times 3.72 + 0.094 = 3.45 \text{ L/kg}$
$K_{oc} = 2.84 \times 10^3 \text{ L/kg}$
$K_d = 2.84 \times 10^3 \text{ L/kg} \times 0.04 = 114 \text{ L/kg}$

Class specific

$\log K_{oc} = 0.904 \log K_{ow} - 0.539$ (Chiou et al., 1983)
$\log K_{oc} = (0.904 \times 3.72) - 0.539 = 2.82 \text{ L/kg}$
$K_{oc} = 667 \text{ L/kg}$
$K_d = 667 \text{ L/kg} \times 0.04 = 26.7 \; 667 \text{ L/kg}$

The class-specific estimate was closer to the log KOC vaule listed in Table 8.7 for lindane (2.96) than the three general equations.

Example 3b) Estimate K_{oc} and K_d for lindane in a soil containing 4% organic carbon using the aqueous solubility value (log S = –6.07 m\ listed in Table 8.7.

Table 8.2 gives both a general and a class specific equation for estimating log K_{oc} from log S:

log K_{oc} = –0.508 log S + 0.953 Fc(Gerstl, 1990)
log K_{oc} = (–0.508 –6.07) + 0.953 – 0= 4.04
K_{oc} = 1.10 × 10^4 L/kg
K_d = 1.10 × 10^4 × 0.04 = 440 L/kg

log K_{oc} = –0.56 log S (µmole/L) (Chiou et al., 1979)
log K_{oc} = –(0.56 × 0.07) + 4.28 = 4.32 667 L/kg
log K_{oc} =2.08 × 10^4 L/kg

The log K_{oc} reported for lindane is 2.96 in Table 8.7.

Example 3c) Assuming no experimental values of K_{ow} or S are available, estimate K_{oc} for lindane (SMILES String: ClC(C(Cl)C(Cl)C1Cl)C(Cl)C1Cl) from MCIs, using the approach Meylan et al. (1992) described.

$$\log K_{oc} = 0.53\ ^1\chi + 0.63 + \Sigma\ P_fN \tag{8}$$

First, draw out the hydrogen suppressed structure for lindane and assign delta values to each of its atoms, as shown below. For $^1\chi$, the delta value for each atom is equal to the number of atoms is bonded to.

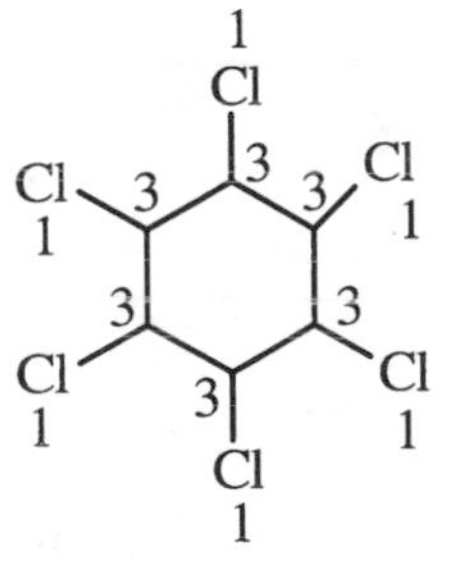

$$^1\chi = \sum \left(\left(6/\sqrt{1\times3}\right) + \left(6/\sqrt{3\times3}\right)\right)$$

$$^1\chi = 5.46$$

Check the table of fragment correction values (Table 8.2) Meylan et al. (1992) provided to see if any appropriate fragments are listed for lindane. No appropriate fragments are listed; therefore $\Sigma\ P_f N = 0$.

Plug the value for $^1\chi$ into Equation (3) and solve for log K_{oc}, as shown below:

log K_{oc} = 0.53 × $^1\chi$ + 0.62 + $\Sigma\ P_f N$
log K_{oc} = 0.53 × 5.46 + 0.62 + 0 = 3.52 L/Kg
K_{oc} = 3.28 × 10^3

Example 4a) Estimate K_{oc} and K_d for 2,6-di(tert-butyl) phenol in a soil containing 4% organic carbon using the K_{ow} value of 5.43 listed in Table 8.7.

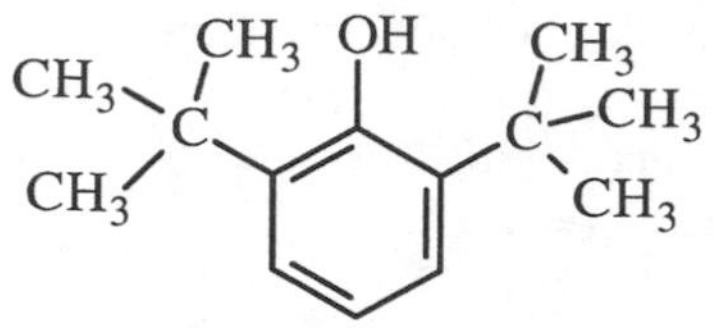

Three general equations for estimating log K_{oc} from log K_{ow} appear in Table 8.1 and below:

log K_{oc} = 0.679 × log K_{ow} + 0.663 (Gerstl, 1990)
log K_{oc} = (0.679 × 5.43) + 0.663 = 4.35 L/kg
K_{oc} = 2.24 × 10^4 L/kg
K_d = 2.24 × 10^4 L/kg × 0.04 = 895 L/kg

Note: K_d is assumed to be a linear sorption coefficient. A non-linear sorption coefficient (K_f) could be calculated if a value for N is known or could be estimated.

log K_{oc} = 0.81 × log K_{ow} + 0.10 (Sabljic et al., 1995)
log K_{oc} = 0.81 × 5.43 + 0.10 = 5.00 L/kg
K_{oc} = 3.15 L/kg
K_d = 234 × 10^4 L/kg × 0.04 = 1260 L/kg

log K_{oc} = 0.903 log K_{ow} + 0.094 (Baker et al., 1997)
log K_{oc} = 0.903 × 5.43 + 0.094 = 5.00 L/kg
K_{oc} = 1 × 10^5 L/kg
K_d = 1 × 10^5 L/kg × 0.04 = 4000 L/kg

Example 4b) Assuming no experimental values of K_{ow} or S are available, estimate K_{oc} for 2,6-di(tert-butyl) phenol from MCIs using the following expression (Meylan et al., 1992):

$$\log K_{oc} = 0.53 \times {}^1\chi + 0.63 + \Sigma\, P_f\, N$$

Calculate the value of ${}^1\chi$ for 2,6-di(tert-butyl) phenol (SMILES string: c1cc(C(C)(C)C)c(O)c(C(C)(C)C)c1)

By hand, you must first draw out its hydrogen suppressed structure and assign delta values to each of its atoms, as shown below. For a first order MCI (1χ), the delta value for each atom is equal to the number of atoms it is bonded to.

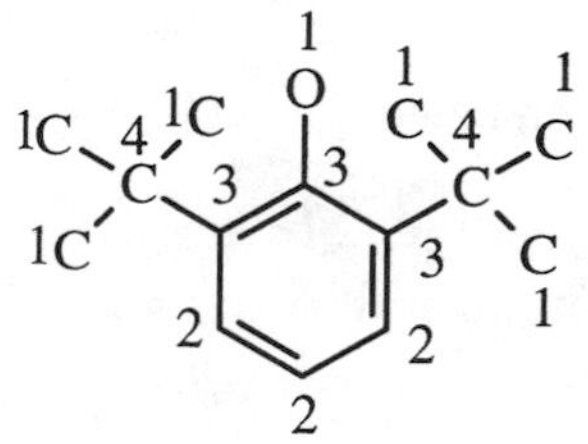

Calculate the 1χ using Equation (2), as follows:

$$^1\chi = \sum \left(\left(6/\sqrt{4 \times 1}\right) + \left(2/\sqrt{4 \times 3}\right) + \left(1/\sqrt{3 \times 1}\right) + \left(2/\sqrt{3 \times 3}\right) + \left(2/\sqrt{3 \times 2}\right) + \left(2/\sqrt{2 \times 2}\right) \right)$$

$$^1\chi = 6.64$$

Check the table of fragment correction values (Table 8.2) Meylan et al. (1992) provided to see if any appropriate fragments are listed for 2,6-di(tert-butyl) phenol. No appropriate fragments are listed; therefore $\Sigma P_f N = 0$.

Plug the value for $^1\chi$ into Equation (3) and solve for log K_{oc} as shown below:

log K_{oc} = (0.53 × 6.64) + 0.62 + 0 = 4.14
$K_{oc} = 1.37 \times 10^4$

Freitag (1982) reported an experimental K_{oc} value of 2600 (log K_{oc} = 3.41).

Example 5a) Estimate K_{oc} and K_d for trichloroethylene in a soil containing 4% organic carbon using the K_{ow} value of 2.53 listed in Table 8.7.

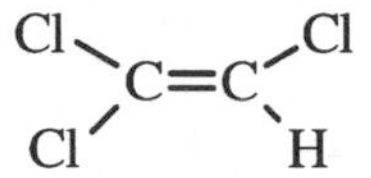

$$\begin{array}{c} Cl \diagdown \\ \diagup C{=}C \diagup \\ Cl \end{array} \begin{array}{c} {\diagup} Cl \\ {\diagdown} H \end{array}$$

Three general equations for estimating log K_{oc} from log K_{ow} appear in Table 8.1 and below:

log K_{oc} = 0.679 × log K_{ow} + 0.663 (Gerstl, 1990)
log K_{oc} = 0.679 × 2.53 + 0.663 = 2.38
K_{oc} = 240 L/kg
K_d = 240 L/kg× 0.04 = 9.61 L/kg

Note: K_d is assumed to be a linear sorption coefficient. A non-linear sorption coefficient (K_f) could be calculated if a value for N is known or could be estimated.

log K_{oc} = 0.81 × log K_{ow} + 0.10 (Sabljic et al., 1995)
log K_{oc} = (0.81 × 2.53) + 0.10 = 2.15 L/kg
K_{oc} = 141 L/kg
K_d = 141 × 0.04 = 5.64 L/kg

log K_{oc} = 0.903 log K_{ow} + 0.094 (Baker et al., 1997)
log K_{oc} = (0.903 x 2.53) + 0.094 = 2.38 L/kg
K_{oc} = 239 L/kg
K_d = 239 × 0.04 = 9.56 L/kg

The average estimated log K_{oc} value is 2.30. This is 0.30 log units higher than the experimental value of 2.00 Seip et al. (1986) reported at 25°C.

Example 5b) Estimate K_{oc} and K_d for trichloroethylene in a soil containing 4% organic carbon using the aqueous solubility value (log S = –158 m) reported in Table 8.7.

Table 8.2 shows no suitable class-specific relationship. However, log K_{oc} can be estimated from log S, using the general equation Gerstl (1990) described.

$log K_{oc} = -0.508 log S + 0.953 - Fc$
$log K_{oc} (-0.508 - 1.88) + 0.953 - = 1.96 L/kg$
$K_{oc} = 91.0 L/kg$
$Kd = 91.0 L/kg ¥ 0.04 = 3.63 L/kg$

The experimental log K_{oc} value Seip et al. (1986) reported is 2.00 at 25°C.

Example 5c) Assuming no experimental values of K_{ow} or S are available, estimate K_{oc} for trichloroethylene from MCIs, using the following expression (Meylan et al. 1992):

$$log K_{oc} = 0.53 \times {}^1\chi + 0.63 + \Sigma P_f N$$

To calculate the value of ${}^1\chi$ for trichloroethylene (SMILES String: ClC(Cl)=CCl), you must first draw out its hydrogen suppressed structure and assign delta values to each of its atoms, as shown below. For 1χ, the delta value for each atom is equal to the number of atoms it is bonded to.

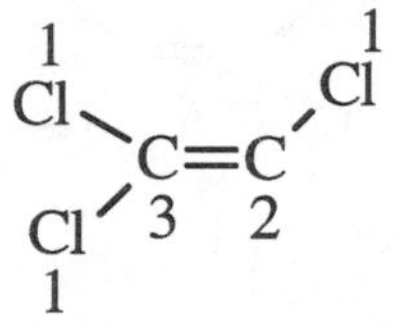

Calculate the 1χ using Equation (2), as follows:

$${}^1\chi = \sum \left(\left(2/\sqrt{1\times3}\right) + \left(1/\sqrt{3\times2}\right) + \left(1/\sqrt{2\times1}\right) \right)$$

$${}^1\chi = 2.27$$

Check the table of fragment correction values (Table 8.2) Meylan et al. (1992) provided to see if any appropriate fragments are listed for trichloroethylene. No appropriate fragments are listed; therefore $\Sigma P_f N = 0$.

Plug the value for ${}^1\chi$ into Equation (3) and solve for log K_{oc}, as shown below:

$log K_{oc} = 0.53 \times {}^1\chi + 0.62 + sum (fragments \times \# of occurrences)$
$log K_{oc} = 0.53 \times 2.27 + 0.62 + 0 = 1.82 L/Kg$
$K_{oc} = 66.5 L/Kg$

The experimental log K_{oc} value Seip et al. (1986) reported is 2.00 at 25°C.

Example 6a) Estimate the K_d for pentachlorophenol in a soil containing 1% organic carbon ($f_{oc} = 0.01$) and having a pH of 7.5. A log K_{ow} value for pentachlorophenol of 5.01 is listed in Table 8.7.

Since pentachlorophenol is a weak acid, and both soil pH and organic carbon content are available, the method Bintein and Devillers (1994) described can be used to estimate K_d. The method uses the following expression:

$$\log K_d = 0.93 \log K_{ow} + 1.09 \log f_{oc} + 0.32 C_{fa} - 0.55 C_{fb'} + 0.25$$

where f_{oc} is the fraction of organic carbon in the soil. For acids, $C_{fa} = \log \dfrac{1}{1+10^{pH-pKa}}$ and C_{fa} is 0 for bases and neutral compounds. Similarly for bases, $C_{fb'} = \log \dfrac{1}{1+10^{pKa-(pH-2)}}$ with C_{fb} is equal to zero for acids and other neutral compounds.

For pentachlorophenol, $C_{fa} = \log \dfrac{1}{1+10^{pH-pKa}} = \log \dfrac{1}{1+10^{7.5-4.75}} = -2.75$

$C_{fb'} = 0$, $\log f_{oc} = -2$
$\log K_d = 0.93\,(5.01) + (1.09 \times -2) + (0.32 \times -2.75) - (0.55 \times 0) + 0.25$
$\log K_d = 4.62 - 2.18 - 0.88 + 0 + 0.25 = 1.85 \text{ L/Kg}$
$K_d = 707 \text{ L/kg}$

Example 7a) Estimate the K_d for quinoline in a soil containing 1% organic carbon ($f_{oc} = 0.01$) and having a pH of 7.5. A log K_{ow} value of 2.0 is listed in Table 8.7 for quinoline.

Since quinoline is a weak base, and both soil pH and organic carbon content are available, the method Bintein and Devillers (1994) described can be used to estimate K_d as shown in the previous example.

For quinoline, $C_{fa} = 0$, $\log f_{oc} = -2$

$C_{fb'} = \log \dfrac{1}{1+10^{pKa-(pH-2)}} = \log \dfrac{1}{1+10^{4.94-(7.5-2)}} = 0.11$
$\log K_d = 0.93\,(2.0) + (1.09 \times -2) + (0.32 \times 0) - (0.55 \times -0.11) + 0.25$
$\log K_d = 1.86 - 2.18 + 0 + 0.06 + 0.25 = 0.010 \text{ L/Kg}$
$K_d = 0.977 \text{ L/kg}$

References

Alexander, M. 1999. *Biodegradation and Bioremediation.* Academic Press, San Diego.

Ames, T.T. and E.A. Grulke. 1995. Group contribution method for predicting equilibria of nonionic organic compounds between soil organic matter and water. *Environ. Sci. Technol.* 29(9):2273-2279.

Anderson, E., G.D. Veith, and D. Wieninger. 1987. SMILES: *A line notation and computerized interpreter for chemical structures.* EPA/600/M-87/021.

ASTM. 1987a. *Standard test method for 24-h batch-type measurement of containment sorption by soils and sediments.* ASTM, D 4646, American Society for Testing and Materials, Philadelphia, PA pp. 120-123.

ASTM. 1987b. *Standard test method for determining a sorption constant (k_{oc}) for an organic chemical in soil and sediments.* ASTM 1195 - 87, American Society for Testing and Materials, Philadelphia, PA, pp. 731-737.

ASTM. 1991. *Annual Book of ASTM Standards, Section 11, Volume 11.04.* American Society for Testing and Materials, Philadelphia, PA.

Bahnick, D.A., and W.J. Doucette. 1988. Use of molecular connectivity indices to estimate soil sorption coefficients for organic chemicals. *Chemosphere* 17(9):1703-1715.

Baker, J.R., J.R. Mihelcic, D.C. Luehrs, and J.P. Hickey. 1997. Evaluation of estimation methods for organic carbon normalized sorption coefficients. *Water Environmental Research* 69(2):136-144.

Bailey, G.W. and J.L. White. 1970. Factors influencing the absorption, desorption, and movement of pesticides of in soils. *Residue Rev.* 32:29-32.

Ball, W.P. and P.V. Roberts. 1991. Long-term sorption of halogenated organic chemicals by aquifer material. 1. Equilibrium. *Environ. Sci. Technol.* 25(7):1223-1237.

Bayard, R., L. Barna, B. Mahjoub, and R. Gourdon. 1998. Investigation of naphthalene sorption in soils and soil fractions using batch and column assays. *Environ. Toxicol. Chem.* 17: 2382-2390.

Benker, E., G.B. Davis, and D.A. Barry. 1998. Estimating the retardation coefficient of trichloroethene for a sand aquifer low in sediment organic carbon—a comparison of methods. *J. Contaminant Hydrol.* 30: 157-178.

Bintein, S. and J. Devillers. 1994. QSAR for organic chemical sorption in soils and sediments. *Chemosphere* 28(6):1171-11188.

Bouchard, D.C., C.G. Enfield, and M.D. Piwoni. 1989. Transport processes involving organic chemicals. In B.L. Sawhney and K. Brown, Eds., *Reactions and Movement of Organic Chemicals in Soils.* Soil Science Society of America, Inc., Madison, pp. 349-371.

Briggs, G.G. 1981. Theoretical and experimental relationships between soil adsorption, octanol-water partition coefficients, water solubilities, bioconcentration factors and the parachor. *J. Agric. Food Chem.* 29:1050-1059.

Brown, D.S., and E.W. Flagg. 1981. Empirical prediction of organic pollutant sorption in natural sediments. *J. Environ. Qual.* 10(3):382-386.

Brownawell, B.J., H. Chen, J.M. Collier, and J.C. Westall. 1990. Adsorption of Organic Cations to Natural Materials. *Environ. Sci. Technol.* 24(8), 1234-1241.

Brusseau, M.L. 1993. Using QSAR to evaluate phenomenological models for sorption of organic compound by soil. *Environ. Toxicol. Chem.* 12:1835-1846.

Brusseau, M.L., and P.S.C. Rao. 1989. Influence of sorbate-organic matter interactions on sorptive nonequilibrium. *Chemosphere* 18:1691-1706.

Brusseau, M.L., R.E. Jessup, and P.S.C. Rao. 1990. Sorption kinetics of organic chemicals – evaluation of gas-purge and miscible-displacement techniques. *Environ. Sci. Tech.* 24(5):727-735.

Brusseau, M.L., R.E. Jessup, and P.S.C. Rao. 1991a. Nonequilibrium sorption of organic chemicals – elucidation of rate-limiting processes. *Environ. Sci. Technol.* 25(1):134-142.

Brusseau, M.L., T. Larsen, and T.H. Christensen. 1991b. Rate-limited sorption and nonequilibrium transport of organic chemicals in low organic carbon aquifer materials. *Water Resour. Res.* 27(6):1137-1145.

Brusseau, M.L., and P.S.C. Rao. 1991a. Influence of sorbate structure on nonequilibrium sorption of organic compounds. *Environ. Sci. Technol.* 25(8):1501-1506.

Brusseau, M.L., and P.S.C. Rao. 1991b. Sorption kinetics of organic chemicals – methods, models and mechanisms. *Rates of Soil Chem. Proces.* 27:281-302.

Burchill, S., M.H.B. Hayes, and D.J. Greenland. 1981. Adsorption. In D.J. Greenland and M.H.B. Hayes, Eds., *The chemistry of soil processes* John Wiley and Sons, Chichester, England, pp. 221-400.

Carski, T.H., and D.L. Sparks. 1985. A modified miscible displacement technique for investigating adsorption-desorption kinetics in soils. *Soil Sci. Soc. Am. J.* 49:1114-1116.

Cheng, H.H. 1990. Organic residues in soils: Mechanisms of retention and extractability. *Intern. J. Environ. Anal. Chem.* 39(2):165-171.

Chiou, C.T. 1989. Theoretical considerations of the partition uptake of nonionic organic compounds by soil organic matter. In B.L. Shawyney and K. Brown, Eds., *Reactions and Movement of Organic Chemicals in Soils,* Soil Science Society of America, Madison, WI.

Chiou, C.T., L.J. Peters, and V.H. Freed. 1979. A physical concept of soil-water equilibria for nonionic organic compounds. *Science* 206:831-832.

Chiou, C.T., P.E. Porter, and D.W. Schmedding. 1983. Partition equilibria of nonionic organic compounds between soil organic matter and water. *Environ. Sci. Technol.* 17(4):227-231.

Chiou, C.T., D.W. Schmedding, and M. Manes. 1982. Partitioning of organic compounds in octanol—water systems. *Environ. Sci. Technol.* 16(1):4-10.

Chiou, C.T., and T.D. Shoup. 1985. Soil sorption of organic vapors and effects of humidity on sorption mechanism and capacity. *Environ. Sci. Technol.* 19:1196-1200.

Davis, J.W. 1993. Physico-chemical factors influencing ethyleneamine sorption to soil. *Environ. Toxicol. Chem.* 27-35.

De Bruijn, J., F. Busser, W. Seinen, and J. Hermens. 1989. Determination of octanol/water partition coefficients for hydrophobic organic chemicals with the "slow-stirring" method. *Environ. Toxicol. Chem.* 8:499-512.

Deeley, G.M., M. Reinhard, and S.M. Stearns. 1991. Transformation and sorption of 1,2-dibromo-3-chloropropane in subsurface samples collected at Fresno, California. *J. Environ. Qual.* 20(3):547-556.

deJonge, H., T.J. Heimovaara, and J.M. Verstraten. 1999. Naphthalene sorption to organic soil materials studied with continuous stirred flow experiments. *Soil Sci. Soc. of Am. J.*

Ditoro, D.M., C.S. Zarba, D.J. Hansen, W.J. Berry, R.C. Swartz, C.E. Cowan, S.P. Pavlou, H.E. Allen, N.A. Thomas, and P.R. Paquin. 1991. Technical basis for establishing sediment quality criteria for nonionic organic chemicals using equilibrium partitioning. *Environ. Toxicol. Chem.* 10(12):1541-1583.

Dobbs, R.A., L. Wang, and R. Govind. 1989. Sorption of toxic organic compounds on wastewater solids: Correlation with fundamental properties. *Environ. Sci. Technol.* 23:1092-1097.

Doucette, W.J., and A.W. Andren. 1988. Estimation of octanol/water partition coefficients: evaluation of six methods for highly hydrophobic aromatic hydrocarbons. *Chemosphere* 17(2):334-359.

Doucette, W.J., and M.S. Holt. 1992. PEP, A microcomputer program for estimating physical/chemical properties. In *Fifth International Workshop on QSAR in Environmental Toxicology (QSAR 92).* Duluth, MN, July 19-23.

Eick, M.J., A. Bar-Tal, D.L. Sparks, and S. Feigenbaum. 1990. Analyses of adsorption kinetics using a stirred-flow chamber: II. Potassium-calcium exchange on clay minerals. *Soil Sci. Soc. Am. J.* 54:1278-1282.

Ellgehousen, H., J.A. Guth, and H.O. Esser. 1980. Factors determining the bioaccumulation potential of pesticides in the individual compartments of aquatic food chains. *Ecotoxicology and Environmental Safety* 4:134-157.

Enfield, C.G. 1985. Chemical transport facilitated by muliphase flow systems. *Water Sci. Technol.* 17:1-12.

Falatko, D.M., and J.T. Novak. 1992. Effects of biologically produced surfactants on the mobility and biodegradation of petroleum hydrocarbons. *Water Environ. Res.* 64(2):163-169.

Fontaine, D.D., R.G. Lehmann, and J.R. Miller. 1991. Soil adsorption of neutral and anionic forms of a sulfonamide herbicide, flumetsulam. *J. Environ. Qual.* 20(4):759-762.

Freitag, D., H. Geyer, A. Kraus, R. Viswanathan, D. Kotzias, A. Attar, W. Klein, and F. Korte. 1982. Ecotoxicological profile analysis. vii. Screening chemicals for their environmental behavior by comparative evaluation. *Ecotox. Environ. Safety* 6:60-81.

Fu, K., and R.G. Luthy. 1986. Effect of organic solvent on sorption of aromatic solutes onto soil. *Journal of Environmental Engineering* 112(2):346-366.

Fugate, H.N. 1989. Using total molecular surface area in quantitative structure activity relationships to estimate environmental fate and transport parameters. Master's thesis, Utah State University, Logan, UT.

Gao, C., R. Govind, and H.H. Tabak. 1996. Prediction soil sorption coefficients of organic chemicals using a neural network model. *Environ. Toxicol. Chem.* 15(7):1089-1096.

Garbarini, D.R., and L.W. Lion. 1985. Evaluation of sorptive partitioning of nonionic pollutants in closed systems by headspace analysis. *Environ. Sci. Technol.* 19:1122-1128.

Garbarini, D.R., and L.W. Lion. 1986. Influence of the Nature of Soil Organics on the Sorption of Toulene and Trichloroethylene. *Environ. Sci. Technol.* 20:1263-1269.

Gauthier, T.D., W.R. Seltz, and C.L. Grant. 1987. Effects of structural and compositional variations of dissolved humic materials on pyrene Koc values. *Environ. Sci. Technol.* 21:243-248.

Gawlik, B.M., N. Sotirious, N. Feicht, S. Schulte-Hostede, and A. Kettrup. 1997. Alternative for the determination of soil adsorption coefficient, k_{oc}, of non-ionicorganic compounds—A review. *Chemosphere* 34(12):2525-2551.

Gerstl, Z. 1990. Estimation of organic chemical sorption by soils. *Journal of Contaminant Hydrology* 6:357-375.

Gerstl, Z., and C.S. Helling. 1987. Evaluation of molecular connectivity as a predicitve method for the adsorption of pesticides by soils. *J. Environ. Sci. Health* B22(1):55-69.

Gerstl, Z., and L. Kliger. 1990. Fractionation of the organic matter in soils and sediments and their contribution to the sorption of pesticides. *J. Environ. Sci. Health B-Pesticides* 25(6):729-741.

Gmehling, J., P. Rasmussen, and A. Fredenslund. 1982. Vapor-liquid equilibria by group contribution. Revision and extension. 2. *Ind. Eng. Chem. Process Des. Dev.* 21:118-127.

Grathwohl, P. 1990. Influence of Organic Matter from soils and sediments from various origins on the sorption of some chlorinated aliphatic hydrocarbons: implications on Koc correlations. *Environ. Sci. Technol.* 24:1687-1692.

Green, R.E., J.M. Davidson, and J.W. Biggar. 1980. An assessment of methods for determining adsorption-desorption of organic chemicals. In A. Banin and U. Kafkefi, Eds., *Agrochemicals in Soils*. New York: Pergamon Press, pp. 73-82.

Green, R.E., and S.W. Karickhoff. 1990. Sorption Estimates for Modeling. In H.H. Cheng, Ed., *Pesticides in the Soil Environment*, Madison, WI: Soil Science Society of America, Inc., pp. 79-101.

Greenland, D.J. and M.H.B. Hayes. 1978. Soils and soil chemistry. In D.J. Greenland and M.H.B. Hayes, Eds., *The Chemistry of Soil Constituents*, John Wiley and Sons, Chichester, England.

Gschwend, P.M. and S.-C. Wu. 1985. On the constancy of sediment-water partition coefficients of hydrophobic organic pollutants. *Environ. Sci. Technol.* 19:90-96.

Guenzi, W.D. 1974. *Pesticides in Soil and Water*. Soil Science Society of America, Madision, WI.

Haky, J.E., and A.M. Young. 1984. Evaluation of a simple hplc correlation method for the estimation of the octanol-water partition coefficients of organic compounds. *Journal of Liquid Chromatography* 7(4):675-689.

Hall, L.H. 1992. *Molconn-X*. Hall Associates Consulting, Quincy, MA.

Hamaker, J.W. and J.M. Thompson. 1972. Adsorption. In C.A.I. Goring and J.W. Hamaker, Eds., *Organic Chemicals in the Soil Environment*, pp. 49-143 Marcel Dekker, New York.

Hance, R.J. 1969. An empirical relationship between chemical structure and the sorption of some herbicides by soils. *J. Agr. Food Chem.* 17(3):667-668.

Hansen, B.G., A.B. Paya-Perez, M. Rahman, and b.R. Larden. 1999. QSARs for K_{ow} and K_{oc} of PCB congeners: a critical examination of data, assumptions, and statistical approaches. *Chemosphere.* 39: 2207-222.

Hassett, J.J., and W.L. Banwart. 1989. The sorption of nonpolar organics by soils and sediments. In B.L. Sawhney and K. Brown, Eds., *Reactions and Movement of Organic Chemicals in Soil.* Soil Science Society of America, Inc, Madison, WI.

Hassett, J.J., W.L. Banwart, S.G. Wood, and J.C. Means. 1981. Sorption of anaphthol: Implications concerning the limits of hydrophobic sorption. *Soil Sci. Soc. Am. J.* 45(1):38-42.

Hassett, J.J., J.C. Means, W.L. Banwart, S.G. Wood, S. Ali, and A. Khan. 1980. Sorption of dibenzothiphene by soils and sediments. *J. Environ. Qual.* 9(2):184-186.

Herbert, B.E., P.M. Bertsch, and J.M. Novak. 1993. Pyrene sorption by water-soluble organic carbon. *Environ. Sci. Technol.* 27(2):398-403.

Hicken, S.T. 1993. Sorption behavior of three pesticides on sorbents with varying clay and organic carbon fractions. Master's thesis, Utah State University, Logan, UT.

Hodson, J., and N.A. Williams. 1988. The estimation of the adsorption coefficient (k_{oc}) for soils by high performance liquid chromatography. *Chemosphere* 17(1):67-77.

Holt, M.S. 1992. Microcomputer program for the estimation of properties for use in environmental fate and transport models. Master's thesis. Utah State University, Logan, UT.

Hong, H., L. Wang, S. Han, and G. Zou. 1996. Prediction of absorption coefficients (K_{oc}) for aromatic compounds by HPLC retention factors (K'). *Chemosphere*, 32(2):343-351.

Huang, P.M., R. Grover, and R.B. McKercher. 1984. Components and the particle size fraction involved in atrazine adsorption by soil. *Soil Sci.* 138:20-24.

Kanazawa, J. 1989. Relationship between the soil sorption constants for pesticides and their physichemical properties. *Environ. Toxicol. Chem.* 8:477-484.

Karickhoff, S.W. 1980. Sorption kinetics of hydrophobic pollutants in natural sediments. In R.A. Baker, ed., *Contaminants and Sediments*, pp. 193-205, Ann Arbor Science, Ann Arbor, MI.

Karickhoff, S.W. 1981. Semi-empirical estimation of sorption of hydrophobic pollutants on natural sediments and soils. *Chemosphere* 10(8):833-846.

Karickhoff, S.W. 1983. Pollutant sorption in environmental systems. (No. EPA-600/D-83-083). Environmental Research Laboratory, U.S. Environmental Protection Agency, July.

Karickhoff, S.W. 1984. Organic pollutant sorption in aquatic systems. *J. Hydraulic Eng.* 110:707-735.

Karickhoff, S.W., D.S. Brown, and T.A. Scott. 1979. Sorption of hydrophobic pollutants on natural sediments. *Water Research*, 13:241-248.

Kenaga, E.E., and C.A.I. Goring. 1980. Relationship between water solubility, soil sorption, octanol-water partitioning, and concentration of chemicals in biota. In J.C. Eaton, P.R. Parrish, and A.C. Hendricks, Eds., *Aquatic Toxicology ASTM STP 707*, pp. 78-115, American Society for Testing and Materials, Philadelphia, PA.

Khan, A., J.J. Hassett, W.L. Banwart, J.C. Means, and S.G. Wood. 1979. Sorption of acetophenone by sediments and soils. *Soil Sci.* 128:297-302.

Kier, L.B. 1980. *Molecular Connectivity as a Description of Structure for SAR Analysis in Physical Chemical Properties of Drugs*. Marcel Dekker, Inc., New York.

Kier, L.B., and L.H. Hall. 1976. Molecular connectivity in chemistry and drug research. In *Medicinal Chemistry, a Series of Monographs*. Academic Press, New York.

Kier, L.B., and L.H. Hall. 1986. *Molecular Connectivity in Structure-Activity Analysis*. Research Studies Press, LTD., Letchworth, Hertfordshire, England, and John Wiley and Sons, Inc., New York.

Kile, D.E., C.T. Chiou, H. Zhou, H. Li, and O. Zu. 1995. Partition of nonpolar organic pollutants from water to soil and sediment organic matters. *Environ. Sci. Technol.* 29: 1401-1406.

Kile, D.E., R.L. Wershaw, and C.T. Chiou. 1999. Correlation of soil and sediment organic matter polarity to aqueous sorption of organic compounds. *Environ. Sci. Technol.* 33: 2053-2056.

Koch, R. 1983. Molecular Connectivity index for assessing ecotoxicological behavior of organic compounds. *Toxicol. Environ. Chem.* 6:87-96.

Kontogeorgis, G.M., A. Fredeslund, and D.P. Tassios. 1993. *Ind. Eng. Chem. Res.* 32:362-372.

Kordel, W., J. Stutte, and G. Kotthoff. 1993. HPLC-screening method for the determination of the adsorption-coefficients in soil-comparison of different stationary phases. *Chemosphere* 27:2341-2352.

Kordel, W., J. Stutte, and G. Kotthoff. 1995. HPLC-screening method for the determination of the adsorption-coefficients on soil-results of a ring test. *Chemosphere* 30:1373-1384.

Koskinen, W.L. and S.S. Harper. 1990. The retention processes: Mechanisms. In *Pesticides in the Soil Environment*, pp. 51-77. Soil Science Society of America, Madison, WI.

Laird, D.A., E. Barriuso, R.H. Dowdy, and W.C. Koskinen. 1992. Adsorption of atrazine on smectites. *Soil Sci. Soc. Amer. J.* 56(1):62-67.

Lambert, S.M. 1967. Functional relationship between sorption in soil and chemical structure. *J. Agr. Food Chem.* 15(4):572-576.

Lee, L.S., P.S.C. Rao, and M.L. Brusseau. 1991. Nonequilibrium sorption and transport of neutral and ionized chlorophenols. *Environ Sci. Technol.* 25(4):722-729.

Lyman, W.J., W.F. Reehl, and D.H. Rosenblatt. 1982. *Handbook of Chemical Property Estimation Methods, Environmental Behavior of Organic Compounds*. McGraw-Hill, Inc., New York.

MacIntyre, W.G., and deFur, P.O. 1985. The effect of hydrocarbon mixtures on adsorption of substituted naphthalenes by clay and sediment from water. *Chemosphere* 14(1):103-111.

MacIntyre, W.G., T.B. Stauffer, and C.P. Antworth. 1991. A comparison of sorption coefficients determined by batch, column, and box methods on a low organic carbon aquifer material. *Ground Water* 29(6):908-913.

Mackay, D., and W.Y. Shiu. 1977. Aqueous solubility of polynuclear aromatic hydrocarbons. *J. Chem. Eng. Data* 22(4):399-402.

Mackay, D., W.Y. Shiu, and K.C. Ma. 1992. *Illustrated Handbook of Physical-Chemical Properties and Environmental Fate for Organic Chemicals. Vol. 1 – 3.* Lewis Publishers, Chelsea, MI.

Maraqua, M.A., X. Zhao, and T.C. Voice. 1998. Retardation coefficients of nonionic organic compounds determined by batch and column techniques. *Soil Sci. of Am. J.*

McDuffie, B. 1981. Estimation of octanol/water partition coefficients for organic pollutants using reverse-phase HPLC. *Chemosphere* 10:73-83.

McGowan, J.C. 1954. The physical toxicity of chemicals. IV. Solubilities, partition coefficients and physical toxicities. *J. Appl. Chem.* 4:41-47.

Meylan, W., P.H. Howard, and R.S. Boethling. 1992. Molecular topology/fragment contribution method for predicting soil sorption coefficients. *Environ. Sci. Technol.* 26(8):1560-1567. PCK-OCWIN software version 1.61, Syracuse Research Corp, Syracuse, NY.

Mihelcic, J.R. and R.G. Luthy. 1991. Sorption and microbial degradation of naphthalene in soil water suspensions under denitrification conditions. *Environ Sci. Technol.* 25(1):169-177.

Miller, G.C., V.R. Hebert, and W.W. Miller. 1989. Effect of sunlight on organic contaminants at the atmosphere-soil interface. In B.L. Sawhney and K. Brown, Eds., *Reactions and Movement of Organic Chemicals in Soils*. Soil Science Society of America, Inc, Madison, WI.

Miller, G.C., V.R. Hebert, and R.G. Zepp. 1987. Chemistry and photochemistry of low-volatility organic chemicals on environmental surfaces. *Environ. Sci. Technol.* 21(12):1164-1167.

Miller, G.C., and R.G. Zepp. 1979. Photoreactivity of aquatic pollutants sorbed on suspended sediments. *Environ. Sci. Technol.* 13:860.

Mingelgrin, U., and Z. Gerstl. 1983. Reevaluation of partitioning as a mechanism of nonionic chemicals adsorption in soils. *J. Environ. Qual.* 12(1):1-11.

Nkedi-Kizza, P. 1985. Influence of organic cosolvents on sorption of hydrophobic organic chemicals by soils. *Environ. Sci. Technol.* 19(10):975-979.

Nkedi-Kizza, P., P.S.C. Rao, and A.G. Hornsby. 1987. Influence of organic cosolvents on leaching of hydrophobic organic chemicals through soils. *Environ. Sci. Technol.* 21:1107-1111.

O'Connor, D.J., and J.P. Connolly. 1980. The effect of concentration of adsorbing solids on the partition coefficient. *Water Res.* 14.

Pearlman, R.S. 1980. Molecular surface areas and volumes and their use in structure/activity relationships. In S.H. Yalkowsky, A.A. Sikula, and S.C. Valvani, Eds., *Physical Chemical Properties of Drugs*, pp. 331-346. Marcel Dekker, Inc., New York.

Perlinger, J.A., S.J. Eisenreich, and P.D. Capel. 1993. Application of headspace analysis to the study of sorption of hydrophobic organic chemicals to alpha-AL2O3. *Environ. Sci. Technol.* 27:928-937.

Pignatello, J.J. 1989. Sorption dynamics of organic compounds in soils and sediments. In B.L. Sawhney and K. Brown, Eds., *Reactions and Movement of Organic Chemicals in Soils*. Soil Science Society of America, Inc., Madison, WI.

Pionke, H.B., and R.J. DeAngelis. 1980. Method for distributing pesticide loss in field runoff between the solution and adsorbed phase. In *CREAMS, A Field Scale Model for Chemicals, Runoff, and Erosion from Agricultural Management Systems*, pp. 607-643. USDA, Washington, DC.

Podoll, R.T., K.C. Irvin, and S. Brendlinger. 1987. Sorption of water soluble oligomers on sediments. *Environ. Sci. Technol.* 21:562-568.

Podoll, R.T., and K.C. Irwin. 1988. Sorption of organic oligomers on sediment. *Environ. Toxicol. Chem.* 7:405-415.

Pussemier, L., R. DeBorger, P. Cloos, and R. VanBladel. 1989. Relation between the molecular structure and the adsorption of arylcarbamate, phenylurea and anilide pesticides in soil and model organic adsorbents. *Chemosphere* 18(9/10):1871-1882.

Pussemier, L., G. Szabo and R.A. Bulman. 1990. Prediction of the soil adsorption coefficient Koc for aromatic pollutants. *Chemosphere* 21:1199-1212.

Quayle, O.R. 1953. *The parachors of organic compounds. An interpretation and catalogue.* Department of Chemistry, Emory University, GA.

Randic, M. 1975. On characterization of molecular branching. *J. Amer. Chem. Soc.* 97(23):6609-6615.

Rao, P.S.C., and J.M. Davidson. 1980. Estimation of pesticide retention and transformation parameters required in non-point source pollution models. In M.R. Overcashand and J.M. Davidson, Eds., *Environmental Impact of Nonpoint Source Pollution*, pp. 23-67. Ann Arbor Sci. Publ., Ann Arbor, MI.

Rao, P. S. C., and J. M. Davidson. 1982. *Retention and transformation of selected pesticides and phosphorus in soil-water systems: A critical review.* U.S. EPA 600/3-82-060.

Rao, P.S.C., L.S. Lee, and R. Pinal. 1990. Cosolvency and sorption of hydrophobic organic chemicals. *Environ. Sci. Technol.* 24(5):647-654.

Rutherford, D.W., C.T. Chiou, and D.E. Kile. 1992. Influence of soil organic matter composition on the partition of organic compounds. *Environ. Sci. Technol.* 26:336-340.

Ruthven, D.M. 1984. Physical adsorption and the characterization of porous adsorbents. In *Principles of Adsorption and Adsorption Processes*, pp. 29-61. John Wiley and Sons, New York.

Sabljic, A. 1984. Predictions of the nature and strength of soil sorption of organic pollutants by molecular topology. *J. Agric. Food Chem.* 32:243-246.

Sabljic, A. 1987. On the prediction of soil sorption coefficients of organic pollutants from molecular structure: application of molecular topology model. *Environ. Sci. Technol.* 21:358-366.

Sabljic, A., and D. Horvatic. 1993. A computer program for calculating molecular connectivity indices on microcomputers. *J. Chem. Inf. Comput. Sci.* 33:292-295.

Sabljic, A., H. Gusten, H. Verhaar, and J. Hermans. 1995. QSAR modeling of soil sorption, improvements and systematics of log K_{oc} vs. log K_{ow} correlations. *Chemosphere* 31(11/12):4489-4514.

Sacan, M.T., and I.A. Balcioglu. 1996. Prediction of the soil sorption coefficient of organic pollutants by the characteristic root index model. *Chemosphere* 32(10):1993-2001.

Sacan, T., and Y. Inel. 1995. Application of the characteristic root index model to the estimation of octanol water partition coefficients: Polychlorinated biphenyls. *Chemosphere* 30:30-39.

Santana-Casiano, J.M. and M. Gonzalez-Davila. 1992. Characterization of the sorption and desorption of lindane to chitin in seawater using reversible and resistant components. *Environ. Sci. Technol.* 26(1):90-95.

Santos-Buelga, M.D., M.J. Sanchez-Martin, and M. Sanchez-Camazano. 1992. Effect of dissolved organic matter on the adsorption of ethofumesate by soils and their components. *Chemosphere* 25(5):727-734.

Sarna, L.P. 1984. Octanol-water partition coefficients of chlorinated dioxins and dibenzofurans by reversed-phase HPLC using several C18 columns. *Chemosphere* 13(9):975-983.

Schellenberg, K., C. Leuneberger, and R.P. Schwarzenbach. 1984. Sorption of chlorinated phenols by natural sediments and aquifer materials. *Environ. Sci. Technol.* 18:652-657.

Schnitzer, M., and S.U. Khan. 1972. *Humic Substances in the Environment*. Marcel Dekker, New York.

Schwarzenbach, R.P., and J. Westall. 1981. Transport of nonpolar organic compounds from surface water to groundwater. Laboratory sorption studies. *Environ. Sci. Technol.* 15(11):1360-1367.

Schwarzenbach, R.P., P.M. Gschwend, and D.M. Imboden. 1993. *Environmental Organic Chemistry*, p. 13. Wiley-Interscience, New York.

Scow, K.M. and M. Alexander. 1992. Effect of diffusion on the kinetics of biodegradation – experimental results with synthetic aggregates. *Soil Sci. Soc. Amer. J.* 56(1):128-134.

Scow, K.M. and C. Johnson. 1997. Effection of sorption on biodegradation of soil pollutants. In D.L. Sparks, Ed., *Advance in Agronomy*. Academic, San Diego, pp. 1-56.

Seip, H.M., J. Alstad, G.E. Carlberg, K. Martinsen, and R. Skaane. 1986. Measurement of mobility of organic compounds in soils. *The Science of the Total Environment* 50 (1986):87-101.

Seth, R., D. Mackay, and J. Muncke. 1999. Estimating the organic carbon partition coefficient and its variability for hydrophocbic chemicals. *Environ. Sci. Technol.* 33: 2390-2394.

Shelton, D.R., and R.B. Parkin. 1991. Effect of moisture on sorption and biodegradation of carbofuran in soil. *J. Agr. Food Chem.* 39(11):2063-2068.

Shimizu, Y., S. Yamazaki, and Y. Terashima. 1992. Sorption of anionic pentachlorophenol (PCP) in aquatic environments: The effect of pH. *Wat. Sci. Tech.* 25(11):41-48.

Singh, G., W.F. Spencer, M.M. Cliath, and M.T. van Genuchten. 1990. Sorption behavior of s-triazine and thicarbamates on soils. *J. Environ. Qual.* 19:520-525.

Smith, E.H. 1991. Modified solution of homogeneous surface diffusion model for adsorption. *J Environ. Eng-ASCE* 117(3):320-338.

Somasundaram, L., J.R. Coats, K.D. Racke, and V.M. Shanbhag. 1991. Mobility of pesticides and their hydrolysis metabolites in soil. *Environ. Toxicol. Chem.* 10(2):185-194.

Sposito, G. 1984. *The Surface Chemistry of Soils*. Oxford University Press, New York.

Stevenson, F.J. 1982. Role and function of humus in soil with emphasis on adsorption of herbicides and chelation of microorganisms. *Bioscience* 22:643-650.

Su, P.-H. 1994. Sorption studies of atrazine using continuous stirred-flow and batch techniques. Master's thesis. Utah State University, Logan, UT.

Swanson, R.A., and G.R. Dutt. 1973. Chemical and physical processes that affect atrazine and distribution in soil systems. *Soil Sci. Soc. Am. Proc.* 37:872-876.

Szabo, G., S.L. Prosser, and R.A. Bulman. 1990a. Adsorption coefficient (K$_{oc}$) and HPLC retention factors of aromatic hydrocarbons. *Chemosphere* 21(4-5):495-505.

Szabo, G., S.L. Prosser, and R.A. Bulman. 1990b. Prediction of the adsorption coefficient (K$_{oc}$) for soil by a chemically immobilized humic acid column using RP-HPLC. *Chemosphere* 21(6), 729-739.

Syracuse Research Corporation (SRC). 1997. PHYSPROP(c) Database, Sept 1997 (ISIS/Base version for IBM-PC, Windows operating system). Syracuse, NY: Syracuse Research Corp.

Talbert, R.E. and O.H. Fletchall. 1965. The adsorption of some s-triazines in soils. *Weeds* 46-52.

Tate, R.L. 1987. *Soil Organic Matter: Biological and Ecological Effects*. John Wiley & Sons, New York.

Tao, S., H.S. Piao, R. Dawson, X.X. Lu, and H.Y. Hu. 1999. Estimation of organic carbon normalized sorption coefficient (K-oc) for soils using the fragment constant method. *Environ. Sci. and Technol.*

ten Hulscher, E.M., and G. Cornelissen. 1996. Effect of temperature on sorption equilibrium and sorption kinetics of organic micropollutants – A Review. *Chemosphere* 32(4):609.

Tewari, Y.B., D.E. Martire, S.P. Wasik, and M.M. Miller. 1982. Aqueous solubilities and octanol-water partition coefficients of binary liquid mixtures of organic compounds at 25 C. *J. Solution Chemistry* 11(6):435-445.

Theng, B.K.G. 1974. *The Chemistry of Clay-Organic Reactions*. John Wiley and Sons, New York.

van Gestel, C.A.M., W.C. Ma, and C.E. Smit. 1991. Development of QSARs in terrestrial ecotoxicology – earthworm toxicity and soil sorption of chlorophenols, chlorobenzenes and dichloroaniline. *Sci. Total Envir.* 109(DEC):589-604.

VanHoof, P.L., and A.W. Andren. 1990. Partitioning and sorption kinetics of a PCB in aquarous suspensions of model particles: Solids concentration effect. In *Organic Substances and Sediments in Water*.

Voice, T.C., C.P. Rice, and W.J. Weber Jr. 1983. Effect of solids concentration on the sorptive partitioning of hydrophobic pollutants in aquatic systems. *Environ. Sci. Technol.* 17(9):513-518.

Von Oepen, B., W. Kordel, W. Klein, and G. Schuurmann. 1991. Predictive QSPR models for estimating soil sorption coefficients: Potential and limitation based on dominating processes. *The Science of the Total Environment*, 109/110:343-354.

Vowles, P.D., and R.F.C. Mantoura. 1987. Sediment-water partition coefficients and hplc retention factor of aromatic hydrocarbons. *Chemosphere* 16(1):109-116.

Walters, R.W., and A. Guiseppi-Elle. 1988. Sorption of 2,3,7,8-tetrachlorodibenzo-p-dioxin from water/methanol mixtures. *Environ. Sci. Technol.* 22(7):819-825.

Webb, O.F., T.J. Phelps, P.R. Bienkowski, P.M. Digrazia, G.D. Reed, B. Applegate, D.C. White, and G.S. Sayler. 1991. Development of a differential volume reactor system for soil biodegradation studies. *Appl. Biochem. Biotechnol.* 28-9(SPR):5-19.

Weber, W.J., P.M. McGinley, and L.E. Katz. 1991. Sorption phenomena in subsurface systems – concepts, models and effects on contaminant fate and transport. *Water Res.* 25(5):499-528.

Weber, W.J., P.M. McGinley, and L.E. Katz. 1992. A distributed reactivity model for sorption by soils and sediments-1: Conceptual basis and equilibrium assessments. *Environ. Sci. Technol.* 26:1955-1962.

Weininger, D. 1988. SMILES, a chemical language and information system. 1. Introduction to methodology and encoding rules. *J. Chem. Inf. Comput. Sci.*, 28:31-36.

Weissenfels, W.D., H.J. Klewer, and J. Langhoff. 1992. Adsorption of polycyclic aromatic hydrocarbons (PAHs) by soil particles — influence on biodegradability and biotoxicity. *Appl. Microbiol. Biotechnol.* 36(5):689-696.

Wood, A.L., D.C. Bouchard, M.L. Brusseau, and P.S.C. Rao. 1990. Cosolvent effects on sorption and mobility of organic contaminants in soils. *Chemosphere* 21(4-5):575-587.

Wu, S., and P.M. Gschwend. 1986. Sorption kinetics of hydrophobic organic compounds to natural sediments and soils. *Environ. Sci. Technol.* 20:717-725.

Xing, B., J.J. Pignatello, and B. Gigliotti. 1996. Competitive sorption between atrazine and other organic compounds in soils and model sorbents. *Environ. Sci. Technol.* 31: 2442.

Xing, B. and J.J. Pignatello. 1998. Competitive sorption between 1,3-dichlorobenzene or 2,d-dichlorobenzene and natural aromatic acids in soil organic matter. *Environ. Sci. Technol.* 32: 6j14-619.

Xu, F., X.M. Liang, B.C. Lin, F. Su, K.W. Schramm, and A. Kettrup. 1999. Soil column chromatography for correlation between capacity factors and soil organic partition coefficients for eight pesticides. *Chemosphere* 39: 2239-2248.

9

Bioconcentration and Biomagnification in the Aquatic Environment

Frank A.P.C. Gobas and
Heather A. Morrison

CONTENTS

1-56670-456-1/00/$0.00+$.50
© 2000 by CRC Press LLC

9.1 Introduction

Environmental toxicology is based on the premise that the amount of chemical reaching an active site in, or on the surface of, an organism determines whether a compound will exert a specific beneficial or adverse effect or pose an ecological or health risk. This basic toxicological principle is embodied in the expression attributed to Paracelsus that it is "the dose that makes the poison."

To interpret the toxicological responses of organisms, it is necessary to understand the relationship between the concentration of chemical in environmental media (e.g., water, sediment, or air) and the concentration of chemical in an organism, or in the target organ(s) of an organism, in contact with that media. In environmental toxicology literature, environment-organism concentration relationships often are discussed in terms of bioaccumulation, bioconcentration, and biomagnification and quantified in the form of bioconcentration factors, bioaccumulation factors, and biomagnification factors. These factors can be multiplied by the ambient concentration in the medium, for which the factor was derived, to arrive at the internal concentration in the organism, which through a dose-response relationship can be related to a toxicological response. Hence, information about the bioaccumulation behavior of substances is a crucial first step in determining the toxicological impact of the substance.

For regulatory purposes, bioconcentration and bioaccumulation factors often are used to assess the potential of new and existing substances for uptake in biological organisms and to derive of environmental quality criteria and standards. For example, in regulations under the Canadian Environmental Protection Act (1995), bioaccumulation factors play an important role in decisions related to the virtual elimination of discharges of certain substances, chemical ranking, and the derivation of water quality guidelines. Reliable information about the bioaccumulation behavior of chemical substances is therefore of great importance in making sound environmental quality management decisions.

This chapter summarizes the current state of knowledge about the bioaccumulation, bioconcentration, and biomagnification of hydrophobic organic chemicals in aquatic organisms, including an overview of:

(i) terminology and scientific definitions

(ii) the mechanism of bioaccumulation

(iii) experimental methods to determine bioconcentration and bioaccumulation factors

(iv) methods for the derivation of bioconcentration factors and other bioconcentration kinetic data from laboratory bioconcentration tests

(v) methods to assess the bioaccumulation of organic chemical substances under field conditions, including sample calculations

(vi) key bioaccumulation texts and databases

The aim is to provide a summary of the most relevant information about bioaccumulation of chemical substances and to provide references to other publications for more detail.

9.2   Definitions

9.2.1   Bioaccumulation

This is the process by which the chemical concentration in an aquatic organism achieves a level that exceeds that in the water, as a result of chemical uptake through all possible routes of chemical exposure (e.g., dietary absorption, transport across the respiratory surface, dermal absorption, inhalation).

Bioaccumulation takes place under field conditions. It is a combination of chemical bioconcentration and biomagnification.

9.2.2   Bioaccumulation Factor

The extent of chemical bioaccumulation usually is expressed in the form of a bioaccumulation factor (BAF), which is the ratio of the chemical concentrations in the organism (C_B) to those in water (C_W):

$$BAF = C_B/C_W \tag{1}$$

Because chemical sorption to particulate and dissolved organic matter in the water column can reduce substantially the fraction of chemical in water that can be absorbed by aquatic organisms (see discussion on bioavailability), the BAF also can be expressed in terms of the freely dissolved chemical concentration (C_{WD}):

$$BAF = C_B/C_{WD} \tag{2}$$

Defining BAF this latter way keeps it independent of the concentrations of particulate and dissolved organic matter in the water phase and thus makes it more universally applicable from site to site.

The chemical concentration in the organism usually is expressed in units of mass of chemical per kg of organism, whereas the concentration in water is expressed in mass per litre. The weight of the organism can be expressed on a wet weight (WW), dry weight (DW), or lipid weight (LW) basis.

Most commonly, the weight of the organism is expressed on a wet weight basis, and the units of the BAF are L/kg. However, when concentration measurements are made in specific tissues of the organism (rather than the whole organism), it is preferable to report the concentration on a lipid weight basis, as organs and tissues can vary substantially in their lipid content. The lipid content is an important factor controlling the extent of chemical bioaccumulation of organic substances.

9.2.3   Bioavailability

Definitions of chemical bioavailability vary widely among environmental chemists, pharmacologists, physiologists, and ecologists. In this chapter, bioavailability of a chemical substance in a particular environmental media such as water, sediment, and food is defined as "the fraction of chemical in a medium that is in a state which can be absorbed by the organism." Bioavailability usually is expressed as a fraction or a percentage and is specific to the medium and route of exposure.

9.2.4 Bioconcentration

The process in which the chemical concentration in an aquatic organism exceeds that in water as a result of exposure to waterborne chemical. Bioconcentration refers to a condition, usually achieved under laboratory conditions, where the chemical is absorbed only from the water via the respiratory surface (e.g., gills) and/or the skin.

9.2.5 Bioconcentration Factor

Bioconcentration can be described by a bioconcentration factor (BCF), which is the ratio of the chemical concentration in an organism (C_B) to the concentration in water (C_W):

$$BCF = C_B/C_W \qquad (3)$$

The BCF, like the BAF, also can be expressed in terms of the dissolved chemical concentration (CWD):

$$BCF = C_B/C_{WD} \qquad (4)$$

9.2.6 Biomagnification

This is the process in which the chemical concentration in an organism achieves a level that exceeds that in the organism's diet, due to dietary absorption. The extent of chemical biomagnification in an organism is best determined under laboratory conditions, where organisms are administered diets containing a known concentration of chemical, and there is no chemical uptake through other exposure routes (e.g., respiratory surface, dermis).

   Biomagnification also can be determined under field conditions, based on chemical concentrations in the organism and its diet. Biomagnification factors derived under controlled laboratory conditions, which exclude uptake through routes other than the diet, are different from those determined under field conditions, because field-based biomagnification factors are inevitably the result of chemical uptake by all routes of chemical uptake, rather than dietary absorption alone.

9.2.7 Biomagnification Factor

Biomagnification can be described by a biomagnification factor (BMF), which is the ratio of the chemical concentration in the organism to the concentration in the organism's diet:

$$BMF = C_B/C_D \qquad (5)$$

The chemical concentration in the organism (C_B) and the diet of the organism (C_D) usually are expressed in units of mass of chemical per kg of the organism and mass chemical per kg of food, respectively. Again, the weight of the organism and food can be expressed on a wet weight (WW), dry weight (DW), or lipid weight (LW) basis. Most commonly, the weight of the organism is expressed on a wet weight basis.

9.2.8   Biotransformation

Biotransformation is the process by which chemical substances undergo chemical or bio-chemical reactions in organisms. The rate of transformation usually is expressed in terms of a rate constant or half life.

9.2.9   Biota-Sediment Bioaccumulation Factor (BSAF and BSAF')

The biota-sediment bioaccumulation factor describes bioaccumulation in sediment dwell-ing organisms and fish relative to chemical concentrations in sediment. It is the ratio of chemical concentration in an organism to that in the sediments in which the organism resides. This ratio is usually expressed in one of two ways:

$$\text{BSAF} = \frac{C_B}{C_S} \tag{6}$$

or

$$\text{BSAF}' = \frac{C_B}{C_S} \cdot \frac{\text{TOC}}{L} \tag{7}$$

where C_B is the chemical concentration in the organism (g chemical/kg organism), C_S is the chemical concentration in the sediment (g chemical/kg dry weight sediment), L is the con-centration of lipid in the organism (g lipid/g organism), and TOC is the total organic car-bon content in the sediments (g organic carbon/g dry weight sediment).

Because hydrophobic organic chemicals preferentially bioaccumulate in the lipid tissues of organisms and the organic matter fraction of sediments, the BSAF (expressed in units of kg sediment/kg organism) depends on the lipid content of the organism and the organic carbon content of the sediments. Lipid and organic matter fractions can differ among sites, types of organisms, and individuals of one species of organism. The BSAF' (expressed in units of kg organic carbon/kg lipid) is more universal in its application because it accounts for differences in lipid content between organisms and the organic carbon content of sedi-ments. Both BSAFs are the result of chemical uptake from all routes of exposure.

9.2.10 Dietary Uptake Efficiency or Chemical Assimilation Efficiency

This is the fraction of ingested chemical actually absorbed by the organism via the gastro-intestinal tract. It usually is expressed in terms of a unitless fraction of doses, eg., g/day:

$$E_D = \text{absorbed dose/administered dose} \tag{8}$$

9.2.11 Equilibrium

Chemical equilibrium is achieved when chemical is distributed among environmental media (including organisms) according to the chemical's physico-chemical partitioning behavior. Thermodynamically, an equilibrium is defined as "a condition where the chemi-cal's potentials (also chemical activities and chemical fugacities) are equal in the environ-mental media." At equilibrium, chemical concentrations in static environmental media remain constant over time.

9.2.12 Food-Chain Bioaccumulation

Food-chain bioaccumulation is the process in which chemical concentrations in organisms increase with each step in the food-chain, resulting in chemical concentrations in predators that are greater than those in their prey. Because concentrations of many hydrophobic organic chemicals in organisms increase as the lipid content of the organism increases, the occurrence of food-chain bioaccumulation is detected best by comparing chemical concentrations in predators and prey on a lipid weight basis. An increase in lipid-based concentrations in organisms with increasing trophic level indicates food-chain bioaccumulation.

9.2.13 Food-Chain Multiplier

This factor is applied to bioconcentration factors to account for chemical biomagnification and bioaccumulation in the food web. The US-EPA uses it to derive bioaccumulation factors of very hydrophobic organic chemicals.

9.2.14 Half-Life or Half-Time

The time (hours, days or years) required for the chemical concentration in a medium to be reduced by half. If the elimination rate involves transport and transformation processes that follow first order kinetics, the half-life time is related to the total elimination rate constant k by 0.693/k.

9.2.15 Octanol-Water Partition Coefficient

Because 1-octanol is a good surrogate phase for lipids in biological organisms, the octanol-water partition coefficient, a ratio of concentrations in 1-octanol and water, represents how a chemical would thermodynamically distribute between the lipids of biological organisms and water. It further represents the lipophilicity and the hydrophobicity of the chemical substance. It usually is referred to as K_{ow} or P, or in its 10-based logarithmic form as log K_{ow} or log P, and is unitless. For more detail see Chapter 5.

9.2.16 Rate Constant

Rate constants describe the fraction of the total chemical mass or concentration in a particular medium or organism that is transported from and/or transformed per unit of time. It has units of 1/day or 1/hour or 1/year.

9.2.17 Rates of Uptake and Elimination

The rates of uptake (or elimination) are the amount of chemical (in grams or moles) absorbed (or eliminated) by the organism per unit of time.

9.2.18 Steady-State

This situation applies when the total flux of chemical into an organism equals the total flux out with no net change in mass or concentration of the chemical. Steady-state differs from equilibrium in that it is achieved as a result of a balance of transport and transformation processes acting upon the chemical, whereas an equilibrium is the end result of a physical-chemical partitioning process.

9.3   Theoretical Background: Mechanisms Controlling Bioaccumulation

This section summarizes the current state of knowledge about the mechanisms controlling the bioaccumulation of organic substances in aquatic organisms. It focuses on a set of general mechanisms that apply to the bioaccumulation of many organic chemicals in many aquatic organisms. It does not discuss bioaccumulation for individual organisms or classes of organisms because numerous aquatic species are of interest.

Most studies show that bioaccumulation predominantly results from chemical absorption directly from water via the respiratory surface (e.g., gills and/or skin) of the organism (i.e., bioconcentration) and from diet via the gastro-intestinal tract (i.e., biomagnification). Exceptions occur; e.g., phytoplankton species generally do not absorb chemical by dietary ingestion, organisms obtain chemical via maternal transfer, and some species contain unique mechanisms for chemical uptake and elimination.

9.3.1   Bioavailability

The term *bioavailability* is used frequently by environmental toxicologists to express the extent of bioaccumulation of chemicals from environmental matrices (e.g., water, sediment). Differences in bioaccumulation factors often are explained by differences in bioavailability. Here, bioavailability is defined more narrowly as "the fraction of the total concentration, in a specific medium or matrix, that can be absorbed by the organisms via a specific route of intake." Bioavailability here does not refer to the rate or the extent to which the chemical is absorbed and accumulated.

The merit of this definition is that it distinguishes factors controlling the toxicokinetics of the substance (i.e., the rates of uptake, elimination, transformation, and internal distribution) from environmental factors (external to the organism) that control the amount of chemical that can be absorbed by the organism through a specific route of uptake.

9.3.1.1   *Sorption to Particulate and Dissolved Organic Matter*

Chemical substances can cross biological membranes via simple molecular diffusion, facilitated diffusion, mediated transport (Stein, 1981) or a combination of these processes. Facilitated diffusion and mediated transport occur for chemical substances such as ions and nutrients that play a specific role in the organism's physiology. Most environmentally relevant organic substances tend to permeate through biological membranes by simple molecular diffusion due to their lipophilic nature. This process requires that the chemical is in a form that allows individual molecules to move "freely," i.e., as single molecules, through the membrane.

If the molecule is attached or associated with a particle (i.e., "adsorbed" to the outer surface of the particle or "sorbed" into particle's interior matrix), then it is believed to be too large to permeate through membranes. The fraction of total chemical in water that is freely dissolved is referred to, therefore, as the fraction of bioavailable chemical or, simply, the chemical's bioavailability in the water. Particle associated chemical can be absorbed through other routes of uptake and hence become available. Also, particle associated chemical may become available following desorption. For hydrophobic organic chemicals, **sorption** to particulate organic matter (POM) and dissolved organic matter (DOM) in the water phase is the predominant process controlling the fractions of freely dissolved and sorbed chemical in the water phase (e.g., Servos et al., 1989; Black and McCarthy, 1988).

Particulate organic matter (POM) usually refers to organic matter that is present in particles with a diameter exceeding a certain diameter (0.45 μm frequently is used). Organic matter present in truly dissolved form or in particles with a diameter less than 0.45 μm usually is defined as *dissolved organic matter* (DOM). Definitions regarding POM and DOM are a matter of convention and are based on the ability to separate particulate matter from the water column by filtration or centrifugation. Sorption, which is discussed in more detail elsewhere (see Chapters 7 and 8), refers to the process where chemical dissolves into and onto the organic matter of particulate matter in the water column.

Many hydrophobic organic chemicals exhibit a high affinity for the organic fraction of suspended **particulate matter** in the water column. The affinity of organic chemicals for the mineral matter fraction is believed to be low; hence, this fraction contains an insignificant amount of chemical. Chemical sorption to the organic fraction of waterborn particles can be viewed as a partitioning process between water and the organic fraction of the particulate matter. The extent of chemical partitioning in waterborne particulate matter can be expressed by the sorption coefficient K_p, which has units of L water/kg suspended matter. Because the sorption of hydrophobic organic chemicals largely reflects partitioning into the organic fraction of the particle, K_p can be related and expressed in terms of an organic carbon based sorption coefficient K_{oc} (L water/kg organic carbon), which, in turn, can be correlated with K_{ow}:

$$K_p = C_p/C_{WD} = OM.K_{OM} = OM.\varphi_{OC}.K_{OC} = \phi_{OC}.K_{OC} \approx \phi_{OC \cdot 0.41.Kow} \qquad (9)$$

In Equation 9, C_p is the chemical concentration in the particulate matter (g/kg particle), C_{WD} is the freely dissolved chemical concentration in the water (g/L water), OM is the fraction of organic matter in the particulate matter (kg organic matter/kg particle), K_{OM} is the sorption coefficient of the chemical to the organic matter content of the particulate matter (L water/kg organic matter), φ_{oc} is the fraction of organic carbon in the organic matter (kg organic carbon/kg organic matter), and ϕ_{oc} is the fraction of organic carbon in the particulate matter (kg organic carbon/kg particulate matter).

Much research has been devoted to determining the bioavailability of organic chemicals in the presence of **dissolved organic matter (DOM)**. In experiments involving commercially produced humic acids, such as Aldrich humic acids, reported chemical-specific, organic-carbon-based sorption coefficients (K_{OC}) range between 0.1 of K_{ow} for a given compound, to 1.0 of K_{ow}, suggesting that commercially extracted humic acids behave like particulate organic matter with an "octanol-like" composition (McCarthy and Jimenez, 1985; Carter and Suffet, 1982; Landrum et al., 1984; Hassett and Milicic, 1985; Yin and Hassett, 1986; Chiou et al., 1987; Landrum et al., 1985; Resendes et al., 1992).

Laboratory and field experiments with naturally occurring dissolved organic matter show that the chemical binding to natural dissolved organic matter, as K_{OC} expresses, can be considerably less than that for commercially available dissolved organic matter. For example, Landrum et al. (1984) found that K_{OC} values for benzo-a-pyrene and p,p'-DDT are approximately one order of magnitude lower than the K_{OC} for Aldrich Humic Acids, and Evans (1988) showed that K_{OC} for natural dissolved organic matter from different lakes is approximately 2% of K_{OW}. Eadie et al. (1990) found similar results in the Great Lakes.

In addition, field studies show considerable variations in the partition coefficients for several organic chemicals to dissolved organic matter from different sources, although the concentration of dissolved organic matter in the different waters was similar (Evans, 1988;

Eadie et al., 1990; Moorehead, 1986). These findings agree with the observations of Gauthier et al. (1987), who found that chemical-specific K_{OC} values could vary by as much as a factor of 10 depending on the type of humic materials used in the measurement.

These results indicate that the structure and composition of the humic materials are important factors controlling the binding of hydrophobic organic chemicals to dissolved organic matter in the water phase. Experimental difficulties associated with the separation of particulate and freely dissolved organic matter probably also contribute to reported differences in the sorptive capacities among dissolved organic carbon fractions found in the environment (Gobas and Zhang, 1994).

9.3.1.2   *Ionizing Hydrophobic Organic Chemicals*

Organic chemicals that are weak acids or bases may dissociate in water, causing part of the chemical to be in an ionic form. The extent to which the chemical substance is in the ionic form is expressed by the dissociation constant K_d. For example, a weak acid like pentachlorophenol (C_6Cl_5OH) can dissociate to form the pentachlorophenate ion ($C_6Cl_5O^-$) as follows:

$$C_6Cl_5OH + H_2O \Leftrightarrow C_6Cl_5O^- + H_3O^+ \tag{10}$$

where the acid dissociation constant K_d is given by:

$$K_d = [C_6Cl_5O^-]\,[H_3O^+]/[C_6Cl_5OH] \tag{11}$$

If only the non-ionic fraction can be absorbed via the respiratory surface, it follows that the fraction of chemical that is available for uptake is:

$$C_N/C_T = 1/(1 + K_d/[H_3O^+]) = 1/(1 + 10^{(pH-pK_d)}) \tag{12}$$

where C_N is the concentration (mol/L) of the non-ionic species in the water, C_T is the concentration (mol/L) of the non-ionic and ionic species, K_d is the dissociation constant and $[H_3O^+]$ can be deduced from the pH, i.e. $[H_3O^+]$ equals 10^{-pH}. However, there is evidence in fish species that this may not always apply.

Saarikoski et al. (1986) found that Equation (12) gives a satisfactory representation of the fraction of weak acids that can be absorbed by fish at the lower pH levels when most of the chemical is in the easily absorbable non-ionic form. However, with increasing pH, the extent of chemical bioavailability is underestimated (Saarikoski et al., 1986; Erickson and McKim, 1990a). At very high pH levels, when most of the chemical is in the ionic form, the extent of underestimation of the chemical's bioavailability can be as large as an order of magnitude or more.

McKim and Erickson (1991) proposed three explanations to explain this phenomena. First, it is possible that, when the non-ionic chemical is absorbed at the gill surface, rapid acid-base equilibration results in the transformation of ionized compound into non-ionized chemical which can be further absorbed. Second, respiration causes the pH in the gill water near the gills to drop, causing more of the ionic chemical to be converted to the non-ionic form. Third, the chemical in the ionic form may be taken up through the gills but at a much lower rate than the non-ionic form. McKim and Erickson (1991) showed that incorporating any of these three scenarios in Equation (4) improves estimates of the chemical bioavailability of weak acids in fish.

9.3.1.3   Molecular Weight and Size

A high molecular weight or large molecular size often is offered as an explanation for the apparently low bioavailability or uptake efficiency of extremely hydrophobic organic chemicals (McKim et al., 1985; Anliker and Moser, 1987; Bruggeman et al., 1984; Niimi and Oliver, 1987). The rationale is that membrane diffusion coefficients drop with increasing weight.

In the membranes, the molecular weight dependence of the diffusion coefficient is greater than in the water layers that are associated with the membranes, where the diffusion coefficient is inversely proportional to molecular weight to the power of 0.33 to 0.50. However, because of the increase in lipophilicity (e.g., higher K_{ow}) with increasing molecular weight, the membrane permeability usually increases with increasing molecular weight (Stein, 1981).

For high molecular weight lipophilic substances, permeation through the lipophilic bilayer is often so rapid that it is not the rate limiting step in the membrane permeation process. (Flynn and Yalkowsky, 1972; Gobas and Mackay, 1987) and diffusion through the membrane is relatively rapid. Because of their very low solubility in water, high molecular weight substances exhibit a low rate of mass transfer through the aqueous diffusion layers, causing slow membrane permeation kinetics.

9.3.1.4   Molecular Size

Opperhuizen et al. (1985) proposed that molecules with a minimal internal cross section exceeding 0.95 nm cannot cross biological membranes and thus are not bioavailable. This proposition was to account for the apparent lack of bioconcentration of some very hydrophobic chemical substances such as hexabromobenzene, octachloronaphthalene, and octachlorodibenzo-p-dioxin.

In follow-up experiments, Gobas et al. (1989) found that brominated biphenyls which have minimal internal cross sections somewhat greater than 0.95 nm were available for uptake in fish via the gills. In addition, Muir et al. (1986) and Muir and Yarechewski (1988) reported that octachlorodibenzo-p-dioxin, which has a minimal internal cross section greater than 0.95 nm, was present in internal tissues of trout exposed under field conditions.

It is presently unclear how molecular size affects the membrane permeability of simple hydrophobic organic chemicals. A size restriction to membrane permeation may exist, but it appears to be greater than the proposed 0.95 nm and it may be erroneous to view it as a sharp "cut-off."

9.3.2   Bioconcentration

9.3.2.1   General Definition

Bioconcentration in aquatic organisms involves uptake of chemical from water via the respiratory surface of the organism (e.g., gills and skin) (Figure 9.1) and the loss or elimination of chemical from the organism. Elimination of the chemical occurs predominantly via the respiratory surface (e.g., gills in fish), fecal egestion, and metabolic transformation. Chemical elimination also can occur through other mechanisms, such as egg deposition in oviparous organisms or sperm production. Growth of the organism tends to lower or "dilute" the internal concentration in the organism and thus can be viewed as a route of chemical elimination, although there is no net loss of chemical mass from the organism. Because concentrations generally increase when the rate of chemical uptake exceeds the combined rate of elimination and growth dilution, growth can have a considerable effect on bioaccumu-

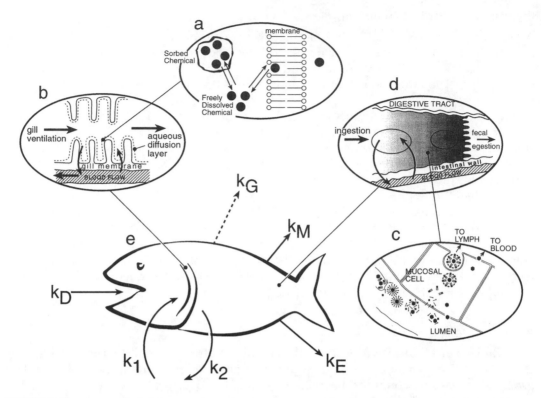

FIGURE 9.1

Bioconcentration and biomagnification process. Schematic diagram of the most important molecular, physiological, and pharmacokinetic processes involved in the bioconcentration and biomagnification of chemical substances in aquatic organisms as illustrated by fish: (a) illustrates the association of the chemical substance with particulate matter in the water phase and the non-associated freely dissolved chemical passing through the gill membranes; (b) illustrates the ventilation of contaminant containing water over the gill membranes and the transport of a fraction (i.e., the gill absorption efficiency) of the chemical through the gill membranes into the blood; (c) illustrates the digestion of the lipid fraction of the diet in the gastro-intestinal tract leading to the absorption of the chemical across the gastro-intestinal tract; (d) illustrates the increase of the chemical activity of the contaminant in the gastro-intestinal tract and the egestion of the chemical in fecal matter; (e) illustrates the kinetic balance between chemical absorption from the water (via the gills) and the diet (via the gastro-intestinal tract) and chemical elimination via gill elimination, fecal egestion, metabolic transformation, and pseudo-elimination resulting from growth.

lation factors and concentrations in organism, in particular for chemicals of high K_{ow} (Thomann and Connolly, 1984; Clark et al., 1990).

9.3.2.2   Toxicokinetic Model (Two-Compartment)

The bioconcentration process can be modeled by an organism–water two-compartment toxicokinetic model where the organism is described as a single compartment in which the chemical is homogeneously distributed.

$$dC_B/dt = k_1.C_{WD} - (k_2 + k_E + k_M + k_G).C_B \qquad (13)$$

Here C_{WD} is the dissolved chemical concentration in water (g/L), C_D is the chemical concentration in food (g/kg), C_B is the chemical concentration in the organism (g/kg organism), t is time in days (Gobas et al., 1989), and dC_B/dt expresses the rate of net chemical accumulation the organism. The first order rate constants are k_1, for uptake from water via the respiratory surface (L water/kg organism/day); k_2, for elimination via the respiratory surface to the water (1/day); k_D, for chemical uptake from food (in kg food/kg organism/day); k_E, for elimination by fecal egestion (1/day); k_G, for growth dilution (1/day); and k_M, for metabolic transformation of the chemical (in 1/day).

If C_{WD} is constant, this equation can be integrated to give the following time dependent relationship between the concentration in the organism and the concentration in the water:

$$C_B = (k_1/(k_2 + k_E + k_M + k_G)).C_{WD} \cdot (1 - \exp\{-(k_2 + k_E + k_M + k_G).t\}) \tag{14}$$

The bioconcentration factor (BCF) is defined under conditions where the chemical reaches a steady-state ($dC_B/dt = 0$) or when t in equation 14 is large, namely;

$$BCF = C_B/C_{WD} = k_1/(k_2 + k_E + k_M + k_G) \tag{15}$$

Figure 9.1 illustrates the processes of chemical uptake and elimination via the respiratory surface for fish. These processes result from the combination of the water ventilation rate, the rate of chemical permeation across the membranes of the respiratory surface, and the rate of distribution of the chemical within the organism.

9.3.2.3   Ventilation Rate

The **ventilation rate**, expressed here in units of litres of water per day, brings waterborn chemical substances into contact with the membranes that make up the respiratory surface. In fish and certain benthic invertebrate species, the ventilation rate is generated predominantly by the gills and is referred to as the **gill ventilation rate**. In organisms that do not contain gills, the ventilation rate usually is generated by natural water movement or by movement of the organism in the water. Gill ventilation rates have been measured for many fish species, and reliable empirical relationships can be used to estimate gill ventilation rates from fish weights (e.g., Randall, 1970; Thurston and Gehrke, 1993) and for benthic invertebrate species (Morrison, 1995).

9.3.2.4   Gill Membranes

Gill membranes consist of a series of lamellae, each consisting of biological membranes associated with "stagnant" water or "diffusion" layers (Stein, 1981). Chemical substances can permeate through the membranes of the respiratory surface by simple molecular diffusion, or facilitated diffusion, or by enzyme(s) mediated transport processes ("active" transport), or a combination of these processes. Simple molecular diffusion is by far the most common membrane permeation process for hydrophobic organic substances because these substances tend to dissolve well in membrane lipids. Several authors (Spacie and Hamelink, 1982; Gobas et al., 1986; Barber et al., 1988, 1991; Erickson and McKim, 1990b; Gobas and Mackay, 1987) have published models for gill membrane transfer of organic substances in fish.

9.3.2.5   Blood Transport

Blood transports chemicals from the respiratory surface throughout the tissues of the organism. Several authors have attempted to elucidate the relative importance of this and other transport processes in regulating the uptake of chemicals via the gills in fish (e.g., Barber et al., 1988 and 1991; Erickson and McKim, 1990b; Gobas and Mackay, 1987). In most cases, it has been concluded that the role of blood flow in regulating the overall rate of chemical uptake is insignificant. Consequently, the gross rate of chemical uptake (U_W, g/day) across the gills can be expressed as the product of the gill uptake efficiency (E_W, unitless fraction), the gill ventilation rate (G_W, L/day), and the bioavailable chemical concentration in the water (C_{WD}, g/L):

$$U_W = E_W.G_W.C_{WD} \tag{16}$$

9.3.2.6 Gill Uptake Efficiency

The **(gill) uptake efficiency** E_W expresses the amount of chemical that actually is absorbed via the respiratory surface per unit of time, relative to the amount of chemical that is brought into contact with the respiratory surface, through the process of (gill) ventilation, per unit of time. It reflects the rates of chemical permeation through the membranes of the respiratory surface (e.g., gills) and the rate of chemical transport to the respiratory area (e.g., gills) via (gill) ventilation.

Gill uptake efficiencies have been observed to vary from 0 to 90% and have shown a relationship with the K_{ow} (Figure 9.2). The gill uptake efficiency increases with increasing K_{ow} for chemical substances with a log K_{ow} between 0.5 and 3. For chemicals with a log K_{ow} between 3 and approximately 6.5, the gill uptake efficiency remains constant at approximately 55% for a 0.75 kg rainbow trout (McKim et al., 1985) but this varies with fish species and size. For some super-hydrophobic chemical, which have a log K_{ow} greater than approximately 6.5, gill uptake efficiencies show a tendency to fall with increasing K_{ow}. This relationship with K_{ow}, which also has been observed for the gill uptake rate constant k_1, has been explained by a reduced membrane permeability and/or low aqueous bioavailability of the chemicals in the experiments.

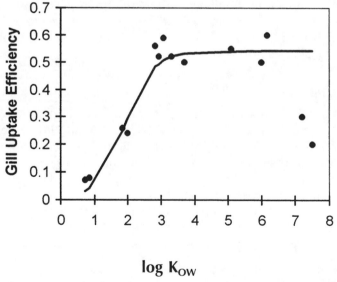

FIGURE 9.2

Gill uptake efficiencies. Gill uptake efficiencies of a range of hydrophobic organic chemicals in rainbow trout (data from McKim et al. (1985)) as a function of the chemical's K_{ow}. The solid line represents the fish gill ventilation model predictions of Gobas and Mackay (1987).

The relationship between E_W and k_1 is

$$k_1 = E_W.G_W/W_B \tag{17}$$

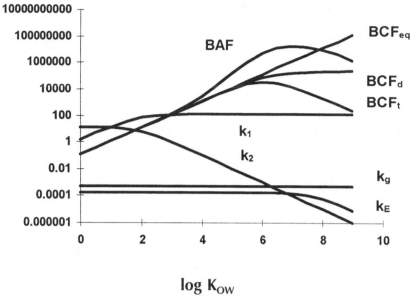

FIGURE 9.3

Gill uptake rate relationships. General relationships between the gill uptake rate constant (k_1), gill elimination rate constant (k_2), fecal egestion rate constant (k_E), and the bioconcentration factor, based on chemical equilibrium (BCF_{eq}) and on the truly dissolved chemical concentration in the water (BCF_d) and the total chemical concentration in the water (BCF_t), the field based bioaccumulation factor based on the truly dissolved chemical concentration in the water (BAF) and the octanol-water partition coefficient K_{ow}.

where W_B is the mass of the organism (kg) and k_1 can be viewed as the number of litres of water from which contaminants are effectively (i.e., 100%) extracted per kg of organism per day. Figure 9.3 presents the general relationship between the uptake rate constant and K_{ow}.

9.3.2.7 Respiratory Elimination

9.3.2.7.1 General Definition

The process of **elimination** by the respiratory surface is essentially the reverse process of uptake, i.e., chemicals from internal tissues are transported to the surface of the gills, where they diffuse through the gill membranes into the gill compartment, from which they are eliminated to water during gill ventilation. Because the rate of chemical transport from the lipids to water decreases with increasing lipophilicity or decreasing water solubility, the gill elimination rate constant declines with increasing K_{ow}.

Fish with a high lipid content have a greater ability to sequester the chemical, and hence it is more difficult for the chemical to transfer from the tissues of the fish to the ambient water. This is reflected in the expression for the **elimination rate constant**, which is:

$$k_2 = E_w.G_w/L_B.K_{ow}.W_B \tag{18}$$

where L_B is the lipid content (g lipid/g wet weight fish).

Figure 9.3 presents the general relationship between the gill elimination rate constant and K_{ow} and shows that k_2 does not vary with K_{ow} for lower K_{ow} chemicals but shows a distinct drop with K_{ow} for higher K_{ow} chemicals. Numerous studies have reported this drop, usually in the form of a log-log relationship. The relationship between k_2 and K_{ow} for lower K_{ow} chemicals ($K_{ow} < 1,000$) rarely has been observed, because bioaccumulation studies usually involve substances with a log K_{ow} greater than 3, but Tolls et al. (1996) show that k_2 is fairly constant for log K_{ow} less than 3.

If uptake and elimination via the respiratory surface are the only significant processes controlling bioconcentration, it follows that, by dividing Equation 17 by Equation 18, the steady-state bioconcentration factor BCF is:

$$BCF = k_1/k_2 \approx L_B.K_{ow} \text{ (for log } K_{ow} > 1) \tag{19}$$

If K_{ow} is less than 10, i.e., $\log K_{ow} < 1$, partitioning into water in the fish can be important, and Equation (19) is expressed more accurately as;

$$BCF = k_1/k_2 = 1 + (L_B.K_{ow}) \text{ (for } \log K_{ow} < 1) \tag{20}$$

Hamelink et al. (1971) first proposed the hypothesis that bioconcentration is essentially a chemical partitioning process. Since then, several relationships between the bioconcentration factor and K_{ow} have been reported. Meylan et al. (1999) reported the following correlation for 694 substances in fish:

$$\log BCF = 0.86.\log K_{ow} - 0.39 \tag{21}$$

which is equivalent to:

$$BCF = 0.41 K_{ow}^{0.86} \tag{22}$$

Mackay (1982) reported:

$$\log BCF = 1.0.\log K_{ow} - 1.31 \tag{23}$$

which is equivalent to:

$$BCF = 0.048.K_{ow} \tag{24}$$

The linear correlation between BCF and K_{ow} apparently breaks down for chemicals with a log K_{ow} greater than approximately 6 (Figure 9.4), resulting in a "parabolic" or "bilinear" type relationship between the BCF and K_{ow} (Bintein 1993, Meylan et al., 1999). For these superhydrophobic chemicals, the BCF appears to be much lower than expected from the chemical's octanol-water partition coefficient. A loss of linear correlation between the BCF and K_{ow} can be caused by a number of experimental artifacts (described in section 9.4.3) and physiological processes, including metabolic transformation, fecal egestion, and growth.

9.3.2.7.2   *Metabolic Transformation*

Metabolic transformation of the absorbed substance by the organism results in a reduction in the BCF below the value that is expected from the chemical's ability to partition into the lipids of the organism's tissues. Figure 9.4 illustrates this by showing the bioconcentration factors of some metabolizable chlorodibenzo-p-dioxins together with those of a range of non-metabolizable substances.

If the rate constant for metabolic transformation (k_m) in Equation (15) is significant relative to the rate of chemical elimination to water (k_2), the steady-state bioconcentration factor will be $k_1/(k_2 + k_m)$ and hence lower than k_1/k_2, which reflects the situation where the chemical substance is chemically partitioning between the organism and the water.

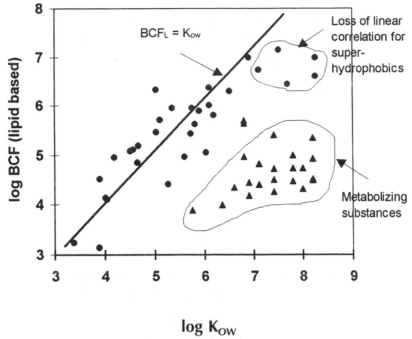

FIGURE 9.4

Bioconcentration factors. Observed Bioconcentration Factors (based on lipid based concentrations in the organism, i.e., L water/L lipid) of non-metabolizable polychlorinated benzenes, biphenyls and naphthalenes (·) and chlorodibenzo-p-dioxins (filled triangles) in Poecilia reticulata as a function of the chemical's octanol-water partition coefficient K_{ow}. The figure illustrates the breakdown of the linear correlation between the BCF and K_{ow} for chemicals with high K_{ow} and the effect of metabolic transformation on the relationship between K_{ow} and the BCF.

Although there is considerable information about metabolic transformation rates of a variety of organic substances in aquatic organisms, it is often difficult to predict metabolic transformation rates in organisms under field conditions.

9.3.2.7.3 Fecal Egestion

In addition to eliminating chemical substances via the respiratory surfaces of organisms, many aquatic organisms can eliminate chemical substances in fecal matter. Elimination by fecal egestion is generally insignificant unless the chemical substance is very hydrophobic (i.e., log $K_{ow} > 5$). As Figure 9.3 illustrates, the rate constant of elimination via the respiratory surface (k_2) drops with increasing chemical hydrophobicity (and hence increasing K_{ow}) but the rate constant of elimination through fecal egestion (k_E) remains relatively constant, with the possible exception of chemicals with extremely high K_{ow} (log $K_{ow} > 7$), for which the rate constant is expected to drop with further increases in K_{ow} (Gobas et al., 1988).

Figure 9.3 illustrates that, for chemicals with log $K_{ow} < 5$, the fecal egestion rate in most organisms is too small to have a significant effect on the total rate of elimination. As the K_{ow} of a chemical increases, fecal egestion becomes increasingly important and ultimately (i.e., for log $K_{ow} > 6$) fecal egestion is the predominant mechanism of elimination. When fecal egestion is a significant route of elimination, the bioconcentration process no longer reflects simple fish–water partitioning, which can be mimicked by octanol-water partitioning. For this reason, the linear correlation between the BCF and K_{ow} breaks down for high K_{ow} chemicals.

9.3.2.7.4   Growth

Because natural increases in the body weight of an organism tend to reduce (or "dilute") the internal concentration in the fish, growth can be viewed as a route of chemical elimination, although the fish loses no chemical mass. If the growth rate is significant relative to that of elimination, growth reduces the bioconcentration factor below that expected based on k_1/k_2. In general, k_2 exceeds k_G for log K_{ow} less than 5, but, for chemicals with logK_{OW}>5, k_G can exceed k_2. However, growth dilution is a pseudo elimination route. It is therefore possible (as observed for very hydrophobic chemicals) that, while the organism grows, the amount of chemical in the organism increases, because the amount absorbed exceeds that which is eliminated; but the concentrations actually fall.

9.3.3   Biomagnification

9.3.3.1   Toxicokinetic Model (Two-Compartment)

Biomagnification or dietary accumulation in aquatic organisms involves the uptake of chemical from the diet (Figure 9.1) and the loss or elimination of chemical by the organism by respiration, fecal egestion, metabolic transformation, growth dilution, and other routes of chemical elimination. A two-compartment organism-diet toxicokinetic model, ignoring uptake from water, yields:

$$dC_B/dt = k_D.C_D - (k_2 + k_E + k_M + k_G).C_B \qquad (25)$$

where C_D is the chemical concentration in the diet of the organism (g/kg), and C_B and the other quantities are as defined earlier.

If C_D is constant, integration gives the following time-dependent relationship between the concentration in the organism and the concentration in the diet:

$$C_B = (k_D/(k_2 + k_E + k_M + k_G)).C_D . (1 - \exp\text{-}(k_2 + k_E + k_M + k_G).t) \qquad (26)$$

At steady-state, the biomagnification factor BMF is defined as:

$$\text{BMF} = C_B/C_D = k_D/(k_2 + k_E + k_M + k_G) \qquad (27)$$

9.3.3.2   Digestion

The rate (g/day) at which the chemical is introduced into the gastro-intestinal tract (GIT) is referred to as the **ingestion rate**, and is the product of the organism's feeding rate and the concentration of chemical in the diet. In the GIT, the ingested food is digested. Digestion causes the chemical to become more readily available for dietary uptake. Digestion involves two major processes, namely changes to the composition of and changes in the volume of the digested material (Gobas et al. 1993, 1999).

9.3.3.2.1   Changes in Composition

Digestion of food changes the matrix in which the chemical resides as lipids, carbohydrates, and proteins are absorbed. Because hydrophobic organic chemicals have the highest affinity for these components of the food, the removal of these food components, as a result of digestion, causes the chemical activity in the food to increase (Gobas et al., 1993, 1999). This increase in chemical activity can be viewed as an increase in the chemical's potential to escape from the food phase, as expressed by the chemical's fugacity, which is effectively the partial pressure that the chemical exerts and is expressed in units of pressure, i.e., Pascal. The chemical's concentration and fugacity in the food are simply related by

$$\text{Concentration C Fugacity} = (\text{mol}/\text{m}^3) \text{ f (Pa)} \div \text{f} = \text{C}/\text{Z} \ (\text{mol}/\text{m}^3 \text{ Pa}) \tag{28}$$

(Mackay and Paterson, 1982; Mackay, 1991). The fugacity capacity (Z) is the ability of the food matrix to "solubilize" the chemical and represents the increase in chemical concentration which results in a one-unit increase in fugacity. When food is digested, its fugacity capacity is reduced, due to the absorption of the lipids and other organic rich molecules. Consequently, the activity and fugacity of the chemical in the gastro-intestinal tract increases beyond that in the ingested food.

9.3.3.2.2 Changes in Volume

The second phenomenon is the reduction in volume of the food as it moves through the intestinal tract. The concentration of the chemical thus increases, resulting in an increase in the chemical activity and fugacity difference between the gastro-intestinal contents and the gut lumen (Gobas et al., 1988, 1993, 1999). This increase in fugacity during food digestion explains why certain chemicals achieve a higher concentration and fugacity in the food-consuming organisms (i.e., the predators) than in the diet of these organisms (i.e., prey) in the absence of an active transport mechanism for gastro-intestinal uptake.

However, gastro-intestinal magnification (i.e., the increase of the chemical activity in the *gastro-intestinal tract* over that in the ingested food) does not always result in biomagnification (i.e., the increase of the chemical fugacity in the *organism* above that of the ingested food). Gastro-intestinal magnification can occur only if the combined rate of chemical elimination via the respiratory surface area, metabolic transformation, growth dilution and other processes is small compared to the rate of fecal egestion (i.e., $k_2 + k_M + k_G < k_E$). Under those conditions, the chemical fugacity in the organism will approach the fugacity in the GIT. If the combined rate of chemical elimination is high (e.g., the chemical substance is metabolized) relative to the rate of chemical elimination into fecal matter, the chemical fugacity in the organism will not be able to achieve the level in the gastro-intestinal tract, and biomagnification will not occur.

Because of the rapid rate of chemical elimination by respiration relative to the rate of fecal excretion, chemicals with log $K_{ow} < 5$ rarely biomagnify in aquatic food-chains. Chemicals with greater K_{ow} (i.e., log K_{ow} between 5 and 7.5) exhibit a significant biomagnification potential unless they are being metabolized at a significant rate. Superhydrophobic chemicals with a log K_{ow} greater than 7.5 also have considerable biomagnification potential, and there is evidence of biomagnification of these substances under field conditions.

9.3.3.3   Food-Chain Bioaccumulation

The increase in fugacity with increasing trophic level has been observed in several food chains (Connolly and Pedersen, 1988; Clark et al., 1988) and is referred to as food chain bioaccumulation. Increases in wet weight chemical concentrations in organisms concomitant with increasing trophic level can occur as a result of increases in the lipid content of organisms as their trophic level increases. This phenomenon is not food chain bioaccumulation but resembles it. The test for food chain bioaccumulation is an increase in either fugacity or concentration *on a lipid weight basis, not* on concentration on a wet weight basis.

Thus a fundamental difference exists between the processes of bioconcentration and biomagnification of hydrophobic organic chemicals. *Bioconcentration* results from the chemical's inherent tendency to thermodynamically partition between the water and the organism, approaching a situation where the activity (and fugacity) of the chemical in the ambient water and organism are equal. *Biomagnification* results from food digestion and absorption, which can elevate the chemical's potential and fugacity above that in the ingested food. Biomagnification therefore acts as a "fugacity pump," where fugacity of the chemical substance is being elevated at each step in the food chain. With significant metabolic transformation, gill elimination, or growth dilution, biomagnification will not occur. This is analogous to a "leaky pump," where the elevated fugacity escapes. If this occurs, the chemical concentration may become smaller with each step in the food-chain, a process referred to as "trophic dilution."

9.3.3.4   Bioaccumulation Factor (BAF)

Bioaccumulation is the result of *simultaneous bioconcentration and biomagnification*. It usually is observed under field conditions but can be measured in laboratory experiments, as well. In terms of a toxicokinetic model, where the organism is presented as a single compartment, addition of equations (13) and (25) can describe the bioaccumulation of a chemical in an organism:

$$dC_B/dt = k_1 . C_{WD} + k_D . C_D - (k_2 + k_E + k_M + k_G) . C_B \qquad (29)$$

The bioaccumulation factor BAF is defined under steady-conditions $(dC_B/dt = 0)$ as:

$$BAF = C_B/C_{WD} = \{k_1 + k_D . (C_D/ C_{WD})\}/(k_2 + k_E + k_M + k_G) \qquad (30)$$

An important consideration is the relative importance of the water and the diet as a source of chemical uptake in the organism. Equation (30) illustrates that, to a large degree, the relative concentrations of the chemical substance in the water and the diet of the organism control the BAF. In laboratory experiments where bioaccumulation factors are measured, it is therefore important to choose appropriate concentrations C_{WD} and C_D, because they affect conclusions drawn about the relative importance of the diet and water as routes of chemical intake. Under field conditions, the bioaccumulation factors in organisms can vary, depending on the relationship between the chemical concentrations in the water and the diet, especially if the chemical is very hydrophobic (log $K_{ow} > 5$). If the BAF measured in one ecosystem is used to assess the BAF in another ecosystem, an error could be made if this diet/water concentration relationship or values of k_D vary between the ecosystems.

9.4   Experimental Methods and Measurements

9.4.1   Freely Dissolved Concentration in the Water

The measurement of the freely dissolved chemical concentration in water C_{WD} is a crucial step in the measurement of bioconcentration factors in laboratory experiments and in the assessment of body burdens under field conditions. This is particularly important for very hydrophobic chemicals (log $K_{ow} > 5$), which exhibit a high tendency to associate with organic matter in the water.

For chemicals of log $K_{OW} < 5$, the total chemical concentration in the water, often derived through solvent or XAD-resin extraction, is usually equal to the freely dissolved chemical concentration. Several techniques have been used to measure freely dissolved chemical concentrations in the water.

Landrum et al. (1984) developed a methodology in which the water is pumped over a SEP-PACK filter system. The premise of their methodology is that sorbed chemical in the water will pass through the filter, while the freely dissolved chemical in the water will adhere to the filter. Yin and Hasset (1988) and Sproule et al. (1991) have applied gas-sparging techniques to measure C_{WD}. The principle of this technique is that only freely dissolved chemical in the water will tend to partition into the gas phase. C_{WD} is then estimated from the gas phase concentration and K_{AW}, the air-water partition coefficient.

9.4.2 Bioconcentration Tests

In Canada, the United States, and Europe, a "flow-through" method is generally used to conduct bioconcentration tests with aquatic biota (in Canada and the U.S., "Standard Practice for Conducting Bioconcentration Tests with Fishes and Saltwater Bivalve Molluscs" American Standards for Testing and Materials (1988), E 1022; in Europe, "Bioaccumulation: Flow-through Fish Test," Organization for Economic Co-operation and Development (1996), 305E). This methodology enables the determination of steady-state bioconcentration factors as well as chemical uptake and depuration (i.e., elimination) rate constants.

Two groups of organisms of the same species are held in separate flow-through aquaria and subjected to a treatment consisting of an uptake phase and a depuration phase. The control group is exposed to water that contains no test chemical during both the uptake and depuration phases, thus providing information on the impact of the test conditions on the test organisms. The other group is subjected to (i) an uptake phase with exposure to water containing a constant concentration of the test chemical until steady-state is achieved or for at least 28-30 days and then (ii) a depuration phase during which organisms are exposed to water containing no test chemical for approximately twice the duration time of the uptake phase. During the experiments, organisms and water are removed from aquaria in geometric time series and analyzed. The resulting time-series data are used to estimate uptake rate constants (k_1), depuration rate constants (k_2), bioconcentration factors (BCFs), and their confidence limits, using mathematical models which best fit the data.

To properly design this bioconcentration test, an estimate of duration of the uptake phase is necessary. An uptake phase approximately equal to 1.6 times the reciprocal of the elimination rate constant ($1.6/k_2$), but no longer than 3.0 times the reciprocal of the elimination rate constant (i.e., $3.0/k_2$), is considered optimal (Reilly et al., 1977). Prior estimates of k_2 can be obtained using water solubility data or octanol-water partition coefficients of the chemical substance (Organization for Economic Co-operation and Development, 305E). Although methods that use physical-chemical properties to estimate exposure durations in bioconcentration tests are useful for estimating the time period for the bioconcentration test, it should be stressed that elimination rate constants (hence, appropriate exposure duration times) are a function of the size and the lipid content of the organism. In general, elimination rate constants drop significantly with increasing size and lipid content of the organism, hence requiring longer exposure durations in bioconcentration tests.

In most cases, uptake and depuration profiles follow the patterns Figure 9.5 illustrates. The BCF then follows as k_1/k_2. Assuming that the organism behaves as a single homogeneous unit, bioconcentration can be described by an organism-water two-compartment model (i.e., the organism is described as a single compartment) consisting of first-order rate constants, as follows:

$$dC_B/dt = k_1.C_{WD} - k_t.C_B \qquad (31)$$

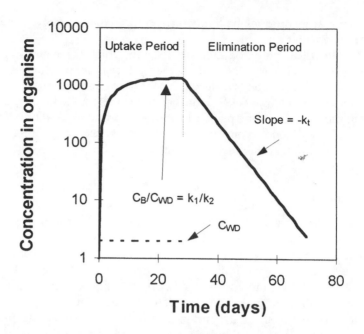

FIGURE 9.5

Uptake and elimination. Illustrative example of the uptake and elimination of a chemical substance in a bioconcentration test consisting of an uptake period, where the organism is exposed to a constant concentration of a chemical substance in the water phase, and an elimination period, where the organism is exposed to clean water. In this example, bioconcentration is described by a first order organism-water two compartment model. The concentration in the organism is C_B, the freely dissolved concentration in the water is C_{WD}.

where k_t is the total elimination rate constant (1/day) which can be calculated from the slope of ln(concentration)-time profile determined in the elimination experiment:

$$C_B(t) = C_B(0).\exp(-k_t.t) \text{ or } \ln C_B(t) = \ln C_B(0) - k_t.t \tag{32}$$

where $C_B(0)$ and $C_B(t)$ are the concentrations of the test chemical in the organism initially and at a given time t. The k_1 can be calculated by fitting the data to the following equation if C_{WD} remained constant throughout the uptake experiment:

$$k_1 = k_2 \cdot (C_B/C_{WD}).(1 - \exp(-k_t.t))^{-1} \tag{33}$$

Non-linear parameter estimation methods can also be used to estimate k_1 and k_t. Two methods are recommended. Both use an organism–water two-compartment model (i.e., Equation (29)). The first method is the BIOFAC method (Blau and Algin 1978). This method is based on the condition that the chemical concentration in the water is constant throughout the duration of the experiment, such that Equation 29 can be integrated to give

$$C_B = (k_1/k_t).C_{WD}.(1 - \exp(-k_t.t)) \tag{34}$$

after which the rate constants k_1 and k_t and the steady-state bioconcentration factor (BCF), i.e., k_1/k_t, are derived from a series of experimental data of C_W and C_B at different times t, through non-linear regression.

Gobas and Zhang (1995) developed a second method for conditions frequently observed in bioconcentration tests with very hydrophobic substances (log $K_{ow} > 3.5$) where the concentration in the water varies throughout the duration of the test. This method, which the author refers to as BIOFIT, combines numerical integration of Equation (29) and a maximum likelihood method to derive the bioconcentration rate constants, k_1 and k_t, and the steady-state bioconcentration factor (BCF), i.e., k_1/k_2, from a time-series of experimental measurements of the chemical concentration in the water and the organism. If a first order two-compartment model does not adequately describe the bioconcentration data, one can use more complex models such as those Blau et al. (1975), Moriarty (1975), Walker (1987), and Nichols et al. (1990) describe.

9.4.3 Common Difficulties with the Bioconcentration Test

A number of physical and biological constraints, as well as experimental problems, often effect proper execution of bioconcentration tests. The most frequently encountered limitations of bioconcentration tests involve duration, measurement of concentration, and variability of concentrations.

9.4.3.1 *Too Short Duration of the Test to Achieve Steady State*

This problem is most prevalent in tests with very high K_{ow} (i.e., log $K_{ow} > 5$) substances, because organisms eliminate these chemicals slowly, and so steady-state is achieved slowly. Insufficiently long test duration times are partly responsible for the frequently observed loss of linear correlation between the BCF and K_{ow} for the high K_{ow} chemicals. Because it is not always possible to conduct tests for an adequate time period while maintaining constant exposure conditions, Blau and Algin (1978) and Gobas and Zhang (1992) have proposed quantitative methods for extrapolating steady-state BCFs measured in laboratory experiments under non-steady-state conditions.

9.4.3.2 *Difficulties with Measuring Bioavailable Chemical Concentrations in the Water*

Measuring the concentration of dissolved chemical in water throughout a bioconcentration experiment is often difficult because test organisms introduce organic matter such as fecal matter, lipids, and oils into the water. Hydrophobic substances bind to this organic matter, and the fraction of the chemical that binds to organic matter in the water increases with increasing K_{ow} of the chemical. If one erroneously assumes that all the chemical is bioavailable, one will underestimate the BCF, because the concentration in the organism will reflect only the bioavailable concentration. Substances which exhibit a reduced bioavailability in laboratory tests also may be partly, but probably not equally, unavailable under field conditions.

9.4.3.3 *Variability of the Chemical Concentration in Water Over Time*

It is difficult to maintain a constant concentration of test chemical in water during the duration of the test. For example, OECD guidelines (1996) for bioconcentration tests recommend a water-flow-through rate of 1.13 L/day or more for a bioconcentration test with 25 goldfish of 4.5 gram in a 100 L tank. However, the gill ventilation rate of a single goldfish is approximately 2.4 L/day, resulting in a gill ventilation rate in the experimental tank of approximately 60 L/day.

If the BCF of the chemical is sufficiently large (e.g., >1,000) for the fish to absorb a significant fraction of the chemical in the tank, then the relatively low renewal rate of the chemical solution into the tank results in a significant drop in concentration, especially early in the test when most uptake occurs. This effect becomes more problematic with increasing K_{ow}, because concentrations of chemical in the water are more dilute and these chemicals have a greater potential to bioconcentrate. Gobas and Zhang (1992) provide a data analysis methodology to account for the effect of time dependent variations in the water concentration of bioconcentration tests on the BCF.

If, during the bioconcentration test, the chemical concentrations in the organism and water reach steady-state, the bioconcentration factor can be calculated from the steady-state concentrations in the organism (C_B) and the water (C_W) as C_B/C_W. However, when steady-state is not achieved during the test because the test was conducted for an insufficiently long period of time or because exposure concentrations were variable during the test, the derivation of the BCF and the rate constant for chemical uptake and elimination require a more specific method of data analysis.

9.4.4   Sediment Bioaccumulation Tests

The U.S. Environmental Protection Agency (Lee et al., 1993) designed Bedded Sediment Bioaccumulation tests to estimate the bioaccumulation of sediment-associated contaminants by benthic organisms. The bioaccumulation potential of a chemical in sediment usually is expressed as a Biota-Sediment Accumulation Factor (BSAF). The simplest test determines the BSAF as the ratio of the steady-state chemical concentration in the test organism and the chemical concentration in the sediment.

Two groups of organisms of the same species are placed in separate exposure chambers (either static or flow-through). One group (the "control" group) is exposed to sediments that contain no test chemical, and the other is exposed to sediments containing a constant concentration of test chemical until steady-state is achieved or for at least 28 days. Organisms are analyzed for lipid content and test chemical concentration. Sediment samples are analyzed for grain size distribution, moisture content, test-chemical concentration, and Total Organic Carbon (TOC) or Loss On Ignition (LOI). Resulting steady-state chemical concentrations are used to estimate BSAFs as described earlier (see Equations 6 and 7).

A second test derives the BSAF as the ratio of the sediment uptake rate constant and the depuration rate constant. This experiment is the same as the one described just above, but both groups of organisms are subjected to a treatment consisting of (i) an uptake phase, during which the organisms are exposed to sediment containing a constant concentration of the test chemical until steady-state is achieved or for at least 28-30 days, and (ii) a depuration phase, during which the organisms are exposed to sediment containing no test chemical for approximately twice the duration time of the uptake phase. During the uptake and depuration experiments, organisms and sediment are removed from the exposure chambers in geometric time series and analyzed as above. The time-series data are used to estimate the uptake rate constant of the chemical from the sediment (k_s) and the depuration rate constants (k_d) and the BSAF as the ratio (k_s/k_d).

9.5 Estimation Methods

9.5.1 Freely Dissolved Chemical Fraction in Water

The bioavailability of non-dissociating chemicals is determined largely by the fraction of the chemical in the water phase that is freely dissolved. In the presence of both particulate and dissolved organic matter, the following equation can be used as a reasonable estimate of the fraction of freely dissolved chemical:

$$F_{DW} = C_{DW}/C_W = 1/(1 + \phi_P . K_{oc} + \phi_D . 0.1 . K_{OC}) \qquad (35)$$

where ϕ_P is the concentration of particulate organic matter in the water (kg organic carbon/L water), ϕ_D is the concentration of dissolved organic carbon in the water (kg organic carbon/L water), and K_{OC} is the organic carbon-water sorption coefficient (L water/kg organic carbon), which can be estimated from the chemical's K_{ow} according to Karickhoff (1981, 1984):

$$K_{OC} = 0.41 . K_{ow} \qquad (36)$$

Dissolved organic carbon is defined operationally as the fraction of the total organic carbon in the water phase that will pass through a 0.45 µm filter. Assuming illustrative values for ϕ_P of 10^{-7} kg/L and ϕ_D of 10^{-7} kg/L, it follows that a chemical with a log K_{ow} of 5 will have F_{DW} of 99.6%, whereas a chemical with a log K_{ow} of 8 will have a F_{DW} of only 18%.

9.5.2 Bioconcentration in Algae and Phytoplankton

Bioaccumulation in algae and phytoplankton often is viewed as an equilibrium partitioning process of the chemical between the phytoplankton's lipid or organic carbon and water (Geyer et al., 1984; Mallhot, 1987; Swackhamer & Skoglund, 1993). Both the lipid-water and the organic carbon-water partition coefficients have been used to approximate the partitioning behavior of hydrophobic organic chemicals between phytoplankton and water. Considerable discussion surrounds whether lipids or organic carbon is the more suitable surrogate phase for the cell structures in which hydrophobic organic substances tend to bioconcentrate.

At equilibrium, the chemical concentration in algae or phytoplankton (C_A, g chemical/kg organism) can be estimated as the product of the freely dissolved chemical concentration in the water (C_{WD}, g chemical/L water); the lipid content of the algae or phytoplankton (L_A, kg lipid/kg organism) or the organic carbon content in algae or phytoplankton (OC_A, kg organic carbon/kg organism); and the lipid-water (K_L, L water/kg lipid) or organic carbon-water partition coefficient (K_{OC}, L water/kg organic carbon):

$$C_A = C_{WD} \cdot L_A \cdot K_L \quad \text{or} \quad C_A = C_{WD} \cdot OC_A \cdot K_{OC} \qquad (37)$$

This assumes that (i) the chemical has achieved a chemical equilibrium between algae or phytoplankton and water, (ii) there is no transformation of the chemical substance in the plankton, and (iii) "growth dilution" is insignificant.

Based on short duration experiments (less than 3 days), several researchers (e.g., Soder-gren, 1968; Reinert, 1972; Rice, 1973; Geyer, 1984; Mallhot, 1987) concluded that equilibrium between organic chemicals and algal cells occurs within hours. In contrast, Swackhammer (1991) found that the initial uptake of chemical to 40-90% of steady state was rapid and occurred within 24 hours, but that subsequent uptake was slow. In this work, most PCB congeners did not reach chemical equilibrium during the 20 day exposure. Swackhammer and Skoglund (1993) reported that bioaccumulation in certain algal species was consistent with equilibrium partitioning between cell lipids and water, but that, during conditions of rapid algal growth, growth dilution prevented achievement of equilibrium. No predictive kinetic bioaccumulation models for algae or phytoplankton have been proposed.

9.5.3 Bioconcentration in Aquatic Macrophytes

Bioconcentration in aquatic macrophytes also has been described by equilibrium partitioning of chemical between the macrophytes' lipid or organic carbon fraction and water (e.g., Gobas et al., 1991). If the lipid-water partition coefficient is used to describe the partitioning of chemical between macrophytes and water, then the chemical concentration in macrophytes (C_M, g chemical/kg organism) is approximately equal to the product of the freely dissolved chemical concentration in water (C_{WD}, g chemical/L water), the lipid content of the macrophytes (L_M, kg lipid/kg organism), and the lipid-water partition coefficient (K_L, L water/kg lipid), which K_{OW} can approximate:

$$C_M = C_{WD} \cdot L_M \cdot K_{OW} \tag{38}$$

This equation provides a relatively simple estimation method and should yield acceptable results when log K_{ow} is less than 5 and there is no metabolism in or on the surface of the macrophytes. For substances with a greater K_{ow}, changes in macrophyte biomass, i.e., growth dilution, can be important. If appropriate data regarding rate constants are available, a kinetic model may be preferable in assessing the bioconcentration of hydrophobic organic substances in aquatic macrophytes (e.g., Gobas et al., 1991). However, in most cases, this information is not available.

9.5.4 Bioconcentration in Zooplankton

Like algae, phytoplankton, and macrophytes, an equilibrium partitioning model commonly is used to estimate the chemical concentration in zooplankton (C_Z, g chemical/kg organism). Hence, C_Z is the product of the freely dissolved chemical concentration in the water (C_{WD}, g chemical/L water), the lipid content of zooplankton (L_Z, kg lipid/kg organism), and the lipid-water partition coefficient (K_L, L water/kg lipid), which the octanol-water partition coefficient approximates (Clayton et al. 1977):

$$C_Z = C_{WD} \cdot L_Z \cdot K_{OW} \tag{39}$$

This assumes no biomagnification, metabolism, or growth dilution. The fact that chemical concentrations achieve equilibrium rapidly because of the large surface area to volume ratio of these organisms justifies use of this equation.

Trowbridge & Swackhammer, (1995) suggested that, under field conditions, certain zooplankton species can biomagnify chemicals of high K_{ow} while feeding. Furthermore, seasonal growth spurts in zooplankton might result in significant discrepancies between equilibrium partitioning model predictions and actual chemical concentrations, due to growth dilution.

Thomann et al. (1992) and Morrison et al. (1997) have developed **kinetic models** employing rate constants to assess the extent of chemical bioaccumulation in zooplankton, as Tables 9.1 and 9.2 summarize. Thomann et al. (1992) list relationships which incorporate organism physiology, bioenergetics, and chemical characteristics to estimate uptake and elimination rate constants which are used to estimate bioaccumulation. Morrison et al. (1997) rely on physiological information to estimate bioaccumulation. Both models provide a potentially more realistic description of bioaccumulation by zooplankton, although, to date, neither model has been tested independently against field data.

TABLE 9.1

Model Equation, Parameters, and their Units of the Zooplankton Bioaccumulation Model of Thomann et al. (1992)

$$v_{zoo} = \frac{k_u \cdot C_W + \alpha_{pl} \cdot I_L \cdot v_{pl}}{K_{zoo} + G_{zoo}} \qquad (39)$$

Parameter	Units	Definition
α_{pl}	g chemical assimilated·g^{-1} chemical ingested	chemical assimilation efficiency from phytoplankton
c_W	µg·L^{-1}	freely-dissolved chemical concentration in overlying water (w)
G_{zoo}	d^{-1}	growth rate
I_L	g lipid ingested·g^{-1} zooplankton lipid·d^{-1}	specific feeding rate on phytoplankton
k_u	L·d^{-1}·g^{-1} lipid	rate of chemical uptake from freely-dissolved chemical in overlying water
K_{zoo}	d^{-1}	excretion rate
v_{pl}	µg·kg^{-1} lipid	lipid-normalized chemical concentration in phytoplankton
v_{zoo}	µg·kg^{-1} lipid	lipid-normalized chemical concentration in zooplankton

9.5.5 Bioaccumulation in Benthic Invertebrates

9.5.5.1 *Equilibrium Partitioning Models*

The models are used most frequently and assume that bioaccumulation is described adequately by equilibrium partitioning between the sediment, the pore-water (or interstitial water), and the organism (e.g. Shea, 1988; Gobas et al., 1989b; Bierman, 1990, DiToro et al., 1990) namely:

$$C_B = \frac{C_S \cdot d_{OC} \cdot L_B}{d_L \cdot OC} = C_{WD} \cdot L_B \cdot K_{LW} \qquad (40)$$

where C_B is the concentration in the benthic invertebrate (g/kg wet weight); C_S is the concentration in the bottom sediments (g/kg dry weight); C_{WD} is the dissolved concentration in the interstitial (or pore) water (g/L); d_{OC} and d_L are the densities of organic carbon and lipid (kg/L); L_B is the mass fraction of lipid in the organism (kg lipid/kg organism); OC is the mass fraction of organic carbon in the sediment (kg organic carbon/ kg sediment); and

TABLE 9.2

Model Equation, Parameters and their Units of the Zooplankton Bioaccumulation Model of Morrison et al. (1997)

$$C_B = \delta_B \cdot \varnothing_B \cdot K_{OW} \cdot$$

$$\left[\frac{C_{WD} \cdot G_W \cdot E_W + C_D \cdot G_D \cdot E_D}{E_W \cdot G_W + E_D \cdot (1-\alpha) \cdot (1-\beta) \cdot G_D \cdot \delta_D \cdot \varnothing_D \cdot K_{OW} + \delta_B \cdot \varnothing_B \cdot G_R \cdot K_{OW} + V_B \cdot G_M \cdot \delta_B \cdot \varnothing_B \cdot K_{OW}} \right] \quad (40)$$

Parameter	Units	Definition
$\varnothing$	unitless fraction	lipid or organic carbon content in biota (B) or diet (D)
α	unitless fraction	fraction of organic carbon or lipid in diet that is removed upon digestion
β	unitless fraction	fraction of ingested diet absorbed by the organism
C_B, C_W, C_D	$\mu g \cdot kg^{-1}$	chemical concentration in biota (C_B), water (C_W), and diet (C_D)
δ	$kg \cdot L^{-1}$	density of biota (B) or diet (D)
E_W, E_D	unitless fraction	efficiency of chemical transfer across the respiratory surface and the organism (E_W), and between gut contents and the organism (E_D)
G_D	$kg \cdot d^{-1}$	ingestion rate of food
G_R	$kg \cdot d^{-1}$	growth rate
G_W	$L \cdot d^{-1}$	rate of water ventilation across the respiratory surface
k_M	d^{-1}	metabolic transformation rate
K_{OW}, K_{OC}		chemical partition coefficient between octanol and water (K_{OW}), and organic carbon and water (K_{OC})
V_B	kg	mass of organism

K_{LW} is the lipid-water partition coefficient (L/kg). This assumes that (i) lipids are the primary site for chemical bioaccumulation in the organism, (ii) the organism has resided in the sediments long enough to have achieved chemical equilibrium, (iii) the organism is unable to metabolize the substance, nor is the substance degraded through abiotic reactions, (iv) biomagnification is unimportant, (v) there is no growth dilution, and (vi) the invertebrate is exposed primarily to chemical in the interstitial water rather than the overlying water.

This model has been tested well against measured data (e.g., Van der Oost et al., 1988; Markwell et al., 1989; Gobas et al., 1989b; Landrum et al., 1992; Lake et al., 1990), with the most thorough validation by Parkerton (1993). The model predicts that the ratio of concentration in benthic invertebrate lipid (C_L) to concentration in sediment organic carbon (C_{oc}) should be between 1 and 3. Parkerton (1993) found, however, that these ratios varied by several orders of magnitude and exhibited a parabolic dependence on the logarithm of the octanol-water partition coefficient (log K_{OW}). Based on these comparisons, Parkerton concluded that the equilibrium partitioning model was not an adequate predictor of chemical accumulation by benthic invertebrates.

Morrison et al. (1996) summarized some of the factors that may be responsible for the lack of predictability of the equilibrium model. For instance, biomagnification can explain why many observed concentrations of hydrophobic organic chemicals are greater than those expected by the partitioning model (Gobas et al., 1993). Differences in feeding strategies, e.g., filter-feeding and detritus consumption, may be responsible for differences between species. For many benthic invertebrate species, uptake of chemical from the overlying water is more important than from the pore-water. Concentrations in sediment pore water and overlying water differ, as a result of temporal changes in chemical inputs (Gobas et al., 1995), relatively slow chemical kinetics between sediment and water (Baker et al., 1986), and changes in the organic carbon pool (Baker et al., 1991). Other factors relating to contaminant kinetics, rates of metabolism, mode of feeding, seasonal changes in feeding, reproductive status, and lipid contents (Landrum et al., 1992) may introduce further errors.

9.5.5.2 Kinetic Models

The models rely on the use of rate constants (usually first order) to quantify the rate of chemical uptake and elimination by organisms. These models have been developed to predict chemical concentrations under non-equilibrium conditions and to account for factors that affect chemical accumulation, such as organism growth, chemical metabolism, and chemical concentrations in diet and water. Although most kinetic-based models are used to predict chemical accumulation under steady-state conditions, these models have the advantage of being adaptable to dynamic conditions (i.e., the equations can be integrated so that they predict the change in biotic chemical concentration through time).

The models rely on the derivation of rate constants through laboratory experiments or field experiments. Because rate constants differ among organisms and chemical substances, the development of kinetic models often requires a substantial effort. In some cases, empirically based correlations can be used to assess the rate constants.

Landrum et al. (1992) developed a kinetic bioaccumulation model for PAHs in the amphipod *Diporeia*, employing first-order kinetic rate constants for uptake of dissolved chemical from the overlying water, uptake by ingestion of sediment, and elimination of chemical via the gills and feces. In this model, diet is restricted to sediment, and chemical metabolism is considered neglagable. The model and its parameters, as Table 9.3 summarizes, treat steady-state and time-variable conditions. Empirically derived regression equations (Landrum and Poore, 1988; and Landrum, 1989) are used to estimate the uptake and elimination rate constants. A field study in Lake Michigan revealed substantial differences between predicted and observed concentrations of PAHs in the amphipod *Diporeia*. Until more robust kinetic rate constant data are available for a variety of benthic invertebrates and chemicals, this model is unlikely to provide accurate estimates of chemical concentrations in benthic invertebrates under field conditions.

TABLE 9.3

Model Equation, Parameters and their Units of the Benthic Invertebrate Bioaccumulation Model of Landrum et al. (1992)

$$dC_b/dt = k_w \cdot C_w + k_s \cdot C_s - k_d \cdot C_b \tag{42}$$

Parameter	Units	Definition
C_{bi}	$ng \cdot g^{-1}$ wet weight	chemical concentration in benthic invertebrate
C_s	$ng \cdot g^{-1}$ dry weight	chemical concentration in sediment
C_w	$ng \cdot mL^{-1}$	chemical concentration in overlying water
k_d	h^{-1}	elimination rate constant
k_s	g sediment dry weight·g^{-1} organism wet weight·h^{-1}	uptake clearance from sediment
k_w	mL water·g^{-1} organism wet weight·h^{-1}	uptake clearance from overlying water
$\log K_{OW}$		octanol-water partition coefficient
t	h	time

9.5.5.3 Physiologically Based Kinetic Models

Thomann et al. (1992) (see Table 9.4) and Morrison et al. (1996) (see Table 9.5) developed models that derive uptake and elimination rate constants based on physiological parameters such as gill ventilation rates, weights and lipid contents. The Thomann model has been

TABLE 9.4

Definitions and Units of Model Parameters Used in the Benthic Bioaccumulation Model of Thomann (1991, 1992).

$$v_{bi} = \frac{k_u\left(b_s \cdot c_s + b_w \cdot c_w\right) + \left[\left(p_s \cdot \alpha_s \cdot I_{LOG}\right) \cdot r_s + \left(p_p \cdot \alpha_p \cdot I_L\right) \cdot v_p\right]}{K_{bi} + G_{bi}} \tag{43}$$

Parameter	Units	Definition
α_s, α_p	g chemical assimilated·g^{-1} chemical ingested	chemical assimilation efficiency from sediment (s) and phytoplankton (p), respectively
b_s, b_w		fraction of chemical uptake from sediment interstitial water (s) and overlying water (w), respectively
c_s, c_w	mg·L^{-1}	freely dissolved chemical concentration in sediment interstitial water (s) and overlying water (w), respectively
G_{bi}	d^{-1}	growth rate
I_L	g lipid ingested·g^{-1} benthic invertebrate lipid·d^{-1}	specific feeding rate for phytoplankton
I_{Loc}	g organic carbon ingested·g^{-1} benthic invertebrate lipid·d^{-1}	specific feeding rate for sediment
K_{bi}	d^{-1}	excretion rate
k_u	L·d^{-1}·g^{-1} lipid	rate of chemical uptake from available freely-dissolved pools (i.e., interstitial water and overlying water)
p_s, p_p		fraction of sediment (s) and phytoplankton (p) in diet, respectively
r_s	mg·kg^{-1} organic carbon	chemical particulate concentration in sediment on an organic carbon basis
v_{bi}	mg·kg^{-1} lipid	lipid-normalized chemical concentration in a benthic invertebrate
v_p	mg·kg^{-1} lipid	lipid-normalized chemical concentration in phytoplankton

applied to amphipods from Lake Ontario to evaluate the relative importance of overlying water, interstitial water, sediment, and phytoplankton as chemical exposure routes. It provided reasonable results and appears to be a useful tool for estimating chemical concentrations in benthic invertebrates.

The Morrison model accounts for chemical uptake from water and food and chemical elimination via gills, feces, growth dilution, and metabolic transformation. It distinguishes between filter-feeding and detritus-feeding strategies. The diets can include any number of items for which contaminant data are available (either calculated or measured). Tests involving observed and predicted polychlorinated biphenyl (PCB) congener data in several benthic invertebrate species indicate that 95% of observed concentrations in filter-feeding and detritus feeding benthic invertebrates are within a factor of approximately two of model predictions (Morrison et al., 1996).

9.5.5.4   *Fugacity-Based Models*

These models use thermodynamic terms to describe the behavior of chemicals in the environment. Fugacity is a thermodynamic quantity that can be viewed as the "escaping tendency of a chemical substance from a phase" (Mackay and Paterson, 1982; Mackay, 1991). At steady-state, Campfens and Mackay (1997) described bioaccumulation in benthic invertebrates as follows:

TABLE 9.5

Model Equations, Parameters and their Units of the Benthic Invertebrate Bioaccumulation Model for Filter Feeders and Benthic Detritovores of Morrison et al. (1996.)

$$C_{Bff} = \left[\frac{C_{WD} \cdot E_W \cdot G_W + C_{SS} \cdot G_W \cdot \varnothing_{SS} \cdot E_D \cdot \sigma}{\left(E_W \cdot G_W / K_{BW}\right) + \left(E_D \cdot G_F / K_{BF}\right) + \left(V_B \cdot k_M\right)} \right] \tag{44}$$

$$C_{Bd} = \left[\frac{C_{WD} \cdot E_W \cdot G_W + C_D \cdot G_D \cdot E_D}{\left(E_W \cdot G_W / K_{BW}\right) + \left(E_D \cdot G_F / K_{BF}\right) + \left(V_B \cdot k_M\right)} \right] \tag{45}$$

Parameter	Units	Definition
σ	unitless fraction	particle scavenging efficiency
ϕ_{SS}	L·L^{-1}	concentration of suspended solids in water column
C_{Bff}, C_{Bd} C_{WD}, C_D, C_{SS}	g·kg^{-1}	chemical concentration in filter feeder (C_{bff}), benthic detritovore (C_{Bd}), concentration of freely dissolved chemical in the water (C_{WD}), the diet of the detritovore (C_D), and suspended solids (C_{SS})
E_W, E_D	unitless fraction	efficiency of chemical transfer across the respiratory surface and the organism (E_W), and between gut contents and the organism (E_D)
G_D	kg·d^{-1}	ingestion rate of food
G_F	kg·d^{-1}	fecal egestion rate
G_W	L·d^{-1}	rate of water ventilation across the respiratory surface
K_{BF}	L/kg	organism-gastro-intestinal content partition coefficient of the chemical substance
K_{BW}	L/kg	organism-water partition coefficient of the chemical substance
k_M	d^{-1}	metabolic transformation rate constant
K_{OW}, K_{OC}		chemical partition coefficient between octanol and water (K_{OW}), and organic carbon and water (K_{OC})
V_B	kg	weight of the organism

$$f_B = \frac{f_W \cdot D_W + f_D \cdot D_D}{D_W + \left(\dfrac{D_D}{Q}\right) + D_M + D_G} \tag{41}$$

where f_B, f_W, and f_D are the chemical fugacities (Pa) in, respectively, the organism, the water, and the organism's diet; D_W, D_D, D_M, and D_G are transport parameters (mol·Pa^{-1}·h^{-1}) representing, respectively, the exchange of chemical between the benthic invertebrate and the water (W), the diet and the organism (D), and the elimination of chemical through metabolism (M) and growth (G). Exchange of chemical between the organism and feces is represented by D_D/Q, where Q is a maximum biomagnification factor. Equations and definitions of the parameters required to calculate these D values and to convert fugacities into concentrations are in Campfens and Mackay (1997).

This model is similar to that of Morrison et al. (1996), the main difference being in the parameterization. The fugacity model provided satisfactory predictions of polychlorinated biphenyl concentrations in *Diporeia* and oligochaetes collected in Lake Ontario. The model is a potentially useful tool for quantifying chemical exposure routes to benthic invertebrates and predicting chemical concentrations in their tissues based on chemical concentrations in water, sediment and diet.

TABLE 9.6

Model Equations, Parameters, and their Units of the Fish Bioaccumulation Model of Thomann et al. (1991, 1992).

$dv_F/dt = k_u.C_{WD} + \Sigma p_i.\alpha_i.I_P.v_i - (K + G).v_B$ (49)

Parameter	Units	Definition
α_i	gram chemical absorbed/gram chemical ingested	chemical assimilation efficiency from diet item i.
C_{WD}	g/L	freely dissolved chemical concentration in water
G	d^{-1}	growth rate
I_i	gram (lipid) prey/gram (lipid) predator/day	lipid specific consumption of the fish
K	d^{-1}	excretion rate constant from fish
k_u	L water.kg^{-1} lipid·d^{-1}	uptake rate from water
v_F	g/kg lipid	chemical concentration in fish
n_i	g/kg lipid	chemical concentration in diet item i
p_i	fraction	fraction of diet consisting a prey i
t	day	time

9.5.6 Bioaccumulation Models for Fish

9.5.6.1 Equilibrium Partitioning Models

The simplest bioaccumulation models for hydrophobic organic chemicals in fish are the **equilibrium partitioning models**, which assume that chemicals reach an equilibrium between the organism and water (e.g., Hamelink et al., 1971; Veith et al., 1979; Mackay, 1982). Dietary uptake and biomagnification, as well as growth dilution and metabolic transformation, are ignored. Most models are empirical and based on a set of measured bioconcentration factors for a particular species. At equilibrium, the bioconcentration factor is correlated to the octanol-water partition coefficient, usually ona logarithmic basis, resulting in correlations such as equations (19) and (21) or, generically:

$$\log BCF = \alpha \cdot \log K_{ow} + \beta \quad (42)$$

where α and β are constants derived by linear regression of the experimental data.

Due to differences in test species among bioconcentration tests as well as differences in the types of chemicals used in the various studies, possible metabolic transformation, and experimental artifacts, empirical relationships between BCF and K_{ow} are known to vary substantially. Because of these differences, assessments of the BCF of chemical substances can differ substantially, depending on the empirical model that is used. Special caution is required when empirical relationships report a value for the slope α that is significantly different from 1.0.

If equilibrium partitioning is assumed, and this equilibrium is successfully mimicked by the chemical's partitioning behavior between octanol and water (Mackay, 1982), and the fish is viewed as a "droplet" of lipid that is extracting the chemical from the water, BCF can be approximated by the product of K_{OW} and lipid content (Gobas and Mackay, 1987):

$$BCF = L.K_{ow} \quad \text{or} \quad \log BCF = 1.0.\log K_{ow} + \log L \quad (43)$$

This simple model provides reliable estimates for bioconcentration factors in fish of non-metabolizable hydrophobic chemical substances with log K_{ow} between 1 and 5. It also reliably estimates bioaccumulation factors of the same substances under field conditions,

TABLE 9.7

Model Equations, Parameters, and their Units of the Fish Bioaccumulation Model of Gobas (1993).

$$dC_F/dt = k_1.C_{WD} - k_2.C_F + k_D. \Sigma P_i.C_{D,i} - k_E.C_F - k_M.C_F - k_G.C_F \tag{50}$$

$$BAF = C_F/C_{WD} = \{k_1 + k_D.(\Sigma P_i.C_{D,i}/C_{WD})\}/(k_2 + k_E + k_M + k_G) \quad \text{at steady-state} \tag{51}$$

Parameter	Units	Definition
$C_{D,i}$	g/kg wet weight	chemical concentration in diet item i
C_F	g/kg wet weight	chemical concentration in fish
C_{WD}	g/mL	freely dissolved chemical concentration in overlying water
k_1	L water.kg^{-1} organism wet weight·d^{-1}	uptake clearance rate (often referred to as uptake rate constant) from water
k_2	d^{-1}	elimination rate constant from fish
k_D	kg food.kg-1 organism wet weight·d^{-1}	dietary uptake clearance rate (often referred to as dietary uptake rate constant)
k_E	d^{-1}	fecal egestion elimination rate constant from fish
k_G	d^{-1}	growth dilution rate constant
k_M	d^{-1}	metabolic transformation rate constant from fish
P_i	fraction	fraction of diet consisting a prey i
t	day	time

because chemical exchange between the fish and water is the predominant mechanism of chemical bioaccumulation. The model should not be applied to assess the bioconcentration and bioaccumulation factors of very hydrophobic substances with a log $K_{ow} > 5$ or to chemicals that are being metabolized in the organism.

9.5.6.2 Kinetic Models

Kinetic Models have been used widely and successfully to describe bioconcentration, biomagnification, and bioaccumulation of hydrophobic organic chemicals in fish. Most describe the fish as a single homogeneous compartment and are referred to as "one-compartment" models or sometimes as "water-fish two-compartment" models. Models treating fish as two compartments also have been developed and distinguish between parts of the fish that the chemical accesses at different rates. The choice of the number of compartments in a fish–water kinetic model is, in some cases, important to accurately describe the bioconcentration kinetics in fish in laboratory experiments. However, there is considerable evidence that indicates that, under field conditions, persistent hydrophobic organic chemicals are generally homogeneously distributed among the internal tissues of the organism as long as the concentrations in the tissues are expressed on a lipid weight basis.

The exception is for chemicals that are being metabolized in certain organs of the organism, such as the liver. If knowledge of differences in concentrations between tissues is required, physiologically based pharmacokinetic models (PBPK) models should be used for the assessment of internal concentrations in the organism.

A number of authors (e.g., Bruggeman et al., 1981 and 1984; Opperhuizen et al., 1985; Gobas et al., 1989) have applied simple kinetic models (i.e., equations (13)-(15), (25)-(27), (29, (30))to fish. In most cases, the models are used in a descriptive sense to describe empirical data derived from bioconcentration or bioaccumulation tests. These models can play an important role in the analysis of the results of bioconcentration tests, but they are generally inapplicable to bioaccumulation under field conditions.

Thomann (1989) and Gobas (1993) have developed physiologically based kinetic models, including rate constants for chemical uptake and elimination based on physiological parameters such as gill ventilation rates, feeding rates and chemical assimilation rates as well as empirical correlations. Tables 9.6 and 9.7 list the model equations and parameters.

Both models include a series of sub-models to estimate the parameters the model uses, based on a set of bioenergetic parameters and empirical correlations. Both models are used widely and have been tested against field data from various locations. Model predictions are generally in good agreement with observations made in the field. Predictions of the BAF by both models are generally within a factor of two to three of observed values. This level of agreement is often acceptable, given the considerable variability in concentrations observed in many field studies. Both models are discussed in section 9.5.7.

9.5.6.3   *Physiological Models*

Physiological Models for chemical bioaccumulation in fish are based on the same mass balance equations as the kinetic models for bioaccumulation, but the rate constants and chemical fluxes that quantify the rates of uptake and elimination of the substance are derived from K_{OW} and a set of physiological parameters. The most well known model in this category is the FGETS (Food and Gill Exchange of Toxic Substances) model Barber et al. (1988, 1991) developed. This is a FORTRAN simulation model that predicts dynamics of a fish's whole body concentration of non-ionic, nonmetabolized, organic chemicals absorbed from the water only, or from water and food jointly.

FGETS considers both the biological attributes of the fish and the physico-chemical properties of the chemical to determine diffusive exchange across gill membranes and intestinal mucosa. The physiological parameters the model uses are the fish's gill and intestinal morphometry, the body weight of the fish, and the fractional aqueous, lipid, and structural organic composition. The physico-chemical properties required are the chemical's aqueous diffusivity, molar volume, and K_{OW}.

FGETS is parameterized for a particular fish species by means of two morphological/physiological databases that delineate the fish's gill morphometry, feeding and metabolic demands, and body composition. Presently, joint water and food exposure is parameterized for salmonids, centrarchids, cyprinids, percids, and ictalurids. The copy of the model can be obtained from the USEPA at web site ftp://ftp.epa.gov/epa_ceam/wwwhtml/software.htm.

9.5.6.4   *Physiologically Based PharmacoKineticModels*

PBPK models segment the fish into a number of compartments representing different organs and tissues. Each compartment receives chemical from the blood compartment, which circulates among all the compartments in the model. In most models, equilibrium partitioning is applied to describe transfer from the blood to the organ or tissue. It is recognized that diffusion can limit the exchange of chemical between the blood and the target organs, but parameterizing this process is difficult. Chemical elimination from each compartment is by a combination of equilibrium partitioning to the blood and organ specific elimination rates, such as urine excretion (from the kidney), metabolic transformation (e.g., from the liver), and fecal excretion.

The advantage of the PBPK models is their ability to assess concentrations in different tissues of the fish. This is important for substances that are readily metabolized and when exposure is of a relatively short duration, such as situations when drugs or vitamins are administered to fish in aquacultures or fish tanks or the fish is exposed to a pulse of contaminant, e.g., following a spill. In these cases, steady-state is rarely achieved, the substance has not been fully distributed among the tissues and organs, and differences in the ability of the various tissues to metabolize the substance may cause substantial differences in tissue concentrations. PBPK models for fish are also applicable to risk assessment involving human consumption of certain edible parts of the fish.

The primary disadvantage of PBPK models is that they require extensive data describing chemical partitioning between blood and the various tissues, as well as physiological data regarding blood flows into all organs, urine and fecal excretion rates, sizes of individual organs, and metabolic transformation rates. Most of these data are both chemical and organism specific and can be difficult to obtain. Even in cases where these data are available, model calibration often is required to produce estimates that conform with observed data. PBPK models generally are developed in conjunction with laboratory experiments, examples being those of Nichols et. al. (1990) and Law et al. (1991).

9.5.7   Foodweb Bioaccumulation Models

To assess the bioaccumulation of hydrophobic organic substances in organisms of aquatic food chains, food web bioaccumulation models can be used. The most frequently used food web bioaccumulation models are those by Thomann et al. (1991, 1992) and Gobas (1993). Campfens and Mackay (1997) recently developed an alternative fugacity based model. All of these models contain sub-models describing bioaccumulation in each organism comprising the aquatic food web. These sub-models are linked to represent chemical transfer between organisms in the food web.

9.5.7.1   *The Thomann Model*

Thomann et al. (1992) developed a steady-state food web bioaccumulation model that combines kinetic and bioenergetic parameters to quantify chemical uptake and elimination by zooplankton, benthic invertebrates and fish. First-order kinetic rate constants quantify uptake of freely-dissolved chemical from interstitial water and overlying water and total chemical elimination from gills and feces. Various physiological and bioenergetic parameters quantify chemical uptake from diet and growth dilution.

The diet of benthic invertebrates is restricted to the detritus fraction of sediment and phytoplankton. Fish can feed on all organisms other than themselves. Phytoplankton and detritus are assumed to be in chemical equilibrium with the overlying and sediment-pore water, respectively.

Because kinetic rate constants are not readily available in the literature, Thomann et al. (1992) used a set of formulas to estimate the gill uptake rate constant and an excretion rate constant. The uptake rate constant is a function of the respiration rate of the organism and the efficiency of chemical transfer across the organism's membrane. The excretion rate constant is related to the uptake rate constant and K_{OW}.

The model has been applied to the Lake Ontario foodweb to evaluate the relative importance of overlying water, interstitial water, sediment, and diet as chemical exposure routes. Additionally, the model was used as a tool for estimating the significance of the sediment/overlying-water partition coefficient for this ecosystem. The model provided reasonable results and insights into bioaccumulation in benthic/pelagic foodwebs. The model is a valuable tool for predicting chemical concentrations in a variety of aquatic biota.

9.5.7.2   *The Gobas Model*

Gobas (1993) published a foodweb bioaccumulation model to predict chemical concentrations in phytoplankton, macrophytes, zooplankton, benthic invertebrates, and fish, based

TABLE 9.8

Illustrative calculations of the organic carbon based suspended and bottom sediment partition coefficient K_{oc} (Equation 35); the chemical bioavailability for uptake from the water via the respiratory surface F_{DW} (Equation 34); the bioconcentration factor BCF in algae and phytoplankton (Equation 36); the bioconcentration factor BCF in aquatic macrophytes (Equation 37); the bioconcentration factor BCF in zooplankton (Equation 38); the biota-sediment accumulation factor BSAF for a benthic detritovore consuming phytoplankton according to the equilibrium partitioning model (Equation 41), and the Morrison et al. (1996) model (Equation 45); the uptake rate constant k_1, the dietary uptake rate constant k_D, the elimination rate constant k_2, the fecal egestion rate constant k_E and the growth dilution rate constant k_G in a 1 kg fish (following Gobas 1993), the metabolic transformation rate constants k_M in a 1 kg fish (derived from literature data); the bioconcentration factor BCF in a 1 kg fish (Equation 14), the bioconcentration factor in a 1 kg fish when assuming a chemical equilibrium (Equation 48), the biomagnification factor BMF in 1 kg fish (Equation 26) and the bioaccumulation factor BAF in 1 kg fish (Equation 29) of lindane, anthracene, Polychlorinated Biphenyl (PCB) 52, 2,3,7,8-tetrachlorodibenzo-p-dioxin (2,3,7,8-TCDD), PCB 153 and PCB 209. The model parameters that were used in the calculations are summarized in the footnotes.

Chemical	log K_{ow}	log K_{oc}	F_{DW}	log BCF Algae	log BCF Macrophytes	log BCF Zooplankton	BSAF(41) Benthic Invertebrate	BSAF(45) Benthic Invertebrate
Lindane	3.70	3.31	100%	1.70	1.70	2.18	2	1.22
Anthracene	4.54	4.15	98%	2.53	2.53	3.02	2	1.23
Pentachlorobenzene	5.03	4.64	95%	3.01	3.01	3.51	2	1.25
PCB52	6.10	5.71	64%	3.90	3.90	4.58	2	1.53
2,3,7,8-TCDD	6.85	6.46	24%	4.23	4.23	5.33	2	2.22
PCB153	6.90	6.51	22%	4.24	4.24	5.38	2	2.28
PCB209	8.25	7.86	1%	4.34	4.34	6.73	2	3.09

Chemical	k_1	k_2	k_E	k_M	k_G	k_D	BCF(14) Fish	BCF(48) Fish, Equil.	BMF Fish	BAF Fish
Lindane	86.6	$1.73 \cdot 10^{-1}$	$1.24 \cdot 10^{-3}$	0	$2.0 \cdot 10^{-3}$	$1.55 \cdot 10^{-2}$	2.69	2.70	0.09	2.71
Anthracene	88.0	$2.54 \cdot 10^{-2}$	$1.24 \cdot 10^{-3}$	0	$2.0 \cdot 10^{-3}$	$1.54 \cdot 10^{-2}$	3.49	3.54	0.54	3.60
Pentachlorobenzene	88.2	$8.23 \cdot 10^{-3}$	$1.23 \cdot 10^{-3}$	0	$2.0 \cdot 10^{-3}$	$1.54 \cdot 10^{-2}$	3.89	4.03	1.34	4.18
PCB52	88.3	$7.00 \cdot 10^{-4}$	$1.20 \cdot 10^{-3}$	0	$2.0 \cdot 10^{-3}$	$1.50 \cdot 10^{-2}$	4.26	5.10	3.06	5.41
2,3,7,8-TCDD	88.3	$1.25 \cdot 10^{-4}$	$1.06 \cdot 10^{-3}$	0.001	$2.0 \cdot 10^{-3}$	$1.33 \cdot 10^{-2}$	4.44	5.85	4.18	6.44
PCB153	88.3	$1.11 \cdot 10^{-4}$	$1.05 \cdot 10^{-3}$	0	$2.0 \cdot 10^{-3}$	$1.31 \cdot 10^{-2}$	4.45	5.90	4.14	6.49
PCB209	88.3	$4.97 \cdot 10^{-6}$	$2.43 \cdot 10^{-4}$	0	$2.0 \cdot 10^{-3}$	$3.03 \cdot 10^{-3}$	4.59	7.25	1.35	7.48

TABLE 9.8 (CONTINUED)

Footnotes for Table 9.8: Model parameters used in the illustrative calculations of chemical bioaccumulation factors.

Density of lipids (kg/L)	0.9
Density of organic carbon (kg/L)	0.9
Dietary uptake efficiency of benthic detritovore	0.72
Feeding rate (kg/day)	0.0002
Fish weight (kg)	1
Fraction of dissolved organic matter in water column (kg/L)	0.000001
Fraction of fish diet consisting of benthic invertebrates	1
Fraction of particulate organic matter in water column (kg/L)	0.000001
Gill uptake efficiency for benthic detritovore	0.75
Gill ventilation rate (L/day)	20
Organism-gastro-intestinal content partition coefficient of the chemical substance	3
Organism-water partition coefficient of the chemical substance	$0.05 \cdot K_{ow}$
Metabolic transformation rate constant for 2,3,7,8-TCDD in fish (1/day)	0.001
Metabolic transformation rate constant for all chemicals in benthic detritovore (1/day)	0
Lipid content fish (kg/kg)	0.1
Lipid content of algae (fraction)	0.01
Lipid content of benthic detritovore (fraction)	0.05
Lipid content of macrophytes (fraction)	0.01
Lipid content of zooplankton (fraction)	0.03
Organic carbon content of bottom sediments (fraction)	0.025
Sediment-water Disequilibrium in this illustrative ecosystem	10
Weight of benthic detritovore (kg)	0.002

on chemical concentrations in water and sediments. A set of kinetic rate constants is derived from a combination of empirical correlations, physiological data and bioenergetic relationships to determine concentrations at steady-state. Biomagnification in the model is based on gastro-intestinal magnification, (Gobas et al. 1993).

The model applies equilibrium partitioning to estimate chemical concentrations in phytoplankton, macrophytes, zooplankton, and benthic invertebrates. Chemical concentrations in sediment and water, along with environmental and trophodynamic information, are used to quantify chemical concentrations in all aquatic biota. This model can be applied to many aquatic food webs and relies on a relatively small set of input parameters which are readily accessible.

The performance of the model was satisfactory when tested by comparing predictions with observed concentrations of a range of organochlorine chemicals in organisms of the Lake Ontario food-chain (Gobas, 1993). The 95% confidence intervals of the ratio of observed and predicted concentrations of persistent organic chemicals is a factor of 2 to 3.

The EPA (1994) has reviewed the model and is applying it in its Great Lakes Water Quality Initiative (EPA 1995). The model is available in a self-contained MicroSoft Windows-based program which also contains a method for uncertainty analysis by Monte Carlo simulation. This model can be obtained from the Internet web site http://fas/sfu.ca/rem/era/era.html".

9.5.7.3   The Campfens and Mackay Model

Campfens and Mackay (1997) developed a steady-state fugacity based foodweb bioaccumulation model to estimate chemical concentrations in aquatic biota from chemical concentrations in sediment and water. The foodweb can consist of any number or classes of organisms including zooplankton, benthic invertebrates, and fish, with all organisms potentially consuming all organisms, including themselves. The submodels treat chemical uptake from water, sediment, and food, and chemical elimination via respiration, egestion, metabolism, and growth dilution. Chemical concentrations in phytoplankton are estimated from equilibrium partitioning relationships with water. Predictions of this model were compared with observed PCB concentrations in the Lake Ontario food chain, yielding satisfactory results. It is available from www.trentu.ca/envmodel.

9.6   Illustrative Calculations

Table 9.8 provides examples of some of the methods for assessing of the bioconcentration and biomagnification potential of a range of hydrophobic organic chemicals. Included are sample calculations for K_{OC}, the dissolved chemical concentration in water, the BCF in algae, zooplankton, a benthic invertebrate species (i.e., a benthic detritovore, e.g., amphipod), and fish (e.g., rainbow trout).

The BCFs and BAFs are expressed in terms of the dissolved chemical concentration in water. The BCF represents a bioaccumulation estimate under laboratory conditions where organisms are exposed to waterborn chemical only. The BAF represents a field-based bioaccumulation estimate, where organisms are exposed to chemical through water and their diet. In this example, the benthic invertebrates are consuming algae, and the fish are consuming the benthic invertebrates.

9.7   Bioaccumulation Databases and Key References

Bioavailability: Physical, Chemical and Biological Interactions. 1994. J.L. Hamelink, P.F. Landrum, H.L. Bergman, and W.H. Benson, Eds., SETAC Special Publication. CRC Press, Boca Raton, FL.
Bioaccumulation of Xenobiotic Compounds. 1990. D.W. Connell. CRC Press, Boca Raton, FL.
Chemical Dynamics in Fresh Water Ecosystems. Gobas, F.A.P.C. and J.A. McCorquodale, Eds.; Lewis Publishers: Ann Arbor, Michigan.

Illustrated Handbook of Physical-Chemical Properties and Environmental Fate for Organic Chemicals. Volumes I, II, III, IV, V. 1992, 1993, 1995, 1997. Mackay, D., W.Y. Shiu, and K.C. Ma. Lewis Publishers, Chelsea, Michigan, USA.

Handbook of Environmental Data on Organic Chemicals. 1983. Verschueren K. Van Norstrand Reinhold Company, New York, New York, USA.

ChemFate. Howard, P.H. Syracuse Research Corporation. Lewis Publishers, Chelsea, Michigan, USA, ISBN 0-87371-785-6.

Environmental Fate Database (EFDB) : http://esc.syrres.com/~ESC/efdb.htm

---

References

American Standards for Testing and Materials. 1988. Standard practice for conducting bioconcentration tests with fishes and saltwater bivalve molluscs. E 1022 – 84. *Annual Book of ASTM Standards* Philadelphia, PA.

Anliker, R. and P. Moser. 1987. The limits of bioaccumulation or organic pigments in fish: Their relation to the partition coefficient and the solubility in water and octanol. *Ecotoxicol. Environ. Safety* 13: 43-52.

Baker, J.E., P.D. Capel, and S.J. Eisenreich. 1986. Influence of colloids on sediment-water partition coefficients of polychlorobiphenyl congeners in natural waters. *Environ. Sci. Technol.* 20, 1136-1143.

Baker, J.E., S.J. Eisenreich, and B.J. Eadie. 1991. Sediment trap fluxes and benthic recycling of organic carbon, polycyclic aromatic hydrocarbons and polychlorobiphenyl congeners in Lake Superior. *Environ. Sci. Technol.* 25, 500-509.

Barber, M.C., L.A. Suárez, and R.R. Lassiter. 1988. Modeling bioconcentration of nonpolar organic pollutants by fish. *Environ. Toxicol. and Chem.* 7:545-558.

Barber, M.C., L.A. Suárez, and R.R. Lassiter. 1991. Modelling bioaccumulation of organic pollutants in fish with an application to PCBs in Great Lakes salmonids. *Can. J.of Fisheries and Aquatic Sciences* 48:318-337.

Bierman, V.J., Jr. 1990. Equilibrium partitioning and biomagnification of organic chemicals in benthic animals. *Environ. Sci. Technol.* 24:1407-1412.

Bintein, S., J. Devillers, and W. Karcher. 1993. Non-linear dependence of fish bioconcentration on n-octanol/water partition coefficients. *Env. Res.* 1: 29–39.

Black M.C. and McCarthy J.F. 1988. Dissolved organic macromolecules reduce the uptake of hydrophobic organic contaminants by the gills of the rainbow trout (Salmo gairdneri). *Environ. Toxicol. Chem.* 7, 593-600.

Blau, G.E. and G.L. Algin. 1978. *A User Manual for BIOFAC: A Computer Program for Characterizing the Ratio of Uptake and Clearance of Chemicals in Aquatic Organisms.* Dow Chemical Corporation, Midland, MI.

Blau, G.E., W.B. Neely, and D.R. Branson. 1975. Ecokinetics: a study of the fate and distribution of chemicals in laboratory ecosystems. *AIChE J.* 21:854-861.

Bruggeman, W.A., L.B.J.M. Matron, D. Kooiman, and O. Hutzinger. 1981. Accumulation and elimination kinetics of di-, tri-, and tetra chlorobiphenyls by goldfish after dietary and aqueous exposure. *Chemosphere* 10, 811-832.

Bruggeman, W.A., A. Opperhuizen, A. Wijbenga, and O. Hutzinger. 1984. Bioaccumulation of superlipophilic chemicals in fish. *Toxicol. Environ. Chem.* 4:779-788.

Campfens, J. and D. Mackay. 1997. Fugacity-based model of PCB bioaccumulation in complex aquatic food webs. *Environ. Sci. Technol.* in press.

Carter, C.W. and I.H. Suffet. 1982. Binding of DDT to dissolved humic materials. *Environ. Sci. Technol.* 16:735-740.

Chiou, C.T., D.E. Kile, T.I. Brinton, R.L. Malcolme, J.A. Leenheer, and P. MacCarthy. 1987. A comparison of water solubility enhancements of organic solutes by aquatic humic materials and commercial humic acids. *Environ. Sci. Technol.* 21:1231-1234.

Clark, K.E., F.A.P.C. Gobas, and D. Mackay. 1990. Model of Organic Chemical Uptake and Clearance by Fish from Food and Water. *Environ. Sci. Technol.* 24, 1203-1213.

Clark, T., K. Clark, S. Paterson, R. Norstrom, and D. Mackay. 1988. Wildlife monitoring, modelling and fugacity. *Environ. Sci. Technol.* 22:120-127.

Clayton, J.R., S.P. Pavlou, and N.F. Breitner. 1977. Polychlorinated biphenyls in coastal marine zooplankton. *Environ. Sci. Technol.* 18, 676-682.

Connolly, J.P. and C.J. Pedersen. 1988. A thermodynamically based evaluation of organic chemical bioaccumulation in aquatic organisms. *Environ. Sci. Technol.* 22:99-103.

DiToro, D.M., C.S. Zarba, D.J. Hansen, W.J. Berry, R. Swartz, C.E. Cowan, S.P. Pavlou, H.E. Allen, N.A. Thomas, and P.R. Paquin. 1991. Technical basis for establishing sediment quality criteria for nonionic organic chemicals using equilibrium partitioning. *Environ. Toxicol. Chem.* 10:1541-1586.

Eadie, B.J., N.R. Moorehead, and P.F. Landrum. 1990. Three-phase partitioning of hydrophobic organic compounds in Great Lakes waters. *Chemosphere* 20:161-178.

Erickson, R.J. and J.M. McKim. 1990a. A simple flow limited model for the exchange of organic chemicals at fish gills. *Environ. Toxicol. Chem.* 9:159-165.

Erickson, R.J. and J.M. McKim. 1990b. A model for exchange of organic chemicals at fish gills: flow and diffusion limitations. *Aquat. Toxicol.* 18:175-198.

Evans, H.E. 1988. The binding of three PCB congeners to dissolved organic carbon in fresh waters. *Chemosphere* 17:2325-2338.

Flynn, G.L. and S.H. Yalkowsky. 1972. Correlation and prediction of mass transport across membranes I. Influence of alkyl chain length on flux determining of barrier and diffusant. *J. Pharm. Sci.* 61:838-851.

Gauthier T.D., W.R. Seltz, and C.L. Grant. 1987. Effects of structural and compositional variations of dissolved humic materials on pyrene K_{OC} values. *Environ. Sci. Technol.* 21:243-248.

Geyer, H., G. Politzki, and D. Freitag. 1984. Prediction of ecotoxicological behaviour of chemicals: relationship between n-octanol-water partition coefficient and bioaccumulation of organic chemicals by alga Chlorella. *Chemosphere* 13, 269-284.

Gobas, F.A.P.C. 1993. A model for predicting the bioaccumulation of hydrophobic organic chemicals in aquatic food-webs: application to Lake Ontario. *Ecol. Modelling* 69:1-17.

Gobas, F.A.P.C. and D. Mackay. 1987. Dynamics of Hydrophobic Organic Chemical Bioconcentration in Fish. *Environ. Toxicol. Chem.* 6:495-504.

Gobas, F.A.P.C. and X. Zhang. 1995. Measuring bioconcentration factors and rate constants of chemicals in aquatic organisms under conditions of variable water concentrations and short exposure time. *Chemosphere.* 25:1961-1971.

Gobas, F.A.P.C. and X. Zhang. 1994. Interactions of Organic Chemicals with Organic Matter in the Aquatic Environment. In *Environ. Toxicol. Chem.* Special Publication, "Bioavailability: Physical, Chemical and Biological Interactions, J.L. Hamelink, Peter F. Landrum, Harold L. Bergman, and W.H. Benson, Eds., pp. 83-91, Lewis Publishers, Chelsea, MI.

Gobas, F.A.P.C., D.C. Bedard, J.J.H. Ciborowski, and G.D. Haffner. 1989b. Bioaccumulation of Chlorinated Hydrocarbons by the Mayfly Hexagenia limbata in Lake St. Clair. *J. Great Lakes Res.* 15, 581-588.

Gobas, F.A.P.C., K.E. Clark, W.Y. Shiu, and D. Mackay. 1989a. Bioconcentration of Polybrominated Benzenes and Biphenyls and Related Superhydrophobic Chemicals in Fish : Role of Bioavailability and Faecal Elimination. *Environ. Toxicol. Chem.* 8, 231-247.

Gobas, F.A.P.C., E.J. McNeil, L. Lovett-Doust, and G.D. Haffner. 1991. Bioconcentration of chlorinated aromatic hydrocarbons in aquatic macrophytes. *Environ. Sci. Technol.* 25:924-929.

Gobas, F.A.P.C., J.R. McCorquodale, and G.D. Haffner. 1993. Intestinal Absorption and Biomagnification of Organochlorines. *Environ. Toxicol. Chem.* 12:567-576.

Gobas, F.A.P.C., M.N. Z'Graggen, and X. Zhang. 1995. Time response of the Lake Ontario Ecosystem to Virtual Elimination of PCBs. *Environ. Sci. Technol.* 29:2038-2046.

Gobas, F.A.P.C., E.J. McNeil, L. Lovett-Doust, and G.D. Haffner. 1991. Bioconcentration of Chlorinated Aromatic Hydrocarbons in Aquatic Macrophytes (*Myriophyllum spicatum*). *Environ. Sci. Technol.* 25, 924-929.

Gobas, F.A.P.C., D.C.G. Muir, and D. Mackay. 1989. Dynamics of Dietary Bioaccumulation and Faecal Elimination of Hydrophobic Organic Chemicals in Fish. *Chemosphere* 17, 943-962.

Gobas, F.A.P.C., A. Opperhuizen, and O. Hutzinger. 1986. Bioconcentration of Hydrophobic Chemicals in Fish : Relationship with Membrane Permeation. *Environ. Toxicol. Chem.* 5:637-646.

Gobas, F.A.P.C., X. Zhang, and R.J. Wells. 1993. Gastrointestinal magnification: the mechanism of biomagnification and food-chain accumulation of organic chemicals. *Environ. Sci. Technol.* 27:2855-2863.

Gobas, F.A.P.C., J.W.B. Wilcockson, R.W. Russell, and G.D. Haffner. 1999. Mechanism of biomagnifiction in fish under laboratory and field conditions. *Environ. Sci. Technol.* 33: 133–141.

Hamelink, J.L., R.C. Waybrandt, and R.C. Ball. 1971. Proposal: Exchange equilibriums control the degree chlorinated hydrocarbons are biologically magnified in lentic environments. *Trans. Am. Fish. Soc.* 100:207-214.

Hassett, J.P. and E. Milicic. 1985. Determination of equilibrium and rate constants for the binding of a polychlorinated biphenyl congener by dissolved humic substances. *Environ. Sci. Technol.* 19:638-643.

Karickhoff, S.W. 1981. Semi-empirical estimation of sorption of hydrophobic pollutants on natural sediments and soils. *Chemosphere* 10:833-846.

Karickhoff, S.W. 1984. Organic Pollutant Sorption in Aquatic Systems. *J. Hydraulic Eng. ASCE* 110, 707-735.

Lake, J.L., N.I. Rubinstein, H. Lee II, C.A. Lake, J. Heltshe, and S. Pavignano. 1990. Equilibrium partitioning and bioaccumulation of sediment-associated contaminants by infaunal organisms. *Environ. Toxicol. Chem.* 9:1095-1106.

Landrum P.F., M.D. Reinhold, S.R. Nihart, and B.J. Eadie. 1985. Predicting the bioavailability of organic xenobiotics to Pontoporeia hoyi in the presence of humic and fulvic materials and natural dissolved organic carbon. *Environ. Toxicol. Chem.* 4, 459-467.

Landrum P.F., S.R. Nihart, B.J. Eadie, and W.S. Gardner. 1984. Reverse-phase separation method for determining pollutant binding to Aldrich humic acid and dissolved organic carbon of natural waters. *Environ. Sci. Technol.* 18:187-192.

Landrum, P.F. and R. Poore. 1988. Toxicokinetics of selected xenobiotics in Hexagenia limbata. *J. Great Lakes Res.* 14:427-437.

Landrum, P.F., T.D. Fontaine, W.R. Faust, B.F. Eadie, and G.A. Lang. 1992. Modeling the accumulation of polycyclic aromatic hydrocarbons by the amphipod *Diporeia* (spp.). In *Chemical Dynamics in Fresh Water Ecosystems*, F.A.P.C. Gobas and J.A. McCorquodale, Eds., pp. 111-128, Lewis Publishers: Ann Arbor, Mi.

Law, F.C. P. S. Abedini, and C.J. Kennedy. 1991. A biologically based toxicokinetic model for pyrene in rainbow trout. *Toxicol. App. Pharmacol.* 110:390-402.

Lee, H. II., B.L. Boese, J. Pelletier, M. Winsor, D.T. Specht, and R.C. Randall. 1993. *Guidance Manual: Bedded Sediment Bioaccumulation Tests.* EPA/600/R-93/183.

Mackay, D. 1982. Correlation of Bioconcentration Factors. *Environ. Sci. Technol.* 16: 274–278.

Mackay, D. and S. Paterson. 1982. Fugacity revisited. *Environ. Sci. Technol.* 16:655a-661a.

Mackay, D. 1991. *Multimedia Environmental Models: The Fugacity Approach*, CRC Press/Lewis Publishers. Boca Raton, FL.

Mallhot, H. 1987. Prediction of algal bioaccumulation and uptake rate of nine organic compounds by ten physicochemical properties. *Environ. Sci. Technol.* 21:1009-1013.

Meylan, W.M., P.H. Howard, R.S. Boethling, D. Aronson, H. Printup, and S. Gouchie. 1999. Improved method for estimating bioconcentration/bioaccumulation factor from octanol/water partition coefficient. *Environ. Toxicol. Chem.* 18: 664–672.

Markwell, R.D., D.W. Connell, and A.J. Gabric. 1989. Bioaccumulation of lipophilic compounds from sediments by oligochaetes. *Water. Res.* 23:1443-1450.

McCarthy J.F. and B.D. Jimenez. 1985. Reduction in bioavailability to bluegills of polycyclic aromatic hydrocarbons bound to dissolved humic material. *Environ. Toxicol. Chem.* 4, 511-521.

McKim, J.M. and R.J. Erickson. 1991. Environmental impacts on the physiological mechanisms controlling xenobiotic transfer across fish gills. *Physiol. Zool.* 64:39-67.

McKim, J.M., P.K. Schnieder, and G. Veith. 1985. Absorption dynamics of organic chemical transport across trout gills as related to octanol-water partition coefficient. *Toxicol. Appl. Pharmacol.* 77, 1-10.

Moorehead, N.R., B.J. Eadie, B. Lake, P.F. Landrum and D. Berner. 1986. The sorption of PAH onto dissolved organic matter in Lake Michigan. *Chemosphere* 15:403-412.

Moriarty, F. 1975. Exposure and Residues. In F. Moriarty, ed., *Organochlorine Insecticides – Persistent Organic Pollutants*, Ch. 2. Academic Press. London.

Morrison, H.A. 1995. Canadian Data Report of Fisheries and Aquatic Sciences No. 955. GLIFAS, Department of Fisheries & Oceans.

Morrison, H.A., F.A.P.C. Gobas, R. Lazar, and G.D. Haffner. 1996. Development and verification of a bioaccumulation model for organic contaminants in benthic invertebrates. *Environ. Sci. Technol.* 30:3377-3384.

Morrison, H.A., F.A.P.C. Gobas, R. Lazar, and G.D. Haffner. 1997. Derivation of a benthic/pelagic foodweb bioaccumulation model for organic contaminants. *Environ. Sci. Technol.* Under review.

Muir, D.C.G. and A.L. Yarechewski. 1988. Dietary accumulation of four chlorinated dioxin congeners by rainbow trout and fathead minnows. *Environ. Toxicol. Chem.* 7:227-236.

Muir, D.C.G., A.L. Yarechewski, A. Knoll, and G.R.B. Webster. 1986. Bioconcentration and disposition of 1,3,6,8-tetrachlorodibenzo-p-dioxin and octachlorodibenzo-p-dioxin by rainbow trout and fathead minnows. *Environ. Toxicol. Chem.* 5:261-272.

Nichols, J.W., J.M. McKim, M.E. Andersen, M.L. Gargas, H.J. Clewell III, and R.J. Erickson. 1990. A physiology based toxicokinetic model for the uptake and disposition of waterborne organic chemicals in fish. *Toxicol. Appl. Pharmacol.* 106:433-447.

Neely, W.B., D.R. Branson, and G.E. Blau. 1974. Partition coefficient to measure bioconcentration potential of organic chemicals in fish. *Environ. Sci. Technol.* 8:1113-1115.

Niimi, A.J. and B.G. Oliver. 1987. Influence of molecular weight and molecular volume on the dietary absorption efficiency of chemicals by fish. *Canadian Journal of Fisheries and Aquatic Sciences* 45:222-227.

Opperhuizen, A., E.W. Van der Velde, F.A.P.C. Gobas, D.A.K. Liem, J.M.D. Van der Steen, and O. Hutzinger. 1985. Relationship between Bioconcentration in Fish and Steric Factors of Hydrophobic Chemicals. *Chemosphere* 14, 1871-1896.

Organization for Economic Co-operation and Development. 1996. Bioaccumulation: Flow-through Fish Test, 305 E. OECD Guideline for Testing Chemicals.

Parkerton, T.F. 1993. Estimating toxicokinetic parameters for modelling the bioaccumulation of non-ionic organic chemicals in aquatic organisms. Ph.D. thesis. Rutgers The State University of New Jersey, New Brunswick, New Jersey.

Randall, D.J. 1970. In *Fish Physiology. Vol. IV.* W.S. Hoar, and D.J. Randall, Eds., Academic Press, New York.

Reilly, P.M., R. Bajramovic, G.E. Blau, D.R. Branson, and M.W. Sauerhoff. 1977. Guidelines for the optimal design of experiments to estimate parameters in first order kinetic models. *Can. J. Chem. Eng.* 55:614-622.

Reinert, R.E. 1972. Accumulation of dieldrin in algae (Scenedesmus obliquus), Daphnia magna, and the guppy (Poecilia reticulata). *J. Fish. Res. Bd. Can.* 29:1413.

Resendes, J.W., Y. Shiu, and D. Mackay. 1992. Sensing the fugacity of hydrophobic organic chemicals in aqueous systems. *Environ. Sci. Technol.* 26:2381-2387.

Rice, C.P. and H.C. Sikka. 1973. Uptake and metabolism of DDT by six species of marine algae. *J. Agr. Food Chem.* 21:148-152.

Saarikoski, J., M. Lindstrom, M. Tyynila, and M. Viluksela. 1986. Factors affecting the absorption of phenolics and carboxylic acids in the guppy (poecilia reticulata). *Ecotoxicol. Environ. Saf.* 11, 158-173.

Servos, M.R., D.C.G. Muir, and G.R.B. Webster. 1989. Effect of dissolved organic matter on bioavailability of polychlorinated dibenzo-*p*-dioxins. *Aquatic Toxicology* 14:169-184.

Shea D. 1988. Developing national sediment quality criteria. *Environ. Sci. Technol.* 22, 1256-1261.

Sodergren, A. 1968. Uptake and accumulation of C^{14}-DDT by chlorella sp. (Chlorophyceae). *Oikos.* 19:126-138.

Spacie, A., and J.L Hamelink. 1982. Alternative models for describing the bioconcentration of organics in fish. *Environ. Toxicol. Chem.* 1:309-320.

Sproule, J.W., W.Y. Shiu, D. Mackay, W.H. Schroeder, R.W. Russell, and F.A.P.C. Gobas. 1991. In-Situ Measurement of the Truly Dissolved Concentration of Hydrophobic Chemicals in Natural Waters. *Environ. Toxicol. Chem.* 10(1), 9-20.

Stein, W.D. 1981. Permeability for lipophilic molecules. In *Membrane Transport*, S.L. Bonting, and J.J.H.H.M. de Pont, Eds., pp. 1-28. Elsevier, Amsterdam Holland.

Swackhamer, D.L. 1991. Bioaccumulation of toxic hydrophobic organic compounds at the primary trophic level. *J. Environ. Sci.* (China) 3:15-21.

Swackhamer, D.L. and R.S. Skoglund. 1993. Bioaccumulation of PCBs by algae: Kinetics versus equilibrium. *Environ. Toxicol. Chem.* 12:831-838.

Thomann R.V. and J.P. Connolly. 1984. Model of PCB in the Lake Michigan lake trout food chain. *Environ. Sci. Technol.* 18, 65-71.

Thomann, R.V. 1989. Bioaccumulation model of organic chemical distribution in aquatic food chains. *Environ. Sci. Technol.* 23, 699-707.

Thomann, R.V., J.P. Connolly, and T. Parkerton. 1991. Modeling accumulation of organic chemicals in aquatic food-webs. In *Chemical Dynamics in Fresh Water Ecosystems*, F.A.P.C. Gobas and J.A. McCorquodale, Eds., CRC Press/Lewis Publishers.

Thomann, R.V., J.P. Connolly, and T.F. Parkerton. 1992. An equilibrium model of organic chemical accumulation in aquatic food webs with sediment interaction. *Environ. Toxicol. Chem.* 11:615-629.

Thurston, R.V. and P.C. Gehrke. 1993. In *Fish Physiology, Toxicology, and Water Quality management. Proceedings of an International Symposium, Sacramento, California, USA, September 18-20, 1990.* R.C. Russo and R.V. Thurston, Eds., p. 95. Environmental Research Laboratory Office of Research and Development. EPA, Athens, Georgia. EP/600/R-93/157.

Tolls, J., M. Haller, M.H.C. Thijssen, and D.T.H.M. Sijm. 1996. LAS Bioconcentration Is Isomer Specific. *Abstracts of 17th Annual Meeting of the Society of Environmental Toxicology and Chemistry,* Washington DC.

Trowbridge, A. and D.L. Swackhammer. 1995. Accumulation of polychlorinated biphenyl congeners in the Lake Michigan lower pelagic foodweb. In: *Abstracts of the Society of Environmental Toxicology and Chemistry, Second SETAC World Congress, November 5-9, Vancouver, B.C. Canada,* pp. 35.

USEPA. 1995. Great Lakes Water Quality Initiative Technical Support Document for the Procedure to Determine Bioaccumulation Factors. EPA-820-B-95-005.

USEPA, 1994. *Great Lakes Water Quality Initiative Technical Support Document.* EPA-822-R-94-002.

Van der Oost, R., H. Heida, and A. Opperhuizen. 1988. Polychlorinated biphenyl congeners in sediments, plankton, molluscs, crustaceans, and eel in a freshwater lake: Implications of using reference chemicals and indicator organisms in bioaccumulation studies. *Arch. Environ. Contam. Toxicol.* 17:721-729.

Van Hoogen, G. and A. Opperhuizen. 1988. Toxicokinetics of chlorobenzenes in fish. *Environ. Toxicol. Chem.* 7, 213-219.

Veith, G.D., D.L Defoe, and B.V. Bergstaedt. 1979. Measuring and estimating the bioconcentration factor of chemicals in fish. *J. Fish. Res. Board Can.* 36:1040-1048.

Walker, C.H. 1987. Kinetic models for predicting bioaccumulation of pollutants in ecosystems. *Environmental Pollution.* 44:227-240.

Yin, C. and J.P. Hassett. 1986. Gas-partitioning approach for laboratory and field studies of mirex fugacity in water. *Environ. Sci. Technol.* 20:1213-1217.

Yoshida, T., F. Takashima, and T. Watanabe. 1973. Distribution of (C14)PCBs in carp. *Ambio* 2:111-113.

10

Sorption to Aerosols

Terry F. Bidleman and
Tom Harner

CONTENTS

10.1 Introduction

Many persistent organic pollutants (POPs) are semivolatile compounds which undergo exchange between air and atmospheric particles, soils, and vegetation. Classes of POPs include polychlorinated biphenyls (PCBs), pesticides, polycyclic aromatic hydrocarbons (PAHs), and polychlorinated dibenzo-p-dioxins and dibenzofurans (PCDDs, PCDFs). The distribution of these substances between the particle and gas phases in the atmosphere is a key factor which controls their deposition to water bodies and vegetation (Bidleman, 1988; Eitzer and Hites, 1989a; Dickhut and Gustafson, 1995; Duinker and Bouchertall, 1989; Golomb et al., 1997; Gustafson and Dickhut, 1997; Hoff et al., 1996; Koester and Hites, 1992a; Ligocki et al., 1985a,b; Lorber et al., 1994; Mackay et al., 1986; McLachlan, 1996; Pirrone et al., 1995; Poster and Baker, 1996a,b; Seiber et al., 1993; Welsch-Paulsch et al., 1995).

PCBs in ambient air are largely gaseous at moderate temperatures, yet even a small fraction on coarse particles (which have high deposition velocities) can lead to large dry deposition fluxes, especially near urban areas (Holsen et al., 1991, 1992). Precipitation scavenging of particulate PCBs in the "urban plume" from Chicago results in enhanced PCB deposition into nearshore Lake Michigan (Offenberg and Baker, 1997). Knowledge of the particulate — and therefore the gaseous — fraction is also critical for estimating the gas exchange of POPs between air and water (Bidleman and McConnell, 1995; Hoff et al., 1996).

The physical state of POPs also influences their reactivity with oxidants in air. Atmospheric lifetimes due to OH radical reactions in the gas phase range from 3 to 40 days for PCBs and 3 to 18 days for PCDD/Fs containing 1 to 4 chlorines (Anderson and Hites, 1996; Kwok et al., 1995). PCDD/Fs sorbed to fly ash are photochemically unreactive (Koester and Hites, 1992b). Atmospheric removal of PCDD/Fs appears to result from a combination of gas-phase reactions for lower chlorinated species and particle deposition for higher chlorinated ones (Brubaker, Jr., and Hites, 1997). Gaseous PAHs are labile photochemically and in the presence of oxidants (Arey et al., 1989; Bunce et al., 1997; Jang and McDow, 1995; Kamens et al., 1994; Kwok et al., 1994; Lane and Tang, 1994), whereas particulate PAHs demonstrate a wide range of reactivities, depending on the particle composition and relative humidity (Beyhmer and Hites, 1988; Kamens et al., 1988,1989; McDow et al., 1994,1995). Certain constituents of atmospheric particulate matter, particularly methoxyphenols, polycyclic aromatic quinones, and substituted benzaldehydes and furans, enhance the photodegradation of labile PAHs (Jang and McDow, 1995; McDow et al., 1994, 1995).

This chapter gives an overview of the experimental and modelling approaches which commonly are used to estimate the particle/gas distribution of POPs in air. More detailed reviews appear in a recent book devoted to gas and particle partitioning measurements (Lane, 1999).

10.2 Experimental Methods for Measuring Particle/Gas Distributions

10.2.1 Filtration Samplers

Most monitoring networks collect POPs with a glass or quartz fiber filter, followed by a vapor trap such as polyurethane foam (PUF), organic resins (Tenax, XAD, Chromasorb), or PUF-resin combinations (Bidleman, 1985; Chuang et al., 1987; Hornbuckle et al., 1993; Monosmith and Hermanson, 1996; Zaranski et al., 1991). These "high volume" (hi-vol) samplers draw air at approximately 0.2–0.6 m^3/min, providing volumes of 300–600 m^3 in a 24-hour period. Longer collection times sometimes are used, particularly in remote locations and for compounds (e.g., PCDD/Fs) having very low atmospheric concentrations. The filters for hi-vol sampling retain particles ≥0.3 μm with 99+ % efficiency, and the quantities of POPs found on the filter and vapor trap often are taken as estimates of the particulate and gaseous fractions.

10.2.2 Impactors

Impactors separate aerosols into several size classes, the smallest being typically 0.5–1.0 µm for high-volume and 0.04–0.05 µm for low-volume samplers (Poster et al., 1995; Umlauf, 1999; Venkataraman et al., 1994a). Measurements with impactors show that PAH emissions from vehicle sources peak sharply in the submicrometer range (Venkataraman et al., 1994a). The distribution broadens in ambient air, with a prominent "tail" above 1 µm, due to gas-to-particle conversion as the particles age (Baek et al., 1991; Venkataraman et al., 1994b). The gas phase is defined operationally by placing a sorbent trap after the last impactor stage or using a separate sampler.

10.2.3 Artifacts in Filtration and Impactor Sampling

Sampling artifacts which compromise measurements of the particle/gas distribution include loss of POPs from the particles on the filter by volatilization ("blow-off"), sorption of gaseous compounds onto the particles on the filter and onto the filter itself, and reaction of compounds with oxidants during sampling (Lane, 1999). Blow-off losses are well recognized, but there is disagreement as to the cause of the problem and how seriously it will bias particle/gas speciation estimates. The pressure drop created across impactor plates and filters during sampling has been suggested as one reason for blow-off, but a theoretical treatment of the problem predicted insignificant volatilization of organic compounds from particles on filters and only moderate losses during impactor sampling (Zhang and McMurry, 1991). Side-by-side collections of POPs, using a hi-vol and a low-pressure cascade impactor, showed that average particulate fractions were higher with *the impactor* (Kaupp and Umlauf, 1992; Umlauf, 1999). This was surprising, since blow-off losses were expected to be more significant with *the impactor*.

In a laboratory simulation of volatilization artifacts (Poster et al., 1995), urban air particulate matter was exposed to PAH-free air for 20 hours in a Berner impactor. Losses of 4-ring PAHs were 21–33% at a pressure of 0.09 atm and <15% at higher pressures. Volatilization of perdeuterated PAHs which were surface-coated onto the particles was greater, amounting to 50% at 0.09 atm. These volatilization losses were expected to be upper limits, since the clean air used for the experiments would have tended to strip PAHs from the particles.

Temperature fluctuation during the sampling period may be a more serious concern than pressure drop (Umlauf, 1999). The liquid-phase vapor pressure of POPs increases by about three-fold for a 10°C rise in temperature (Falconer and Bidleman, 1994; Hinckley et al., 1990). During a normal 24-hour collection run, particle-sorbed species which are deposited on the filter of a hi-vol sampler at night may be revolatilized in the heat of the next day. In some studies, collection periods have been kept to 11–12 hours to avoid the diurnal temperature cycle (Cotham and Bidleman, 1995; Foreman and Bidleman, 1990; Harner and Bidleman, 1998a).

However, shortening the sampling time increases the likelihood of a second (positive) artifact — adsorption of gaseous compounds by the filter itself (McDow and Huntzicker, 1990; Cotham and Bidleman, 1992; Hart and Pankow, 1994; Turpin et al., 1994; McDow, 1999). Attempts have been made to correct for this "positive" artifact by using two filters backed up by a PUF trap. Compounds retained by the front filter (FF) are sorbed onto the particles and onto the filter matrix itself. The back filter (BF) is assumed to sorb gaseous POPs to the same extent as the front filter. Values of C_p and C_g are calculated from quantities of POPs found on the filters and PUF trap:

$$C_p = (FF - BF)/\mu g \text{ particles} \tag{1}$$

$$C_g = (PUF + 2\ BF)/m^3\ air \tag{2}$$

Back filter corrections are typically 10–40% of front filter values when both filters are glass or quartz fiber (Cotham and Bidleman, 1992; Hart and Pankow, 1994; Ngabe and Bidleman, 1992). In general, higher positive artifacts are found for the more volatile compounds, such as 3-ring PAHs and HCHs. The relative amounts on back filters decrease at higher face velocities and longer sampling times (McDow and Huntzicker, 1990; Turpin et al., 1994; McDow, 1999). Sorption of gaseous POPs is substantially less onto Teflon membrane than onto quartz fiber filters (Hart and Pankow, 1994; Ligocki and Pankow, 1989; Turpin et al., 1994).

A suggested means of correcting for the positive artifact is to deploy two samplers, the primary one containing a quartz filter and a gas trap and a second one containing a quartz filter behind a Teflon membrane filter (operated at the same face velocity as the primary hi-vol). Analytes found on the backup quartz filter are used as BF quantities in equations (1) and (2)(Hart and Pankow, 1994; Turpin et al., 1994). However, this method is inconvenient and involves extra analytical work, and the double glass- or quartz-fiber technique (Equation 1-2) is judged to be satisfactory for most purposes (Hart and Pankow, 1994).

10.2.4 Denuders

Several types of denuders have been developed to collect particulate POPs without artifacts. Examples of the design and application of these samplers are in Eatough et al., 1993, Tang et al., 1994, the primary studies cited below, and several chapters (Eatough, Febo et al., Gundel and Lane, Kamens et al., Lane and Johnson, Lewis and Coutant) of a recent book (Lane, 1999) on gas and particle phase measurements of organic compounds in the atmosphere. In some investigations, side-by-side trials compared the particulate fractions denuders and hi-vol samplers estimated. Most of these experiments suggest that volatilization artifacts occur with the hi-vol, as lower particulate percentages occur with the hi-vol than with the denuder (Fan et al., 1995; Kamens et al., 1994; Gundel et al., 1995; Lane and Gundel, 1996; Subramanyam et al., 1994; Wilson et al., 1995; Coutant et al., 1988, 1989, 1992).

In light of these comparisons, it is relevant to ask whether a hi-vol measurement of particulate POPs has any meaning. The hi-vol is the "workhorse" of monitoring programs and, for reasons of simplicity, ruggedness and cost, it probably will be used for years to come. Particulate percentages measured with the hi-vol correlate well with the deposition characteristics of PCBs (Duinker and Bouchertall, 1989) and PCDD/Fs (Eitzer and Hites, 1989a; Koester and Hites, 1992a).

Furthermore, it has not been demonstrated that denuders are entirely artifact-free. Some trials have shown breakthrough of gaseous compounds through the denuder section (Coutant et al., 1992), which can bias the estimated particulate fraction on the high side. This is especially serious for compounds which are largely gaseous with only a few percent of particles. For example, if a compound is split 95-5 between the gas and particle phases, only a 2% vapor breakthrough artifact in the denuder will cause a 40% increase in the observed particulate fraction (i.e, from 5% to 7%). If the 5% were measured with the hi-vol and 7% with the denuder, one could infer a 40% "blow off artifact" for the hi-vol, when in fact the problem is a small degree of vapor breakthrough from the denuder!

A diffusion separator has been described as an alternative to the denuder for eliminating filtration artifacts (Eisenreich and Hornbuckle, 1997; Turpin et al., 1993,1997, 1999). The separator, which uses neither filters nor a denuder section, relies on the different mobilities of gases and particles to separate them. The technique was used to speciate particulate and gaseous PAHs in Minneapolis air (Eisenreich and Hornbuckle, 1997). Comparison of filter-adsorbent and diffusion separation methods for speciating PAHs showed a negative bias (blow off) for compounds that equipartition between the two phases (chrysene and benz[a]anthracene) and a positive bias (filter adsorption of the gas phase) for lower molecular weight PAHs (Turpin et al., 1999)

10.2.5 Analytical Problems

The analytical methodology complicates determining the particulate fraction of POPs. A portion of the compound in particulate matter appears freely exchangeable with its vapor phase, while another portion is strongly sorbed or occluded (Pankow, 1988; Pankow and Bidleman, 1991; 1992). Most analytical schemes for particulate matter on filters call for extraction with dichloromethane, acetone-dichloromethane, or toluene, and these solvents may access some of the "non-exchangeable" POPs (Ligocki and Pankow, 1989; Pankow and Bidleman, 1991; Rounds et al., 1993).

The presence of this non-exchangeable pool for PAHs is implied by the results of experiments which compared the extraction rates of native PAHs and spiked perdeuterated compounds from SRM 1649 urban dust (Burford et al., 1993). Extraction with supercritical CO_2 quantitatively recovered the deuterated PAHs within 30 min. CO_2 alone removed native PAHs more slowly and incompletely, CO_2/methanol gave improved yields. Differences in volatilization of PAHs from urban particles into a clean airstream occurred, depending on whether the PAHs were native to the particles or added by surface coating (Poster et al., 1995).

10.3 Methods for Estimating the Particle/Gas Distribution in Ambient Air

10.3.1 Junge-Pankow Adsorption Model

Despite many years of interest in the phase distribution of POPs, few predictive models are available. A Langmuir-type relationship, which Junge (1977) first proposed and Pankow (1987) later reviewed and critically evaluated, is the most popular model for estimating adsorption onto aerosols. The Junge-Pankow equation relates the fraction of particulate POPs (ϕ) to the saturation liquid-phase vapor pressure of the compound (P_L^S, Pa) and the surface area of particles per unit volume of air (θ, cm^2 aerosol/cm^3 air).

$$\phi = c\theta/(P^S_L + c\theta) \tag{3}$$

The parameter c (Pa cm) is related to the heat of desorption from the particle surface (Q_d, J/mol), the heat of vaporization of the compound (Q_v, J/mol), and the moles of adsorption sites on the aerosol (N_s, mol/cm^2)

$$c = 10^6 RT N_s \exp[(Q_d-Q_v)/RT] \tag{4}$$

where T is temperature (K), R, the gas constant, is 8.314 Pa m^3/mol K or J/mol K, and the factor 10^6 converts m^3 to cm^3. Junge (1977) proposed a value of c = 17.2 Pa cm.

Pankow (1987) argued that if N_s is on the order of 4×10^{-10} mol/cm^2 and Q_d–Q_v is 6300 J/mol, the resultant value for c = 13 Pa cm at 298K is close to Junge's estimate. He also suggested that different values of (Q_d-Q_v) (and therefore c) may be appropriate for different classes of compounds, a point which we discuss later. Junge (1977) did not specify the physical state of the sorbing compound in Equation 3, but the vapor pressure of the sub-cooled liquid (P_L^S) is probably the controlling factor rather than the solid-phase vapor pressure (P_S^S) (Bidleman et al., 1986; Cotham and Bidleman, 1992; Foreman and Bidleman, 1987; Mackay et al., 1986), hence the appearance of P_L^S in Equation (3).

For compounds that are solids at ambient temperatures, P_L^S can be estimated by exploiting the fugacity ratio–melting point relationship discussed in the introduction to this book, namely,

$$P_L^S / P_S^S = \exp\left[\Delta S_f \left(T_m - T\right)/RT\right] \tag{5}$$

In Equation 5 ΔS_f is the entropy of fusion (J/mol K), T_m (K) is melting point of the compound, and T is the ambient temperature. Equation 5 assumes that the heat capacity of the solid and liquid phases are equal. Values of ΔS_f have been summarized for PCDD/Fs (Mackay et al., 1992a,b; Rordorf, 1989), PCBs and chlorinated pesticides (Donnelly et al., 1990; Hinckley et al., 1990; Mackay et al., 1992a,b) and other compounds (Dannenfelser et al., 1993). In the absence of an experimental result, an average $\Delta S_f / R = 6.79$ is often used (Hinckley et al., 1990). Capillary gas chromatography (GC) can estimate liquid-phase vapor pressures for *non-polar* compounds such as PCBs, PAHs, PCNs, PCDD/Fs, and organochlorine pesticides (Eitzer and Hites, 1988; Falconer and Bidleman, 1994; Fischer et al., 1992; Hinckley et al., 1990; Lei et al., 1999; Yamasaki et al., 1984). However, the GC method is not successful for polar compounds such as organophosphate insecticides and herbicides (Hinckley et al., 1990). The hypothesis that the vapor pressure of the liquid phase controls sorption onto aerosols is especially significant for high-melting compounds such as PCDD/Fs, hexachlorobenzene, and anthracene, which have large differences between the solid and liquid-phase vapor pressures.

Estimates of θ for urban, rural, and clean background air are based on a study by Whitby (1978) of the size distribution of accumulation mode aerosols (i.e, aerosols of 0.1 to 1.0 μm diameter) and the average total volume of particles per unit volume of air V_T , (cm³ aerosol/cm³ air). Table 10.1 gives average values of θ and V_T, but note that the surface properties of urban aerosols can change by an order of magnitude from day to day. From V_T and an assumed particle density of 2 g/cm³ (Corn et al., 1971; Mackay et al., 1986), the average total suspended particle (TSP) concentrations in urban and average background air are 140 and 60 μg/m³. These are somewhat higher than the mean TSP concentrations of 80 and 30 μg/m³ measured in 46 U.S. cities and 20 rural locations in 1975 (Shah et al., 1986).

TABLE 10.1

Values of θ, V_T, and TSP for Different Air Regimes.[a,b]

Air Type	θ cm² Aerosol/cm³ Air	V_T cm³ Aerosol/cm³ Air	TSP μg/m³
Clean continental background	0.42×10^{-6}	6.5×10^{-12}	13
Average background	1.5×10^{-6}	30×10^{-12}	60
Background plus local sources	3.5×10^{-6}	43×10^{-12}	86
Urban	11×10^{-6}	70×10^{-12}	140

[a] Bidleman, 1988; Whitby, 1978.
[b] TSP calculated from V_T, assuming a particle density of 2 g/cm³.

Calculations of the aerosol-specific surface area ($A_{tsp} = 10^6\theta/TSP$, cm²/μg) from Table 10.1 data yield estimates of 0.079 in urban air and 0.025 in average background air. The few experimentally measured values of A_{tsp} are similar to these estimates: 0.019–0.031 cm²/μg in Pittsburgh, Pennsylvania (Corn et al., 1971), and 0.023–0.087 cm²/μg in Portland, Oregon (Sheffield and Pankow, 1994). These surface areas were measured on particles which were collected on glass fiber filters, not freely suspended in air. The latter authors noted that particles tended to agglomerate to a greater extent on Teflon membrane filters than on glass fiber filters, and that experimentally determined specific surface areas were higher on glass filters. This suggests that A_{tsp} estimates on filtered particles may be biased on the low side, although more information is needed in this regard.

The fraction of chemical on the particulate phase ϕ also can be expressed by:

$$\phi = C_p(TSP)/[C_g + C_p(TSP)] \tag{6}$$

C_p is the concentration of POPs associated with aerosols (ng/μg particles), C_g is the gas-phase concentration (ng/m³), and TSP is the total suspended particle concentration (μg/m³). Equation 6 is a general relationship that applies to any experimental or model estimate of C_g and C_p. Combining equations (3), (4), and (6):

$$K_p = C_p/C_g = c\theta/P_L^S(TSP) = \left(1/P_L^S\right)RTN_s A_{tsp} \exp\left[(Q_d - Q_v)/RT\right] \tag{7}$$

$$\text{Log } K_p = \left(\text{Log } c\theta/TSP\right) - \text{Log } P_L^S \tag{8}$$

$$\phi = K_p(TSP)/[1 + K_p(TSP)] \tag{9}$$

Here and in other work (Cotham and Bidleman, 1995; Falconer et al., 1995; Gustafson and Dickhut, 1997; Hart and Pankow, 1994; Kamens et al., 1995; Liang and Pankow, 1996; Pankow and Bidleman, 1992; Pankow, 1999), C_p/C_g is referred to as the particle/gas partition coefficient, K_p, with units of m³/μg. Its inverse, $C_g/C_p = 1/K_p$, also has been used for these correlations (Cotham and Bidleman, 1992; Ligocki and Pankow, 1989). Equation 7 differs from that in other publications (Pankow, 1991; Pankow and Bidleman, 1992; Sheffield and Pankow, 1994), where R does not appear in the preexponential term because it is incorporated into other constants. Also, here we use consistent units for R (Pa m³/mol K = J/mol K) and vapor pressure (Pa), whereas the above authors used Torr for vapor pressure and varying units for R. According to Equation (8), the expected slope of log K_p vs. log P_L^S is –1 and the intercept (log cθ/TSP) is related to A_{tsp}. Plots of log K_p vs. log P_L^S for experimental particle/gas partitioning data are usually well correlated and follow the general relationship:

$$\text{Log } K_p = m_r \text{Log } P_L^S + b_r \tag{10}$$

Slopes (m_r) of these plots derived from ambient air sampling data are often different from the expected value of –1 (Equation 8) possibly because of kinetic limitations and/or sampling artifacts or thermodynamic factors (Pankow and Bidleman, 1992). Goss and Schwartzenbach (1998) and Simcik et al. (1998) argued that slopes differing from –1 do not necessarily mean that the aerosols are out of equilibrium with the gas phase. In these situations, the intercept (b_r) partly depends on the slope and cannot be used to estimate θ (Pankow and Bidleman, 1992). Table 10.2 lists reported values of these parameters.

Limitations of the Junge-Pankow model include uncertainties in the parameters c and θ. Pankow (1987) suggested that optimal values of c might be chosen for different classes of

TABLE 10.2

Parameters of Equation 10.8 for Filtration Sampling of POPs

	Location	m_r	b_r	Reference
Urban				
PAHs	Portland, Oregon	−0.882	−5.38	a
PAHs	Denver, Colorado	−0.760	−5.10	b
PCBs	Denver, Colorado	−0.946	−5.86	b
PAHs	Chicago, Illinois	−0.694	−4.61	c
PCBs	Chicago, Illinois	−0.726	−5.18	c
PCBs	Chicago, Illinois	−0.715	−5.14	d
PAHs	Chicago, Illinois	−0.745	-4.66	d
PAHs	Hampton, Virginia	−1.09	−5.75	e
PAHs	London, U.K.	−0.631	−4.61	f
PAHs	Osaka, Japan	−1.04	−5.95	g
PAHs	Brazzaville, Congo	−0.810	−5.31	h
OC pesticides	Brazzaville, Congo	−0.740	−5.76	g
PCDD/Fs	Bloomington, Indiana	−0.775	−5.72	i,j
Rural				
PAHs	Coastal Oregon	−0.724	−4.94	a
PAHs	Lake Superior	−0.614	−4.25	k
PAHs	Lake Superior	−0.586	−3.83	l
PAHs	Green Bay	−1.00	−5.47	c
PAHs	Haven Beach, Virginia	−0.649	−4.43	d
PAHs	Lakes Erie & Ontario	−0.580	−4.14	m
OC pesticides	Lakes Erie & Ontario	−0.688	−5.06	m
PCBs, OC pesticides	Bayreuth, Germany	−0.610	−4.74	n

a) Ligocki and Pankow, 1989; b) Foreman and Bidleman, 1990; c) Cotham and Bidleman, 1995; d) Harner and Bidleman, 1999; e)Gustafson and Dickhut, 1997; f) Baek et al., 1991; g) Yamasaki et al., 1982; h) Ngabe and Bidleman, 1992; i) Bidleman et al., 1997; j) Eitzer and Hites, 1989a; k) Baker and Eisenreich, 1990; l) McVeety and Hites, 1988; m) Hoff et al., 1996; n) Kaupp and Umlauf, 1992.

compounds. His reasoning was that the excess heat of desorption (Q_d-Q_v) in Equation 4 appeared to be a smaller term for OCs than for PAHs. Equation (11) gives the approximate relationship between log K_p and ambient temperature (Pankow, 1987; Yamasaki et al., 1982). Plotting experimentally measured values of log K_p vs. 1/T determines Q_d:

$$\text{Log } K_p = (-Q_d/2.303RT) + b \qquad (11)$$

The intercept of such plots is related to A_{tsp} and should be the same for all classes of compounds (Pankow, 1987, 1991). Pankow (1991, 1992) proposed that Q_d is best estimated from plots of log K_p vs. 1/T by assuming a "common y-intercept" to minimize the residuals in these plots of field data.

It is difficult to extract precise Q_d values from ambient air sampling data. Plots of Equation (11) are constructed by measuring K_p on days when temperatures vary (Bidleman et al., 1986; Foreman and Bidleman, 1990; Gustafson and Dickhut, 1997; Pankow, 1991; Yamasaki et al., 1982). Unfortunately, these experimental K_p estimates also reflect the day-to-day variations in relative humidity and aerosol properties. As a result, confidence intervals around Q_d are typically large (Bidleman et al., 1986; Pankow, 1991). Moreover, differences in the rates at which gases equilibrate with particles on hot and cold days (Kamens et al., 1995) make it problematic whether true sorption equilibrium is attained over the entire temperature range of the field experiments.

Excess heats of desorption (Q_d-Q_v) for PAHs in urban air studies ranged from -6.3 to 8.9 kJ/mol (Krieger and Hites, 1994; Pankow, 1991; Yamasaki et al., 1982). In another investigation, Gustafson and Dickhut (1997) reported Q_d to be 90 to 208 kJ/mol for 3-5 ring PAHs. The highest values of Q_d were found at a rural site, followed by semi-urban, urban, and industrial locations. For the 3-5 ring PAHs, the average Q_v was 82 kJ/mol (Hinckley et al., 1990). Thus, Gustafson and Dickhut's Q_d-Q_v results spanned a large range, from 8 to 126 kJ/mol. Determinations of Q_d for 3-5 ring PAHs at an island site in Lake Superior ranged from 43-114 kJ/mol (McVeety, 1986). (Q_d-Q_v) for organochlorine pesticides varied from -16 to 19 kJ/mol in a field study (Bidleman et al., 1986) and 1.4 to 19 kJ/mol in controlled laboratory experiments (Cotham and Bidleman, 1992). Considering these wide ranges, it is questionable whether (Q_d-Q_v) values for PAHs and OCs are different.

The θ-parameter is also subject to uncertainty. It is likely that Whitby's (1978) estimates of θ (Table 10.1) do not reflect the true surface area distribution, since they were based on the average size spectrum of aerosols and assumed spherical particles. Adsorption/desorption kinetics (Kamens et al., 1995; Rounds and Pankow, 1990, 1993) and relative humidity (Goss and Eisenreich, 1997; Lee and Tsai, 1994; Pankow et al., 1993; Storey et al., 1995; Thibodeaux et al., 1991) influence the adsorption of POPs onto aerosols, and the Junge-Pankow model does not take these factors into account.

10.3.2 Mackay Adsorption Model

Mackay et al. (1986) defined a dimensionless particle/gas partition coefficient

$$K'_P = \left(ng/m^3 \text{ particles}\right)/\left(ng/m^3 \text{ air}\right) = K_P (TSP)/V_T \tag{12}$$

where V_T is the volume fraction of particles in air (Table 10.1). By assuming that semivolatile compounds partition onto aerosols according to the distribution Yamasaki et al. (1982) found for benz[a]anthracene in Osaka, Japan, they showed that the data could be fitted approximately by the simple one parameter model.

$$K'_P = 6 \times 10^6/P_L^S \tag{13}$$

$$K_P = \left(6 \times 10^6\right) V_T/P_L^S \, TSP \tag{14}$$

In their model, TSP and V_T for urban air were taken as 100 μg/m³ and 5×10^{-11}, giving:

$$K_P = 3 \times 10^{-6}/P_L^S \tag{15}$$

10.3.3 Pankow Absorption Model

Pankow (1994a,b) proposed that gaseous POPs also absorb into a liquid-like organic film on the particles. Liang and Pankow (1996), and Pankow (1999) found the sorptive capacity of environmental tobacco smoke (almost entirely organic matter) and urban air particles (about 20% organic matter) for n-alkanes and PAHs was nearly the same when normalized to the organic content of the particles. Gas absorption is expressed by:

$$P_L = \gamma_{om} X_{om} P_L^S \tag{16}$$

In Equation (16), P_L^S is the partial pressure in equilibrium with a solute having mole fraction X_{om} and activity coefficient γ_{om} in the organic film ($\gamma_{om} \to 1$ as $X_{om} \to 1$). Pankow's expression for K_p based on absorption considerations is:

$$K_p = 10^{-6} \, RTf_{om}/M_{om}\gamma_{om} \, P_L^S \tag{17}$$

$$Log \, K_p = Log \, 10^{-6} \, RTf_{om}/M_{om}\gamma_{om} - Log \, P_L^S \tag{18}$$

where f_{om} is the fraction of the particle mass that consists of absorbing organic matter having molecular weight M_{om} (g/mol). A substantial portion of this organic matter may be "secondary organic aerosol," which is formed by oxidation of hydrocarbons and is therefore polar (Forstner et al., 1997; Jang et al., 1997; Pankow, 1994b). Both the adsorption (Equation 8) and absorption (Equation 18) models predict the same functional dependence of K_p on P_L^S; that is, the slope of a plot of log K_p vs. P_L^S should be unity. The combined relationship for both adsorption and absorption is the sum of Equation 7 and 17.

$$K_p = \left(1/P_L^S\right)\left\{RT \, N_s A_{tsp} \, exp\left[(Q_d - Q_v)/RT\right] + 10^{-6} RT \, f_{om}/M_{om} \, \gamma_{om}\right\} \tag{19}$$

However, it is questionable whether a particle which is coated with an absorbing liquid film will have any unoccupied active sites for adsorption to occur. It may be preferable to use either the absorption or adsorption model but not both additively.

An advantage of the absorption model is that the fraction of organic matter on aerosols (f_{om}) is more easily measureable than aerosol surface area (θ). The difficulty in applying the model as in Equation (17) or (18) is that the activity coefficient of the compound in the organic film (γ_{om}) is not known, nor easily measured, and is likely to vary substantially among classes of POPs. From experiments with environmental tobacco smoke, Liang and Pankow (1996) estimated that γ_{om} was 1.8 for n-alkanes and 12.7 for PAHs. Jang et al. (1997) estimated γ_{om} for different combinations of aerosol compositions and absorbing solutes using group contribution methods. These models were successful in explaining the differences in partitioning of PAHs, alkanes, and oxygenated compounds onto particles which contained largely non-polar vs. polar secondary organic aerosol. The authors concluded that equilibrium partitioning of POPs into almost any organic layer on an aerosol can be predicted using the activity coefficient approach.

10.3.4   Octanol-Air Partition Coefficient Model

Absorption also may be viewed according to an alternate logarithmic form of Equation (17):

$$Log \, K_p = Log \, f_{om} + Log \, 10^{-6} \, RT/M_{om}\gamma_{om} \, P_L^S \tag{20}$$

where the term $10^{-6}RT/M_{om}\gamma_{om} \, P_L^S$ is related to the partition coefficient of the compound between the aerosol organic matter and air. This suggests using the octanol/air partition coefficient (K_{oa}) to replace vapor pressure for describing absorption onto aerosols (Finizio et al., 1997).

$$K_{oa} = C_{oct}/C_{air} = 10^3 \rho_{oct} \, RT/M_{oct}\gamma_{oct} \, P_L^S \qquad (21)$$

C_{oct} and C_{air} are the moles of compound per m^3 of octanol and air, ρ_{oct} and M_{oct} are the density (820 kg/m^3 at 20°) and molecular weight (130 g/mol) of octanol, and γ_{oct} is the activity coefficient of the solute in octanol. Eliminating P_L^S from equations (17) and (21) gives

$$K_p = (1/\rho_{oct})[10^{-9}K_{oa}f_{om} \, (\gamma_{oct}/\gamma_{om})(M_{oct}/M_{om})] \qquad (22)$$

The molecular weights of octanol and the organic matter on aerosols are likely to be similar $(M_{oct}/M_{om} \sim 1)$ and $\rho_{oct} = 820$ kg/m^3:

$$\text{Log } K_p = \text{Log } K_{oa} + \text{Log } f_{om} \, (\gamma_{oct}/\gamma_{om}) - 11.91 \qquad (23)$$

The improvement in the absorption model from introducing K_{oa} is twofold. Use of K_{oa} avoids the problem of converting solid to subcooled liquid vapor pressures. The liquid-phase vapor pressure is not directly accessible for compounds that are solids, including many POPs. Values of P_L^S for these substances must be calculated from solid-phase vapor pressures by Equation 5 or estimated by extrapolating gas chromatographic data obtained at higher temperatures (Eitzer and Hites, 1988; Fischer et al., 1992; Hinckley et al., 1990/ Lei et al., 1999). K_{oa} is directly measureable at ambient temperatures and is reported for several classes of POPs (Harner and Bidleman, 1996, 1998b; Harner and Mackay, 1995; Kömp and McLachlan, 1997). Also, the ratio γ_{oct}/γ_{om} should be less variable among classes of compounds than γ_{om} alone (Pankow, 1998). With the assumption that $\gamma_{oct}/\gamma_{om} \sim 1$, Equation 23 reduces to a simple form:

$$\text{Log } K_p = \text{Log } K_{oa} + \text{Log } f_{om} - 11.91 \qquad (24)$$

Shah et al. (1986) reported that the organic carbon content of airborne particles in U.S. cities averaged 8.4%. More recent estimates are higher. Total carbon (organic + elemental) accounted for 23% of the urban aerosol mass in Chicago, Illinois (Cotham and Bidleman, 1995), or 15% organic carbon if an organic/elemental ratio of 1.8 is assumed (Shah et al., 1986). Measurements of organic carbon in Portland, Oregon, ranged from 13-16% of TSP (Sheffield and Pankow, 1994). For an oxidized secondary organic aerosol, the ratio of organic matter to organic carbon is about 2.0, compared to 1.2 for a primary organic aerosol containing mainly heavy alkanes and aldehydes (Liang et al., 1997). If we assume that urban aerosols contain 15% organic carbon, present as compounds with the average molecular formula of octanol (74% carbon), then urban particles contain about 20% organic matter. Inserting 0.2 for f_{om} in Equation 24 gives

$$\text{Log } K_p = \text{Log } K_{oa} - 12.61 \qquad (25)$$

10.3.5 Applying the Models

Illustrative calculations appear below for estimating the particulate fraction (ϕ) of p,p'-DDT in urban air at 20°C using the Junge-Pankow adsorption model (Equation 3), the Mackay adsorption model (Equation 15), and the octanol-air partition coefficient model (Equation 25). Table 10.3 lists values of P_L^S and K_{oa} for p,p'-DDT and other POPs of different chemical classes. All model calculations are for an urban air TSP = 80 μg/m^3 (Shah et al., 1986).

TABLE 10.3

Predicted Sorption of POPs to Urban Aerosols.

Compound	Properties at 20°C[1]		A J–P Model[2] ϕ	B Mackay Model[3] ϕ	C K_{oa} Model[4] ϕ
	Log P^s_L (Pa)	Log K_{oa}			
Pesticides					
Hexachlorobenzene	−1.12 (a)	7.11 (d)	0.0014	0.0032	0.00025
Lindane	−1.41 (a)	8.25 (c)	0.0028	0.0062	0.0035
p,p′–DDT	−3.58 (a)	10.29 (d)	0.29	0.48	0.28
PCBs					
2,2′,4,5′–TeCB (49)	−2.02 (b)	8.57 (e)	0.011	0.024	0.0072
3,3′,4,4′–TeCB (77)	−2.93 (b)	9.96 (e)	0.084	0.17	0.15
2,2′,4,5,5′–PeCB (101)	−2.74 (b)	9.31 (e)	0.056	0.12	0.039
2,3′,4,4′,5–PeCB (118)	−3.36 (b)	10.10 (e)	0.20	0.35	0.20
3,3′,4,4′,5–PeCB (126)	−3.60 (b)	10.61 (e)	0.30	0.49	0.44
2,2′,3,4,4′,5′–HxCB (138)	−3.57 (b)	10.09 (e)	0.29	0.47	0.19
2,2′,3,4,4′,5,5′–HpCB (180)	−4.18 (b)	10.75 (e)	0.62	0.78	0.52
PAHs					
Fluorene	−0.10 (a)	7.04 (f)	0.00014	0.00030	0.00023
Anthracene	−1.25 (a)	7.94 (g)	0.0019	0.0043	0.0017
Phenanthrene	−1.22 (a)	7.83 (f)	0.0018	0.0040	0.0013
Pyrene	−2.09 (a)	9.04 (f)	0.013	0.029	0.021

1. References to properties:
    a) Hinckley et al., 1990. b) Falconer and Bidleman, 1994; c) Estimated from Henry's law constant of 0.52 Pa m^3/mol (Kucklick et al., 1991) and K_{ow} of 6310 (Suntio et al., 1988); d) Harner and Mackay, 1995; e) Harner and Bidleman, 1996; f) Harner and Bidleman, 1998b; g) Estimated from the regression relationship between log K_{oa} and log p^s_L in Harner and Bidleman, 1998b.
2. Equation 1, $\theta = 6.3 \times 10^{-6}$ and c = 17.2 Pa–cm. 3. Equation 15 and 9, TSP = 80 $\mu g/m^3$. 4. Equation 25 and 9, TSP = 80 $\mu g/m^3$.

In the Junge-Pankow model, the surface area parameter θ for urban air usually is taken as 1.1×10^{-5} (Table 10.1), but, as explained earlier, this corresponds to a TSP concentration of about 140 $\mu g/m^3$. We assumed that θ could be scaled to TSP and adjusted θ to $1.1 \times 10^{-5}(80/140) = 6.3 \times 10^{-6}$. The value of c was 17.2 Pa-cm.

The Mackay adsorption and octanol-air partition coefficient models were applied by first using Equation 15 or 25 to estimate K_p from P^S_L or K_{oa} and then calculating ϕ from Equation 9 using TSP = 80 $\mu g/m^3$. As noted above, Equation 25 assumes that 20% of urban aerosols is composed of organic matter which is capable of sorbing POPs.

Results of these calculations for p,p′-DDT (below) and other POPs (Table 10.3) show that (except for hexachlorobenzene) the model estimates of particulate fractions are within a factor of about 2–3. The agreement among the three approaches is remarkable, considering the different assumptions that were used in deriving and applying the models.

Junge-Pankow Adsorption Model

$$P^S_L \ (\text{p,p′-DDT, 20°C}) = 2.6 \times 10^{-4} \ \text{Pa}$$

$$c\theta = 17.2(6.3 \times 10^{-6}) = 1.9 \times 10^{-4}$$

Equation 3: $\phi = c\theta/(P_L^S + c\theta) = 1.08 \times 10^{-4}/(2.6 \times 10^{-4} + 1.08 \times 10^{-4}) = 0.29$

Mackay Adsorption Model

P_L^S (p,p'-DDT, 20°C) = 2.6×10^{-4} Pa

TSP = 80 μg/m³

Equation 15: $K_p = 3 \times 10^{-6}/ P_L^S = 0.0115$; $K_p(TSP) = 0.923$

Equation 9: $\phi = K_p(TSP)/[1 + K_p(TSP)] = 0.923/(1 + 0.923) = 0.48$

Octanol-Air Partition Coefficient Model

K_{oa} (p,p'-DDT, 20°C) = 1.95×10^{10}

TSP = 80 μg/m³

Equation (25): Log K_p = Log K_{oa} − 12.61 = 10.29 − 12.61 = −2.32; $K_p = 4.79 \times 10^{-3}$

$K_p(TSP) = 0.383$

Equation(9): $\phi = K_p(TSP)/1 + K_p(TSP)] = 0.383/(1 + 0.383) = 0.28$

10.4 Comparison of Approaches

10.4.1   Filtration Sampling vs. Junge-Pankow Adsorption Model

Most estimates of particle/gas distributions have been made with the hi-vol sampler. Table 10.2 presents a summary of data from several urban and rural locations, expressed as the regression parameters (m_r and b_r) of Equation 10. Values of K_p and TSP were sometimes not reported in the original paper, and in these cases K_p was calculated from the particle/gas partitioning data, assuming a TSP of 80 μg/m³ for urban air and 30 μg/m³ for rural air (Shah et al., 1986). Correlations to liquid-phase vapor pressure were made using reported values of P_L^S for PAHs (Yamasaki et al., 1984), PCBs (Falconer and Bidleman, 1994), and OC pesticides (Hinckley et al., 1990). In the case of PCDD/Fs, the phase distribution and vapor pressure data from Eitzer and Hites (1988, 1989b) were recalculated to take into account temperature effects, and replotted according to Equation (10) (Bidleman et al., 1999).

Figure 10.1 shows the results of a study in Chicago that measured PCBs, PAHs, and TSP (Cotham and Bidleman, 1995). Double glass fiber filters were used to correct for filter adsorption artifacts in calculation of C_p and C_g (Equation 1 and 2). Log-log plots of $K_p = C_p/C_g$ vs. P_L^S at the ambient temperature were well correlated, but slopes differed from the −1 value expected from Equation (8), common in other studies.

Figures 10.2 and 10.3 compare the particulate percentages obtained from the hi-vol sampling correlations in Table 10.2 with those estimated from the Junge-Pankow adsorption model (Equation 3). The model was applied using a value for c of 17.2 Pa-cm, surface area parameters (θ, cm² aerosol/cm³ air) of 6.3×10^{-6} for urban air, and 7.5×10^{-7} for rural (average background) air. These θ values were scaled downward from those in Table 10.1, to adjust for the lower TSP concentrations, e.g., urban air = 1.1×10^{-5} (80/140) = 6.3×10^{-6}; rural air = 1.5×10^{-6} (30/60) = 7.5×10^{-7}. The fraction on aerosols (ϕ) was calculated from K_p and TSP by Equation 9.

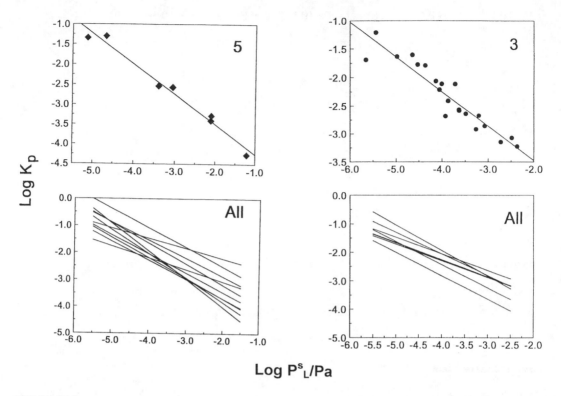

FIGURE 10.1

Plots of Equation (8) for PAHs (squares) and PCB congeners (dots). The upper panels show two individual air samples from Chicago, Illinois (sample 5 for PAHs and 3 for PCBs). Lower panels show the regression lines for all samples. Data from Cotham and Bidleman (1995).

The agreement between the measured and predicted phase distribution of PAHs is quite good for urban air (Figure 10.2a). At $\log P_L^S = -4.0$ where the model predicts equipartitioning, the range of particulate percentages estimated from hi-vol sampling is 40–62%. PAHs speciation in rural ("average background," Table 10.1) air is compared to the J-P model in Figure 10.2b. Most experimental PAH distributions are higher than those predicted. Equipartitioning in rural air is estimated to occur at $\log P_L^S = -4.9$. At this vapor pressure, the measured particulate percentages range from 55 to 85%.

In all cases the hi-vol curves for OCs in urban (Figure 10.3a) and rural (Figure 10.3b) air fall below the model predictions, indicating that the sampler underestimates (or the model overestimates) the particulate fraction of these compounds. The discrepancy might be due to blow-off artifacts during hi-vol sampling, although this problem should be manifest with PAHs, for which measured and modeled phase distributions in urban air are in better agreement (Figures 10.2a and 10.2b). It is also possible that OCs sorb onto particles less strongly than PAHs, as suggested by Pankow (1987).

Only a few reports have compared the sorption of different compound classes onto the same aerosol, or the same compound onto different aerosol types. However, these studies demonstrate that vapor pressure alone does not explain the interaction of POPs with aerosols.

In four urban investigations, PAHs were associated with urban air particles to a greater extent than OCs of the same vapor pressure (Bidleman et al., 1986; Cotham and Bidleman, 1995; Ngabe and Bidleman, 1992; Harner and Bidleman, 1998a), but another study in Denver, Colorado found about the same particle/gas distributions for alkanes, PAHs, PCBs,

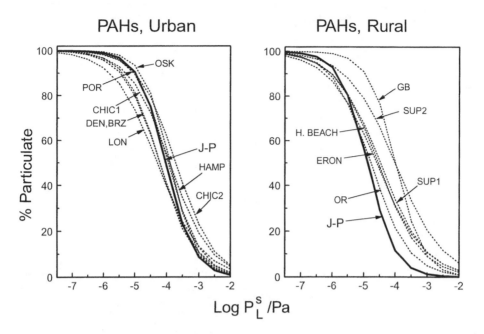

FIGURE 10.2

Comparison of measured and predicted particulate percentages of PAHs in urban and rural air. Urban field values were normalized to TSP = 80 μg/m³, and rural values to 30 μg/m³. The J-P model was run with the following values of θ : urban = 6.3×10^{-6}, average background (rural) = 7.5×10^{-7} (obtained by adjusting values in Table 10.1 for the lower TSP content of today's ambient air). POR = Portland, Oregon (Ligocki and Pankow, 1989); CHIC1 and CHIC2 = Chicago, Illinois (1 = Cotham and Bidleman, 1995; 2 = Harner and Bidleman, 1998a); DEN = Denver, Colorado (Foreman and Bidleman, 1990); BRZ = Brazzaville, Congo (Ngabe and Bidleman, 1992); LON = London, England (Baek et al., 1991); OSK = Osaka, Japan (Yamasaki et al., 1982); HAMP = Hampton, Virginia (Gustafson and Dickhut, 1997); SUP1 and SUP 2 = Lake Superior (1 = Baker and Eisenreich, 1990; 2 = McVeety and Hites, 1988); ERON = lakes Erie and Ontario (Hoff et al., 1996); OR = coastal Oregon (Ligocki and Pankow, 1989), GB = rural Green Bay, Wisconsin (Cotham and Bidleman, 1995); H. BEACH = Haven Beach, Virginia (Gustafson and Dickhut, 1997).

and OC pesticides (Foreman and Bidleman, 1990). Indoor sampling of environmental tobacco smoke (ETS) showed that, when normalized for vapor pressure, values of K_p increased in the order: n-alkanes < PAHs < nitrogenous aromatics (Figure 10.4) (Liang and Pankow, 1996; Pankow et al., 1994). Differences in partitioning also are observed within a compound class. PCBs that are more nearly planar (e.g., "coplanar" PCBs that have no ortho-substituted chlorines) are sorbed preferentially onto urban air particulate matter, compared to multi-ortho congeners of the same volatility (Falconer et al., 1995; Harner and Bidleman, 1998a).

Liang et al. (1997) compared sorption of PAHs and n-alkanes onto three types of organic aerosols: dioctyl phthalate mist (DOP), environmental tobacco smoke (ETS), and secondary organic aerosol generated from gasoline combustion. Sorption of PAHs onto all three aerosol types was similar. For alkanes, partition coefficients normalized for organic matter content were higher on the DOP than the other two aerosols, possibly because of the structural similarity between the n-alkanes and the octyl chains on DOP. Sorption of PAHs to ETS and secondary aerosol was greater than for alkanes.

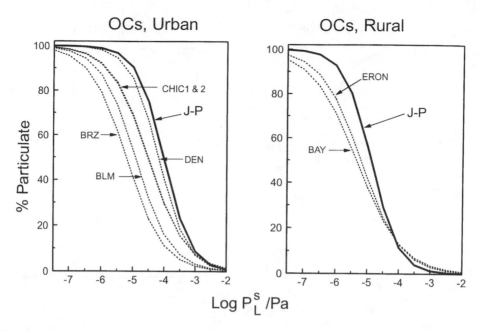

FIGURE 10.3

As in Figure 10.2, for PCBs, PCDD/Fs and OC pesticides. BLM = Bloomington, Indiana (Eitzer and Hites, 1989a,b); DEN = Denver, Colorado (Foreman and Bidleman, 1990); BRZ = Brazzaville, Congo (Ngabe and Bidleman, 1992); CHIC1 and CHIC2 = Chicago, Illinois (1 = Cotham and Bidleman, 1995; 2 = Harner and Bidleman, 1998a); BAY = Bayreuth, Germany (Kaupp and Umlauf, 1992); ERON = lakes Erie and Ontario (Hoff et al., 1996).

10.4.2 Vapor Pressure vs. Octanol-Air Partition Coefficient for Correlating Experimental Particle/Gas Distributions

Particle-gas partitioning can be described by a variation of the Pankow absorption model, in which liquid-phase vapor pressure is replaced by the octanol-air partition coefficient as a fitting parameter (equations (22)–(25)). Section 10.3.4 states the advantages of doing so. K_{oa} has been reported as a function of temperature for several PCB congeners, chlorobenzenes, PAHs, polychloronaphthalenes (PCNs), and p,p′-DDT (Harner and Bidleman, 1996, 1998b; Harner and Mackay, 1995; Kömp and McLachlan, 1997). For others, K_{oa} can be estimated from the ratio of the octanol-water partition coefficient to the Henry's law constant (H = Pa m³/mol) (Finizio et al., 1997; Harner and Mackay, 1995; Simonich and Hites, 1995).

$$K_{oa} = K_{ow}RT/H \qquad (26)$$

Limitations of using Equation 26 include disagreement among literature values for K_{ow} and H, and the effect of the mutual solubility of octanol and water on the determination of K_{ow}. It is preferrable to measure K_{oa} directly (Harner and Bidleman, 1996; Harner and Mackay, 1995).

Log–log plots of K_{oa} vs. P_L^S for PCBs show a close correlation between the two properties, but differences appear between two classes of congeners which have four or more total chlorines. Partitioning of coplanar (non-ortho) and mono-ortho PCBs from air into octanol is enhanced compared to the less planar multi-ortho PCBs. No distinction appears for the lower molecular weight congeners (Harner and Bidleman, 1996, 1998b) (Figure 10.5). Reasons for this behavior are unclear but may relate to differences in γ_{oct}.

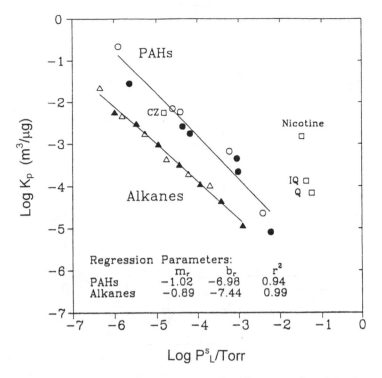

FIGURE 10.4

Comparison of the particle/gas partition coefficient (K_p) among compound classes when normalized to liquid-phase vapor pressure (P^S_L). Filled and open triangles = alkanes, filled and open circles = PAHs from Liang and Pankow (1996) or Pankow et al. (1994). Figure from Liang and Pankow (1996).

Hi-vol estimates of K_p have been correlated to K_{oa} on a log-log scale for several classes of compounds, demonstrating the general utility of K_{oa} as a descriptor of particle/gas partitioning (Finizio et al., 1997; Figure 10.6). Moreover, K_{oa} may be able to resolve the observed differences in speciation among compound classes that vapor pressure does not explain. The laboratory experiments of PCB partitioning to Chicago aerosol, supported this (Falconer et al., 1995).

Values of K_p at 25°C for "particle set A" from Falconer et al. are plotted in log–log format vs. P^S_L or K_{oa} of the PCB congeners (Figure 10.7). Three distinct lines for non-, mono- and multi-ortho PCB congeners are seen in the plot of log K_p vs log P^S_L (Figure 10.7a), indicating preferential sorption of the more nearly planar molecules onto aerosols when the congeners are compared on a vapor pressure basis. The same trend occurs in hi-vol measurements of PCBs in the ambient air of Chicago (Falconer et al., 1995; Harner and Bidleman, 1998a). Differences among the ortho-chlorine groups are largely resolved when K_{oa} is used as a fitting parameter (Figure 10.7b).

Figures 10.8a and 10.8b compare the P^S_L — based J-P and Mackay models to the K_{oa} model, using actual air data from Chicago at 0°C (Harner and Bidleman, 1998b). K_p values were determined from particle phase and gas phase concentrations (equations (1) and (2)) for PCBs, PAHs, and PCNs (polychlorinated naphthalenes) and expressed as particulate fractions, ϕ, using Equation (9). The model estimates were calculated as described in the example (section 3-5) except that Equation (24) instead of (25) was used for the K_{oa} model, to allow for variation in the organic matter fraction, f_{om}, of the aerosol. In Figure 10.8a, the J-P model overpredicts the particulate fraction, ϕ, for PCBs but is in good agreement for PAHs. The Mackay model produces somewhat higher values. Another point of interest in

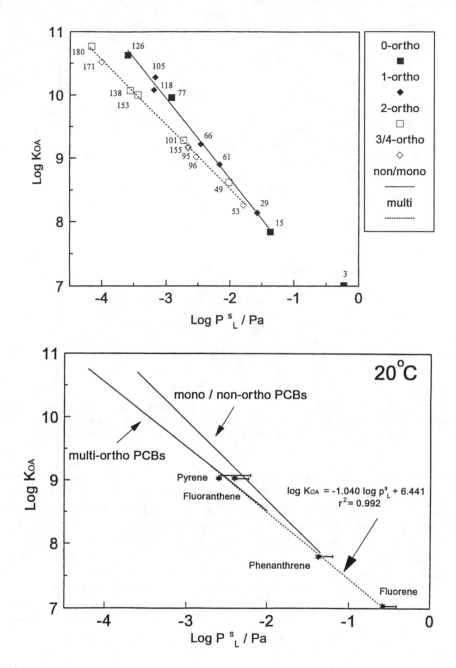

FIGURE 10.5

Top: Octanol-air partition coefficient (K_{oa}) vs. liquid-phase vapor pressure (P^S_L) for PCB congeners (indicated by IUPAC numbers). Black squares = mono- and non-ortho substituted congeners, white squares = multi-ortho congeners (Harner and Bidleman, 1996). Bottom: PAHs compared to PCBs (Harner and Bidleman, 1998a).

the field data is the higher particulate fractions observed for the PAHs and non-ortho (coplanar) PCBs relative to mono-/multi-ortho PCBs of the same vapor pressure.

The selective partitioning of coplanar PCBs is resolved by plotting the observed ϕ vs. log K_{oa} (Figure 10.8b). Results for PCNs also appear in Figure 10.8b since K_{oa} values recently have been determined for many congeners (Harner and Bidleman, 1998b). PCN partition-

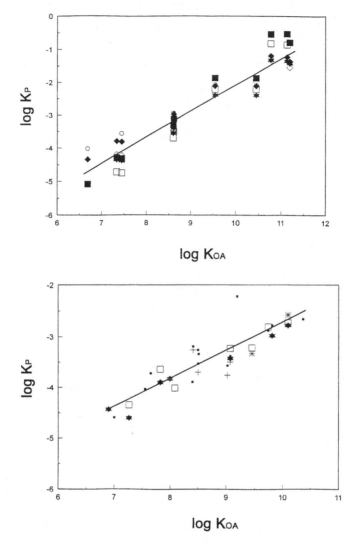

FIGURE 10.6
Log-log plots of the particle/gas partition coefficient (K_p) for PAHs (top) and organochlorine compounds (PCBs, pesticides) (bottom) vs. the octanol-air partition coefficient (K_{oa}). Points denote individual field studies. Figures and data from Finizio et al. (1997).

ing is also well described by K_{oa}. The fit between the experimental data and the K_{oa} model is better if the value of f_{om} = 0.05–0.1 rather than 0.2. This suggests that only a portion of the total organic matter in aerosols is effective for scavenging gas-phase POPs. PAHs are still enriched relative to the multi-ortho PCBs, and K_{oa} does not explain the field distribution well. Indeed, Pankow (1998) estimated that g oct/gom ~ 11 for PAHs, rather 1.0 assumed in our model. A possible explanation is the presence of bound PAHs, which are non-exchangeable with the surrounding air (Harner and Bidleman, 1998a).

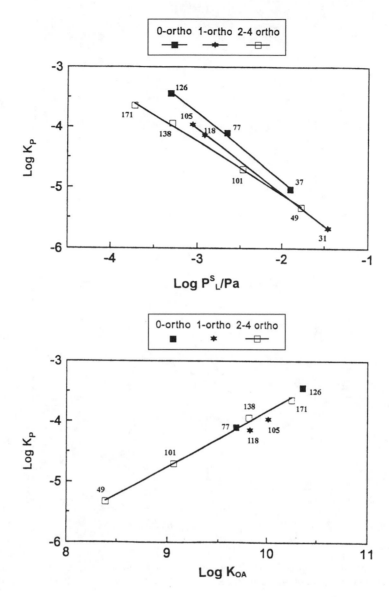

FIGURE 10.7
Values of log K_p for PCBs having different ortho-chlorine substitutions (laboratory equilibration experiments with Chicago aerosols at 5, 11, 18 and 25°C; Falconer et al., 1995) plotted vs. log P_L^S or log K_{oa}. Congeners used in these experiments were: 77,126 (non-ortho); 105,118 (mono-ortho); 49,101,138 (di-ortho) and 171 (tri-ortho).

10.5 Conclusions

Several predictive approaches are available for estimating the fraction of POPs sorbed onto aerosols. The Junge-Pankow model assumes that gaseous POPs are adsorbed onto active sites on the aerosol. Key parameters in the model are particle surface area per unit volume of air (θ), the liquid-phase vapor pressure of the compound (P_L^S) and a factor (c) which depends on the excess heat of desorption from the particle surface.

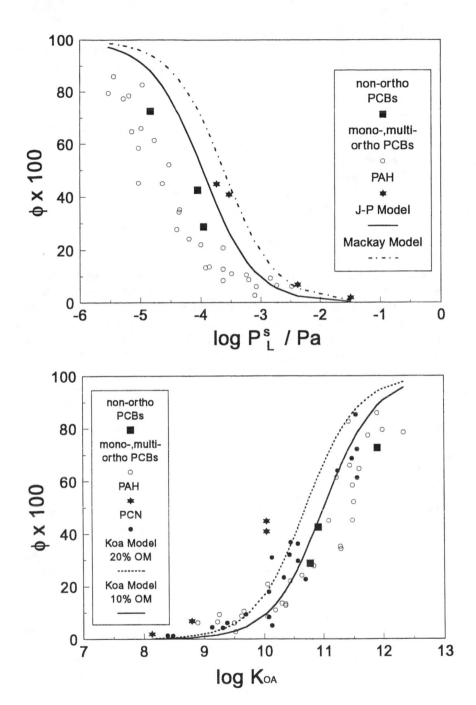

FIGURE 10.8

Comparison of Junge-Pankow and Mackay adsorption models (top) and K_{oa} model (bottom) with observed particulate percentages ($\phi \times 100$) of PCBs, PAHs and PCNs (bottom only) in Chicago (Harner and Bidleman, 1998a).

The Mackay simple one-parameter adsorption model assumes that POPs partition onto aerosols in a manner similar to the observed speciation of benz[a]anthracene in urban air; it also uses P_L^S as a descriptor.

The Pankow absorption model supposes that gaseous POPs partition into a liquid-like organic film on the aerosol. It uses P_L^S, the fraction of organic matter in the particle (f_{om}), and the activity coefficient of the solute in the liquid film (γ_{om}). The latter is apt to vary substantially among different chemical classes of POPs.

Replacing P_L^S with the octanol-air partition coefficient (K_{oa}) improves the absorption model. Advantages of the K_{oa} – based model are twofold: K_{oa} is directly measureable as a function of temperature, and the model depends on the ratio of activity coefficients of the solute in octanol and the organic film on the particle (γ_{oct}/γ_{om}). This ratio is more likely to be constant among classes of compounds than γ_{om} alone.

Comparison of the Junge-Pankow, Mackay, and K_{oa} models for predicting the particulate fraction of POPs in urban air show reasonable agreement, with the K_{oa} model giving somewhat lower and the Mackay model higher results. An unanswered question is: does octanol adequately represent the polar and non-polar classes of organic matter on aerosols; i.e., is the assumption that γ_{oct} and γ_{om} are equal valid?

Hi-vol air samplers are deployed routinely in air monitoring programs and, for reasons of simplicity and cost, probably will continue to be used for estimating particle/gas distributions. Particulate fractions obtained with the hi-vol often are compared to those predicted by models, but neither necessarily give the "correct" result. Further work is required to improve the models, and experimental methods for speciating particulate and gaseous POPs in air are needed, as are additional measurements of P_L^S and K_{oa} for different chemical classes. Our understanding of the phase distribution also will improve with fundamental investigations of the sorption mechanism onto aerosols.

References

Anderson, P.N. and R.A. Hites. 1996. OH radical reactions: the major removal pathway for polychlorinated biphenyls from the atmosphere. *Environ. Sci. Technol.* 30: 301-306.

Arey, J., R. Atkinson, B. Zielinska, and P.A. McElroy. 1989. Diurnal concentrations of volatile polycyclic aromatic hydrocarbons and nitroarenes during a photochemical pollution episode in Glendora, California. *Environ. Sci. Technol.* 23: 321-327.

Baek, S.O., M.E. Goldstone, P.W.W. Kirk, J.N. Lester, and R. Perry. 1991. Phase distribution and particle size dependency of polycyclic aromatic hydrocarbons in the urban atmosphere. *Chemosphere* 22: 503-520.

Baker, J.E. and S.J. Eisenreich. 1990. Concentrations and fluxes of PAHs and PCB congeners across the air-water interface of Lake Superior. *Environ. Sci. Technol.* 24: 342-352.

Beyhmer, T.D. and R.A. Hites. 1988. Photolysis of polycyclic aromatic hydrocarbons adsorbed on fly ash. *Environ. Sci. Technol.* 22: 1311-1319

Bidleman, T.F. 1988. Atmospheric processes: wet and dry deposition of organic compounds are controlled by their vapor-particle partitioning. *Environ. Sci. Technol.* 22: 361-367.

Bidleman, T.F. 1985. High volume collection of organic vapors using solid adsorbents. In J.F. Lawrence, Ed., *Trace Analysis*, Vol. 4 (Academic Press, NY), pp. 51-100.

Bidleman, T.F., W.N. Billings, and W.T. Foreman. 1986. Vapor-particle partitioning of semivolatile organic compounds: estimates from field collections. *Environ. Sci. Technol.* 20: 1038-1043.

Bidleman, T.F., D. Cleaverly, and M. Lorber. 1999. *Methodology for Assessing Health Risks Associated with Indirect Exposure to Combustor Emissions*, U.S. Environmental Protection Agency, Office of Research and Development, in press.

Bidleman, T.F. and L.L. McConnell. 1995. A review of field experiments to determine air-water gas exchange of persistent organic pollutants. *Sci. Total Environ.* 159: 101-117.

Brubaker, W.W. Jr. and R.A. Hites. 1997. Polychlorinated dibenzo-p-dioxins and dibenzofurans: gas-phase hydroxyl radical reactions and related atmospheric removal. *Environ. Sci. Technol.* 31: 1805-1810.

Bunce, N.J., L. Liu, J. Zhu, and D.A. Lane. 1997. Reaction of naphthalene and its derivatives with hydroxyl radicals in the gas phase. *Environ. Sci. Technol.* 31: 2252-2259.

Burford, M.D., S.B. Hawthorne, and D.J. Miller. 1993. Extraction rates of spiked vs. native PAHs from heterogeneous environmental samples using supercritical fluid extraction and sonication in methylene chloride. *Anal. Chem.* 65: 1497-1505.

Chuang, J.C., S.W. Hannan, and N.K. Wilson. 1987. Field comparison of polyurethane foam and XAD-2 resin for air sampling of polynuclear aromatic hydrocarbons. *Environ. Sci. Technol.* 21: 798-804.

Corn, M., T.L. Montgomery, and N.A. Esman. 1971. Suspended particulate matter: seasonal variation in specific surface areas and densities. *Environ. Sci. Technol.* 5: 155-158.

Cotham, W.E. and T.F. Bidleman. 1995. Polycyclic aromatic hydrocarbons and polychlorinated biphenyls in air at an urban and rural site near Lake Michigan. *Environ. Sci. Technol.* 29: 2782-2789.

Cotham, W.E. and T.F. Bidleman. 1992. Laboratory investigations of the partitioning of organochlorine compounds between the gas phase and atmospheric aerosols on glass fiber filters. *Environ. Sci. Technol.* 26: 469-478.

Coutant, R.W., L. Brown, J.C. Chuang, R.M. Riggin, and R.G. Lewis. 1988. Phase distribution and artifact formation in ambient air sampling for polynuclear aromatic hydrocarbons. *Atmos. Environ.* 22: 403-409.

Coutant, R.W., P.J. Callahan, M.R. Kuhlman, and R.G. Lewis. 1989. Design and performance of a high volume compound annular denuder. *Atmos. Environ.* 23: 2205-2211.

Coutant, R.W., P.J. Callahan, J.C. Chuang, and R.G. Lewis. 1992. Efficiency of silicone grease coated denuders for collection of polynuclear aromatic hydrocarbons. *Atmos. Environ.* 26A: 2831-2834.

Dannenfelser, R.M., N. Surendran, and S.H. Yalkowsky. 1993. Molecular symmetry and related properties. *SAR and QSAR in Environ. Res.* 1: 273-292.

Dickhut, R.M. and K.E. Gustafson. 1995. Atmospheric washout of polycyclic aromatic hydrocarbons in the southern Chesapeake Bay. *Environ. Sci. Technol.* 29: 1518-1525, correction *ibid.* 2904.

Donnelly, J.R., L.A. Drewes, R.L. Johnson, W.D. Munslow, K.K. Knapp, and G.W. Sovocool, 1990. Purity and heat of fusion data for environmental standards as determined by differential scanning calorimetry. *Thermochim. Acta* 167: 155-187.

Duinker, J.C. and F. Bouchertall. 1989. On the distribution of atmospheric polychlorinated biphenyl congeners between the vapor phase, aerosols and rain. *Environ. Sci. Technol.* 23: 57-62.

Eatough, D.J. 1999. BOSS, The Brigham Young University organic sampling system: determination of particulate carbonaceous material using diffusing denuder sampling technology In: Lane, D.A., Ed., *Gas and Particle Measurements of Atmospheric Organic Compounds*. Gordon and Breach Science Publishers, Amsterdam, pp. 233–285.

Eatough, D.J., A. Wadsworth, D.A. Eatough, J.W. Crawford, L.D. Hansen, and E.A. Lewis, 1993. A multi-system, multi-channel diffusion denuder sampler for the determination of fine particulate organic matter in the atmosphere. *Atmos. Environ.* 27: 1213-1219.

Eisenreich, S.J., K.C. Hornbuckle, and D.A. Achman. 1997. Air-water exchange of semivolatile organic chemicals (SOCs) in the Great Lakes. In J.E. Baker, Ed. *Atmospheric Deposition of Contaminants to the Great Lakes and Coastal Waters,* Society of Environmental Toxicology and Chemistry, SETAC Press, Pensacola, Florida, 109-135.

Eitzer, B.D. and R.A. Hites. 1989a. Atmospheric transport and deposition of polychlorinated dibenzo-p-dioxins and dibenzofurans. *Environ. Sci. Technol.* 23: 1396-1401.

Eitzer, B.D. and R.A. Hites. 1989b. Polychlorinated dibenzo-p-dioxins and dibenzofurans in the ambient atmosphere of Bloomington, Indiana. *Environ. Sci. Technol.* 23: 1389-1395.

Eitzer, B.D. and R.A. Hites. 1988. Vapor pressures of chlorinated dioxins and furans. *Environ. Sci. Technol.* 22: 1362-1364

Falconer, R.L. and T.F. Bidleman. 1994. Vapor pressures and predicted particle/gas distributions of polychlorinated biphenyl congeners as functions of temperature and ortho-chlorine substitution. *Atmos. Environ.* 28: 547-554.

Falconer, R.L., T.F. Bidleman, and W.E. Cotham. 1995. Preferential sorption of non- and mono-ortho polychlorinated biphenyls to urban aerosols. *Environ. Sci. Technol.* 29: 1666-1673.

Fan, Z.H., D. Chen, P. Birla, and R.M. Kamens, 1995. Modeling of nitro-polycyclic aromatic hydrocarbon formation and decay in the atmosphere. *Atmos. Environ.* 29: 1171-1181.

Febo, A., C. Perrino, and I. Allegrini. 1999. Selective gas and particle sampling: diffusion denuders. In: Lane, D.A., Ed., *Gas and Particle Measurements of Atmospheric Organic Compounds.* Gordon and Breach Science Publishers, Amsterdam, pp. 127-175

Finizio, A., D. Mackay, T.F. Bidleman, and T. Harner. 1997. Octanol-air partition coefficient as a predictor of partitioning of semi-volatile organic chemicals to aerosols. *Atmos. Environ.* 31: 2289-2296.

Fischer R.C., R. Wittlinger, and K. Ballschmiter. 1992. Retention-index based vapor pressure estimation for polychlorobiphenyl (PCB) by gas chromatography. *Fres. J. Analyt. Chem.* 342: 421-425.

Foreman, W.T. and T.F. Bidleman. 1987. An experimental system for investigating vapor-particle partitioning of trace organic pollutants. *Environ. Sci. Technol.* 21: 869-875.

Foreman, W.T. and T.F. Bidleman. 1990. Semivolatile organic compounds in the ambient air of Denver, Colorado. *Atmos. Environ.* 24A: 2405-2416.

Forstner, H.L., R.C. Flagan, and J.H. Seinfeld. 1997. Molecular speciation of secondary organic aerosol from photooxidation of higher alkenes: 1-octene and 1-decene. *Atmos. Environ.* 31: 1953-1964.

Golomb, D., D. Ryan, J. Underhill, T. Wade, and S. Zemba. 1997. Atmospheric deposition of toxics onto Massachusetts Bay. II. Polycyclic aromatic hydrocarbons. *Atmos. Environ.* 31: 1361-1368.

Goss, K.-U. and S.J. Eisenreich. 1997. Sorption of volatile organic compounds to particles from a combustion source at different temperatures and relative humidities. *Atmos. Environ.* 31: 2827-2834.

Goss, K.-U. and R.P. Schwartzenbach. 1998. Gas/solid and gas/liquid partitioning of organic compounds: critical evaluation of the interpretation of equilibrium constants. *Environ. Sci. Technol.* 32: 2025-2032.

Gundel, L.A. and D.A. Lane. 1999. Sorbent-coated diffusion denuders for direct measurement of gas/particle partition by semivolatile organic compounds. In: Lane, D.A., Ed., *Gas and Particle Measurements of Atmospheric Organic Compounds.* Gordon and Breach Science Publishers, Amsterdam, pp. 287-332.

Gundel, L.A., V.C. Lee, K.R.R. Mahanama, R.K. Stevens, and J.M. Daisey. 1995. Direct determination of the phase distributions of semi-volatile polycyclic aromatic hydrocarbons using annular denuders. *Atmos. Environ.* 29: 1719-1733.

Gustafson, K.E. and R.M. Dickhut. 1997. Particle/gas concentrations and distributions of PAHs in the atmosphere of southern Chesapeake Bay. *Environ. Sci. Technol.* 31: 140-147.

Harner, T. and T.F. Bidleman. 1998b. Measurements of octanol-air partition coefficients for polycyclic aromatic hydrocarbons and polychloronaphthalenes. *J. Chem. Eng. Data,* 43: 40-46.

Harner, T. and T.F. Bidleman. 1998a Octanol-air partition coefficient (K_{oa}) for describing particle-gas partitioning of aromatic compounds in urban air. *Environ. Sci. Technol.,* 32: 1494-1502.

Harner, T. and T.F. Bidleman. 1996. Measurements of octanol-air partition coefficients for polychlorinated biphenyls. *J. Chem. Eng. Data* 41: 895-899.

Harner, T. and D. Mackay. 1995. Measurements of octanol-air partition coefficients for chlorobenzenes, PCBs and DDT. *Environ. Sci. Technol.* 29: 1599-1606.

Hart, K.M. and J.F. Pankow. 1994. High volume air sampler for particle and gas sampling. 2. Use of backup filters to correct for adsorption of gas-phase polycyclic aromatic hydrocarbons to the front filter. *Environ. Sci. Technol.* 28: 655-661.

Hinckley, D.A., T.F. Bidleman, W.T. Foreman, and J. Tuschall. 1990. Determination of vapor pressures for nonpolar and semipolar organic compounds from gas chromatographic retention data. *J. Chem. Eng. Data* 35: 232-237.

Hoff, R.M., W.M.J. Strachan, C.W. Sweet, C.H. Chan, M. Shackleton, T.F. Bidleman, K.A. Brice, D.A. Burniston, S. Cussion, D.F. Gatz, K. Karlin, and W.H. Schroeder. 1996. *Atmos. Environ.* 30: 3505-3527.

Holsen, T.M. and K.E. Noll. 1992. Dry deposition of atmospheric particles: application of current models to ambient data. *Environ. Sci. Technol.* 26: 1807-1815.

Holsen, T.M., K.E. Noll, S.-P. Liu, and W-J. Lee. 1991. Dry deposition of polychlorinated biphenyls in urban areas. *Environ. Sci. Technol.* 25: 1075-1081.

Hornbuckle, K.C., D.R. Achman, and S.J. Eisenreich. 1993. Over-water and over-land polychlorinated biphenyls (PCBs) in Green Bay, Lake Michigan. *Environ. Sci. Technol.* 27: 75-87.

Jang, M. and S.R. McDow. 1995. Benz[a]anthracene photodegradation in the presence of known organic constituents of atmospheric aerosols. *Environ. Sci. Technol.* 29: 2654-2660.

Jang, M., R.M. Kamens, K.B. Leach, and M.R. Strommen. 1997. Thermodynamic approach using group contribution methods to model the partitioning of semivolatile organic compounds on atmospheric particulate matter. *Environ. Sci. Technol.* 31: 2805-2811.

Junge, C.E. 1977. Basic considerations about trace constituents in the atmosphere as related to the fate of global pollutants. In I.H. Suffet, Ed., *Fate of Pollutants in the Air and Water Environments.* Adv. in Environ. Sci. Technol. Ser., Vol. 8, Part I, Wiley-Interscience, NY. pp. 7-26.

Kamens< R., Z. Fan, M. Jang, J. Odum, J. Hu, D. Coe, J. Zhang, S. Chen, and K. Leach. 1999. The use of denuders for semivolatile characterization studies in outdoor chambers. In: Lane, D.A., Ed., *Gas and Particle Measurements of Atmospheric Organic Compounds.* Gordon and Breach Science Publishers, Amsterdam, pp. 333-368.

Kamens, R.M., Z-H. Fan, Y. Yao, D. Chen, S. Chen, and M. Vartiainen. 1994. A methodology for modeling the formation and decay of nitro-PAH in the atmosphere. *Chemosphere* 28: 1623-1632.

Kamens, R.M., Z. Guo, J.N. Fulcher, and D.A. Bell. 1988. Influence of humidity, sunlight and temperature on the daytime decay of polyaromatic hydrocarbons on atmospheric soot particles. *Environ. Sci. Technol.* 22: 103-108.

Kamens, R.M., H. Karam, J. Guo, J.M. Perry, and L. Stockburger. 1989. The behavior of oxygenated polycyclic aromatic hydrocarbons on atmospheric soot particles. *Environ. Sci. Technol.* 23: 801-806.

Kamens, R., J. Odum, and Z.-H. Fan. 1995. Some observations on times to equilibrium for semivolatile polycyclic aromatic hydrocarbons. *Environ. Sci. Technol.* 29: 43-50.

Kaupp, H. and G. Umlauf. 1992. Atmospheric gas-particle partitioning of organic compounds: comparison of sampling methods. *Atmos. Environ.* 26A: 2259-2267.

Koester, C.J. and R.A. Hites. 1992a. Wet and dry depositon of chlorinated dioxins and furans. *Environ. Sci. Technol.* 26: 1375-1382.

Koester, C.J. and R.A. Hites. 1992b. Photodegradation of polychlorinated dioxins and dibenzofurans adsorbed to fly ash. *Environ. Sci. Technol.* 26: 502-507.

Kömp, P. and M. McLachlan. 1997. Octanol-air partitioning of polychlorinated biphenyls. *Environ. Toxicol. Chem.* 16: 2433-2437.

Krieger, M.S. and R.A. Hites. 1994. Measurement of polychlorinated biphenyls and polycyclic aromatic hydrocarbons in air with a diffusion denuder. *Environ. Sci. Technol.* 28: 1129-1133.

Kucklick, J.R., D.A. Hinckley, and T.F. Bidleman. 1991. Determination of Henry's law constants for hexachlorocyclohexanes in distilled water and artificial seawater as functions of temperature. *Marine Chem.* 34: 197-209.

Kwok, E.S.C., R. Atkinson, and J. Arey, 1995. Rate constants for the gas-phase reactions of the OH radical with dichlorobiphenyls, 1-chlorodibenzo-p-dioxin, 1,2-dimethoxybenzene and diphenyl ether: Estimation of OH radical reaction rate constants for PCBs, PCDDs and PCDFs. *Environ. Sci. Technol.* 29: 1591-1598.

Kwok, E.S.C., W.P. Harger, J. Arey, and R. Atkinson. 1994a. Reactions of gas-phase phenanthrene under simulated atmospheric conditions. *Environ. Sci. Technol.* 29: 1591-1598.

Lane, D.A., 1999. *Gas and Particle Phase Partition Measurements of Atmospheric Organic Compounds.* Gordon and Breach Publishers, in press.

Lane, D.A. and L. Gundel, 1996. Gas and particle sampling of airborne polycyclic aromatic compounds. *Polycyclic Aromatic Compounds* 9: 67-73.

Lane, D.A. and N.D. Johnson. 1999. Gas/particle measurements with the gas and particle (GAP) sampler. In: Lane, D.A., Ed., *Gas and Particle Measurements of Atmospheric Organic Compounds.* Gordon and Breach Science Publishers, Amsterdam, pp. 177-200.

Lane, D.A. and H. Tang. 1994. Photochemical degradation of polycyclic aromatic compounds. 1. Napthalene. *Polycyclic Aromatic Compounds* 5: 131-138.

Lee, W.-M. G. and L.-Y. Tsay. 1994. The partitioning model of polycyclic aromatic hydrocarbon between gaseous and particulate (PM-10) phases in urban atmosphere with high humidity. *Sci. Total Environ.* 145: 163-171.

Lei, Y. D., F. Wania, and W.-Y. Shu. 1999. Determination of the vapor pressures of polychlorinated naphthalenes. *J. Chem. Eng. Data* 44: 577-582.

Lewis, R.G. and R.W. Coutant. 1999. Determination of phase-distributed polycyclic aromatic hydrocarbons in air by grease-coated denuders. In: Lane, D.A., Ed., *Gas and Particle Measurements of Atmospheric Organic Compounds*. Gordon and Breach Science Publishers, Amsterdam, pp. 201-231.

Liang, C. and J.F. Pankow. 1996. Gas-particle partitioning of organic compounds to environmental tobacco smoke: partition coefficient measurements by desorption and comparison to urban particulate material. *Environ. Sci. Technol.* 30: 2800-2805.

Liang, C., J.F. Pankow, J.R. Odum, and J.H. Seinfeld. 1997. Gas/particle partitioning of semivolatile organic compounds to model inorganic, organic and ambient smog aerosols. *Environ. Sci. Technol.* 31: 3086-3092.

Ligocki, M.P., C. Leuenberger, and J.F. Pankow. 1985a. Trace organic compounds in rain. II. Gas scavenging of neutral organic compounds. *Atmos. Environ.* 19: 1609-1617.

Ligocki, M.P., C. Leuenberger, and J.F. Pankow. 1985b. Trace organic compounds in rain. III. Particle scavenging of neutral organic compounds. *Atmos. Environ.* 19: 1619-1626.

Ligocki, M.P. and J.F. Pankow. 1989. Measurement of the gas/particle distribution of atmospheric organic compounds. *Environ. Sci. Technol.* 23: 75-83.

Lorber, M., D. Cleverly, J. Schaum, L. Phillips, G. Schweer, and T. Leighton. 1994. Development and validation of an air-to-beef food chain model for dioxin-like compounds. *Sci. Total Environ.* 156: 39-65.

Mackay, D., S. Paterson, and W.H. Schroeder. 1986. Model describing the rates of transfer processes of organic chemicals between atmosphere and water. *Environ. Sci. Technol.* 20: 810-816.

Mackay, D., W.-Y. Shiu, and K.C. Ma. 1992a. *Illustrated Handbook of Physical - Chemical Properties and Environmental Fate of Organic Compounds. I. Monoaromatic Hydrocarbons, Chlorobenzenes and Polychlorinated Biphenyls.* Lewis Publishers, Boca Raton, Florida, 697 pages.

Mackay, D., W.-Y. Shiu, and K.C. Ma. 1992b. *Illustrated Handbook of Physical - Chemical Properties and Environmental Fate of Organic Compounds. II. Polynuclear Aromatic Hydrocarbons, Polychlorinated Dibenzodioxins and Dibenzofurans.* Lewis Publishers, Boca Raton, Florida, 597 pages.

McDow, S.R. 1999. Sampling artifact errors in gas/particle partitioning measurements. In: Lane, D.A., Ed., *Gas and Particle Measurements of Atmospheric Organic Compounds*. Gordon and Breach Science Publishers, Amsterdam, pp. 105-126.

McDow, S.R. and J.J. Huntzicker. 1990. Vapor adsorption artifact in the sampling of organic aerosol: face velocity effects. *Atmos. Environ.* 24A: 2563-2571.

McDow, S.R., Q Sun, M. Vartiainen, Y Hong, Y Yao, T. Fister, R. Yao, and R. Kamens, 1994. Effect of composition and state of organic components on polycyclic aromatic hydrocarbon decay in atmospheric aerosols. *Environ. Sci. Technol.* 28: 2147-2153.

McDow, S.R., M. Vartiainen, Q. Sun, Y. Kong, and R.M. Kamens. 1995. Combustion aerosol water content and its effect on polycyclic aromatic hydrocarbon reactivity. *Atmos. Environ.* 29: 791-797.

McLachlan, M.S. 1996. Bioaccumulation of hydrophobic chemicals in agricultural food chains. *Environ. Sci. Technol.* 30: 252-259.

McVeety, B.D. 1986. Atmospheric deposition of polycyclic aromatic hydrocarbons to water surfaces: a mass balance approach. Ph.D. Thesis, Indiana University, 455 pages.

McVeety, B.D. and R.A. Hites. 1988. Atmospheric deposition of polycyclic aromatic hydrocarbons to water surfaces: a mass balance approach. *Atmos. Environ.* 22: 511-536.

Monosmith, C.L. and M.H. Hermanson. 1996. Spatial and temporal trends of atmospheric organochlorine vapors in the central and upper Great Lakes. *Environ. Sci. Technol.* 30: 3464-3472.

Ngabe, B. and T.F. Bidleman. 1992. Occurrence and vapor-particle partitioning of organic compounds in ambient air of Brazzaville, Congo. *Environ. Pollut.* 76: 147-156.

Offenberg, J.H. and J.E. Baker. 1997. Polychlorinated biphenyls in Chicago precipitation: enhanced wet deposition to nearshore Lake Michigan. *Environ. Sci. Technol.* 31: 1534-1538.

Pankow, J.F. 1987. Review and comparative analysis of the theories on partitioning between the gas and aerosol particulate phases in the atmosphere. *Atmos. Environ.* 21: 2275-2283.

Pankow, J.F. 1988. The calculated effects of non-exchangeable material on the gas-particle distribution of organic compounds. *Atmos. Environ.* 22: 1405-1409.

Pankow, J.F. 1991. Common y-intercept and single compound regressions of gas-particle partition coefficient vs. 1/T. *Atmos. Environ.* 25A: 2229-2239.

Pankow, J.F. 1992. Application of the common y-intercept regression parameters for log K_p vs. 1/T for predicting gas-particle partitioning in the urban environment. *Atmos. Environ.* 26A: 2489-2497.

Pankow, J.F. 1994a. An absorption model of gas-particle partitioning in the atmosphere. *Atmos. Environ.* 21: 185-188.

Pankow, J.F. 1994b. An absorption model of the gas-aerosol partitioning involved in the formation of secondary organic aerosol. *Atmos. Environ.* 21: 189-193.

Pankow, J.F. 1998. Further discussion of the octanol-air partition coefficient, K_{oa}, as a correlating parameter for gas/particle partitioning coefficients. *Atmos. Environ.* 32: 1493-1497.

Pankow, J.F. 1999. Fundamentals and mechanisms of gas/particle partitioning in the atmosphere. In: Lane, D.A., Ed., *Gas and Particle Measurements of Atmospheric Organic Compounds.* Gordon and Breach Science Publishers, Amsterdam, pp. 25-37.

Pankow, J.F. and T.F. Bidleman. 1992. Interdependence of the slopes and intercepts from log-log correlations of measured gas-particle partitioning vs. subcooled liquid vapor pressure. 1. Basic considerations and review of available data. *Atmos. Environ.* 25A: 2241-2249.

Pankow, J.F. and T.F. Bidleman. 1991. Effects of temperature, TSP and percent non-exchangeable material in determining the gas-particle distributions of organic compounds. *Atmos. Environ.* 26A: 1071-1080.

Pankow, J.F., L.M. Isabelle, D.A. Buchholz, W. Luo, and B.D. Reeves. 1994. Gas-particle partitioning of polycyclic aromatic hydrocarbons and alkanes to environmental tobacco smoke. *Environ. Sci. Technol.* 28: 363-365.

Pankow, J.F., J.M.E. Storey, and H. Yamasaki. 1993. Effects of relative humidity on gas-particle partitioning of semivolatile organic compounds to urban particulate matter. *Environ. Sci. Technol.* 27: 2220-2226.

Pirrone, N., G.J. Keeler, and T.M. Holsen. 1995. Dry deposition of semivolatile organic compounds to Lake Michigan. *Environ. Sci. Technol.* 29: 2123-2132.

Poster, D.L. and J.E. Baker. 1996a. Influence of submicron particles on hydrophobic organic contaminants in precipitation. 1. Concentrations and distributions of polycyclic aromatic hydrocarbons and polychlorinated biphenyls in rainwater. *Environ. Sci. Technol.* 30: 341-348.

Poster, D.L. and J.E. Baker. 1996b. Influence of submicron particles on hydrophobic organic contaminants in precipitation. 2. Scavenging of polycyclic aromatic hydrocarbons by rain. *Environ. Sci. Technol.* 30: 349-354.

Poster, D.L., R.M. Hoff, and J.E. Baker. 1995. Measurement of the particle size distributions of semivolatile organic contaminants in the atmosphere. *Environ. Sci. Technol.* 29: 1990-1997.

Rordorf, B.F. 1989. Prediction of vapor pressures, boiling points and enthalpies of fusion for twenty-nine halogenated dibenzo-p-dioxins and fifty-five dibenzofurans by a vapor pressure correlation method. *Chemosphere* 18: 783-788.

Rounds, S.A. and J.F. Pankow. 1990. Application of a radial diffusion model to describe gas-particle sorption kinetics. *Environ. Sci. Technol.* 24: 1378-1386.

Rounds, S.A., B.A. Tiffany, and J.F. Pankow. 1993. Description of gas-particle sorption kinetics with an intraparticle diffusion model: desorption experiments. *Environ. Sci. Technol.* 27: 366-377.

Seiber, J.N., B.W. Wilson, and M.M. McChesney. 1993. Air and fog deposition residues of four organophosphate insecticides used on dormant orchards in the San Joaquin Valley, California. *Environ. Sci. Technol.* 27: 2236-2243.

Shah, J.J., R.L. Johnson, E.K. Heyerdahl, and J.J. Huntzicker. 1986. Carbonaceous aerosol at urban and rural sites in the United States. *J. Air Pollut. Cont. Assoc.* 36: 254-257.

Sheffield, A.E. and J.F. Pankow. 1994. Specific surface area of urban atmospheric particulate matter in Portland, Oregon. *Environ. Sci. Technol.* 28: 1759-1766.

Simcik, M.F., T.P. Franz, H. Zhang, and S. J. Eisenrich. 1998. Gar-particle partitioning of PCBs and PAHs in the Chicago urban and adjacent coastal atmosphere: states of equilibrium. *Environ. Sci. Technol.* 32: 251-257.

Simonich, S. and R.A. Hites. 1995. Organic pollutant accumulation in vegetation. *Environ. Sci. Technol.* 29: 2905-2914.

Storey, J.M.E., W. Luo, L.M. Isabelle, and J.F. Pankow. 1995. Gas-particle partitioning of semivolatile organic compounds to model atmospheric solid surfaces as a function of relative humidity. 1. Clean quartz. *Environ. Sci. Technol.* 29: 2420-2428.

Subramanyam, K.T. Valsaraj, L.J. Thibodeaux, and D.D. Reible. 1994. Gas-to-particle partitioning of polycyclic aromatic hydrocarbons in an urban atmosphere. *Atmos. Environ.* 28: 3083-3091.

Suntio, L.R., W.-Y. Shiu, D. Mackay, J.N. Seiber, and D. Glotfelty. 1988. Critical review of Henry's law constants for pesticides. *Rev. Environ. Contam. Toxicol.* 103: 1-59.

Tang, H., E.A. Lewis, D.J. Eatough, R.M. Burton, and R.J. Farber. 1994. Determination of the particle size distribution and chemical composition of semi-volatile organic compounds in atmospheric fine particles with a diffusion denuder sampling system. *Atmos. Environ.* 28: 939-947.

Thibodeaux, L.J., K.C. Nadler, K.T. Valsaraj, and D.D. Reible. 1991. The effect of moisture on volatile organic chemical gas-to-particle partitioning with atmospheric aerosols — competitive adsorption theory predictions. *Atmos. Environ.* 25A: 1649-1656.

Turpin, B.J. , S.-P. Liu, P.H. McMurry, and S.J. Eisenriech. 1999. Definitive measurement of semivolatilve PAHs with a diffusion separator: Design and investigation of sampling artifacts in filter-adsorbent samplers. In: Lane, D.A., Ed., *Gas and Particle Measurements of Atmospheric Organic Compounds*. Gordon and Breach Science Publishers, Amsterdam, pp.369-392 .

Turpin, B.J., J.J. Huntzicker, and S.V. Hering. 1994. Investigation of organic aerosol sampling artifacts in the Los Angeles Basin. *Atmos. Environ.* 28: 3061-3071.

Turpin, B.J., S-P. Liu, K.S. Poldoske, M.S.P. Gomes, S.J. Eisenreich, and P.H. McMurray. 1993. *Environ. Sci. Technol.* 27: 2441-2449.

Umlauf, G. 1999. Measuring gas/particle fractions of trace organic compounds in ambient air. Evaluation of electrostatic precipitation, impaction and cnoventional filtration for separating particles. In: Lane, D.A., Ed., *Gas and Particle Measurements of Atmospheric Organic Compounds*. Gordon and Breach Science Publishers, Amsterdam, pp. 73-103.

Venkataraman, C. and S.K. Friedlander. 1994. Size distributions of polycyclic aromatic hydrocarbons and elemental carbon. 2. Ambient measurements and effects of atmospheric processes. *Environ. Sci. Technol.* 26: 563-572.

Venkataraman, C., J.M. Lyons, and S.K. Friedlander. 1994. Size distributions of polycyclic aromatic hydrocarbons and elemental carbon. 1. Sampling, measurement methods and source characterization. *Environ. Sci. Technol.* 26: 555-562.

Welsch-Paulsch, K., M. McLachlan, and G. Umlauf. 1995. Determination of the principal pathways of polychlorinated dibenzo-p-dioxins and dibenzofurans to *Lolium multiflorum* (Welsh ray grass). *Environ. Sci. Technol.* 29: 189-194.

Whitby, K.T. 1978. The physical characteristics of sulfate aerosols. *Atmos. Environ.* 12: 135-159.

Wilson, N.K., T.R. McCurdy, and J.C. Chuang. 1995. Concentrations and phase distributions of nitrated and oxygenated polycyclic aromatic hydrocarbons in ambient air. *Atmos. Environ.* 29: 2575-2584.

Yamasaki, H., K. Kuwata, and Y. Kuge. 1984. Determination of vapor pressures of polycyclic aromatic hydrocarbons in the supercooled liquid phase and their adsorption on airborne particulate matter. *Nippon Kagaku Kaishi* 8: 1324-1329 (in Japanese), *Chem. Abst.* 101: 156747p.

Yamasaki, H., K. Kuwata, and H. Miyamoto. 1982. Effects of ambient temperature on aspects of atmospheric polycyclic aromatic hydrocarbons. *Environ. Sci. Technol.* 16: 189-194.

Zaranski, M.T., G.W. Patton, L.L. McConnell, T.F. Bidleman, and J.D. Mulik. 1991. Collection of nonpolar organic compounds from ambient air using polyurethane foam – granular adsorbent sandwich cartridges. *Anal. Chem.* 63: 1228-1232.

Zhang, X. and P.H. McMurry. 1991. Theoretical aspects of evaporative losses of adsorbed or absorbed species during atmospheric aerosol sampling. *Environ. Sci. Technol.* 25: 456-459.

11

Absorption Through Cellular Membranes

Stephen C. DeVito

CONTENTS

11.1 Introduction

Substances that are capable of producing a pharmacological or toxicological effect in humans or other life forms do so because they can be absorbed into and distributed throughout the body, and they contain structural features that bestow these biological properties. *Absorption* is the movement of a chemical substance from its site of exposure into the systemic circulation (bloodstream). *Distribution* is the movement of a chemical substance from the systemic circulation to various tissues or organs throughout the body. The ability of a substance to be absorbed and enter the systemic circulation following exposure, and subsequently be distributed via the circulation to specific organs or areas of the body where it can exert a pharmacological or toxicological effect, is referred to as *bioavailability*.

Bioavailability is the amount of substance that reaches bodily tissues in its non-metabolized form following exposure. Most substances are metabolized (i.e., undergo enzyme-catalyzed transformations) by the body to other substances. The bioavailability of a substance is affected greatly by its metabolism. While metabolism usually occurs following absorption, it also can occur prior to or during absorption, depending on the substance and the route of exposure. Bioavailability is clearly an important factor that can affect the therapeutic efficacy or toxicity of chemical substances. A candidate drug substance that contains structural features necessary for a specific pharmacologic property will have limited, if any, therapeutic value if it has low bioavailability. Similarly, a chemical substance that contains structural components or features known to cause a toxic effect will not be toxic if it is not bioavailable. Thus, to accurately assess a substance's biological properties— whether one is designing a new drug substance to control or cure an illness, or assessing human health or ecological risks of a known or planned chemical substance intended for commercial use — one needs to consider the substance's propensity to be absorbed in humans and other species.

This chapter provides an overview of the factors that govern the absorption of chemical substances in humans. Emphasis is on those physicochemical properties of the substance and the anatomical and physiological characteristics of the skin, gastrointestinal tract, and lung that are most significant in relation to absorption. No discussion of the other factors (i.e., extent and duration of exposure, metabolism, distribution, excretion, etc.) important to bioavailability of chemical substances appears here (for additional discussion of these factors, see Benet et al., 1996; Bronaugh, 1990; Hrudey et al., 1996; Medinsky and Klaassen, 1996; Rozman and Klaassen, 1996; Wright, 1995). Unless stated otherwise, the terms "chemical substance" or "substance" refer to non-electrolyte organic substances. The goal of this chapter is to provide an approach to accurately estimate absorption of substances for which little or no absorption data exist.

11.2   Theoretical Background: The Cellular Membrane

Regardless of the route of exposure, absorption of a chemical substance from any site of exposure involves its passage across cellular membranes. Absorption of a substance includes the following processes in succession:

- passage of the substance across the portion of cell membranes that compose the surface of the exposure site (e.g., surface of the skin, inner lining of the gastrointestinal tract, or alveolar sacs ([air sacs] of the lungs
- entry into the cellular cytoplasm
- passage through the cytoplasm and into the portions of the cellular membranes that interface with blood capillaries (these ultimately lead to the major blood vessels)

- passage into the membranes of the cells composing the blood capillaries, and entry into the blood

Once in the blood, substances are transported throughout the circulatory system and enter other organs and tissues via passage across the membranes of the cells composing the organs and tissues. Hence, a critical determinant of a substance's bioavailability is its ability to cross cellular membranes.

11.2.1   Cell Membrane Composition

All mammalian cells contain a membrane as their outer surface. The membrane surrounds and contains the cytoplasm and cellular constituents therein, separating them from the extracellular environment. The primary purpose of the membrane is to protect the cell by regulating intracellular entrance of substances, allowing passage of only those substances that the cell uses either as a food source or to perform biochemical processes necessary for a physiological purpose or function of the organ or tissue to which the cell belongs. However, many substances foreign to the cell (e.g., drug substances and other man-made chemicals) also may cross cell membranes and enter cells.

The key structural component of the cell membrane is a phospholipid bilayer in which proteins are embedded. Some of the proteins protrude from the outer membrane surface. These proteins typically serve as receptor sites for endogenous substances such as hormones or neurotransmitters, for purposes of eliciting a biological response from the tissue to which the cell belongs. Other proteins traverse the lipid bilayer, forming aqueous pores or channels that connect the exterior of the cell with its interior. These pores vary from about 4 to 10 Å in diameter, being predominantly about 7 Å.

The lipid bilayer is such that the polar heads (often phosphatidylcholine or phosphatidylethanolamine) of the phospholipids are juxtaposed on the external and internal surfaces of the membrane, causing the ends of the hydrophobic (i.e., long-chained alkyl) portions of the phospholipids to extend inside the membrane. Also contained within the lipid bilayer are cholesterol and other sterols.

Because of their high lipid content, cell membranes are not permeable to highly polar substances and are fluid or "water bed-like" rather than firm or rigid. The relative fluidity is determined largely by the type and abundance of unsaturated fatty acids in the phospholipids. The greater the abundance of unsaturated fatty acids, and the greater the amount of unsaturation within the acids, the greater the fluidity of the membrane.

Cell membranes differ mainly in their thickness, the type of lipid that composes their membrane bilayer, the diameters of their aqueous pores, and, therewith, their specificity in allowing substances to enter the cell. The thickness of cellular membranes varies greatly, but most range from 5 to 9 nm, depending upon the type of cell to which they belong. Figure 11.1 shows a general representation of a cell membrane.

11.2.2 How Substances Cross Cell Membranes

Chemical substances can cross a cell membrane to enter or leave a cell in several ways. These are:

- Passive permeation (diffusion) through the lipid bilayer
- Passive transport through membrane channels or pores
- Active transport

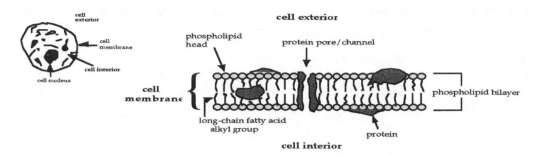

FIGURE 11.1
An enlarged cross-sectional representation of a segment of the cell membrane. Also shown (upper left) is a general depiction of the cell.

- Facilitated transport (carrier-mediated transport)
- Phagocytosis (also pinocytosis and endocytosis)

The specific process(es) by which a substance crosses a membrane depends largely upon the substance's structural and physicochemical properties (Benet et al., 1996; Hrudey et al., 1996; Medinsky and Klaassen, 1996; Rozman and Klaassen, 1996; Wright, 1995). Note that the individual *molecules* of a substance cross cell membranes, not the aggregated forms (e.g., crystals or powder) or formulations of the substance to which exposure may occur. Hence, for a substance to cross cell membranes and be absorbed, it must first leave its aggregated form and reach a state of molecular disaggregation, i.e, individual molecules. The relative ease with which a substance can disaggregate into individual molecules depends largely on its structural and physicochemical properties.

11.2.2.1 Passive Permeation (Diffusion) through the Membrane Lipid Bilayer

Most drug substances and substances of interest to health and environmental risk assessors enter cells by passive permeation (diffusion). In this process, a substance dissolves in the membrane lipid bilayer, permeates through the membrane, and enters into the cytoplasm of the cell. The substance thus must be soluble in lipids. The process is passive because the rate and extent to which a substance will enter a cell by this means depends on its concentration outside and inside the cell. The net movement is from the region of higher concentration to that of lower concentration. Unlike the cell membrane, which is chiefly lipid, the extracellular and intracellular spaces separated by the membrane are aqueous. The higher the concentration of substance outside of the cell, and the more soluble the substance in the membrane lipid bilayer, the greater will be the tendency for the substance to diffuse across the membrane and enter the cytoplasm. The rate and extent of diffusion will decrease as the concentration of the substance inside the cell increases until, eventually, equilibrium is reached.

If the water solubility of a substance is very low, an extracellular concentration sufficient to establish an adequate concentration gradient is unlikely. Consequently, little of the substance will permeate the membrane, even if it is lipid soluble. Hence, in addition to lipid solubility, water solubility is an important factor that controls the extent to which a substance can enter a cell by passive permeation. In fact, lipid-to-water partitioning, rather than lipid solubility or water solubility alone, is the more important factor governing a substance's ability to diffuse through cell membranes. Other factors that control the rate and extent to which a substance permeates a cell membrane are the thickness and surface area of the membrane. Figure 11.2 illustrates a chemical substance entering a cell via passive diffusion across the cell membrane.

The rate at which a substance enters a cell via passive diffusion is represented by Fick's First Law of Diffusion (Equation (1)):

$$\frac{dQ}{dt} = \frac{DA(C_1 - C_2)}{x} \tag{1}$$

where Q is the net quantity of substance that diffuses across the membrane; t is time; C_1 and C_2 are the concentrations on the outside and inside of the cell, respectively; D is the diffusion coefficient of the substance in the membrane lipids; A is the surface area of the membrane; and x is the thickness of the membrane. The diffusion coefficient is related to the lipid/water partition coefficient of the substance (Wright, 1995). At equilibrium, C_1 and C_2 are equal and there is no flux of the substance.

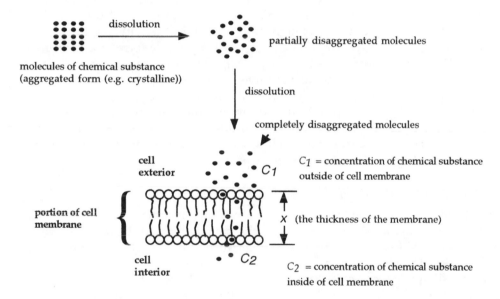

FIGURE 11.2
A Chemical Substance Entering a Cell via Passive Diffusion. Note: dissolution of the substance into a state of complete molecular disaggregation is a prerequisite step for crossing a cellular membrane by any means of transport.

Organic substances that are weak acids or weak bases exist as charged species when ionized. The charged species exist in equillibrium with the un-ionized species. Only the un-ionized species will permeate cellular membranes. The extent to which weak organic acids or bases are ionized under physiological conditions depends upon the pH and their dissociation constants. Strong organic acids or bases are typically highly ionized under physiologic conditions, and hence have low lipid solubility and permeate the lipid bilayer poorly. Organic acids that contain sulfonic acid ($-SO_3^{-?}$) moieties, for example, generally are poorly absorbed.

11.2.2.2 Passive Transport through Membrane Channels or Pores

Transport through membrane channels or pores requires the substance to be water soluble and to have a cross-sectional diameter less than the diameter of the pore or channel. This transport process is passive; transport occurs with the concentration gradient, and the rate and direction of movement of the substance depends on its extracellular and intracellular concentrations. Transport through pores is generally possible only for water and small, water-soluble organic molecules that have molecular weights less than about 200 daltons (e.g., ethanol and urea). Theoretically, an elongated molecule such as alkane could enter a cell via a pore; however, the probability that such a molecule would be oriented properly is low. While it would seem that most inorganic ions are sufficiently small to pass through membrane channels, their hydrated ionic radi are too large to permit their passage by this mechanism. Important cellular ions such as sodium, potassium, calcium, chloride, and others cross cellular membranes by active transport processes.

11.2.2.3 Active Transport

Active transport is a process by which membrane-bound proteins carry ions or other substances across the membrane and into (or out of) the cell (Rozman and Klaassen, 1996; Wright, 1995). These carrier proteins serve as "pumps" that move important ions such as sodium, potassium, calcium, chloride, and others against their electrochemical gradients (i.e., from a region of lower to one of higher electrochemical activity). As the word "active" implies, this type of transport requires energy, usually in the form of adenosine triphosphate (ATP). Active transport is a process designed for rapid movement of important endogenous substances across cellular membranes, whose transport across membranes by other processes would be too slow. In mammalian cells, for example, the intracellular concentrations of sodium and potassium are maintained by specific membrane bound sodium/potassium pumps. Some nonionic endogenous organic substances also require active transport in order to enter a cell. Certain neurotransmitters, for example, are believed to be taken up into neural tissue by active transport.

11.2.2.4 Facilitated Transport

In facilitated transport (also known as carrier-mediated membrane transport), a substance combines with a specific carrier protein on the membrane, and the resultant protein-substance complex diffuses to the other side of the membrane, where it dissociates to release the substance. The absorption of glucose from the intestines into the blood, for example, requires facilitated transport of glucose across the cellular membranes of the epitheleal lining of the intestines. Many amino acids cross cellular membranes by facilitated transport.

  Facilitated transport is not energy dependent, and therefore, cannot move a substance against a concentration gradient, as in active transport. The direction in which the substance moves across the membrane depends upon the difference in the substance's concentration between the exterior and interior of the cell; transport across the membrane occurs in the direction of higher to lower concentration of substance. Facilitated transport is stereospecific.

11.2.2.5 Phagocytosis, Pinocytosis, Endocytosis

Phagocytosis is a term used to describe engulfment and subsequent ingestion of a substance or particulate matter by a cell. While not a membrane transport process *per se*, phagocytosis nonetheless represents a means by which substances can enter cells. In phagocytosis, the cell changes its shape and forms a cavity that surrounds the substance or particulate matter. **Pinocytosis** differs from phagocytosis only in that a comparatively smaller portion of the cell and cell membrane are involved in the engulfment. In pinocytosis, a segment of the cellular membrane invaginates around the substance, and the two sides of the invagination merge to form an intracellular vacuole or vesicle that contains the substance. **Endocytosis** involves binding of the substance to a receptor protein on the outersurface of the cell membrane, which subsequently triggers cellular engulfment.

Substances ingested into cells by any of these three mechanisms either are stored in vesicles inside the cell or are degraded by intracellular enzymes, and only certain cells are capable of either process. The endothelial cells of capillaries are capable of pinocytosis. Neutrophils and macrophages (specific types of white blood cells) are capable of phagocytosis. Phagocytosis, endocytosis, and pinocytosis are slow and inefficient when compared to the other processes by which substances enter cells.

While not fully understood, these three processes appear to occur with substances that cannot enter cells by other means. Macromolecules such as immunoglobulins, transferrin, and low-density lipoproteins, for example, are important endogenous substances that are too large to enter the cells by the other processes and enter by phago-, pino-, or endocytosis. Absorption of macromolecules from the gastrointestinal tract may involve phagocytosis. The body also often uses phagocytosis as a defense mechanism to trap and destroy foreign substances, bacteria, viruses, and other matter that may be harmful. Phagocytic cells may remove inhaled particulate matter from the lung (Benet et al., 1996; Rozman and Klaassen, 1996).

11.3   Estimating Absorption of Chemical Substances

Most substances that can be absorbed enter cells by passive diffusion, but the extent to which a substance is or can be absorbed from the skin, gastrointestinal tract, or lung is not necessarily uniform among these sites and, in fact, may vary substantially. Typically for a substance well absorbed from one site, less is absorbed (if absorbed at all) from one or both of the other possible sites of exposure. For example, a substance may be completely absorbed from the gastrointestinal tract, moderately absorbed from the lung, and poorly absorbed from the skin. On the other hand, there are some substances are well absorbed from each of the sites of exposure, and others are poorly absorbed from each site. Whether a given substance can be absorbed depends on its physicochemical properties, their effect on the substance's ability to cross cellular membranes, and the site's anatomical and physiological characteristics.

11.3.1   Physicochemical Factors that Affect Absorption

Physicochemical properties are critical determinants of a substance's ability to be absorbed (Dethloff, 1993; DeVito, 1996; Klaassen and Rozman, 1991). Table 11.1 lists the physicochemical properties that most significantly govern passage across cellular membranes and absorption.

TABLE 11.1

Physicochemical Properties Most Important for Absorption

Physical State	Vapor Pressure
Molecular Mass/Size	Water Solubility
Melting Point	Octanol/Water Partition Coefficient
Boiling Point	Polarity

11.3.1.1   Physical State

Substances exist either as solids, liquids, or gases. The physical state of a substance can greatly influence exposure and bioavailability. For example, inhalation exposure and subsequent absorption from the lung is more likely to occur with gases. Liquids generally are absorbed more readily from the skin and gastrointestinal tract than are solids because liquid substances cover more surface area and also are in a finer state of molecular subdivision (dissagregation).

11.3.1.2 Molecular Mass/Size

Molecular size greatly affects transport across cellular membranes. As molecular size increases, transport across membranes decreases, because increases in molecular size will increase "frictional resistance" and decrease the diffusivity through the cell membrane. Since molecular size is generally directly proportional to molecular mass, and molecular mass is easily calculated; molecular mass is often used as a descriptor of molecular size. Hence, as a general rule, the lower the molecular mass, the smaller are the molecules composing the substance, and the more easily the substance can cross membranes and be absorbed from the gastrointestinal tract, lung, and skin.

11.3.1.3 Melting Point

In general, low-melting non-ionic organic substances are more likely to be absorbed than non-ionic organic substances that melt at higher temperatures. This is because substances that melt at lower temperatures require less energy to separate from their crystalline lattice and, hence, will disaggregate into free molecules more readily than compounds that melt at higher temperatures. For this reason substances that melt at lower temperatures tend to be more water soluble than those that melt at higher temperatures (see chapter on estimation of water solubility) and therefore, are more likely to achieve higher extracellular concentrations, which establishes a concentration gradient and favors passive diffusion into the cell. Substances that are liquid at ambient temperature generally are absorbed much better from all routes of exposure than analogous substances that are solids.

11.3.1.4 Boiling Point/Vapor Pressure

These properties, which characterize volatility, also determine the concentration of the substance which can be achieved in the vapor phase and is thus available for uptake by inhalation. Volatile, low boiling point substances are thus more readily absorbed in the lung. Anesthetic gases are examples.

11.3.1.5 Water Solubility

Water solubility is one of the most important physicochemical properties affecting the potential for exposure and bioavailability of chemical substances. For any substance to be absorbed, it must have sufficient water solubility to achieve a concentration gradient with respect to the cell exterior/interior. Also, to be transported across the cellular membrane, a substance must be in its finest state of molecular disaggregation (Figure 11.2). When in solution, a substance is completely disaggregated.

Even if other physicochemical properties are optimal for absorption, substances that are poorly soluble in water generally are not well absorbed because they do not dissaggregate readily and, hence, do not achieve extracellular concentrations that are sufficiently high for significant absorption to take place. Chemicals that are fairly water soluble are generally absorbed into the body from all absorption sites, assuming that the other physicochemical properties are suitable to enable absorption.

11.3.1.6 Octanol/Water Partition Coefficient

Partition coefficient represents the equilibrium ratio of the molar concentrations of a chemical substance (the solute) in a system containing two immiscible liquids. The octanol/water partition coefficient is expressed as either K_{ow} or P and is a descriptor of a substance's relative affinity for lipids and water. For purposes of simplification, K_{ow} is usually reported as its common logarithm (log K_{ow} or log P). A large log K_{ow} value for a chemical (relative to other substances) indicates that the chemical has a greater affinity for the *n*-octanol phase and, hence, is more hydrophobic (lipophilic). A negative log K_{ow} value indicates that a chemical has a greater affinity for the water phase and, hence, is more hydrophilic.

Biological membranes and systems (e.g., organs, cell membranes, capillaries, blood-brain barrier, skin, intestines) typically contain various combinations of lipid and aqueous components. For a chemical substance to gain entry into and distribute throughout a biological system, it must have a certain amount of both lipid and water solubility. Since the substance experiences resistances to diffusion in both lipid and water phases and these resistances are controlled in part by the concentrations which can be reached in water and lipid, the distribution of resistances between these two phases is strongly influenced by the lipid–water partition coefficient or K_{ow}, which is essentially the ratio of these concentrations. Substances with a high K_{ow} are capable of achieving only low relative concentrations in water, thus the concentration gradients are also necessarily low and the resistance is high in the water phase. Conversely, substances of low K_{ow} tend to experience most resistance in the lipid phase. Generally, substances with a logK_{ow} greater than 4 are not well absorbed by any route.

11.3.1.7 Particle Size

For a given solid substance, the tendency for absorption will be greatest for the smallest particles. This is because smaller particles have greater surface area, which favors dissolution. Larger particles have less surface area and dissolve more slowly. There are also anatomical reasons that larger particles are less readily absorbed. In the lung, larger particles may not be able to penetrate into the narrow air passages that lead to the air sacs where oxygen-carbon dioxide exchange takes place and where absorption is most favored.

11.3.1.8 Polarity

Polarity is the extent to which a substance, at molecular level, is characterized by a non-symmetrical distribution of electron density. Polarity is often expressed as dipole moment, which is a function of the magnitude of the partial charges on the molecule, and the distance between the charges. Substances that have larger dipole moments have greater polarity than substances with lower dipole moments. Water and acetone, for example, have dipole moments of 1.85 and 2.80, respectively. Benzene and carbon tetrachloride are nonpolar and have dipole moments of zero.

In general, the greater the polarity of a substance within a series of analogous substances, the lower is its lipophilicity and the greater is its water solubility with respect to the other substances in the series. Substances that are nonpolar (i.e., have a dipole moment of zero) are absorbed more readily through the skin than are polar substances. Polar substances, on the other hand, generally are absorbed more readily from the lung and gastrointestinal tract than are nonpolar substances. Substances that are ionic (e.g., weak organic acids and bases) dissociate to form ions in aqueous media and can exist in both their ionized (salt) and un-ionized forms depending upon pH. This can greatly affect absorption, especially from the gastrointestinal tract where pH varies considerably.

11.3.2   Anatomical and Physiological Factors Affecting Absorption

The anatomical and physiological factors that most significantly affect absorption are surface area of the absorption site, thickness of the cellular membranes composing the site of exposure, and blood flow to the exposure site (Table 11.2). A brief discussion of these factors follows.

TABLE 11.2

Anatomical and Physiological Factors Affecting Absorption[a]

Route of Exposure	Surface Area (m²)	Thickness of Absorption Barrier (μm)[b]	Blood Flow (L/min)
Skin	1.8	100–1000	0.5
Gastrointestinal Tract	200	8–12	1.4
Lung	140	0.2–0.4	5.8

[a] Adapted from DeVito (1996).
[b] This refers to the approximate distance that the molecules of a chemical substance travel from the site of exposure to the capillaries.

11.3.2.1   Absorption from the Gastrointestinal Tract

The purpose of the gastrointestinal tract is to digest and absorb food. As a result, it is a major site for absorption of xenobiotic chemical substances. Many environmental toxicants enter the food chain and are absorbed together with food. In occupational settings, airborne toxic substances enter the mouth from breathing and, if not inhaled, can be swallowed and absorbed from the gastrointestinal tract.

The gastrointestinal tract is composed of the esophagus, stomach, small intestine, and large intestine. The esophagus is approximately 30 cm long. The small intestine is composed of three segments: the duodenum (which is attached directly to the stomach and is 30 cm long), the jejunum (ranges from 100 to 250 cm long), and the ileum (ranges from 180 to 360 cm long). The large intestine is about 150 cm long.

The gastrointestinal tract is a major site site of absorption primarily because of its very large surface area and extensive blood flow (Table 11.2). Most of the absorptive surface area of the gastrointestinal tract is in the small intestine and thus most absorption from the gastrointestinal tract occurs there.

The large internal surface area of the small intestine is attributable to its length, folding, and the presence of villi and microvilli within its lumen. The villi contain capillaries and protrude into the lumen of the small intestine. There are approximately four to five million villi in the small intestine. Each villus has many microvilli as its outer surface (Figure 11.3). The microvilli represent the absorptive barrier of the small intestine. The stomach and large intestine do not contain villi and, therefore, have a small absorptive surface area compared with the small intestine.

The pH of the gastrointestinal tract varies, being 1–2 in the stomach, 5–6 in the duodenum, 6–7 in the jejunum, 7–8 in the ileum, and 8–9 in the large intestine. This variation in pH influences the extent to which acidic or basic chemical substances are ionized, which influences the extent of their absorption. The degree of ionization of an acidic substance or a basic substance at a given pH can be expressed by the Henderson-Hasselbalch equations (equations (2) and (3), respectively):

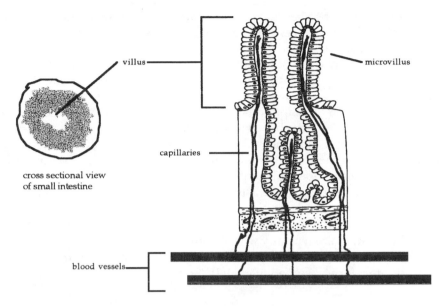

villus

microvillus

capillaries

cross sectional view
of small intestine

blood vessels

FIGURE 11.3

The lumen of the small intestine. The surface area of the small intestine is large due to its length, folding, millions of villi, and tens of millions of microvilli (shown in close-up view on right).

for weak acids (Equation 2)   for weak bases (Equation 3)

$$\log \frac{[\text{un-ionized}]}{[\text{ionized}]} = pKa - pH \qquad \log \frac{[\text{ionized}]}{[\text{un-ionized}]} = pKa - pH$$

Hence, even without the larger surface area of the small intestine, acidic substances would be absorbed much more readily from the small intestine than the large intestine, because they are less dissociated in the small intestine. The opposite is true for basic substances. While acidic substances are less dissociated in the stomach than in the small intestine, absorption is greater in the small intestine because of the much larger surface area.

The physicochemical properties that most significantly affect the extent to which a substance is absorbed from the gastrointestinal tract include physical state; particle size (for solids); octanol-water partition coefficient; dissociation constant (for ionizable substances); and molecular weight and size. A substance must be sufficiently water soluble to undergo the requisite dissolution to its free molecular form. Substances that are liquid in their pure form, or are already dissolved in a solvent, are absorbed more quickly from the gastrointestinal tract than solids. Generally, substances that are in the form of salts (e.g., hydrochloride salt or sodium salt) often are absorbed more quickly and more completely than their unionized (neutral) forms, because the salts undergo dissolution in the gastric juices more quickly than the neutral forms. Once in solution, the salts rapidly equilibrate to their ionized and un-ionized species. The un-ionized species is quickly absorbed, and equilibrium is re-established. For solids, particle size also affects the rate of dissolution and, thus, overall absorption. The smaller the particle size, the larger the surface area per unit mass and the faster the dissolution and absorption of the substance. Larger particle size means less surface area per unit mass, and therefore, a slower dissolution in the gastric fluids, and slower or even less absorption.

Lipid solubility is more important than water solubility in regard to absorption from the gastrointestinal tract. The more lipid soluble a substance is, the better it is absorbed. Highly lipophilic substances ($\log K_{ow} > 5$), however, are usually very poorly water soluble and generally are not well absorbed because of their sparing dissolution in the gastric juices. On the other hand, chemical substances with very high water solubility and very low lipid solubility are also not readily absorbed (D'Souza, 1990).

Substances that have low water solubility dissolve poorly in gastrointestinal fluids and, consequently, are poorly absorbed from the gastrointestinal tract, even if they have optimal lipophilicity. Noteworthy exceptions to this generalization are triglycerides, fatty acids and cholesterol. Because of their high lipophilicity (log K_{ow} > 6), these substances are practically insoluble in water, yet they are absorbed appreciably from the gastrointestinal tract, because the body uses these substances as a source of energy (triglycerides, fatty acids) or for steroidal hormone synthesis (cholesterol), and the gastrointestinal tract has specialized mechanisms for their absorption. Substances that are structurally related to fats or cholesterol also might be absorbed from the gastrointestinal tract by the same mechanisms.

The higher the molecular mass, the slower a substance is absorbed from the gastrointestinal tract. Assuming sufficient aqueous and lipid solubility, substances with molecular mass less than 300 daltons are typically well absorbed. Those with molecular mass ranging from 300 to 500 daltons are not as readily absorbed, and substances with molecular mass in the thousands are poorly absorbed, if at all (D'Souza, 1990).

11.3.2.2 Absorption from the Lung

The function of the lung is to exchange oxygen for carbon dioxide. The lung has an enormous surface area for absorption, due to the continuous, repetitive branching of the airways from the trachea to the many terminal alveoli (at least 150 million in each lung), where oxygen/carbon dioxide exchange takes place. The lung also receives 100% of the blood pumped from the right ventricle of the heart. The thickness of the cellular membranes of the alveoli (the absorption barrier of the lung) is only 0.2–0.4 µm. Each alveolus is surrounded by blood capillaries. These anatomical and physiological characteristics of the lung (see Figure 11.4) enable the rapid and efficient absorption of oxygen and favor the absorption of other substances as well (Dethloff, 1993; Klaassen and Rozman, 1991; Overton, 1990; Valberg, 1990).

Absorption of chemical substances from the lung differs from intestinal and dermal absorption in that lipid solubility is less important. Solubility in water is the more important factor because the cellular membranes of the alveoli are very thin, and the distance a substance must travel to cross the alveolar membrane and enter the bloodstream is correspondingly short. As a result, chemicals absorbed through the lung can enter the blood within seconds (Klaassen and Rozman, 1991). As a general rule, substances with solubility in water equal to or greater than their solubility in lipids are likely to be absorbed from the lung. Polar substances generally are absorbed better from the lung than non-polar substances, due to greater water solubility.

Gases generally are absorbed well from the lung. For solid substances, particles of 1 µm and smaller may be absorbed particularly well from the lung because they have a large surface area per unit mass, and also can penetrate deep into the lung's narrow alveolar sacs. Particles of 2 to 5 µm are deposited mainly in the tracheobronchiolar regions, from where they are cleared by retrograde movement of the mucus layer in the ciliated portions of the respiratory tract. These particles eventually are removed from the lung and may be swallowed and absorbed from the gastrointestinal tract. Particles of 5 µm or larger usually are deposited in the nasopharyngeal region and are too large for absorption from the lung but also may be swallowed and absorbed from the gastrointestinal tract (Klaassen and Rozman, 1991).

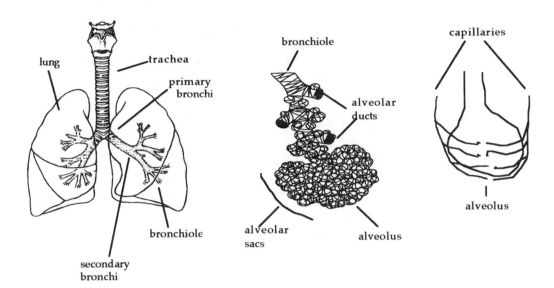

FIGURE 11.4
The lung. The middle diagram is a close-up view of the end of a bronchiole. Each bronchiole terminates in many alveolar sacs. The diagram on the right is a close-up view of an alveolus. The cell membrane of each alveolus is very thin and surrounded by many blood capillaries, thereby allowing for rapid exchange of oxygen for carbon dioxide, as well as absorption of other substances.

11.3.2.3 Absorption from the Skin

The primary purpose of the skin is to serve as a covering that protects the body from the external environment. Compared with the lung and gastrointestinal tract, the skin has much lower surface area and blood flow, as well as a considerably thicker absorption barrier (Table 11.1). Nonetheless, the skin can represent a significant pathway for exposure and absorption.

The skin has two basic layers, the epidermis and the dermis. The epidermis (outer layer) is composed of five sublayers: the stratum corneum (the outermost sublayer); stratum lucidum; stratum granulosum; stratum spinosum; and stratum germinativum. The dermis has two sublayers, the papillary and reticular sublayers, and contains connective tissue, nerves, and capillary networks. The epidermis is 0.05 to 0.1 mm thick, and the dermis 2 to 5 mm thickness (Figure 11.5). The skin also contains appendages that include hair follicles, sebaceous glands and sweat glands (not shown in Figure 11.5.), which provide aqueous channels into the skin. The surface of the skin is weakly acidic (pH ranges from about 4.2 to 5.6), whereas the lower layers are closer to neutral in pH (ca. 7.4).

The stratum corneum consists of densely packed, dead, keratinized cells. Being the outer most sublayer, the stratum corneum is always wearing away, but is formed and replenished approximately every three to four weeks from the continuous outward movement of the other layers of the epidermis. As stratum corneum sloughs off, these layers move toward the outer surface of the skin, where they dehydrate, become keratinized, and form new stratum corneum. The stratum corneum has essentially no water and is highly lipophilic, containing about 75 to 80% lipophilic material. The other layers of the epidermis contain some water and are progressively less lipophilic, the closer they are to the dermis. The dermis layer is more hydrophilic than the epidermis.

The thickness of the stratum corneum varies considerably, depending upon the area of the body and age of the person. It is thicker in areas most subject to friction ranging from 400 to 600 μm on the palms of the hands and soles of the feet, but much thinner in other

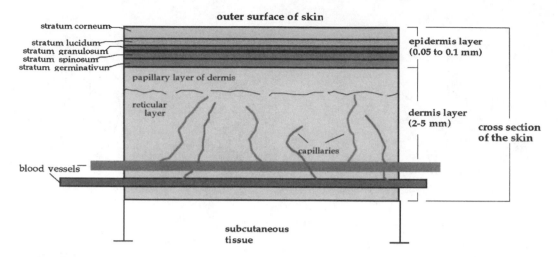

FIGURE 11.5

Cross-sectional view of the skin. The skin has two layers, the epidermis and dermis. The epidermis and dermis have 6 and 2 sublayers, respectively.

areas (back [11 µm], scrotum [5 µm], forehead [13 µm], abdomen [15 µm], back of hand [40 µm]. The stratum corneum sublayer is thinner in infants and young children than in adults because the skin of infants and children is not fully developed.

For a substance to be absorbed into the body following dermal exposure, it must initially dissolve in the stratum corneum sublayer, then diffuse through the remaining sublayers of the epidermis and into the dermis, where it will eventually diffuse into the blood capillaries. This absorption barrier ranges in thickness from 100 to 1000 µm, depending on area of the body (Klaassen and Rozman, 1991).

The stratum corneum is the rate-determining barrier in dermal absorption. Diffusion through the stratum corneum can occur only by passive diffusion. Passage through the remaining sub-layers of the skin is much more rapid. In general, dermal absorption of substances tends to be greater in areas of the body where the stratum corneum is thinner, and in infants and young adults.

Substances also may be absorbed into the body through the appendages of the skin. Such absorption is much faster because it bypasses the stratum corneum. However, the appendages, have considerably less surface area and thus, generally do not represent significant sites of absorption on a mass basis (ECETOC, 1993; Flynn, 1990; Klaassen and Rozman, 1991; USEPA 1992; Wester and Maibach, 1997).

The most important physicochemical properties affecting dermal absorption are physical state, octanol-water partition coefficient, water solubility, and molecular mass or size. Substances that are liquid in their pure form tend to be absorbed more readily from the skin than solids, because liquids cover more dermal surface area and are in a greater state of molecular disaggregation. Non-ionic solids with higher melting points (> 125°C) and substances (particularly solids) that are ionic (e.g., salts of organic acids or bases) or highly polar are generally not well absorbed from the skin.

Substances with greater lipophilicity (higher log K_{ow}) are absorbed more readily from the skin than are less lipophilic substances. As a general rule, substances that have log K_{ow} values between 1 and 2 are well absorbed dermally (ECETOC, 1993; Walsh, 1990). Chemicals having log K_{ow} values less than –1 usually are absorbed poorly through the skin. Dermal absorption generally increases as log K_{ow} increases from –1 to 3.5. Highly lipophilic substances (log K_{ow} > 5) can pass through the stratum corneum but are generally too water insoluble to pass through the remaining sub-layers and enter the bloodstream. These substances are poorly absorbed from the skin (ECOTOC, 1993; Flynn, 1990; Klaassen and Rozman, 1991; USEPA, 1992; Wester and Maibach, 1997).

The rate of absorption of substances through the skin is inversely proportional to molecular mass and size. In general, small molecules (< 200 daltons) that are both lipid- and water-soluble are the most readily absorbed, and molecules of mass greater than 1000 daltons are not well absorbed (Benet et al., 1996; Flynn, 1990; USEPA, 1992).

11.3.3   Strategy for Assessing Absorption

The preceding sections of this chapter discuss the factors that are most important in controlling the absorption of chemical substances following oral, inhalation, and dermal exposure.

Figure 11.6 illustrates a process to follow to assess a substance's absorption potential. The first step is to establish the types of individuals (i.e., workers, consumers, or general population) at greatest risk of exposure and the known or likely routes (dermal, inhalation, or oral) by which exposure will take place. The second step is to determine whether measured absorption data for the substance are available. Such data are often not available, but animal absorption data can be used as surrogates for human data in many cases. If no measured absorption data are available, toxicity data from studies involving humans or animals exposed to the substance may be useful. For example, if systemic toxic effects were noted in humans or animals following dermal (or oral or inhalation) exposure to a substance, especially at low doses, then obviously the substance is absorbed via this route.

Absorption must be estimated when human or animal absorption (or toxicity) data are not available (Figure 11.6). Estimates should be based on consideration of the anatomy and physiology of the exposure site; the structure and physicochemical properties of the substance; and (if available), measured absorption or toxicity data on substances that are analogs of the substance under consideration. If data pertaining to an analogous substance suggest that the analog is well absorbed, it might be possible to conclude that the substance of interest is also well absorbed. In drawing such conclusions, however, one needs to exercise caution. As discussed in other chapters in this book, even subtle structural differences between two substances can result in large differences in physicochemical properties, particularly melting point and water solubility. If such differences are noted, there is a high probability that there are large differences in absorption as well.

While the flow chart in Figure 11.6 is sufficient for assessing the absorption of essentially any chemical substance, a number of mathematical, models also can enable quantitative estimation of membrane diffusion or absorption of a substance from either the skin or gastrointestinal tract. References pertaining to these models appear in the Suggested Readings section of this chapter. These models are useful mainly for estimating membrane diffusion or absorption of substances belonging to certain specific classes of substances. Their use should be reserved for substances that belong to the classes for which the models are known to predict accurately.

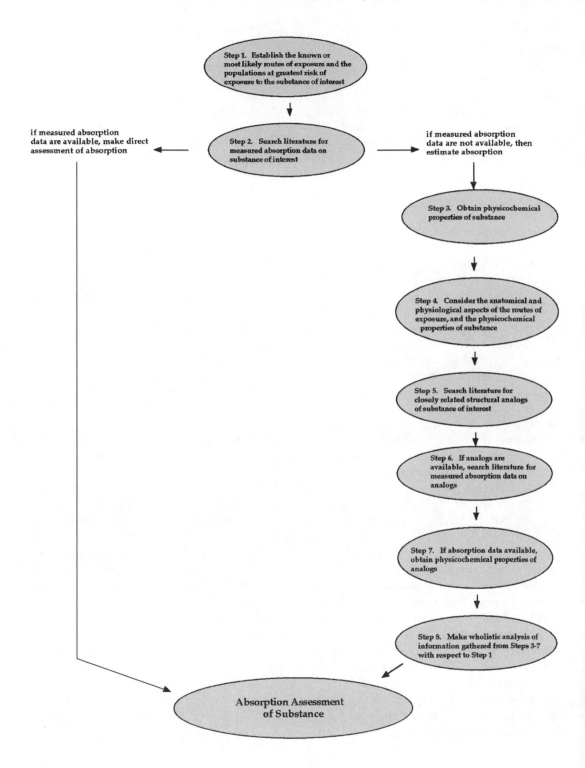

FIGURE 11.6
Scheme for assessing absorption of a chemical substance.

References

Benet, L.Z., D.L. Kroetz, and L.B. Sheiner. 1996. Pharmacokinetics: The Dynamics of Drug Absorption, Distribution, and Elimination. In J.G. Hardman, L.E. Limbird, P.B. Molinoff, R.W. Ruddon, and A.G. Goodman, Eds., *Goodman and Gilman's The Pharmacological Basis of Therapeutics*, Ninth Edition, pp 3-27. McGraw-Hill, New York, N.Y.

Bronaugh, R.L. 1990. Metabolism in Skin. In T.R. Gerrity and C.J. Henry, Eds., *Principles of Route-to-Route Extrapolation for Risk Assessment*, pp 185-191. Elsevier: New York, N.Y.

Dethloff, L.A. 1993. In P.G. Welling and F.A. De La Iglesia, Eds., *Drug Toxicokinetics*, pp 195-219. Marcel Dekker, Inc., New York, N.Y.

DeVito, S.C. 1996. General Principles for the Design of Safer Chemicals, Toxicological Considertions for Chemist. In S.C. DeVito and R.L. Garrett, Eds., *Designing Safer Chemicals Green Chemistry for Pollution Prevention*, ACS Symposium series 640, pp 16-59. American Chemical Society, Washington D.C.

D'Souza, R.W. 1990. Modeling Oral Bioavailability: Implication for Risk Assessment. In T.R. Gerrity and C.J. Henry, Eds., *Principles of Route-to-Route Extrapolation for Risk Assessment*, pp 173-183. Elsevier: New York, N.Y.

ECETOC. 1993. "Percutaneous Absorption." European Centre for Ecotoxicology and Toxicology of Chemicals (ECETOC), monograph no. 20 (August, 1993), pp 5-9. ECETOC, Brussels, Belgium.

Flynn, G.L. 1990. Physicochemical Determinants of Skin Absorption. In T.R. Gerrity and C.J. Henry, Eds., *Principles of Route-to-Route Extrapolation for Risk Assessment*, pp 93-127. Elsevier: New York.

Hrudey, S.E., W. Chen, and C.G. Rousseaux. 1996. Mechanism of Absorption. In *Bioavailability in Risk Assessment*, pp 19-50. Lewis Publishers, New York, N.Y.

Klaassen C.D., and K. Rozman. 1991. Absorption, Distribution, and Excretion of Toxicants. In M.O. Amdur, J. Doull, and C.D. Klaassen, Eds., *Casarett and Doull's Toxicology, The Basic Science of Poisons, Fourth Edition*, pp 50-87. Pergamon Press, New York, N.Y.

Medinsky, M.A. and C.D. Klaassen. 1996. Toxicokinetics. In C.D. Klaassen, ed., *Casarett & Doull's Toxicology, the Basic Science of Poisons*, pp 187-198. Fifth Edition, McGraw-Hill, New York, N.Y.

Overton, J.H. 1990. Respiratory Tract Physiological Processes and the Uptake of Gases. In T.R. Gerrity and C.J. Henry, Eds., *Principles of Route-to-Route Extrapolation for Risk Assessment*, pp 71-91. Elsevier: New York, N.Y.

Rozman, K.K. and C.D. Klaassen. 1996. Absorption, Distribution, and Excretion of Toxicants. In C.D. Klaassen, ed., *Casarett & Doull's Toxicology, the Basic Science of Poisons*, Fifth Edition, pp 91-112. McGraw-Hill, New York, N.Y.

USEPA. 1992. "Dermal Exposure Assessment: Principles and Application." United States Environmental Protection Agency (USEPA), report no. EPA/600/8-91/011B (January, 1992). U.S. EPA, Washington, D.C.

Valberg, P.A. 1990. The Respiratory Tract as a Portal of Entry for Toxic Particles. In T.R. Gerrity and C.J. Henry, Eds., *Principles of Route-to-Route Extrapolation for Risk Assessment*, pp 61-70. Elsevier: New York, N.Y.

Walsh, C.T. 1990. Anatomical, Physiological, Biochemical Characteristics of the Gastrointestinal Tract. IIn T.R. Gerrity and C.J. Henry, Eds., *Principles of Route-to-Route Extrapolation for Risk Assessment*, pp 33-50. Elsevier: New York, N.Y.Wester, R.C., and H.I. Maibach. 1997. Toxicokinetics: Dermal Exposure and Absorption of Toxicants. In I.G. Sipes, C.A. McQueen, and A.J. Gandolfi, Eds., *Comprehensive Toxicology: Volume !; General Principles* (J. Bond, volume editor), pp. 99-114. Pergamon Press, New York, N.Y.

Wright, E.M. 1995. Membrane Transport. In P.G. Welling and F.L.S. Tse, Eds., *Pharmacokinetics; Regulatory, Industrial, Academic Perspectives* Volume 67 of the *Drugs and the Pharmaceutical Sciences* series, pp 89-118. Marcel Dekker, New York, N.Y.

Suggested Reading

Bronaugh, R.L. and H.I. Maibach, Eds., *Percutaneous Absorption: Mechanisms, Methodology, Drug Delivery,* second edition, pp 13-26. Marcel Dekker, New York, N.Y.

Bunge, A.L. and R.L. Cleek. 1995. A new method for estimating dermal absorption from chemical exposure: 2. Effect of molecular weight and octanol-water partitioning. *Pharmaceutical Research* 12(1):88-95.

Dermal Exposure Assessment: Principles and Application. United States Environmental Protection Agency, report no. EPA/600/8-91/011B (January, 1992). U.S. EPA, Washington, D.C.

Guy, R.H. and J. Hadgraft. 1989. Mathematical Models of Percutaneous Absorption. In R.L. Bronaugh and H.I. Maibach, Eds., pp 13-26. *Percutaneous Absorption: Mechanisms, Methodology, Drug Delivery,* second edition, Marcel Dekker, New York, N.Y.

Habucky, K. 1995. Methods to Assess Absorption in Drug Discovery. In P.G. Welling and F.L.S. Tse, Eds., *Pharmacokinetics; Regulatory, Industrial, Academic Perspectives,* Volume 67 of the *Drugs and the Pharmaceutical Sciences* series, pp 21-37. Marcel Dekker, New York, N.Y.

Houk, J. and R.H. Guy. 1988. Membrane models for skin penetration. *Chem. Reviews* 88(3):455-471.

Kao, J. and M.P. Carver. 1990. Cutaneous metabolism of xenobiotics. *Drug Metab. Reviews* 22(4):363-410.

Maitani, Y., A. Coutel-Egros, Y. Obata, and T. Nagai. 1993. Prediction of skin permeabilities of diclofenac and propranolol from theoretical partition coefficients determined from cohesion parameters. *J. Pharm. Sci.* 82(4): 416-420.

Percutaneous Absorption. European Centre for Ecotoxicology and Toxicology of Chemicals (ECETOC) monograph no. 20. (August, 1993). ECETOC, Brussels, Belgium.

Scherrer, R.A. and S.M. Howard. 1977. Use of distribution coefficients in quantitative structure-activity relationships. *J. Med. Chem.* 20(1): 53-58.

Shah, P.V., R.E. Grissom, Jr., and F.E. Guthrie. 1990. Environmental Exposure to Chemicals through Dermal Contact. In J. Saxena, ed., *Hazard Assessment of Chemicals,* Vol. 7 pp 111-155. Hemisphere Publishing, New York, N.Y.

Tayar, N.E., R.-Y. Tsai, B. Testa, P.-A. Carrupt, C. Hansch, and A. Leo. 1991. Percutaneous penetration of drugs: a quantitative structure-permeability releationship study. *J. Pharm. Sci.* 80(8): 744-749.

Tayar, N.E., R.-Y. Tsai, B. Testa, P.-A. Carrupt, and A. Leo. 1991. Partitioning of solutes in different solvent systems: The contribution of hydrogen bonding capacity and polarity. *J. Pharm. Sci.* 80(6): 590-598.

Young, R.C. and R.C. Mitchell. 1988. Development of a new physicochemical model for brain penetration and its application to the design of centrally acting H_2 receptor histamine antagonists. *J. Med. Chem.* 31:656-671.

Part III

Reactivity of Persistence

12

Biodegradation

Philip H. Howard

CONTENTS

12.1 Introduction

Chemicals that are released into the environment are subject to transport (bioconcentration, leaching, volatilization) and degradation (biodegradation, photolysis, hydrolysis, oxidation) processes. The transport processes have a tendency to distribute and dilute or concentrate, but not destroy, the contaminant. On the other hand, chemical and biological reactions in the environment result in alterations and, frequently, degradation of the material to innocuous substances. Of the degradation processes, biological degradation of organic compounds is the most desirable because it results generally in end-products that have been completely mineralized to inorganic compounds.

In 1982, Scow (1982) reviewed information on biodegradation rate estimation methods and concluded that methods were not available. Since that time, considerable research has been reported in the area of quantitative and qualitative structure-biodegradability relationships (QSBR), which this chapter will discuss, along with recommended estimation methods. Detailed discussion of the principles of biodegradation is beyond the scope of this chapter and covered well in numerous recent books (Alexander, 1994; Pitter and Chudoba, 1990; Hurst, 1997). However, this chapter includes some background information on the process of biodegradation, standard test procedures, rules of thumb for biodegradation, and kinetics of biodegradation to illustrate the difficulty of predicting biodegradation rates.

12.2   Theoretical Background: Principles of Biodegradation

12.2.1   Definitions

Biodegradation is a process in which the destruction of a chemical is accomplished by the action of a living organism. In 1967, the Biodegradability Subcommittee of the Water Pollution Control Federation (WPCF, 1967) categorized biodegradation into the following three types:

1. *Primary biodegradation* — Biodegradation to the minimum extent necessary to change the identity of the compound
2. *Ultimate biodegradation* — Biodegradation to (a) water, (b) carbon dioxide, and (c) inorganic compounds (if elements other than C, H, and O are present)
3. *Acceptable biodegradation* — Biodegradation to the minimum extent necessary to remove some undesirable property of the compound such as foaminess or toxicity.

Although biological degradation conceivably might be accomplished by any living organism, available information indicates that by far the most significant biological systems involved in ultimate biodegradation are bacteria and fungi (Howard et al., 1975; Alexander, 1994). Higher life forms, including wildlife, fish, animals, and plants, have highly developed excretory mechanisms, which slightly alter a chemical into water-soluble, easily excreted forms. In contrast, the metabolic versatility of bacteria and fungi suggests that they play a major role in the ultimate biodegradation of natural and synthetic organic chemicals. A number of studies have found chemicals that were degraded by microorganisms, but not by higher organisms. Also, an analysis of the metabolic activity of different organisms in a soil community (Macfadyen, 1957) indicated that bacteria alone accounted for 65% of the total metabolism.

12.2.2   Organism and Environmental Conditions Necessary for Biodegradation

Alexander (1994) indicated several conditions necessary for biodegradation to occur:

(a) An organism must exist that has the necessary enzymes to bring about the biodegradation.

(b) That organism must be present in the environment containing the chemical.

(c) The chemical must be accessible to the organism having the requisite enzymes.

(d) If the initial enzyme bringing about the degradation is extracellular, the bonds acted upon by that enzyme must be exposed for the catalyst to function.

(e) Should the enzymes catalyzing the initial degradation be intracellular, that molecule must penetrate the surface of the cell to the internal sites where the enzyme acts. Alternatively, for the transformation to proceed further, the products of an extracellular reaction must penetrate the cell.

(f) Because the population or biomass of bacteria or fungi acting on many synthetic compounds is initially small, conditions in the environment must be conducive to allowing for proliferation of the potentially active microorganisms.

The initial concentration of the microbial population and the chemical somewhat affects the importance of the last condition and a lag period often occurs between addition of a chemical and the onset of biodegradation. This lag period, usually attributed to the need for acclimation (Alexander, 1994; Spain and Van Veld, 1983), could result from enzyme induction, gene transfer or mutation, predation by protozoa, or growth in the population of responsible organisms.

The initial species present, their relative concentrations, the induction of their enzymes, and their ability to acclimate once exposed to a chemical are likely to vary considerably, depending upon such environmental parameters as temperature, salinity, pH, oxygen concentration (aerobic or anaerobic), redox potential, concentration and nature of various substrates and nutrients, concentration of heavy metals (toxicity), and effects (synergistic and antagonistic) of associated microflora (Howard and Banerjee, 1984). Many of the parameters affect the biodegradation of chemicals in the environment as well as in biodegradation test systems used to simulate the environment.

One important parameter is the chemical substrate concentration. A number of chemicals have biodegradation rates proportional to substrate concentration, but there are also examples of thresholds and inhibitions (Alexander, 1985). Recently, the bioavailability of the chemical to the enzyme (item c above) has been identified as a major factor in determining biodegradability in nature. Several studies have demonstrated that, although a chemical freshly added to soil is biodegraded at a moderate rate, the biodegradation rate for the same chemical that has been present in the soil sample for a long time is very low (Alexander, 1994). Thus, the longer a chemical remains in a sample, the greater the potential for it to become sequestered and less bioavailable.

Some microorganisms are capable of biodegrading chemicals without population growth. In this process, which is known as cometabolism (Alexander, 1994), the microorganism degrades a compound from which it derives no carbon or energy; instead, it is being sustained on other organic nutrients. Alexander (1994) has suggested that cometabolism has the following two characteristics:

1. A chemical subject to cometabolism ... is transformed slowly, and the rate of conversion does not increase with time [low starting population in soils and water and no growth].

2. Many organic products accumulate as a result of cometabolism, and these products tend to persist. [Result of single organism that cannot further metabolize

the product — true in pure culture, but may not be true in nature where another microorganism may be able to use the metabolite].

Soil, water, sediment, and wastewater microenvironments have different microbial populations and different available nutrients which may affect considerably the rate of biodegradation. For example, wastewater treatment plants have high levels of nutrients and high microbial populations that may have been preexposed to the test chemical (acclimated microbial population), but the contact (retention) time is relatively short.

In contrast, marine waters are usually fairly low in nitrogen and phosphorous which may limit the biodegradation of chemicals (e.g., oil spills). Sediment often has high levels of organic nutrients, but often is anaerobic, while surface waters tend, by comparison, to have low levels of organic nutrients. Digestor sludge from wastewater treatment plants has high organic nutrients and is anaerobic. Surface soils have high concentrations of organic nutrients (depending on the type of soil), but this usually decreases with depth.

In the past, it was believed that groundwater aquifers were devoid of life. However, a number of studies (e.g., Ladd et al., 1982) have demonstrated that microorganisms are quite plentiful in certain aquifers, and, in some instances, the bacterial concentration and activity in aquifers may be higher than those in surface waters. In addition, availability of the chemical to the microbial population can be affected considerably by the conditions of the microenvironment (e.g., organic concentration or clay content may bind the chemical tightly).

Thus, the rates of biodegradation are likely to vary considerably, depending on the environment to which a chemical is released, as tests from various media have demonstrated (Boethling et al., 1995; Federle et al., 1997). Also, the rates under different conditions may vary depending upon the type of chemical structure. For example, nitro aromatic compounds are usually fairly resistant to biodegradation under aerobic conditions but are reduced rapidly to amines under anaerobic conditions.

12.2.3 Physical/Chemical Property Effects on Biodegradation

In addition to chemical substrate concentration, chemical structure and physical/chemical properties have considerable impact on the rate and pathways of biodegradation. The chemical structure determines the possible pathways that a substrate may undergo, generally classified as oxidative, reductive, hydrolytic, or conjugative.

Table 12.1 provides some examples of common microbial degradation pathways taken mostly from a review by Alexander (1981). Recently, Klopman et al. (1995) developed a computer program that will predict the most probable metabolites, and Punch et al. (1996) developed a computer program that "simulates the biodegradation of synthetic chemicals through the sequential application of plausible biochemical reactions." The latter program focuses mostly on structures of interest for use as surfactants. Limited application of these programs to biodegradability predictions have been attempted by postulating that chemicals that biodegrade fast should have an identifiable metabolism pathway and those that biodegrade slowly will not have a metabolism pathway.

Over the years, structure/biodegradability "rules of thumb" (Alexander, 1973; Howard et al., 1975; Scow, 1982) have been developed. Table 12.2 summarizes these. Most of the information is from Alexander (1994), Howard et al. (1975), and Scow (1982). Some of these structure/biodegradability relationships have some biochemical mechanistic underpinnings. For example, highly branched compounds frequently are resistant to biodegradation because increased substitution hinders β-oxidation (see Table 12.1), the process by which alkyl chains and fatty acids usually are biodegraded. This structural relationship

TABLE 12.1

Examples of Common Microbial Degradation Pathways.

Type of Reactions (not all steps are given)	Examples of Chemicals Subject to Reaction
β-oxidation 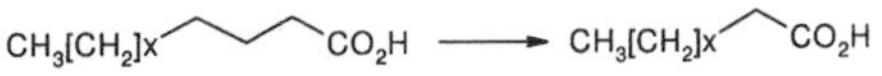	Fatty acids and straight chain hydrocarbons (after oxidation of chain to carboxylic acid - see methyl oxidation)
Methyl oxidation	Aromatic and aliphatic methyl groups
Epoxide formation 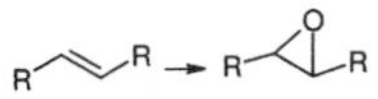	Olefins
Hydroxylation and ketone formation 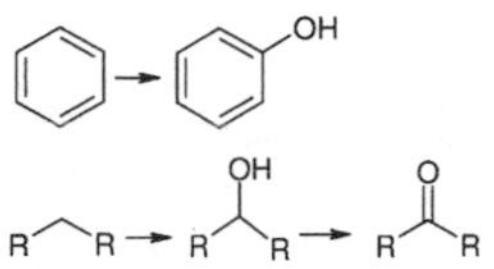	Aromatics to form phenols and hydrocarbons to alcohols and then ketones
Nitrogen oxidation	Aromatic amines to nitroaromatic
Nitro reduction	Nitroaromatics aromatic amines (e.g., parathion) especially fast under anaerobic conditions
Nitrile/amide metabolism 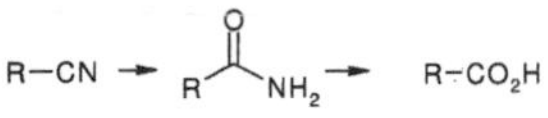	Bromoxynil, Dichlobenil
Sulfur oxidation 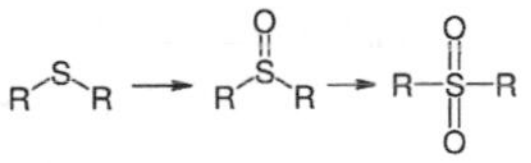	Sulfides such as aldicarb
Thiophosphate ester oxidation 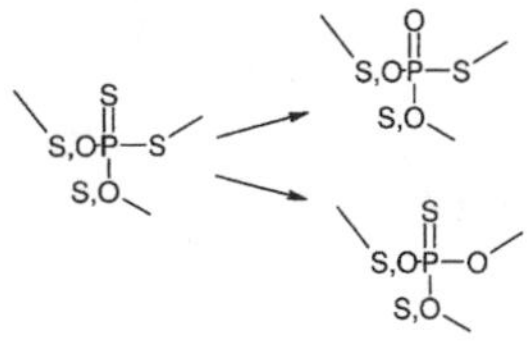	Thiophosphate pesticides
Dehalogenation	Aromatic and aliphatic halogens
Hydrolysis 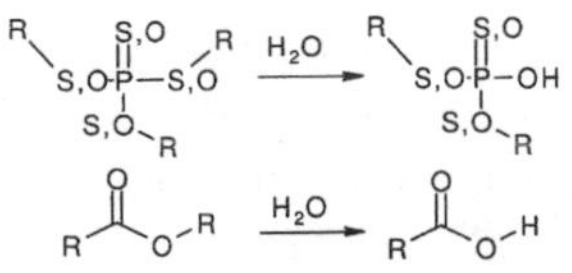	Phosphate and carboxylic esters

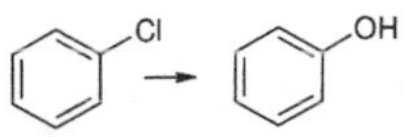

β-oxidation

$CH_3[CH_2]x$ ⟶ $CH_3[CH_2]x$ CO_2H

Methyl oxidation

$R—CH_3 \rightarrow R—CH_2OH \rightarrow R—CHO \rightarrow R—CO_2H$

Epoxide formation

Hydroxylation and ketone formation

Nitrogen oxidation

$R—NH_2 \rightarrow R—NHOH \rightarrow R—N=O \rightarrow R—NO_2$

Nitro reduction

$R—NO_2 \rightarrow R—N=O \rightarrow R—NHOH \rightarrow R—NH_2$

Nitrile/amide metabolism

$R—CN \rightarrow$ ... $\rightarrow R—CO_2H$

Sulfur oxidation

Thiophosphate ester oxidation

Dehalogenation

$R—CH_2Cl \longrightarrow R—CH_2OH$

$R—CCl_3 \longrightarrow R—CO_2H$

Hydrolysis

TABLE 12.2

Relationship Between Chemical Structure and Biodegradability.

More Biodegradable (Less Persistent)		Less Biodegradable (More Persistent)

Branching

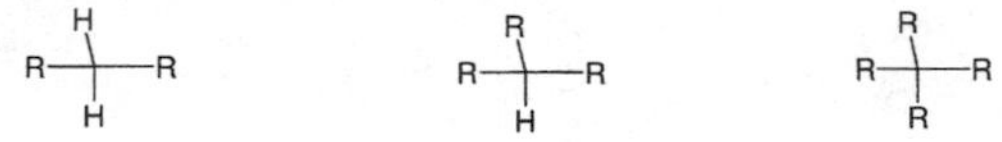

Aliphatic functional groups

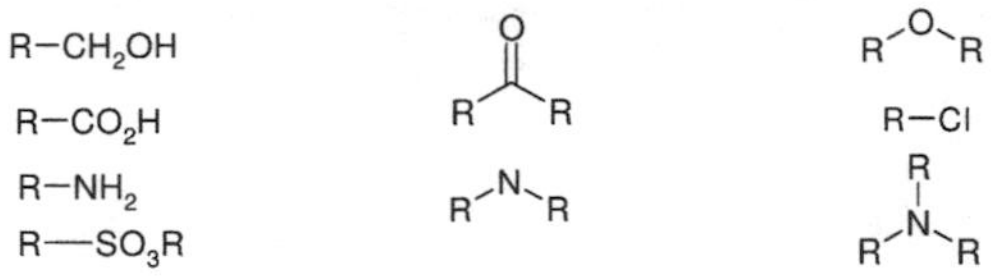

Aromatic functional groups (benzene, naphthalene, pyridine rings)

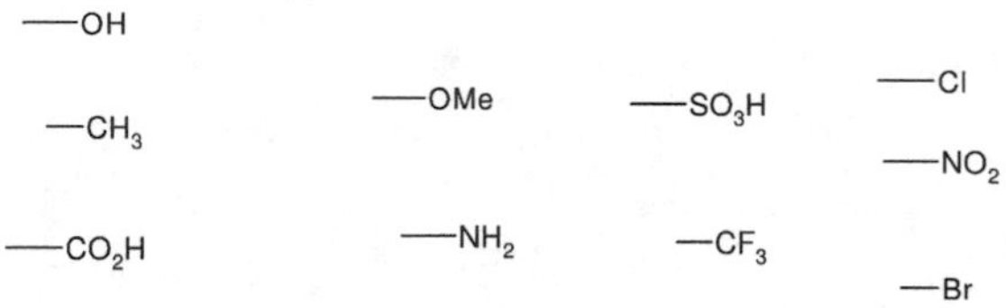

Aliphatic amines

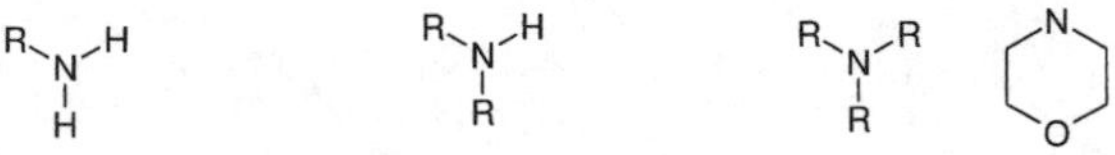

Halophenols

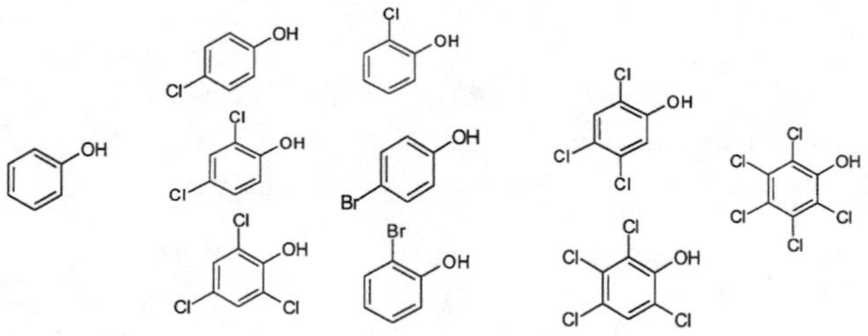

Polycyclic aromatics

≤3 rings   >3 rings

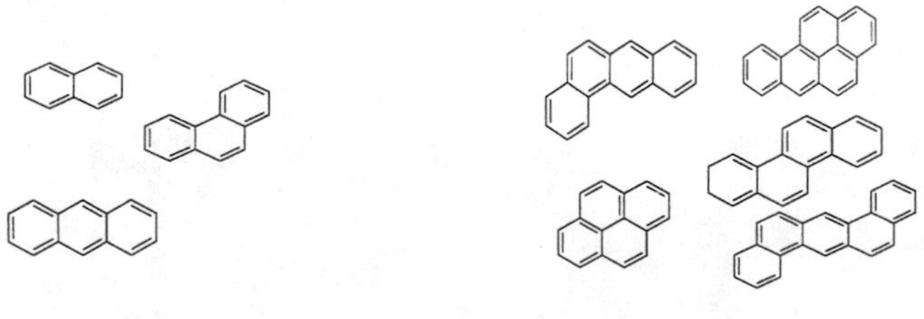

Triazines

was discovered in the 1950s, when detergent scientists found that alkylbenzene sulfonate (ABS) detergents passed through wastewater treatment plants causing foaming problems in rivers and streams. This problem was solved by switching from the highly branched ABS detergents to linear alkylbenzene sulfonate (LAS) detergents, thus illustrating the importance of understanding the relationship between structure and biodegradability.

Few other "rules of thumb" have such mechanistic bases, but there are some general trends. Functional groups commonly seen by microorganisms in natural products usually are degraded easily, probably because the microbes have had eons to develop the required enzymes systems in order to gain carbon and energy from the metabolism. Conversely, functional groups less common in nature or newly synthesized by man usually make a chemical more resistant to biodegradation. Aromatic substituents that are electron withdrawing (e.g., nitro groups and halogens) increase the persistence of a chemical, possibly by making it more difficult for enzymes to attack the aromatic ring, whereas electron donating functionalities (e.g., carboxylic acids, phenols, amines) generally increase biodegradation rates.

Physical/chemical properties affect the rate of biodegradation mostly by affecting bioavailability. Compounds which are sparingly soluble in water tend to be more resistant to biodegradation, possibly due to an inability to reach the microbial enzyme site, a reduced rate of availability due to solubilization, or sequestration due to adsorption or trapping in inert material (Alexander, 1973; Alexander, 1994).

Banerjee et al. (1984) developed a model based on the premise that penetration of the bacterial cell membrane is rate determining, assuming that enzymes capable of degrading the compounds of interest are available. They found classes of chemicals (alcohols) where biodegradation rates increased with the octanol/water partition coefficient (K_{ow}), and classes (phenols) where biodegradation rates decreased with K_{ow}. Their model explained both results: direct correlation of biodegradation rate and K_{ow} - rate (the determining step is adsorption to the cell wall); indirect correlation of biodegradation rate and K_{ow} - rate (the determining step is penetration of the cell wall).

Although this model explains results after the fact, it has little predictive power since a new chemical class could have either a direct or indirect correlation with K_{ow}. However, if experimental biodegradation rates are available for some members of a chemical class, the model could be used for the other members of the class by assuming that they behave in a similar manner.

From the above, it is apparent that variables related to the substrate (test chemical), organism(s), or environment may affect rates of biodegradation. Table 12.3 summarizes the variables.

12.3 Standard Test Methods

Biodegradation studies may use either pure or mixed cultures of microorganisms. Pure cultures can be obtained either from culture collections or by enrichment from natural samples (sewage, soil, river water, etc.). In the latter case, cultures usually are isolated by serial transfer in media containing the test chemical as the sole carbon source.

Pure culture studies are helpful in identifying possible pathways of degradation, but organism(s) isolated by enrichment are not necessarily active in the environment. The organism isolated under high nutrient conditions with the chemical as the only carbon source may not be the organism(s) responsible for the chemical's biodegradation in the environment.

TABLE 12.3

Variables Potentially Affecting Rates of Biodegradation.

Organism Related
- Population (concentration) of individual, viable microorganisms capable of degrading the substrate or its metabolites under the environmental conditions
- Acclimation (previous history of exposure to substrate and organic compounds)
- Intra- and interspecies interaction (predation of viable microorganisms; mutual metabolism – ultimate biodegradation accomplished by several microbial species or strains)

Environment Related
- Temperature
- pH
- Nutrients
- Oxygen concentration/redox potential (aerobic or anaerobic)
- Salinity/ionic strength
- Other organic substrates that may affect microbial population and acclimation
- Hydrostatic pressure
- Presence of sorbing media, especially organic matter, which may render the substrate less available

Substrate Related
- Chemical structure
- Physical/chemical properties
- Concentration

Therefore, mixed microbial cultures from natural samples should be used as an inoculum or as the whole test sample for determining biodegradability in the environment.

Mixed culture biodegradation studies are of four types: (1) screening tests, (2) biological treatment simulations, (3) grab sample tests, and (4) field studies (Howard et al., 1987).

12.3.1   Screening Tests

Screening tests usually employ a defined mineral salts medium and the rates of degradation are determined either directly, by measuring for the disappearance of the parent chemical (UV absorption loss, methylene blue activated substance (MBAS) loss with detergents, etc.) or, more frequently, indirectly, by measuring such parameters as biochemical oxygen demand (BOD) consumption, CO_2 evolution, chemical oxygen demand (COD), consumption, or dissolved organic carbon (DOC) disappearance. Indirect analytical methods usually result in a less expensive screening biodegradation test but provide less information on the loss of the chemical substrate.

Often, different results come from screening tests. This may arise from the source of inoculum, whether acclimation has occurred or can occur during the test period (Cowan et al., 1996), and from the microbial population (concentration) in the test medium (Blok and Booy, 1984), depending on whether the screening test is designed to simulate more closely the natural environment or a biological treatment plant.

12.3.2   Biological Treatment Simulations

Biological treatment simulations consist of either continuous (CAS) or semicontinuous (SCAS) activated sludge tests that use either synthetic or natural sewage. The sewage or initial sewage inoculum usually is taken from a domestic wastewater treatment plant in order to reduce variability (industrial wastewater plants may contain very special chemical loadings) and to simulate treatment of consumer products. If treatability information in an industrial wastewater plant is desired, samples from the plant may be used.

The CAS or SCAS units are aerated and contain a sludge separation step to simulate the sludge separation step in pilot and full scale units. These laboratory biological treatment simulation units are thought to give more realistic results, similar to a pilot or full scale unit. The SCAS units are less expensive to operate and appear to give similar results to the CAS units if SCAS removal is >90% (Boethling et al., 1997). The Organization for Economic Cooperation and Development (OECD) Coupled Units test (OECD Guideline 303A) has been used more frequently, especially in Europe, and is more interpretable than SCAS results of <90%.

12.3.3   Grab Sample Tests

Grab sample tests consist of disappearance or die-away tests of the chemical in a natural sample of soil, aquifer sediments, groundwater, sediment/groundwater slurries, surface water, or surface water/sediment. Often, these studies will use low concentrations of the chemical substrate to more accurately simulate low level contamination.

Because natural samples often have background levels of organic material and the chemical substrate concentration is low, indirect measurement techniques can not be used and ^{14}C, other radiolabelled techniques, or sensitive specific analytical methods are required. If radiolabelled methods are used, the cost of the method increases, due to the cost of synthesis of the radiolabelled material. Use of radiolabelled chemicals is required for registration studies on environmental fate testing of pesticides.

12.3.4   Field Studies

Field studies consist of die-away studies of a chemical in a soil plot, pond, stream, or groundwater and may consist of a number of monitoring sites, with use of models or stable chemicals of similar physical properties to control for dilution. Field studies are considerably less controlled than grab samples, and greater potential exists for other degradation or transport processes that may reduce the chemical substrate concentration but not be a result of biodegradation. However, field studies give perhaps the most realistic biodegradation results, if properly controlled.

Wiedemeier and coworkers (1996) have suggested two methods to approximate biodegradation rates in groundwater field studies: (a) use a biologically recalcitrant tracer (e.g., three isomers of trimethylbenzene) in the groundwater to correct for dilution, sorption, and/or volatilization and calculate the rate constant by using the downgradient travel time; or (b) assume that the plume has evolved to a dynamic steady-state equilibrium and develop a one-dimensional analytical solution to the advection-dispersion equation.

12.3.5   Test Protocol

Frequently, results are highly dependent upon the test protocol. This is especially true for screening tests. For example, many screening tests do not employ an acclimation step prior to the test start and/or may not run long enough to allow for acclimation during the test. In addition, the reproducibility of individual tests often is poor, especially between laboratories, and in some cases even within the same laboratory.

For example, ring tests conducted by the OECD (1981) showed a rather poor reproducibility between laboratories, and various methods gave considerably different results. Gerike and Fischer (1979, 1981) studied 44 chemicals using seven different methods and found only 19 chemicals for which results were consistent with all methods (7 slow degraders and 12 fast degraders). Protocol modifications have been made (e.g., running a test for 28 days instead of 14 days to allow for acclimation) and have resulted in some improvement of reproducibility, but biodegradation tests often still give divergent results (Cowan et al., 1996).

OECD divided their aerobic screening tests into less vigorous "ready" aerobic biodegradability tests, more vigorous "inherent" biodegradability tests, and "simulation" biodegradability tests. A chemical that passes a ready test is expected to biodegrade fairly rapidly under most aerobic conditions. A methodology for translating OECD tests to degradation rates in realistic environmental situations has been presented but applied to only 20 chemicals (Struijs and Van den Berg, 1995) or limited numbers of chemicals (Federle et al., 1997).

Test guidelines are developed by various national and international organizations. Table 12.4 lists the OECD and U.S. Environmental Protection Agency (EPA) Office of Pollution Prevention and Toxics (OPPT) guidelines. It also lists experimental conditions, analytical methods, and criteria for whether a chemical is considered to be biodegradable (pass) and non-biodegradable (fail), when applicable.

12.4 Biodegradation Rate Constants

Biodegradation kinetics have been reviewed in detail (Simkins and Alexander, 1984; Schmidt et al., 1985; Alexander, 1994) and number of kinetic models proposed, including the use of screening tests for generating biodegradation kinetics (Hales et al., 1997). Biodegradation rates typically are interpreted through the Monod equation (Equation (1)), which is analogous to the Michaelis-Menten equation used in enzyme kinetics:

$$\mu = \mu_m \frac{[S]}{K_s + [S]} \tag{1}$$

The parameters μ and μ_m refer to the growth rate in the presence of substrate concentration S and the maximum growth rate, respectively, and K_s is the half-velocity coefficient, i.e., the value of S at which $\mu = 0.5 \mu_m$. Equation 2 can express the degradation rate of a substrate:

$$-\frac{d[S]}{dt} = \frac{\mu_m [S][B]}{Y(K_s + [S])} \tag{2}$$

where B represents biomass and Y is the growth yield factor. The Monod equation assumes that the compound of interest sustains growth and is the only source of carbon and thus, its applicability to the environment may be limited. For example, for cometabolic processes with μ_m and Y defined as zero, the equation would not apply. The equation also ignores toxicity and makes no provision for acclimation. Experimentally, rates are measured either at low substrate concentrates where $K_s > [S]$ and Equation (2) simplifies to Equation (3), or at high substrate concentrations where $[S] > K_s$ and Equation (4) follows from Equation (2):

$$-\frac{d[S]}{dt} = \frac{\mu_m}{YK_s}[S][B] \tag{3}$$

TABLE 12.4

Biodegradation Test Guidelines (Boethling et al., 1993; Cowan et al., 1996; OECD, 1981)

Title	CFR[a]	OECD[b] ISO[c]	OPPTS 835[d]	Concn[e] mg/L	Inocul. Concn.	Inocul. Source	Anal.	Pass/ Fail[f]
Aerobic aquatic biodegradation[g]	3100		3100	10 C	100 ml/L	acclim. medium	CO_2	60%
Closed Bottle[g]	3200	301D 10707	3110	2–10	few drops	effluent	O_2	60%
Two-Phase Closed Bottle		10708					O_2	
Modified AFNOR[h]	3180	301A 7827	3110	40	few drops	effluent	DOC	70%
Modified MITI[h]	3220	301C	3110	100	30 mg/L	sludge solids	O_2	60%
Modified OECD[h]	3240	301E 7827	3110	10–40	5 mL	effluent	DOC	70%
Modified Sturm[h]	3260	301B 9439	3110	10–20	30 mg/L	sludge supernat.	CO_2	60%
Manometic Respirometry[h]		301F 9408	3110					
Modified SCAS[i]	3340	302A 9887	3210	20	66%	mixed liquor	DOC	20% 70%
Modified Zahn-Wellens[i]	3360	302B 9888	3200	50–400 DOC/L	0.2–1.0 g dry matter/L	sludge	DOC COD	70%
Modified MITI (II)[i]		302C						
Coupled Units[j]		303A 11733						
Porous Pot			3220	10–20	100%	sewage	DOC or spec. anal	% rem.
Shake-flask die-away			3170	10mg/L if no [14]C	100%	water and/or sedim.	CO_2 ^{14}C spec. anal	Calc. rate
Sediment/water microcosm (Ecocore)			3180		100%	water/ sedim.	[14]C	Calc. rate
Biodegradation in soil[j] (aerobic/anaerobic)	3400	304A	3300	low[k]	100%	soil	[14]C	Calc. rate
Biodegradation in seawater		306	3160					
Anaerobic biodegradation	3140		3400	50 C	400ml/4 L	digestor sludge	CH_4 + CO_2	50%/56 days
Anaerobic biodegradation		11734				digestor sludge	CH_4 + CO_2	rate/60 days
Anaerobic biodegradation in the Subsurface	795. 54		5154	3 concn	100%	Aquifer material samples (6 sites)	Spec. anal.	calc. rate/ 64 days

[a] 40 Code of Federal Regulations 796.
[b] OECD Guideline Number
[c] Pagga (1997)
[d] USEPA Office of Preventing Pesticides and Toxic Substances (OPPTS) Fate, Transport, etc., Transport and Transformation Test Guidelines 835 Series
[e] Substrate test concentration
[f] Pass/fail criteria for whether the chemical is considered biodegradable or non-biodegradable usually in 28 days
[g] Sealed-Vessel CO_2 Production Test (OPPT 835.3120) is very similar to Aerobic Aquatic Biodegradation Test
[h] Ready biodegradability test
[i] Inherent biodegradability test
[j] Simulation test
[k] Concentration similar to that encountered in environment

$$-\frac{d[S]}{dt} = \frac{\mu_m}{Y}[B] \qquad (4)$$

For the former case (Equation (3)), which is environmentally more relevant for low contamination situations, the rate obeys first-order kinetics with respect to substrate and biomass (second-order overall), whereas in the latter case (Equation (4)), the kinetics have a first-order relationship to biomass but are independent of substrate concentration. Methods for measuring of biomass, B, have varied widely, and, for studies involving mixed populations, in which only a fraction of the organisms can degrade the substrate, a means for quantifying the responsible fraction is not available.

Nevertheless, the Monod equation is useful for comparing biodegradation rates. For example, Paris et al. (1981) used a second-order approach (Equation (3)) to analyze a large body of data on the microbial hydrolysis of three pesticides in numerous natural bodies of water. Under their conditions (surface water die-away at low substrate concentration), the rate constants were independent of the source of microorganisms. This surprising result implies that all the organisms (or a constant fraction thereof) degrade a given compound at the same rate. While this may occur for readily degradable compounds (e.g., pesticides with hydrolyzable functionalities), it is not likely to be true for more recalcitrant compounds, and several investigators have shown that biodegradation rates vary appreciably with differences in the nature of the organism.

Another experimental approach used to develop rates of biodegradation is the heterotropic uptake kinetics approach (Pfaender and Bartholomew, 1982). This approach also is based on Michaelis-Menten kinetics and requires several assumptions: (a) the amount of enzyme remains constant during the experiment, (b) the concentration of substrate does not change significantly during the measurement, (c) the transport systems are responding only to the substrate, and (d) the concentration of a specific pollutant substrate present in the environmental sample is insignificant compared to the amount added. Some of the assumptions (a, b) require that the measurement be made in a short incubation period (<24 hours) before changes can occur in the container, and therefore, acclimation is not considered. The approach usually also requires radiolabeled substrate ($^{14}CO_2$ evolution measured), which may be a drawback in instances in which synthesis of radiolabelled material is difficult. Using this approach, the maximum growth rate, μ_m, and half-velocity coefficient, K_s, can be determined, and a rate constant that has the same units as a first-order rate constant can be calculated by dividing μ_m by K_s:

$$K_1 = \frac{\mu_m}{K_s} \qquad (5)$$

Simkins and Alexander (1984) integrated the above into six kinetic biodegradation models, three for growth and three for non-growth situations. Figures 12.1 and 12.2 illustrate the disappearance curves for the six models and indicate which models apply, given substrate concentration and microbial population of degraders, respectively (Alexander, 1985). These models assume that the substrate is available. If diffusion or desorption are rate determining or the microbial population is metabolizing one substrate while growing on another, the biodegradation kinetics may become even more complicated (Schmidt et al. 1985; Alexander, 1985).

For example, Schmidt et al. (1985) developed 12 kinetic models for the metabolism of organic chemicals that are not supporting bacterial growth. Assignment of the appropriate kinetic model requires measurement of sufficient experimental points on the disappearance curve. Often insufficient data points are collected to assign a kinetic model, especially with screening studies. Temperature definitely has an impact on the biodegradation kinet-

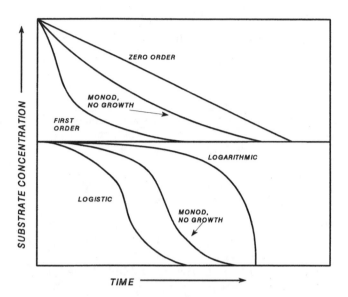

FIGURE 12.1
Disappearance curves for mineralized chemicals as related to individual kinetic models (From Alexander, 1985. Reprinted with permission from the American Chemical Society).

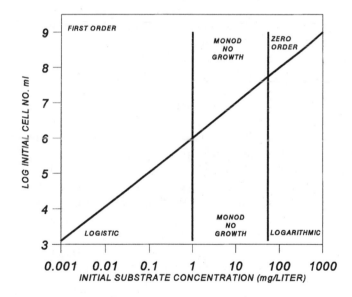

FIGURE 12.2
Kinetic models as a function of initial substrate concentration and bacterial cell density (From Alexander, 1985. Reprinted with permission from the American Chemical Society).

ics and may be modeled by the familiar Arrhenius equation, although few studies have examined the kinetics at different temperatures.

12.5 Estimation Methods

In a recent review of structure-biodegradability relationships, Mani et al. (1991) discussed the chemical properties and descriptors that may be useful in modeling different stages of the biodegradation process. Table 12.5 summarizes the usefulness of these properties and

TABLE 12.5

Examples of Properties and Descriptors Potentially Useful with Various Stages of the Biodegradation Processes (modified from Mani et al., 1991).

Process	Molecular Descriptors	Physicochemical Properties	Thermodynamic and Quantum Chemical Descriptors
Uptake by organism	Molecular weight	Water solubility; Log K_{ow}	Not appropriate
Internal distribution and bioaccumulation	Molecular surface area	Log K_{ow}	Not appropriate
Biodegradation by metabolic processes	Molecular connectivity index (topological indices - shape)	Molar refractivity	Field constants; atomic charges

descriptors and provides a framework for considering the presently available structure-biodegradability relationships. Although one might argue that whatever chemical descriptors work for prediction of biodegradability should be used, it is reasonable to assume that descriptors like bond strength and atom charge probably have little direct relationship to uptake and distribution in microorganisms.

12.5.1 Previous Structure – Biodegradability Relationships of Structurally Related Chemicals

Prediction of quantitative rates of biodegradability is severely limited because of the lack of reproducibility of biodegradation data (Howard et al.,1987), the numerous protocols used for biodegradation tests (Howard and Banerjee, 1984), and the limited number of chemicals with kinetic biodegradation data determined by the same protocol. As a result, quantitative prediction of biodegradation rates or rate constants has been attempted only on a very limited number of structurally related chemicals (Howard et al., 1992).

For example, as noted previously, Banerjee et al. (1984) studied 19 chlorophenols and chloroanisoles and developed both direct and indirect linear correlations between the biodegradation rates (pure culture and river die-away) and K_{ow}. Urushigawa and Yonezawa (1979) studied eight n-alkyl phthalates, and Yonezawa and Urushigawa (1979) studied 22 alcohols and developed correlationships between K_{ow} and the rates of biodegradation in an activated sludge die-away test. Paris et al. (1984) also developed correlations between K_{ow} and rates of biodegradation for 2,4-D esters. Paris et al. (1983) and Paris and Wolfe (1987) developed linear correlations between the biodegradation rates of eight para-substituted phenols and six meta-substituted anilines (river die-away tests) and van der Waal's radii. Wolfe et al. (1980) developed a correlation between the biodegradation rates of several esters and the alkaline chemical hydrolysis rate constants.

Boethling (1986) developed a number of correlations between various molecular connectivity indexes (MCIs) and biodegradation rates for several groups of chemicals, including 2,4-D esters, carbamates, ethers, aliphatic alcohols, aliphatic acids, and alkyl phthalates. Cantier et al. (1986) developed a correlation between the biodegradation rates in soil of propyzamide and nine of its analogs and two variables, the Hammett constant and a hydrophobic constant. Desai et al. (1990) used a group contribution method to predict the biodegradation rate of organic chemicals with a training set of 18 chemicals, 8 structural groups, and a validation set of 11 chemicals. Vaishnav et al. (1987) developed correlations of K_{ow} with 5-day biochemical oxygen demand (BOD_5) of 17 alcohols and 11 ketones.

12.5.2 Previous Structure-Biodegradability Relationships for Structurally Diverse Chemicals

Geating (1981) who used discriminant-function analysis and substructural keys to study a diverse group of 430 chemicals, made perhaps the earliest correlation between structure and biodegradability. The substructural keys were selected statistically, but the criteria for classification of chemicals as biodegradable or non-biodegradable were not stated. Mudder (1981) used multiple linear regression analysis to correlate the biodegradation rates generated by Pitter (1976) on 123 chemicals of widely varying chemical structure with a combination of several variables, including K_{ow} and structural and electronic parameters.

Several investigators have used BOD_5 to develop structure-biodegradability correlationships. Although BOD_5 data are not kinetic data, since only one point is measured at day 5, quantitative correlationships to the BOD_5 values are possible. Both Niemi et al. (1987) and Deardeon and Nicholson (1987) used the same set of BOD_5 data.

Niemi et al. (1987) used MCIs to identify substructures that correlated to biodegradation and non-biodegradation, and their method is discussed in detail below. Dearden and Nicholson (1987) correlated biodegradability to atomic charge difference at a functional group. The method can not be applied to multifunctional chemicals.

Table 12.6 reviews a number of comprehensive approaches that attempted to qualitatively predict biodegradability. Many of the models have used a weight-of-evidence biodegradation database — BIODEG (Howard et al., 1986) — that was developed specifically for structure-biodegradability correlations (Howard et al., 1987). Boethling et al. (1994) used both the experimental BIODEG database and results of an expert survey to develop four models that all used the same structural fragments; these structural fragments were selected from previously known "rules of thumb" (e.g., increasing the number of chlorines on aromatic ring results in increased persistence (Howard et al., 1992)). These models were theoretically satisfying because most of the fragments that were known to increase biodegradability had positive coefficients, and most of the fragments that were known to decrease biodegradability had negative coefficients.

The results of Langenberg et al. (1996) demonstrate the variability of biodegradation tests. They obtained fairly accurate predictions for two models that were both trained and validated on MITI biodegradation test data. In contrast, the model of Boethling et al. (1994), when applied to MITI data (Langenberg et al., 1996), provided accurate predictions only for nonbiodegradable chemicals, which was the opposite of reliability for the BIODEG database results (i.e., more reliable for biodegradable chemicals). Only the method of Boethling et al. (1994) is available as a computer model, although a computer model from unpublished data is available from Case-Western Reserve that uses an approach similar to that of Klopman et al. (1993).

Klopman et al. (1995) developed a method for predicting microbial metabolites. Although this model will not provide much insight into biodegradability, it could provide the user with pathway information and possible metabolites of toxicological concern.

12.5.3 Recommended Biodegradation Models

Three models are recommended for comprehensive prediction of biodegradation of a variety of organic chemicals. These models were selected because they:

(1) are well documented in terms of the chemical descriptors and equations, and

(2) were trained and validated on large numbers of chemicals.

TABLE 12.6

Comparison of Comprehensive Estimation Methods for Biodegradation.

Method	Methodology	Statistical Results
Niemi et al. (1987)	28 Structural fragments Biodegradable or nonbiodegradable	Training: n = 284 Overall 91.9% correct (90.7% for degrad. fast) (95.5% for degrad. slow) MITI Test (Degner et al., 1993) Validation: 756 Overall 76% (80.7% for easily degrad.) (68.7% for non-easily degrad.) (2% – 18 chemicals not classified)
Howard et al. (1992)	35 Structural fragments Biodegradability probability	Training (linear): n = 264 Overall 90.5% correct (97.1% for degrad. fast) (78.5% for degrad. slow) Training (nonlinear): n = 264 Overall 89.8% correct (94.7% for degrad. fast) (80.6% for degrad. slow) Validation (linear): n = 27 Overall 81.5% correct (100% for degrad. fast) (64.3% for degrad. slow) Validation (nonlinear): n = 27 Overall 88.8% correct (92.3% for degrad. fast) (85.7% for degrad. slow)
Boethling et al. (1994)	36 Structural fragments + MW Biodegradability probability + Numerical score for hrs, days, wks, months, longer	Experimental Validation (linear): n = 295 Overall 89.5% correct (97.3% for degrad. fast) (76.1% for degrad. slow) Validation (nonlinear): n = 295 Overall 93.2% correct (97.3% for degrad. fast) (86.2% for degrad. slow) Expert Survey (17 experts; 200 chem.) Validation (primary): n = 200 r^2 = 0.71 me = 0.173 Validation (ultimate): n = 200 r^2 = 0.72 me = 0.206 MITI-Test (Langenberg et al., 1996) Validation (linear): n = 488 Overall 56% correct (91% for degrad. fast 127 chem.) (44% for degrad. slow 361 chem.) Validation (nonlinear): n = 488 Overall 63% correct (87% for degrad. fast 127 chem.) (54% for degrad. slow 361 chem.) 50/50 Validation set - Linear 84%, Nonlinear 80% Correct for Nondegradable
Degner et al. (1993)	5 Descriptors (Acyclic) (Model 74)	Training (acyclic): n = 47 Overall 85% correct (92% for easily degrad.) (79% for non-easily degrad.) Validation (acylic): n = 240 (MITI test) Overall 89% correct (97% for easily degrad.) (61% for non-easily degrad.) Training (acyclic+): n = 65 Overall 92% correct
	7 Descriptors (Acyclic + Phosphoric acids + Tertiary amines) (Model 75)	Validation (acyclic+): n = 222 (MITI test) Overall 91% correct (95% for easily degrad.) (78% for non-easily degrad.)
	9 Descriptors (Monocyclic aromatic) (Model 78)	Training (monocyclic arom.): n = 60 Overall 85% correct Validation (monocyclic arom.): n = 57 (MITI test) Overall 84% correct (87% for easily degrad.) (83% for non-easily degrad.)
Langenberg et al. (1996)	6 Descriptors + MW (Acyclic model) 8 Descriptors + MW (Monocyclic aromatic)	MITI-Test Training (acyclic): n = 65 Training (monocyclic arom.): n = 60 "Validation (acyclic)": n = 222 Overall 91% correct (94% - degrad. fast 175 chem.) (81% - degrad. slow 47 chem.) "Validation (monocyclic aromatic)": n = 57 Overall 84% correct (75% - degrad. fast 24 chem.) (91% - degrad. slow 33 chem.)
Gombar and Enslein (1991)	Aromatic model: 31 descriptors (22 Structural fragments + 5 MCI or Charge calculations)	Training (aromatic) model: n = 149 Overall 91% correct

TABLE 12.6 (CONTINUED)

Comparison of Comprehensive Estimation Methods for Biodegradation.

Method	Methodology	Statistical Results
	Aliphatic model: 22 descriptors (21 Structural fragments and 1 MCI)	Training (aliphatic) model: n = 134 Overall 91% correct
Klopman et al. (1993)	DB 1 - 26 Biophores (degrad. rapid), 11 Biophores (degrad. slow)	Training (DB1): n = 283 (BIODEG chem.) Correctly predicted 74%
		Training (DB3): n = 153 (Pitter chem.)
	DB3 - 8 Biophores, 5 similar to DB1	Validation (Using DB1 and DB3): n = 27 Correctly predicted 74%
Klopman et al. (1995)	Predict metabolites	Training: n = 385
		Validation: n = 27 (Of 13 chemicals known to biodeg., at least one transformation predicted)
Devillers (1993)	20 Structural descriptors - Neural network, Only benzene derivatives	Training: n = 113 Correctly predicted 92% (104/113)
		Validation: n = 14 Correctly predicted 86%
Devillers et al. (1996)	Atomic and group contributions	Training: n = 172 (Expert survey data of 200 chem.) r^2 = 0.76 Residuals <0.5 = 94.8%
		Validation I: n = 12 Residual <0.5 except 1 chem.
		Validation II: n = 57 (19 Survey + 38 Experimental) Correctly predicted 82.5%.

For each model, the descriptors and equations are described, and sample calculations are presented for five benchmark chemicals.

12.5.3.1 Qualitative Substructure Model (Niemi et al., 1987)

The Qualitative Substructure Model of Niemi and coworkers was developed by using a BOD database with BOD data converted to half-lives by assuming, for example, that a theoretical BOD of 50% or greater in 5 days had a half-life of 5 days and a 25% theoretical BOD in 10 days had a half-life of 20 days. Using these data, they identified structural features associated with chemicals that degraded or were persistent (see Table 12.7).

If a chemical has any non-degradable substructures (Table 12.7), the chemical is concluded to be persistent. If a chemical has no non-degradable substructures and does have degradable substructures, the chemical is concluded to be degradable. Chemicals without either degradable or non-degradable substructures are concluded to be unclassified.

Degner et al. (1993) used the model with a dataset of 774 chemicals with MITI test data and found 80.7% correct estimates for easily degradable/non-degradable classifications. They were unable to classify 18 chemicals due to a lack of indicator variables.

Table 12.8 provides sample classifications for five "benchmark" chemicals, some of which are used in other chapters of this book. The table lists the substructure from Table 12.7 that determine the classification for the five chemicals.

12.5.3.2 *Biodegradability Probability Program (Howard et al., 1992; Boethling et al., 1994)*

The Biodegradability Probability Program has been developed over a period of years and is still being refined. As indicated in Table 12.6, the original model was developed with 35 structural fragments whose coefficients were developed by linear and non-linear regression to an experimental "weight-of- evidence" evaluation for 264 chemicals from the BIO-DEG database (Howard et al., 1987). Chemicals were used only if two or more

TABLE 12.7

Qualitative Substructure Model (Niemi et al., 1987).

No.	Description of Association	Range T$^{1}/_{2}$ (days)	Correctly Predicted Chemicals	Incorrectly Predicted Chemicals
Criteria for degradable substances[a]				
1	One halogen substitution on an unbranched chem.	<12	4	0
2	One cyano substitution on an unbranched chem.	<10	3	0
3	Aldehydes	2–11	15	0
4	Hydrocarbons	3–17	6	0
5	Alcohols, esters, and amines	2–16	64	13
6	Acids	3–12	29	2
7	Amino acids	2–5	13	0
8	Sulfonates	2–17	15	3
9	Benzene rings with various substitutions and log K_{ow} < 2.18	2–16	21	0
10	Biphenyl and 2 or less hydroxy-substituted polyaromatics	<15	4	0
11	Cyclic chemicals consisting only of C, O, N, and H	2–15	21	2
12	Two aromatic rings (e.g., naphthalene and amino-naphthalene)	<15	2	0
Criteria for non-degradable substances				
1	At least one *tert*-butyl terminal branch	>15	11	0
2	Epoxides	>20	6	0
3	Aliphatic chemicals with fused rings & no branches	>35	3	0
4	Two terminal isopropyl subgroups for a noncyclic chemical	>35	3	0
5	Aliphatic cyclic chemicals without branches	>40	5	0
6	One or more halogen substitution on a branched, noncyclic or cyclic chemical	>5	12	1
7	At least one isopropyl or dimethyl amine substitution on a cyclic chemical without other "degradable" substituents	>25	4	2
8	Two halogen substitutions on an unbranched, noncyclic chemical	>15	3	0
9	More than two hydroxyl substitutions on an aromatic ring	>15	2	0
10	Two or more rings (with a few exceptions as noted above)	>20	8	0
11	Two terminal diamino groups on a noncyclic chemical	>35	1	0
12	More than 1 amino branch on a ring with nitrogen a member of the ring	>100	2	0
13	Two terminal double-bond carbons on an unbranched chemical	>100	1	0
14	Benzene ring with >2 substitutions (non-hydroxyl) and log K_{ow} > 2.18	>100	1	0
15	Cyano group on a chain consisting of > 8 atoms	>100	1	0
16	Highly branched chemicals	>100	1	0

[a] All the degradable descriptors are only used if other subgroups associated with persistence are not present.

biodegradation studies offered consistent data about them (biodegraded "fast" or "slow") and the data suggest that acclimation would not play a major role.

Two years later, a revised version was developed (Boethling et al., 1994) that included five new or redefined substructures and molecular weight, and the coefficients were developed by linear and non-linear regression with 295 chemicals from the BIODEG database as well as primary and ultimate biodegradation estimates from an expert survey of 200 chemicals (total of four models). The survey database was built from the opinions a panel of 17 experts who were asked to provide semi-quantitative (approximate time for complete deg-

TABLE 12.8

Sample Qualitative Substructure Model (Niemi et al., 1987) Biodegradation
Classifications for Benchmark Chemicals

Structure	Descriptors	Classification (Biodegradable/ Persistent)
Anthracene		
	Two or more rings	Persistent
Lindane		
	Aliphatic cyclic chemicals without branches One or more halogen substitution on a branched, noncyclic or cyclic chemical [not really a branched chemical]	Persistent
2,6-Di(tert)butylphenol		
	At least one *tert*-butyl terminal branch Highly branched chemicals	Persistent
Chlorpyrifos		
	One or more halogen substitution on a branched, noncyclic or cyclic chemical	Persistent
Trichloroethylene		
	Two halogen substitutions on an unbranched noncyclic chemical	Persistent

radation of days, weeks, months, longer) estimates for primary and ultimate aerobic biodegradation on 200 chemicals.

For the experimental database and the primary and ultimate model, a linear equation is defined as:

$$Y_j = a_0 + a_1 f_1 + a_2 f_2 + \dots + a_{36} f_{36} + a_m M_w + e_j \qquad (6)$$

where

Y_j = probability that chemical j will biodegrade fast (experimental data), orthe primary or ultimate biodegradation rate for the survey models
f_n = number of nth substructure in jth chemical
a_0 = intercept (equation constant in Table 12.9)
a_n = regression coefficient for nth substructure
M_w = molecular weight
a_m = regression coefficient for M_w
e_j = error term; mean value is zero

The logistic equation was used as the basis for the nonlinear model. This model,

$$Y_j \frac{\exp(a_0 + a_1 f_1 + a_2 f_2 + \dots a_{36} f_{36} + a_m M_w)}{1 + \exp(a_0 + a_1 f_1 + a_2 f_2 + \dots a_{36} f_{36} + a_m M_w)} \qquad (7)$$

estimates probabilities near 0.0 when the linear combination in the exponent takes large negative values, near 0.5 when that linear combination is near 0.0, and close to 1.0 when the linear combination takes a large positive value.

Table 12.9 lists these fragments and their regression-derived coefficients. Atoms were used only once; that is, if an atom was part of one fragment, it cannot be part of another. The standard errors and the test statistics (or their p values) were used only as an approximate indication of the contribution of a particular fragment rather than as a basis for eliminating the fragment from the model. As a result, there are some collinearities among some of the fragments.

Table 12.10 provides sample calculations for the five "benchmark" chemicals. The results presented for anthracene are unaltered output from the Biodegradation Probability Program (Syracuse Research Corporation, 1996).

12.5.3.3   *Acyclic Aliphatic and Monocyclic Aromatic Model (Degner et al., 1993)*

The Acyclic Aliphatic and Monocyclic Aromatic models were developed during an evaluation of other biodegradation prediction models. As Table 12.6 indicates, the two models were developed with seven structural fragments (acyclic aliphatic) and nine structural fragments (monocyclic aromatic) by means of discriminant analysis with MITI test data.

For all the Degner et al. (1993) models, if the biodegradation factor (B) is > 0, the chemical is considered an easily degradable substance, and if it is < 0, the chemical is considered a non-easily degradable substance.

The Acyclic Aliphatic model (#74) consists of the following equation:

$$\text{Biodegradation factor} = \Sigma(\text{Substructures factor} \times \text{No. factors}) + \text{intercept} \qquad (8)$$

with the five terminal substructures and factors given in Table 12.11. This equation should only be used if the structure is non-aromatic and does not contain a phosphoric acid or a tertiary amine.

TABLE 12.9

Biodegradability Probability Program Structural Fragments and Coefficients
(Boethling et al., 1994).

Fragment or Parameter	BIODEG models			Survey models		
	Freq[a]	Linear Coefficient	Nonlinear Coefficient	Freq[a]	Primary Coefficient	Ultimate Coefficient
Equation constant	–	.748	3.01	–	3.848	3.199
M_W	295	–.000476	–.0142	200	–.00144	–.00221
Unsubstituted aromatic (≤ 3 rings)	2	.319	7.191	1	–.343	–.586
Phosphate ester	5	.314	44.409	6	.465	.154
Cyanide/nitrile (–C≡N)	5	.307	4.644	11	–.065	–.082
Aldehyde (–CHO)	4	.285	7.180	5	.197	.022
Amide (–C(=O)–N or – C(=S)–N)	9	.210	2.691	13	.205	–.054
Aromatic –C(=O)OH	24	.177	2.422	6	.0078	.088
Ester (–C(=O)–O–C)	23	.174	4.080	25	.229	.140
Aliphatic –OH	34	.159	1.118	18	.129	.160
Aliphatic –NH₂ or –NH–	13	.154	1.110	7	.043	.024
Aromatic ether	11	.132	2.248	11	.077	–.058
Unsubstituted phenyl group (–C₆H₅)	25	.128	1.799	22	.0049	.022
Aromatic –OH	46	.116	.909	21	.040	.056
Linear C4 terminal alkyl (–CH₂CH₂CH₂CH₃)	44	.108	1.844	26	.269	.298
Aliphatic sulfonic acid or salt	4	.108	6.833	4	.177	.193
Carbamate	4	.080	1.009	6	.194	–.047
Aliphatic –C(=O)OH	33	.073	.643	10	.386	.365
Alkyl substituent on aromatic ring	36	.055	.577	36	–.069	–.075
Triazine ring	5	.0095	–5.725	4	–.058	–.246
Ketone (–C–C(=O)–C–)	12	.0068	–.453	10	–.022	–.023
Aromatic –F	1	–.810	–10.532	1	.135	–.407
Aromatic –I	2	–.759	–10.003	2	–.127	–.045
Polycyclic aromatic hydrocarbon (≥ 4 rings)	6	–.657	–10.164	2	–.702	–.799
N–nitroso (–N–N=O)	4	–.525	–3.259	1	.019	–.385
Trifluoromethyl (–CF₃)	1	–.520	–5.670	2	–.274	–.513
Aliphatic ether	11	–.347	–3.429	16	–.0097	–.0087
Aromatic –NO₂	14	–.305	–2.509	13	–.108	–.170
Azo group (–N=N–)	2	–.242	–8.219	3	–.053	–.300
Aromatic –NH₂ or –NH–	32	–.234	–1.907	23	–.108	–.135
Aromatic sulfonic acid or salt	11	–.224	–1.028	8	.022	.142
Tertiary amine	10	–.205	–2.223	10	–.288	–.255
Carbon with 4 single bonds & no H	9	–.184	–1.723	32	–.153	–.212
Aromatic –Cl	40	–.182	–2.016	27	–.165	–.207
Pyridine ring	18	–.155	–1.638	8	–.019	–.214
Aliphatic –Cl	12	–.111	–1.853	14	–.101	–.173
Aromatic –Br	5	–.110	–1.678	4	–.154	–.136
Aliphatic –Br	5	–.046	–4.443	2	.035	.029

[a] Number of compounds in the training set containing the fragment

For aliphatic chemicals with phosphoric acid or a tertiary amine, the Acyclic Aliphatic + (#75) model should be used, which consists of the following equation:

$$\text{Biodegradation factor} = \Sigma(\text{standardized Substructures} \times \text{No. factors}) + \text{intercept} \quad (9)$$

where standardized substructure = (substr.-MV of substructure)/SD and MV is the mean value of the substructure and SD is the standard deviation of the substructure (see Table 12.12)

TABLE 12.10

Sample Biodegradation Probability Program Calculations for Benchmark Chemicals.

Anthracene (full printout)

```
SMILES: c(c(ccc1)cc(c2ccc3)c3)(c1)c2
CHEM:   Anthracene
MOL FOR:C14 H10
MOL WT: 178.24
```

```
Linear Model Prediction : Biodegrades Fast
Non-Linear Model Prediction: Biodegrades Fast
Ultimate Biodegradation Timeframe: Months
Primary Biodegradation Timeframe: Weeks
```

Type	Num	Biodeg Fragment Description	Coeff	Value
Frag	1	Unsubstituted aromatic (3 or less rings)	0.3192	0.3192
MolWt	*	Molecular Weight Parameter		-0.0849
Const	*	Equation Constant		0.7475
Result		Linear Biodegradation Probability		0.9819

Type	Num	Biodeg Fragment Description	Coeff	Value
Frag	1	Unsubstituted aromatic (3 or less rings)	7.1908	7.1908
MolWt	*	Molecular Weight Parameter		-2.5309
Result		Non-Linear Biodegradation Probability		0.9995

```
A Probability Greater Than or Equal to 0.5 indicates --> Biodegrades Fast
A Probability Less Than 0.5 indicates --> Does NOT Biodegrade Fast
```

Type	Num	Biodeg Fragment Description	Coeff	Value
Frag	1	Unsubstituted aromatic (3 or less rings)	-0.5859	-0.5859
MolWt	*	Molecular Weight Parameter		-0.3939
Const	*	Equation Constant		3.1992
Result		Survey Model - Ultimate Biodegradation		2.2194

Type	Num	Biodeg Fragment Description	Coeff	Value
Frag	1	Unsubstituted aromatic (3 or less rings)	-0.3428	-0.3428
MolWt	*	Molecular Weight Parameter		-0.2572
Const	*	Equation Constant		3.8477
Result		Survey Model - Primary Biodegradation		3.2478

```
Result Classification:    5.00 -> hours    4.00 -> days    3.00 -> weeks
(Primary & Ultimate)      2.00 -> months   1.00 -> longer
```

g-Hexachlorocyclohexane (Lindane) (abbreviated output)

```
SMILES : C(C(C(C(C(C1CL)CL)CL)CL)(C1CL)CL
CHEM :   Cyclohexane, 1,2,3,4,5,6-hexachloro-, (1.alpha., 2.alpha., 3.beta.,
         4.alpha, 5.alpha.,6.beta.)-
MOL FOR: C6 H6 CL6
MOL WT : 290.83
```

```
Linear Model Prediction : Does Not Biodegrade Fast
Non-Linear Model Prediction: Does Not Biodegrade Fast
Ultimate Biodegradation Timeframe: Recalcitrant
Primary Biodegradation Timeframe: Weeks
```

TABLE 12.10 (CONTINUED)

Sample Biodegradation Probability Program Calculations for Benchmark Chemicals.

Type	Num	Biodeg Fragment Description	Coeff	Value
Frag	6	Aliphatic chloride [-CL]	-0.1114	-0.6683
MolWt	*	Molecular Weight Parameter		-0.1385
Const	*	Equation Constant		0.7475
Result		Linear Biodegradation Probability		-0.0593
Result		Non-Linear Biodegradation Probability		0.0000
Result		Survey Model - Ultimate Biodegradation		1.5174
Result		Survey Model - Primary Biodegradation		2.8245

2,6-Di(tert-butyl) phenol (abbreviated output)

```
SMILES :  Oc(c(ccc1)C(C)(C)C)c1C(C)(C)C
CHEM :   Phenol, 2,6-bis(1,1-dimethylethyl)-
MOL FOR: C14 H22 O1
MOL WT : 206.33
```

Linear Model Prediction : Does Not Biodegrade Fast
Non-Linear Model Prediction: Does Not Biodegrade Fast
Ultimate Biodegradation Timeframe: Weeks-Months
Primary Biodegradation Timeframe: Days-Weeks

Type	Num	Biodeg Fragment Description	Coeff	Value
Frag	1	Aromatic alcohol [-OH]	0.1158	0.1158
Frag	2	Carbon with 4 single bonds & no hydrogens	-0.1839	-0.3679
MolWt	*	Molecular Weight Parameter		-0.0982
Const	*	Equation Constant		0.7475
Result		Linear Biodegradation Probability		0.3973
Result		Non-Linear Biodegradation Probability		0.0788
Result		Survey Model - Ultimate Biodegradation		2.3753
Result		Survey Model - Primary Biodegradation		3.2829

Chlorpyrifos (abbreviated output)

```
SMILES :  CCOP(=S)(OCC)Oc1nc(CL)c(CL)cc1CL
CHEM :   Chlorpyrifos
MOL FOR: C9 H11 CL3 N1 O3 P1 S1
MOL WT : 350.59
```

Linear Model Prediction : Does Not Biodegrade Fast
Non-Linear Model Prediction: Biodegrades Fast
Ultimate Biodegradation Timeframe: Recalcitrant
Primary Biodegradation Timeframe: Days-Weeks

Type	Num	Biodeg Fragment Description	Coeff	Value
Frag	3	Aromatic chloride [-CL]	-0.1824	-0.5473
Frag	1	Pyridine ring	-0.1546	-0.1546
Frag	1	Phosphate ester	0.3139	0.3139
MolWt	*	Molecular Weight Parameter		-0.1669
Const	*	Equation Constant		0.7475
Result		Linear Biodegradation Probability		0.1928
Result		Non-Linear Biodegradation Probability		1.0000
Result		Survey Model - Ultimate Biodegradation		1.7442
Result		Survey Model - Primary Biodegradation		3.2925

TABLE 12.10 (CONTINUED)

Sample Biodegradation Probability Program Calculations for Benchmark Chemicals.

Trichloroethylene (abbreviated output)

```
SMILES : C(=CCL)(CL)CL
CHEM :    Ethene, trichloro-
MOL FOR: C2 H1 CL3
MOL WT : 131.39


Linear Model Prediction : Does Not Biodegrade Fast
Non-Linear Model Prediction: Does Not Biodegrade Fast
Ultimate Biodegradation Timeframe: Weeks-Months
Primary Biodegradation Timeframe: Days-Weeks
```

Type	Num	Biodeg Fragment Description	Coeff	Value
Frag	3	Aliphatic chloride [-CL]	-0.1114	-0.3342
MolWt	*	Molecular Weight Parameter		-0.0626
Const	*	Equation Constant		0.7475
Result		Linear Biodegradation Probability		0.3508
Result		Non-Linear Biodegradation Probability		0.0119
Result		Survey Model – Ultimate Biodegradation		2.3893
Result		Survey Model – Primary Biodegradation		3.3563

TABLE 12.11

Acyclic Aliphatic Model #74.

Substructures	Factor
(1) CO	+0.07
(2) CH_2-NH_2	+0.14
(3) CH_3	−0.11
(4) OH	−0.13
(5) hal	−0.20
Intercept	+0.39

TABLE 12.12

Acyclic Aliphatic+ Model #75 for Phosphoric Acid and Tertiary Amine.

Structures	Factor	MV	SD
(1) CH_2NH_2	+0.960	0.092	0.341
(2) CH_3	+0.004	1.431	1.571
(3) CO	+0.001	0.431	0.684
(4) OH	−0.002	0.583	0.967
(5) hal	−0.008	0.831	1.842
(6) phosphoric acid	−0.089	0.062	0.242
(7) tert. amine	−0.212	0.108	0.312
Intercept	+0.164		

The Monocyclic Aromatic model (#78) consists of the following equation:

$$\text{Biodegradation factor} = \Sigma(\text{Substructures factor} \times \text{No. factors}) + \text{intercept} \qquad (10)$$

TABLE 12.13

Monocyclic Aromatic Model
#78.

Substructures	Factor
(1) aryl COO(R$_1$)	+0.090
(2) aryl OH	+0.003
(3) aryl CH$_2$(R$_1$)	+0.003
(4) aryl amide	−0.052
(5) aryl NH$_2$	−0.338
(6) aryl NO$_2$	−0.551
(7) aryl hal	−0.480
(8) aryl SO$_2$(R$_2$)	−0.320
(9) aryl (R$_3$)	−0.500
Intercept	+0.39

R$_1$: H or unbranched alkyl chain;
R$_2$: OH or NH$_2$ or SH; aryl (R$_3$): chain
with non-terminal heteroatoms or
branched alkyl chain

with the nine terminal substructures and factors given in Table 12.13.

Table 12.14 provides sample calculations for the five "benchmark" chemicals.

12.6 Available Data

A number of recent books review the relationship of chemical structure to biodegradability (e.g., Alexander, 1994; Pitter and Chudoba, 1990) and present tabulations of specific biodegradation data (e.g., Pitter and Chudoba, 1990, give specific values of BOD). Perhaps the most comprehensive source of biodegradation information is the BIOLOG (~ 6000 chemicals) and BIODEG (~ 800 chemicals) files of the Environmental Fate Data Base (Howard et al., 1986). These are searched easily by the Chemical Abstract Registry (CAS) number, and are available at no charge at http://esc.syrres.com.

A little-known source of biodegradation rates of pesticides is data developed by the California Department of Food and Agriculture (CDFA). They estimated aerobic and anaerobic soil metabolism half-lives from open scientific literature and studies submitted to CDFA from chemical companies in compliance with the data call-in requirements of the Pesticide Contamination Prevention Act. Table 12.15 tabulates these data.

TABLE 12.14

Sample Biodegradation Calculations Using OECD Aliphatic Acyclic and Monocyclic Aromatic Models for Benchmark Chemicals.

Anthracene

Cannot be calculated - not monocyclic and no substituents

γ-Hexachlorocyclohexane (Lindane) (Model 74)

Biodegradation factor (B) = Σ(Substructures factor × No. factors) + intercept
B = Σ(hal substructures factor (−0.20) × No. factors(6)) + intercept(+0.39)
B = (−0.20 × 6) +0.39
B = −1.20 +0.39
B = −1.20 +0.39
B = −0.81 = "non-easily degradable substance"

2,6-Di(tert-butyl)phenol (Model 78)

Biodegradation factor (B) = Σ(Substructures factor × No. factors) + intercept
B = Σ[one aryl OH substructures factor (+0.003) + two aryl R3 − branched alkyl chains (−0.500)] + intercept(+0.380)
B = (+0.003 + (2 × -0.500) +0.38)
B = (0.003 - 1.0) + 0.38
B = −0.997 +0.39
B = −0.607 = "non-easily degradable substance"

Chlorpyrifos (Model 75)
(This compound has an aromatic ring and therefore, use of this model may not be appropriate)

Biodegradation factor (B) = Σ(standardized Substructures × no. factors) + intercept (4)
standardized substructure = (substr.-MV of substructure)/SD
standardized substructure = (phosphoric acid substr. (−0.089)-MV (0.062) of phosphoric acid substructure)/SD (0.242) of phosphoric acid
standardized substructure = (−0.089) − (0.062)/0.242
standardized substructure = −0.151/0.242 = −0.624
B = Σ(standardized Substructures (−0.624) × no. factors(1 Phosphoric acid + intercept (+0.164)
B = (−0.624) + (+0.164)
B = −0.46 = "non-easily degradable substance"

Chlorpyrifos (Model 78)

Biodegradation factor (B) = Σ(Substructures factor × No. factors) + intercept
B = Σ[three aryl halide substructures factor (−0.480)] + intercept(+0.380)
B = (3 × −0.480) +0.38
B = (−1.44) + 0.38
B = -1.06 = "non-easily degradable substance"

Trichloroethylene (Model 74)

Biodegradation factor (B) = Σ(Substructures factor × No. factors) + intercept
B = Σ(hal substructures factor (−0.20) × No. factors(3)) + intercept(+0.39)
B = (−0.20 × 3) +0.39
B = −0.60 +0.39
B = −0.21 = "non-easily degradable substance"

TABLE 12.15

Aerobic and Anaerobic Biodegradation Half-lives of Pesticides (Johnson, 1991).

Pesticide	Aerobic Metabolism $T^1/_2$ (days)	Anaerobic Metabolism $T^1/_2$ (days)
Alachlor	18	5
Aldicarb	14	41
Aldrin	120	130
Ametryne	37	320
Atrazine	190	3400
Bentazon	NA	3500
Bromacil	300	170
Carbaryl	8	76
Carbofuran	23	20
Chloramben	NA	59
Chlordane	54	8200
Chlorothalonil	35	8
Chlorpyrifos	88	140
Chlorthal dimethyl	24	150
Cyanazine	15	110
1,3-D	13	16
2,4-D	8	60
DBCP	180	740
DDD	NA	160
DDT	3800	53
Diazinon	17	35
Dicamba	61	88
Dieldrin	1000	270
Dimethoate	2	1
Dinoseb	NA	NA
Disulfoton	2	2.4
Diuron	NA	1000
EDB	44	230
Endosulfan	32	150
Ethoprop	25	130
Fenamiphos	22	120
Fonofos	120	150
Heptachlor	2000	39
Lindane	790	37
Linuron	78	15
Malathion	1	30
Methiocarb	NA	64
Methyl bromide	35	6
Metolachlor	NA	84
Metribuzin	110	60
Naled	3	2
Oxamyl	180	3
Pendimethalin	1300	50
Phorate	3	7
Picloram	350	5100
Prometon	280	61
Prometryn	150	360
Propachlor	NA	NA
Propylene dichloride	NA	NA
Silvex	16	32
Simazine	110	58
Toxaphene	NA	25
Trifluralin	180	37

References

Alexander, M. 1973. Nonbiodegradable and other recalcitrant molecules. *Biotechnol. Bioeng.* 15:611-647.

Alexander, M. 1981. Biodegradation of chemicals of environmental concern. *Science* 211:132-138.

Alexander, M. 1985. Biodegradation of organic chemicals. *Environ. Sci. Technol.* 19:106-111.

Alexander, M. 1994. Biodegradation and Bioremediation. Academic Press, New York.

Banerjee, S., P.H. Howard, A.M. Rosenberg, A.E. Dombrowski, H. Sikka, and D.L. Tullis. 1984. Development of a general kinetic model for biodegradation and its application in chlorophenols and related compounds. *Environ. Sci. Technol.* 18:416-422.

Blok, J. and M. Booy. 1984. Biodegradability test results related to quality and quantity of the inoculum. *Ecotox. Environ. Safety* 8:410-422.

Boethling, R.S. 1986. Application of molecular topology to quantitative structure-biodegradability relationships. *Environ. Toxicol. Chem.* 5:797-806.

Boethling, R.S. 1993. Biodegradation of xenobiotic chemicals. In Handbook of Hazardous Materials. M. Corn, ed., pp 55-67. Academic Press, New York.

Boethling, R.S., P.H. Howard, W.M. Meylan, W. Stiteler, J. Beauman, and N. Tirado. 1994. Group contribution method for predicting probability and rate of aerobic biodegradation. *Environ. Sci. Technol.* 28:459-465.

Boethling R.S., P.H. Howard, J.A. Beauman, and M.E. Larosche. 1995. Factors for intermedia extrapolation in biodegradation assessment. *Chemosphere* 30:741-752.

Boethling, R.S., P.H. Howard, W. Stiteler, and A. Hueber. 1997. Does the Semi-Continuous Activated Sludge (SCAS) Test Predict Removal in Secondary Treatment. *Chemosphere* 35:2119-2130.

Cantier, J.M., J. Bastide, and C. Coste. 1986. Structure-degradability relationships for propyzamide analogues in soils. *Pestic. Sci.* 17:235-241

Cowan, C.E., T.W. Federle, R.J. Larson, and T.C.J. Feijtel. 1996. Impact of biodegradation test methods on the development and applicability of biodegradation QSARs. *SAR QSAR Environ. Res.* 5:37-49.

Dearden, J.C. and R.M. Nicholson. 1987. Correlation of biodegradability with atomic charge difference and superdelocalizability. In K.L.E. Kaiser, ed., *QSAR Environmental Toxicology – II*, pp. 83-89. Reidel, Dordrecht, The Netherlands.

Degner, P., M. Muller, M. Nendza, and W. Klein. 1993. Structure-activity relationships for biodegradation. *OECD Environmental Monographs* No. 68, p. 103.

Desai, M.D., R. Govind, and H.H. Tabak. 1990. Development of quantitative structure-activity relationships for predicting biodegradation kinetics. *Environ. Toxicol. Chem.* 9:473-477.

Devillers, J. 1993. Neural modelling of the biodegradability of benzene derivatives. *SAR and QSAR in Environ. Res.* 1:161-167.

Devillers, J., D. Domine, and R.S. Boethling. 1996. Use of a backpropagation neural network and autocorrelation descriptors for predicting the biodegradation of organic chemicals. In J. Devillers, ed., *Neural Networks in QSAR and Drug Design*. Academic Press, London, England.

Federle, T.W., S.D. Gasior, and B.A. Nuck. 1997. Extrapolating mineralization rates from the ready CO_2 screening test to activated sludge, river-water and soil. *Environ. Toxicol. Chem.* 16:122-134.

Geating, J. 1981. Literature study of the biodegradability of chemicals in water. PB 82-1000843 (EPA 600/2-81-175). In *Biodegradability Prediction, Advances in and Chemical Interferences with Wastewater Treatment, Vol 1.* National Technical Information Service, Springfield, VA.

Gerike, P. and W.K. Fischer. 1979. A correlation study of biodegradability determinations with various chemicals in various test. *Ecotox. Environ. Safety* 3:159-173.

Gerike, P. and W.K. Fischer. 1981. A correlation study of biodegradability determinations with various chemicals in various test. II. Additional results. *Ecotox. Environ. Safety* 5:45-55.

Gombar, V.K. and K. Enslein. 1991. A structure-biodegradability relationship model by discriminant analysis. In J. Devillers and W. Karcher, Eds., *Applied Multivariate Analysis in SAR and Environmental Studies*, pp. 377-414. Kluwer Academic Publ., Dordrecht, Holland.

Hales, S.G., T. Feijtel, H. King, K. Fox, and W. Verstraete. 1997. *Biodegradation Kinetics: Generation and Use of Data for Regulatory Decision Making. SETAC-Europe Workshop - Port-Sunlight,* UK 4-6 Sept. 1996, SETAC-Europe Brussels.

Howard, P.H., J. Saxena, P.R. Durkin, and L.-T. Ou. 1975. *Review and Evaluation of Available Techniques for Determining Persistence and Routes of Degradation of Chemical Substances in the Environment.* EPA-560/5-75-006. U.S. NTIS PB 243825.

Howard, P.H. and S. Banerjee. 1984. Interpreting Results from Biodegradability Tests of Chemicals in Water and Soil. *Environ. Toxicol. Chem.* 3:551-562.

Howard, P.H., A.E. Hueber, B.C. Mulesky, J.C. Crisman, W.M. Meylan, E. Crosbie, D.A. Gray, G.W. Sage, K. Howard, A. LaMacchia, R.S. Boethling, and R. Troast. 1986. BIOLOG, BIODEG, and fate/expos: new files on microbial degradation and toxicity as well as environmental fate/exposure of chemicals. *Environ. Toxicol. Chem.* 5:977-80. (Available http://esc.syrres.com)

Howard, P.H., A.E. Hueber, and R.S. Boethling. 1987. Biodegradation data evaluation for structure/biodegradation relations. *Environ. Toxicol. Chem.* 6:1-10.

Howard, P.H., R.S. Boethling, W.M. Stiteler, W.M. Meylan, A.E. Hueber, J.A. Beauman, and M.E. Larosche. 1992. Predictive model for aerobic biodegradability developed from a file of evaluated biodegradation data. *Environ. Toxicol. Chem.* 11:593-603.

Hurst, C.J. (Editor-in-chief). 1997. *Manual of Environmental Microbiology.* American Society of Microbiology Press, Washington, DC.

Johnson, B. 1991. *Setting Revised Specific Numerical Values, April 1991, Pursuant to the Pesticide Contamination Prevention Act.* California Department of Food and Agriculture, Environmental Monitoring and Pest Management Branch, EH 91-6 Sacramento, CA.

Klopman, G., D.M. Balthasar, and H.S. Rosenkranz. 1993. Application of the computer automated structure evaluation (CASE) program to the study of structure biodegradation relationships of miscellaneous chemicals. *Environ. Toxicol. Chem.* 12:231-240.

Klopman, G., Z. Zhang, D.M. Balthasar, and H.S. Rosenkranz. 1995. Computer-automated predictions of aerobic biodegradation of chemicals. *Environ. Toxicol. Chem.* 14:395-403.

Ladd, T.I., R.M. Ventullo, P.M. Wallis, and J.W. Costerton. 1982. Heterotropic activity and biodegradation of labile and refractory compounds in groundwater and stream microbial populations. *Appl. Environ. Microbiol.* 44:321-329.

Langenberg, J.H., W.J.G.M. Peijnenburg, and E. Rorije. 1996. On the usefulness and reliability of existing QSBRs for risk assessment and priority setting. *SAR and QSAR in Environ. Res.* 5:1-16.

Macfadyen, A. 1957. *Animal Ecology, Aims and Methods.* Pitman and Sons, London, England.

Mani, S.V., D.W. Connell, and R.D. Braddock. 1991. Structure activity relationships for prediction of biodegradability of environmental pollutants. *Crit. Rev. Environ. Contr.* 21:217-236.

Mudder, T.I. 1981. *Development of empirical structure-biodegradability relationships and testing protocol for slightly soluble and volatile priority pollutants.* Order No. 8123345. University Microfilms International, Ann Arbor, MI.

Niemi, G.J., G.D. Veith, R.R. Regal, and D.D. Vaishnav. 1987. Structural features associated with degradable and persistent chemicals. *Environ. Toxicol. Chem.* 6:515-527.

OECD (Organization for Economic Cooperation and Development). 1981. *Guidelines for Testing of Chemicals.* Paris, France.

Pagga, U. 1997. Standardized tests on biodegradability. In Hales, S.G., T. Feijtel, H. King, K. Fox, and W. Verstraete, Eds., *Biodegradation Kinetics: Generation and Use of Data for Regulatory Decision Making.* SETAC-Europe Workshop - Port-Sunlight, UK 4-6 Sept. 1996, SETAC-Europe Brussels, pp. 69-80.

Paris, D.F., W.C. Steen, G.L. Baughman and J.T. Barnett, Jr. 1981. Second-order model to predict microbial degradation of organic compounds in natural waters. *Appl. Environ. Microbiol.* 41:603-609.

Paris, D.F., N.L.Wolfe, W.C. Steen, and G.L. Baughman. 1983. Effect of phenol molecular structure on bacterial transformation rate constants in pond and river samples. *Appl. Environ. Microbiol.* 45:1153-1155.

Paris, D.F., N.L. Wolfe, and W.C. Steen 1984. Microbial transformation of esters of chlorinated carboxylic acids. *Appl. Environ. Microbiol.* 47:7-11.

Paris, D.F. and N.L. Wolfe. 1987. Relationship between properties of a series of aniline and transformation by bacteria. *Appl. Environ. Microbiol.* 53:911-916

Pfaender, F.K., and G.W. Bartholomew. 1982. Measurement of aquatic biodegradation rates by determining heterotrophic uptake of radiolabelled pollutants. *Appl. Environ. Microbiol.* 44:159-164.

Pitter, P. 1976. Determination of biological degradability of organic substances. *Water Res.* 10:231-235.

Pitter, P., and J. Chudoba. 1990. *Biodegradability of Organic Substances in the Aquatic Environment.* CRC Press, Boca Raton, FL.

Punch, W.F., L.J. Forney, A. Patton, K. Wright, P. Masscheleyn, and R.J. Larson. 1996. BESS, A computerized system for predicting the biodegradation potential of new and existing chemicals. *7th International Workshop on QSARs in Environmental Science.* June 24–28, Elsinore, Denmark.

Schmidt, S.K., S. Simkins, and M. Alexander. 1985. Models for the kinetics of biodegradation of organic compounds not supporting growth. *Appl. Environ. Microbiol.* 50:323-331.

Scow K.M. 1982. Rate of biodegradation. In: W.J. Lyman, W.F. Reehl, and D.H. Rosenblatt, Ed., *Handbook of Chemical Property Estimation Methods: Environmental Behavior of Organic Compounds*, pp 9-1–9-85. McGraw-Hill, New York.

Simkins, S. and Alexander, M. 1984. Models for mineralization kinetics with the variables of substrate concentration and population density. *Appl. Environ. Microbiol.* 47:1299-1306.

Spain, J.C. and P.A. Van Veld. 1983. Adaptation of natural microbial communities to degradation of xenobiotic compounds: Effects of concentration, exposure time, inoculum and chemical structure. *Appl. Environ. Microbiol.* 45:428-435.

Struijs, J. and R. Van den Berg. 1995. Standardized biodegradability tests: Extrapolation to aerobic environments. *Water Res.* 29:255-262.

Syracuse Research Corporation. 1996. BIODEG Program. Estimation of Biodegradation Probability, (computer software for MS-DOS & MS-Windows 3.1). http://esc.syrres.com. North Syracuse, NY.

Urushigawa, Y. and Y. Yonezawa. 1979. Chemicobiological interactions in biological purification system. VI. Relation between biodegradation rate constants of di-n-alkyl phthalate esters and their retention times in reverse phase partition chromatography. *Chemosphere* 3:139-142.

Vaishnav, D.D., R.S. Boethling, and L. Babeu. 1987. Quantitative structure-biodegradability relationships for alcohols, ketones and alicyclic compounds. *Chemosphere* 16:695-703.

Wiedemeier, T.H., M.A. Swanson, J.T. Wilson, D.H. Kampbell, R.N. Miller, and J.E. Hansen. 1996. Approximation of biodegradation rate constants for monoaromatic hydrocarbons (BTEX) in ground water. *Ground Water Monit. Remed.* 16:186-194.

Wolfe, N.L., D.F. Paris, W.C. Steen, and G.L. Baughman. 1980. Correlation of microbial degradation rates with chemical structure. *Environ. Sci. Technol.* 14:1143-1144.

WPCF (Water Pollution Control Federation – Biodegradability Subcommittee). 1967. Required characteristics and measurement of biodegradability. *J. Water Pollut. Control Fed.* 39:1232-1235.

Yonezawa, Y. and Y. Urushigawa. 1979. Chemicobiological interactions in biological purification system. V. Relation between biodegradation rate constants of aliphatic alcohols by activated sludge and their partition coefficients in 1-octanol-water system. *Chemosphere* 3:139-142.

13

Hydrolysis

N. Lee Wolfe and Peter M. Jeffers

CONTENTS

1-56670-456-1/00/$0.00+$.50
© 2000 by CRC Press LLC

13.1 Introduction

Hydrolysis is a bond-making, bond-breaking process in which a molecule, RX, reacts with water, forming a new R-O bond with the oxygen atom from water and breaking the R-X bond in the original molecule (March 1977). One possible pathway is the direct displacement of X with OH, as Equation (1) shows.

$$RX + H_2O \quad \rightarrow \quad ROH + HX \tag{1}$$

Hydrolytic processes provide the baseline loss rate for any chemical in an aqueous environment. Although various hydrolytic pathways account for significant degradation of certain classes of organic chemicals, other organic structures are completely inert. Strictly speaking, hydrolysis should involve only the reactant species water provides — that is, H^+, OH^-, and H_2O — but the complete picture includes analogous reactions and thus the equivalent effects of other chemical species present in the local environment, such as SH^- in anaerobic bogs, Cl^- in sea water, and various ions in laboratory buffer solutions.

Methods to predict the hydrolysis rates of organic compounds for use in the environmental assessment of pollutants have not advanced significantly since the first edition of the Lyman Handbook (Lyman et al., 1982). Two approaches have been used extensively to obtain estimates of hydrolytic rate constants for use in environmental systems. The first and potentially more precise method is to apply quantitative structure/activity relationships (QSARs). To develop such predictive methods, one needs a set of rate constants for a series of compounds that have systematic variations in structure and a database of molecular descriptors related to the substituents on the reactant molecule. The second and more widely used method is to compare the target compound with an analogous compound or compounds containing similar functional groups and structure, to obtain a less quantitative estimate of the rate constant.

Predictive methods can be applied for assessing hydrolysis for simple one-step reactions where the product distribution is known. Generally, however, pathways are known only for simple molecules. Often, for environmental studies, the investigator is interested in not only the parent compound but also the intermediates and products. Therefore, estimation methods may be required for several reaction pathways.

Some preliminary examples of hydrolysis reactions illustrate the very wide range of reactivity of organic compounds. For example, triesters of phosphoric acid hydrolyze in near-neutral solution at ambient temperatures with half-lives ranging from several days to several years (Wolfe, 1980), whereas the halogenated alkanes pentachloroethane, carbon tetrachloride, and hexachloroethane have "environmental" (pH = 7; 25°C) half-lives of about 2 hr, 50 yr, and 1000 millennia, respectively (Mabey and Mill, 1978; Jeffers et al., 1989). On the other hand, pure hydrocarbons from methane through the PAHs are not hydrolyzed under any circumstances that are environmentally relevant.

Hydrolysis can explain the attenuation of contaminant plumes in aquifers where the ratio of rate constant to flow rate is sufficiently high. Thus 1,1,1-trichloroethane (TCA) has been observed to disappear from a mixed halocarbon plume over time, while trichloroethene and its biodegradation product 1,2-dichloroethene persist. The hydrolytic loss of organophosphate pesticides in sea water, as determined from both laboratory and field studies, suggests that these compounds will not be long-term contaminants despite runoff into streams and, eventually, the sea (Cotham and Bidleman, 1989). The oceans also can provide a major sink for atmospheric species ranging from carbon tetrachloride to methyl bromide. Loss of methyl bromide in the oceans by a combination of hydrolysis and Cl⁻ for Br- exchange constitutes a significant contribution to the total degradation and is a key factor in modeling atmospheric concentrations and balance schemes. It is therefore an important part of the assessment of stratospheric ozone depletion potential (Jeffers and Wolfe, 1996a).

13.2 Background: Kinetic Characteristics and Mechanisms of Hydrolysis

Hydrolysis of certain classes of compounds or of certain members of a class occurs by any or all of the processes of acid, neutral, or basic reaction, shown as a kinetic expression in Equation(2) for loss of reactant X:

$$-d[X]/dt = k_{H^+} [H^+][X] + k_{H2O} [H_2O][X] + k_{OH^-} [OH^-][X] \tag{2}$$

The first term, representing acid-"catalyzed" hydrolysis, is important in reactions of carboxylic acid esters but is relatively unimportant in loss of phosphate triesters and is totally absent for the halogenated alkanes and alkenes. Alkaline hydrolysis, the mechanism indicated by the third term in Equation (2), dominates degradation of pentachloroethane and 1,1,2,2-tetrachloroethane, even at pH 7. Carbon tetrachloride, TCA, 2,2-dichloropropane, and other "gem" haloalkanes hydrolyze only by the neutral mechanism (Fells and Molewyn-Hughes, 1958; Molewyn-Hughes, 1953). Monohaloalkanes show alkaline hydrolysis only in basic solutions as concentrated as 0.01–1.0 molar OH- (Mabey and Mill, 1978). In fact, the terms in Equation(2) can be even more complex: both elimination and substitution pathways can operate, leading to different products, and a true unimolecular process can result from initial bond breaking in the reactant molecule.

In addition, as mentioned above, kinetic terms such as:

$$k_{SH^-}[SH^-][X], \ k_{Cl^-}[Cl^-][X], \ k_{B^-} [B^-][X]$$

where B⁻ represents one or a sum of buffer anions, may be required to represent the loss due to nucleophilic attack by these other aqueous reactants. Note that, although the water-promoted process is bimolecular, it appears as a pseudo-first order kinetic process. Another point worthy of mention is that, since all quantities in brackets in Equation (2) represent *concentrations*, the H⁺ or OH⁻ values deduced from pH measurement or buffer calculations must be converted from activities to true molar concentrations, usually a minor correction for dilute solutions.

Different hydrolytic reactions can lead to different products. 1,2-Dichloroethane hydrolyzes in neutral to acidic solutions by nucleophilic attack of H_2O to yield ethylene glycol, but in basic solutions, such as might be found in limestone-rich areas, hydroxyl ions promote the elimination of HCl and the formation of vinyl chloride, a product of considerably greater environmental concern.

Each of the rate constants written above (k_{H^+}, k_{H2O}, k_{OH^-}) can be assumed to show Arrhenius temperature dependence, as in Equation (3):

$$k = A \, e^{-(E/RT)} \ \text{or,} \ \ln (k) = \ln (A) - E/RT \tag{3}$$

where A is called the frequency factor, E is the activation energy and has units consistent with the value of R, the gas constant, and T is the absolute temperature. Theoretical formulations of chemical kinetics relate the frequency factor to entropic effects and the activation energy to enthalpy or bond strengths. Lower reactivity can reflect either high E or low A values or both.

For example, most of the greater reactivity of brominated alkanes and alkenes relative to their chlorinated analogs is due to lower activation energies. However, the environmental half-lives of TCA, 1,1-dichloro-1-fluoroethane, and 1,1-difluoro-1-chloroethane are about 1 yr, 18 yr, and 16,000 yr, respectively, and reflect differences of two to three orders of magnitude in the A factors. Many small fluorinated or partially fluorinated halocarbons form stable hydrates with water, indicating a highly ordered and tightly held hydration sphere that may be responsible for the low observed hydrolysis rates. Note that the interplay of energetics and entropic factors is complex. The differences in bond strengths for C-F, C-Cl, C-Br are far larger than any observed differences in hydrolytic activation energies for analogous compounds.

The next sections discuss the general features of hydrolytic reactions various classes of compounds for which the process is significant are presented below.

13.2.1   Carboxylic Acid Esters

Carboxylic acid esters have been studied widely and detailed mechanisms have been worked out for these reactions. Equation (4) shows the overall reaction of a carboxylic acid ester with hydroxide via the nucleophilic attack of hydroxide at the carbonyl oxygen, with the subsequent loss of the alcohol moiety (HOR). Esters generally have larger k_{OH-} than k_{H+} values, with the result that they hydrolyze by base-promoted reactions at pH 5-6 (Mabey and Mill, 1978).

$$
\begin{array}{ccc}
O & & O \\
\| & & \| \\
R\text{--}C\text{--}OR + OH^- & \longrightarrow & R\text{--}C\text{--}OH + HOR
\end{array}
\qquad (4)
$$

13.2.2   Amides

Hydrolysis of amides results in the formation of a carboxylic acid and an amine (Equation (5)):

$$
\begin{array}{ccc}
O \quad R2 & & O \qquad R2 \\
\| \quad | & & \| \qquad | \\
R1\text{-}C\text{-}N\text{-}R3 + H_2O & \longrightarrow & R1\text{-}C\text{-}OH + HN\text{-}R3
\end{array}
\qquad (5)
$$

In general, amides are much less hydrolytically reactive than esters. Typical hydrolysis half-lives under conditions common to aquatic environments range from hundreds to thousands of years. Hydrolysis of amides generally requires acid or base for the reactions to achieve measurable rates.

13.2.3   Halocarbons

Equation (6) gives the overall reaction of methyl bromide with water to form the corresponding alcohol. In freshwater aquatic systems, water is the most important nucleophile, and methyl bromide is hydrolyzed to methanol. On the other hand, in sea water where there are high concentrations of Cl⁻, nucleophile substitution by Cl⁻ gives methyl chloride in a competing reaction (Equation (7)):

$$CH_3Br + H_2O \longrightarrow CH_3\text{-}OH + HBr \tag{6}$$

$$CH_3Br + Cl^- \longrightarrow CH_3Cl + Br^- \tag{7}$$

For polyhalogenated alkanes, the most general elimination reaction is the 1,2 elimination (Equation (8)). These reactions can be mediated by water, acid, or base. However, base-mediated reactions are studied most widely.

$$
\begin{array}{c}
X \\
| \\
H_2C\text{-}CH_2 + OH^- \longrightarrow H_2C\text{=}CHX + X^- + H_2O \\
| \\
X
\end{array}
\tag{8}
$$

A second type of elimination reaction for some halocarbons is the 1,1 elimination. For example, hydrolysis of the halogenated methanes or haloforms is thought to occur by proton abstraction and subsequent formation of a carbene that reacts with water or hydroxide to form carbon monoxide and water (Equations (9)-(11)).

$$CHCl_3 + HO^- \longrightarrow CCl_3^- + H_2O \tag{9}$$

$$CCl_3^- \longrightarrow :CCl_2 + Cl^- \tag{10}$$

$$:CCl_2 + 2\,HO^- \longrightarrow CO + 2\,Cl^- + H_2O \tag{11}$$

13.2.4 Epoxides

The hydrolysis of epoxides can occur through neutral, acid-, or base- mediated reactions. Because the acid and neutral processes generally dominate over the range of environmental pH, the base-mediated reaction often can be ignored. The products of hydrolysis are usually the corresponding diol and sometimes rearranged products (Equation (12)).

$$
\begin{array}{ccc}
O & HO\ OH & O \\
/\ \backslash \quad H_2O & |||\ ||| & || \\
R1R2C\text{-}CR3R4 \longrightarrow R1R2C\text{-}CR3R4 & + & R1R2R3C\text{-}C\text{-}R4
\end{array}
\tag{12}
$$

13.2.5 *Nitriles*

Nitriles undergo both acid and alkaline hydrolysis to give the corresponding amide first, and then the carboxylic acid ester and ammonia (equations (13) and (14)):

$$
\begin{array}{c}
O \\
|| \\
R\text{-}C\text{≡}N + OH^- \longrightarrow R\text{-}C\text{-}NH_2
\end{array}
\tag{13}
$$

$$RC(\text{=}O)NH_2 + OH\text{-} \longrightarrow R\text{-}C(\text{=}O)\text{-}O\text{-} + NH_3 \tag{14}$$

13.2.6   Carbamates

Carbamates, widely used as pesticides, can undergo facile hydrolysis, depending on the substituents on the N atom. Carbaryl (Ar = napthyl), used as an insecticide, undergoes rapid alkaline hydrolysis at room temperature even at pH 7 (Equation (15)).

$$
\begin{array}{c}
\quad\quad O \\
\quad\quad \| \\
Ar\text{-}N\text{-}C\text{-}O\text{-}R + OH^- \longrightarrow Ar\text{-}N\text{-}H + CO_2 + HO\text{-}R \\
\quad | \quad\quad\quad\quad\quad\quad\quad\quad\quad\quad | \\
\quad H \quad\quad\quad\quad\quad\quad\quad\quad\quad\quad H
\end{array}
\tag{15}
$$

When an alkyl substituent is present on the N atom, hydrolysis is much slower. For example, carbaryl at pH 7 has a half-life of a few hours, depending on pH, whereas chlorpropham (Ar = p-chlorophenyl) has a half-life 6 to 7 orders of magnitude longer (Equation (16)) under the same conditions (Wolfe et al., 1978):

$$
\begin{array}{c}
\quad\quad O \\
\quad\quad \| \\
Ar\text{-}N\text{-}CO\text{-}R + OH^- \longrightarrow Ar\text{-}N\text{-}H + CO_2 + HOR \\
\quad | \quad\quad\quad\quad\quad\quad\quad\quad\quad\quad | \\
\quad CH_3 \quad\quad\quad\quad\quad\quad\quad\quad\quad CH_3
\end{array}
\tag{16}
$$

The difference is in the mechanism of reaction. In the case of carbaryl, the reaction proceeds by an elimination process in which the proton acidity on the nitrogen atom determines the reactivity. On the other hand, the chlorpropham reaction proceeds in a manner analogous to the hydrolysis of carboxylic acid esters. Much like carboxylic acid esters, electron-withdrawing substituents in carbamates accelerate the reaction by an amount that depends on whether the substituents are on N or O. Conversely, electron-donating substituents (methyl in the case of chlorpropham, above) slow the rate of hydrolysis.

13.2.7   Sulfonylureas

Hydrolysis appears to be a very significant pathway for the fate of sulfonylurea pesticides in sediments and natural waters. Although these chemicals often have many different functional groups, the principal cleavage occurs at the sulfonylurea bridge. The reaction is highly pH dependent (Equation (17)):

$$
\begin{array}{c}
\quad O\ H \quad\quad\quad\quad\quad\quad\quad\quad\quad\quad\quad O\ H \\
\quad \|\ | \quad\quad\quad\quad\quad\quad\quad\quad\quad\quad\quad \|\ | \\
R\text{-}SO_2\text{-}C\text{-}N\text{-}R + H^+ \longrightarrow R\text{-}SO_2\text{-}OH + HO\text{-}C\text{-}N\text{-}R
\end{array}
\tag{17}
$$

13.2.8   Organophosphate Esters

Organophosphorus compounds are used widely as pesticides and some also have other industrial uses. These diagrams shows the generic chemical structures for the major classes of organophosphorus pesticides:

R1-O   R1-O

```
        \                              \
O = P–O--R2                    S = P--O--R2
      /                              /
  R1-O                            R1-O
organophosphate            organophosphorothioate


  R1-O                            R1-O
      \                              \
    O = P–S--R2                    S = P–S--R2
      /                              /
  R1-O                            R1-O
organophosphorothionate        organophosphorodithioate
```

Mechanistic studies have shown that hydrolysis of organophosphate esters can occur by direct nucleophilic attack at the P atom without the formation of a pentavalent intermediate, mainly via reaction with OH^- (Equation (18)):

$$
\begin{array}{cc}
R1\text{-}O & R1\text{-}O \\
\backslash & \backslash \\
O = P\text{-}O\text{-}CH_2\text{-}R2 + OH^- \longrightarrow O = P\text{-}OH + HOCH_2R2 \\
/ & / \\
R1\text{-}O & R1\text{-}O
\end{array}
\qquad (18)
$$

Nucleophilic attack by water at one of the methyl or ethyl substituents also may occur (Lacorte et al., 1995) (Equation (19)):

$$
\begin{array}{cc}
R1\text{-}O & R1\text{-}O \\
\backslash & \backslash \\
O{=}P\text{-}O\text{-}CH_2\text{-}R2 + H_2O \longrightarrow O{=}P\text{-}O\text{-}CH_2\text{-}R2 + R1HOR \\
/ & / \\
R1\text{-}O & HO
\end{array}
\qquad (19)
$$

13.3 Test Methods

Mill et al. (1982) developed a protocol for measuring hydrolysis rates for organic compounds. Ellington et al. (1986,1987), following this protocol, measured degradation rate constants at pH values of 3, 7, 9, and 11 (achieved with appropriate buffers) and at temperatures from 20° to 87°C, and they reported rate constants for 80 organic compounds of potential concern for landfill disposal. Jeffers and Wolfe (1996b) developed another approach, using dilute solutions of HCl or NaOH to achieve the desired $[H^+]$ or $[OH^-]$ values, rather than buffers. If reactant concentrations are significantly lower than $[H^+]$ or $[OH^-]$, then the reactions can be treated as pseudo first-order, and the experimental rate constant can be evaluated as the slope of the plot of ln (reactant) versus time. For example, for halocarbons where only neutral and alkaline processes contribute, Equation (2) reduces to Equation (20), which upon integration yields Equation (21):

$$
-d[X]/dt = \{k_{H2O}[H_2O] + k_{OH^-}[OH^-]\}[X] = k_{obs}[X] \qquad (20)
$$

$$\ln [X] = \ln [X]_o - k_{obs} t. \tag{21}$$

In dilute acid, $[OH^-]$ is sufficiently small that neutral hydrolysis dominates, and since $[H_2O]$ is constant, $k_{obs} = k_{H2O}[H_2O] = k_{H2O}$, the neutral hydrolysis rate constant. In basic solutions, $k_{OH^-}[OH^-] \gg k_{H2O}$, so that $k_{obs} = k_{OH^-}[OH^-]$.

13.4 Extrapolation

The test methods just discussed imply that many of the laboratory measurements will be performed under conditions far different from those in the environment. There are questions about the validity of extrapolations, and of the validity of using laboratory numbers, however extrapolated, in real applications as complex as the seas, brackish and turbid surface waters, or aquifers. Haag and Mill (1988) observed augmented hydrolytic degradation of epoxides in contact with subsurface sediments, although they found no change in the rate of hydrolysis of TCA under similar conditions. Jeffers et al. (1994) found no measurable effect of high ionic strength or of the presence of crushed minerals, including sulfides, oxides, hydroxides, and aquifer materials on either the neutral or alkaline hydrolysis rates of typical halogenated hydrocarbons. These observations are consistent with the predictions of Absolute Reaction Rate Theory, that an ionic strength effect should occur only if both reactants involved in forming the activated complex are charged ions (Steinfeld et al., 1989).

For compounds like TCA, for which both high temperature and room temperature hydrolysis rates have been measured and contaminant plumes have been observed over a considerable length of time, extrapolation using the Arrhenius parameters appears valid. In addition, apparent augmentation of reactivity in real environmental situations often can be rationalized in terms of additional reaction pathways, so that it is not necessary to invoke the non-validity of the extrapolation process.

13.5 Overview of Estimation Methods

Quantitative structure/activity relationships (QSARs) for hydrolysis are based on the application of linear free energy relationships (LFERs) (Well, 1968). An LFER is an empirical correlation between the standard free energy of reaction (ΔG_o), or activation energy (E_a) for a series of compounds undergoing the same type of reaction by the same mechanism, and the reaction rate constant. The rate constants vary in a way that molecular descriptors can correlate.

For information about the theoretical underpinnings of LFERs, which we do not cover here, the reader is referred to several comprehensive treatments of the subject. Also, Harris (1982) has reviewed estimation methods related to hydrolysis reactions; in general, the methods discussed in that review remain valid. Here we summarize recent applications of hydrolysis QSARs and focus on their practical application to hydrolysis reactions in water.

13.5.1 QSARs

The Hammett equation (Equation (22)) was developed to correlate reactivity in aromatic compounds. It is

$$\log k = \rho\sigma + \log k_o \tag{22}$$

where k is the first- or second-order rate constant, ρ is the sensitivity of the reaction to substitution, σ is the substituent constant and k_o is the reference compound rate constant. The Taft equation (Equation (23)) was developed to correlate reactivity among aliphatic compounds for which steric as well as electronic effects are important. It is

$$\log k = \rho^*\sigma^* + \delta E_s + \log k_o \tag{23}$$

where k is the first- or second-order rate constant, ρ^* is the sensitivity of the reaction to substitution, δ is the sensitivity of the reaction to steric effects, σ^* is the Taft substituent constant, a measure of the polar effect, E_s is a measure of steric effects, and $\log k_o$ is the reference compound rate constant.

In cases where the steric effect is dominant; i.e., when $\rho^* = 0$, Equation (23) reduces to Equation (24):

$$\log k = \delta E_s + \log k_o \tag{24}$$

Unfortunately, only a few of these correlations were developed using water as a solvent and extrapolation of rate constants from organic solvents to water can be difficult. Most hydrolysis data are from measurements of test compounds in water or water-solvent mixtures (Table 13.1).

Several good correlations have been developed using the pK_a of the leaving group as a molecular descriptor (Table 13.2). For some compounds, particularly more complex molecules, pK_a values are more accessible than substituent constants (Perrin et al. 1981).

13.5.2 Elimination Reactions

Although dehydrohalogenation of polyhalogenated hydrocarbons is not hydrolysis by the strictest definition, it is an environmentally important reaction in water because many halogenated compounds are environmentally important. Equation (25) shows dehydrohalogenation:

$$
\begin{array}{ccc}
\overset{\displaystyle H}{\underset{\displaystyle X}{\overset{\textstyle|}{\underset{\textstyle|}{C-C}}}} & \xrightarrow[\text{or}]{\;\; ^-:OH \;\;} & \overset{\displaystyle\diagdown\;\;\diagup}{\underset{\displaystyle\diagup\;\;\diagdown}{C=C}} + X^- + H_2O \\
& :OH_2 &
\end{array}
\tag{25}
$$

Roberts et al. (1993) have provided an in-depth analysis of QSARs for dehydrohalogenation reactions of polychlorinated and polybrominated alkanes. The QSARs were developed based on a dataset of 28 polychlorinated and polybrominated compounds in aqueous solution at 25° C. The QSARs are for the OH– and water-mediated second-order elimination reactions (E_2), as Equation (25) shows. The first QSAR is for base-promoted dehydrochlorination and is based on the inductive parameter σ_I (Equations (26) and (27)):

TABLE 13.1

Summary of QSARs for Hydrolysis Reactions Based on Hammett and Taft
Correlations (Harris, 1982).

Rate Constant	Type of Reaction	Descriptors	Solvent	Temp
k_H	benzamides	Hammett	water	
	($\log k_H = 0.12\ \sigma - 3.51$)			
	benzamides	Hammet		
	($\log k_H = 0.11\ \sigma + -7.00$)			
	benzsulfonates			
	($\log k_H = 0.60\ \sigma - 3.92$)			
	benzamides	Taft		
	($\log k_H = 0.81 E_s - 4.48$)			
k_O	benzyl halides	Hammett		
	($\log k_O = -1.31\sigma - 5.21$)			
	benzyl halides			
	($\log k_O = -4.48\sigma - 3.95$)			
	benzyl tosylates			
	($\log k_O = -2.32\sigma - 3.58$)			
k_{OH}	methyl benzoates	Hammett		
	($\log k_{OH} = 2.38\sigma - 2.14$)		60% acetone	25°
	($\log k_{OH} = 1.17\sigma - 2.26$)		3% acetone	25°
	ethyl benzoates			
	($\log k_{OH} = 2.47\sigma - 2.62$)		60% acetone	25°
	ethyl benzylates xx			
	($\log k_{OH} = 1.00\sigma - 1.36$)		60% acetone	25°
	ethyl phenylacetates xx			
	($\log k_{OH} = 1.24\sigma - 2.86$)		85% ethanol	25°
	benzamides			
	($\log k_{OH} = 1.40\sigma - 5.12$)		60% ethanol	53°
	methyl phenoxyalates xx			
	($\log k_{OH} = 1.51\sigma - 0.62$)			
	methyl benzylates		60% acetone	25°
	($\log k_{OH} = 1.17\sigma - 2.26$)			
	tert–benzamides		10% ethanol	25°
	($\log k_{OH} = 1.14\sigma - 2.59$)			
	sec–benzamides		10% ethanol	38°
	($\log k_{OH} = 2.69\sigma - 2.44$)			
	benzyl chlorides		water	—
	($\log k_{OH} = -0.33\sigma - 5.48$)			
	ethyl benzosilates xx		50% ethanol	25o
	($\log k_{OH} = 1.4\Sigma\sigma - 0.47$)			

$$\log (k_{OH^-}/k_{H2O}) = \rho_{I*}\Sigma.\sigma_I + constant \tag{26}$$

$$\log(k_{OH^-}/k_{H2O}) = 8.07\ (\pm2.36)\ \Sigma.\sigma_I + 1.40\ (\pm2.40) \tag{27}$$

The second QSAR is for water-promoted dehydrobromination and also is based on σ_I but is much more sensitive to inductive effects of other substitutents ($\Sigma.\sigma_I$), as equations (28) and (29) show:

$$\log (k_{OH^-}/k_{H2O}) = \rho_{I*}\Sigma.\sigma_I + constant \tag{28}$$

$$\log (k_{OH^-}/k_{H2O}) = 11.76\ (\pm1.88)\ \Sigma.\sigma_I - 0.51\ (\pm1.26) \tag{29}$$

Note that log (k_{OH^-}/k_{H2O}) versus $\Sigma.\sigma_I$ produces a better fit to the experimental data for both the dechlorination and debromination reactions than does log k_{OH^-} or log k_{H2O} versus $\Sigma.\sigma_I$. This is probably because the former approach eliminates steric effects.

TABLE 13.2

Summary of Correlations that Use the pK$_a$ of the Leaving Group as the Molecular Descriptor (Harris, 1982).

Rate Constant	Type of reaction	Descriptors	Solvent	Temp
k_{OH}	phthalate esters	pK$_a$	water	25°
	phosphonates		water	65°
	(log k_{OH} = −0.68 pKa + 8.9)			
	N–phenylcarbamates		water	25°
	(log k_{OH} = −1.15 pKa + 13.6)			
	N–methyl–N–phenylcarbamates		water	25°
	(log k_{OH} = −0.26 pKa −1.3)			
	N–methylcarbamates		water	25°
	(log k_{OH} = −.91 pKa + 9.3)			
	N,N–dimethylcarbamates		water	25°
	(log k_{OH} = −0.17 pKa −2.6)			
	dimethyl phosphates		water	27°
	(log k_{OH} = −0.28 pKa + 0.50)			
	diethyl phosphates		water	27°
	(log k_{OH} = −0.28 pKa − 0.22)			
	dimethyl thiophosphates		water	27°
	(log k_{OH} = −0.25 pKa + 3.4)			
	diethyl thiophosphates		water	27°
	(log k_{OH} = −0.21 pKa − 1.6)			

13.5.3 Other Approaches to QSARs

Although the use of substituent effects and pK$_a$s of leaving groups as described above is valuable for predicting hydrolysis rate constants, this approach is often of limited utility because the required parameters have been developed only for groups of compounds with a relatively narrow range of structural features. Moreover, substituent parameters are not always available and determining them can be time consuming.

Collette (1992) proposed a novel approach to developing and applying QSARs for the alkaline hydrolysis of carboxylic acid esters using infrared (IR) spectroscopic data. The concept is to extend QSARs to a larger and more diverse group of organic compounds by including the structural information encoded in IR spectra. Such data are available from spectral data bases, but even in the absence of literature data, IR frequencies are readily measurable.

Equation (30) represents a QSAR for the base-mediated hydrolysis of formates and acetates. The correlation is between the second-order alkaline hydrolysis rate constants and the linear combination of the shifts of the vC=O and vC–O stretching peaks for 12 of the 41 compounds in Table 13.3.

$$\log k_{OH^-} = C^*[\Delta v(C{=}O) - \Delta v(C{-}O)] + k_o \tag{30}$$

where k_{OH^-} is the second-order alkaline hydrolysis rate constant (M^{-1} sec^{-1}), C is the slope, $[\Delta v(C{=}O) - \Delta v(C{-}O)]$ is the difference between the carbonyl and alkoxy stretching frequencies, and k_o is the rate constant for methyl formate. Figure 13.1 is a plot of the log of the alka-

TABLE 13.3

Alkaline Hydrolysis Rate Constants (k_{OH}, M^{-1} s^{-1}, near 25°C) of the Compounds Used in the IR Spectral Frequency Correlations.

Compound Name	$\log^k_{OH}$	Ref
ethyl-n-butyrate	−1.26	8
methyl formate	1.56	11
benzyl acetate	−0.71	11
n-butyl acetate	−1.06	8
ethyl isobutyrate	−1.49	8
ethyl acetate	−0.96	8
ethyl benzoate	−1.50	8
n–propyl acetate	−1.06	8
methyl acetate	−0.74	8
isopropyl formate	1.04	11
ethyl bromoacetate	1.70	8
ethyl iodoacetate	1.21	8
ethyl formate	1.41	8
methyl benzoate	−1.10	8
ethyl chloroacetate	1.56	8
methyl methacrylate	−1.25	12
benzyl benzoate	−2.10	11
isopropyl acetate	−1.52	8
n–butyl formate	1.34	11
n–propyl formate	1.36	11
ethyl acrylate	−1.11	8
sec–butyl acetate	−1.76	11
ethyl 2–bromopropionate	1.00	83
–chlorethyl acetate	−0.41	83
–methoxyethyl acetate	−0.69	8
ethyl *p*–fluorobenzoate	−1.41	8
methyl *p*–fluorobenzoate	−1.15	*b*
ethyl dibromoacetate	2.31	8
methyl *p*–hydroxybenzoate	−1.52	*b*
methyl *p*–aminobenzoate	−2.35	*b*
isopropyl *p*–hydroxybenzoate	−2.23	*b*
ethyl *p*–nitrobenzoate	−0.13	8
ethyl *p*–aminobenzoate	−2.59	8
methyl *m*–aminobenzoate	−1.47	*b*
ethyl trichloroacetate	3.41	8
ethyl pivalate	−2.77	8
isopropyl *p*–aminobenzoate	−3.04	*b*
methyl 2,4–D[a]	1.06	12
ethyl aminoacetate	−0.19	83
–butoxy 2,4–D[a]	1.48	13
n–octyl 2,4–D[a]	0.57	13

[a] 2,4–D, (2,4–dichlorophenoxy)acetate. [b]$\log^k_{OH}$ calculated from chemical structure theory by SPARC (SPARC Performs Automated Reasoning in Chemistry) (Karickhoff et al., 1991).

line hydrolysis rate constant versus difference in carbon-oxygen frequencies for 12 carboxylic acid esters.

13.5.4   Sulfonylurea Herbicides

QSARs for hydrolysis are very limited for complex organic molecules. This is due to the difficulty of obtaining adequate molecular descriptors. The use of quantum-chemical

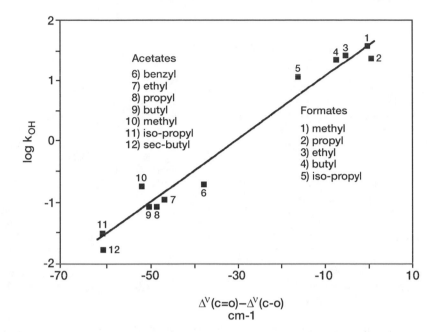

FIGURE 13.1
Log of the alkaline hydrolysis rate constant vs. difference in carbon-oxygen stretching frequency for 12 carboxylic acid esters.

molecular descriptors can sometimes provide the detailed molecular information needed to develop successful correlations, provided a dataset of hydrolysis rate constants is available. In addition, quantum mechanical descriptors have the advantage of being readily accessible for a wide variety of structures and parameters.

Berger and Wolfe (1996) reported a correlation of hydrolysis data for 12 sulfonylurea herbicides. The use of bond strength or Hammett σ constants was impossible because of the complex structures of the compounds. The hydrolysis pathways for this class of compounds are also more complex, but the use of quantum mechanical parameters provided the detailed structural information needed to develop a useful correlation. As a result of the many different functional groups, several reaction pathways are available depending on the substituents. Also, there is a complicating pH effect on the pathways and the kinetics of hydrolysis as shown by product studies. The 12 herbicides used in this study are listed in Table 13.4, and the pseudo first-order hydrolysis rate constants are given in Table 13.5. Figure 13.2 shows the basic structure of these compounds.

Parameters tested for the correlations were bond orders, atomic charges and lowest unoccupied molecular orbital (LUMO) parameters. Of the 12 sulfonylureas, a QSAR was developed for six of the compounds that hydrolyze exclusively at the sulfonylurea bridge. Equation (31) gives the best correlation for this group of compounds:

$$k_{(pH)} = -0.65 * LUMO - 0.62 \tag{31}$$

$$(r^2 = 0.776 \ n = 6)$$

where $k_{(pH)}$ is the acid-catalyzed hydrolysis pseudo first-order rate constant at pH 4.

13.5.5   Benzonitriles

TABLE 13.4

Sulfonylurea Herbicides Investigated for Hydrolysis and Reaction Pathways.

	CAS-RN	X	Y	Z	A	B
Bensulfuron-methyl	83055-99-6	C	H	COOCH$_3$[a]	OCH$_3$	OCH$_3$
Sulfometuron-methyl	74222-97-2	C	H	COOCH$_3$	CH$_3$	CH$_3$
Ethametsulfuron-methyl	97780-06-8	N	H	COOCH$_3$	OCH2CH$_3$	NHCH$_3$
Triasulfuron	82097-50-5	N	H	OCH$_2$CH$_2$Cl	OCH$_3$	CH$_3$
Tribenuron-methyl	101200-48-0	N	CH$_3$	COOCH$_3$	OCH$_3$	CH$_3$
Nicosulfuron	111991-09-4	C	H	CON(CH$_3$)$_2$[d]	OCH$_3$	OCH$_3$
Chlorimuron-ethyl	90982-32-4	C	H	COOCH$_2$CH$_3$	OCH$_3$	Cl
Thifensulfuron-methyl	79277-27-3	N	H	COOCH$_3$b	OCH$_3$	CH$_3$
Primisulfuron-methyl	86209-51-0	C	H	COOCH$_3$	OCF$_2$H	OCF$_2$H
Metsulfunon-methyl	74223-64-6	N	H	COOCH$_3$	OCH$_3$	CH$_3$
Chlorsulfuron	64902-72-3	N	H	Cl	OCH$_3$	CH$_3$
Amidosulfuron	120923-37-7	C	H	c	OCH$_3$	OCH$_3$

[a] Benzyl instead of phenyl ring.
[b] Thiophen instead of phenyl ring.
[c] N(CH$_3$)SO$_2$CH$_3$ instead of phenyl ring.
[d] Pyridine instead of phenyl ring.

Masunaga and Wolfe (1995) developed QSARs for hydrolysis of benzonitriles in aqueous buffers and then extended this approach to more complex systems, Equation (32) gives the QSAR for 14 *para*-substituted benzonitriles in water:

$$\log k = 1.64\ (\pm 0.42) \times \sigma_p - 1.37\ (\pm 0.17) \tag{32}$$

$$(r^2 = 0.0858 \quad n = 14)$$

where log k is the neutral hydrolysis rate constant and σ_p is the Hammett constant. Rate constants for hydrolysis of 12 *para*-substituted benzonitriles in sediments also were correlated with the Hammett constants (Masunaga et al., 1995). Here, correlation analysis explored the mechanisms of reaction of benzonitriles in anaerobic sediment systems. The correlation Equation (33) shows accounts for 73% of the variance of the data and is useful for estimating rate constants for other benzonitriles in anaerobic sediments:

$$\log k = 0.308 \times \log K_{ow} + 0.778 \times \sigma_p - 2.57 \tag{33}$$

$$(r^2 = 0.727 \quad n = 10)$$

where k is the first order rate constant, K_{ow} is the n-octanol/water partition coefficient, and σ_p is the Hammett constant.

13.6   Estimation of Hydrolysis Rate Constants Based on Analogy

Because of the large number of organic compounds and the diversity of their structures and reactivity, it often is not possible to use the more precise and reliable QSARs to estimate hydrolysis rates (Karickhoff et al., 1991). However, even for compounds for which no data

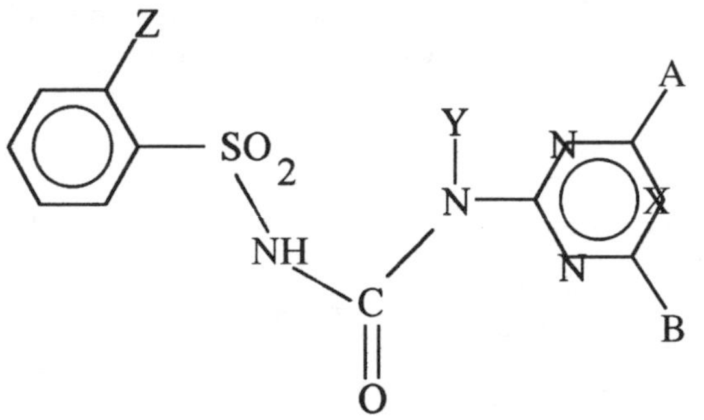

FIGURE 13.2
Generic chemical structure of sulfonylurea herbicides Berger and Wolfe (1996) studied.

TABLE 13.5

Pseudo First-Order Hydrolysis Rate Constants with Standard Errors [(k ± SE) × 10^{-2}] in Aqueous Buffer Solutions for 12 Sulfonylurea Herbicides at Different pHs and Temperatures.

	(k ± SE) × 10^{-2} (d^{-1}) pH 4, 22°C	(k ± SE) × 10^{-2} (d^{-1}) pH 4, 40°C	(k ± SE) × 10^{-2} (d^{-1}) pH 7, 40°C	(k ± SE) × 10^{-2} (d^{-1}) pH 10, 40°C
Bensulfuron-methyl	6.16 ± 0.22	114.91 ± 3.33	4.48 ± 0.19	0.98 ± 0.076
Sulfometuron-methyl	6.58 ± 0.020	129.21 ± 2.67	6.48 ± 0.18	3.76 ± 0.52
Ethametsulfuron-methyl	2.98 ± 0.29	60.35 ± 1.52	2.15 ± 0.028	1.62 ± 0.086
Triasulfuron	7.95 ± 0.23	96.31 ± 2.30	1.83 ± 0.033	1.31 ± 0.18
Tribenuron-methyl	a	a	119.38 ± 0.066	3.66 ± 0.62
Nicosulfuron	12.40 ± 0.29	128.16 ± 2.72	2.13 ± 0.022	1.34 ± 0.069
Chlorimuron-ethyl	15.36 ± 0.39	295.02 ± 20.04	2.47 ± 0.063	2.08 ± 0.070
Thifensulfuron-methyl	13.72 ± 0.47	143.76 ± 2.46	3.69 ± 0.092	21.76 ± 6.60
Primisulfuron-methyl	12.38 ± 0.53	214.49 ± 4.13	1.17 ± 0.040	0.73 ± 0.17
Metsulfuron-methyl	14.72 ± 0.51	133.23 ± 2.08	1.05 ± 0.049	1.31 ± 0.17
Chlorsulfuron	14.89 ± 0.248	138.69 ± 3.08	0.66 ± 0.048	1.26 ± 0.085
Amidosulfuron	9.89 ± 0.29	132.45 ± 17.94	1.44 ± 0.058	1.72 ± 0.072

[a] Reaction too fast for the convenient determination of the hydrolysis rate constant under these conditions.

or QSARs exist, one often can estimate hydrolytic activity by structural analogy to related compounds for which kinetic data exist. A recent EPA report (Kollig et al., 1993) used this approach extensively in assessing hydrolysis rate constants and reaction pathways. In that report, the authors assigned chemicals to one of three categories, NHFG, NLFG, and HG.

1. No hydrolyzable functional groups (NHFG)

   NHFG compounds are those that do not have any heteroatoms that can undergo hydrolysis over the pH range of 5 to 9 at 25°C. Examples include xylenes, carboxylic acids, and polycyclic aromatic hydrocarbons (PAHs).

2. No labile functional groups (NLFG)

   NLFG compounds contain one or more heteroatoms that can react, but they react so slowly over the pH range of 5 to 9 at 25°C that their half-lives will be greater than 50 years, if they react at all. Examples of these compounds include anilines/amines, halogenated aromatics, and ethers.

3. Hydrolyzable groups (HG)

HG compounds have functional groups more labile to hydrolysis. For compounds that can be deduced to be reactive but for which no measured or calculated rate constants can be obtained, rate constants can often be estimated semi-quantitatively by comparison to compounds for which hydrolysis data are available.

The general approach is straightforward. First, the reaction pathway(s) is outlined, based

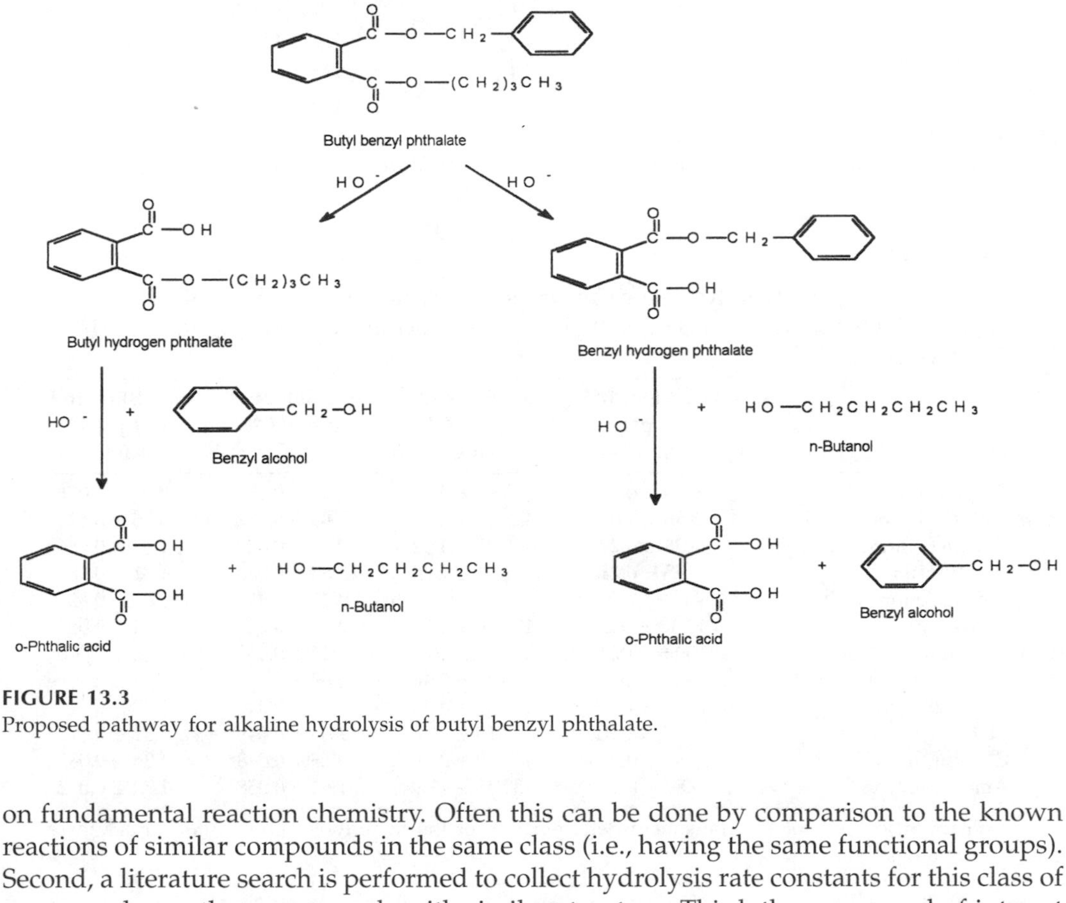

FIGURE 13.3
Proposed pathway for alkaline hydrolysis of butyl benzyl phthalate.

on fundamental reaction chemistry. Often this can be done by comparison to the known reactions of similar compounds in the same class (i.e., having the same functional groups). Second, a literature search is performed to collect hydrolysis rate constants for this class of compounds or other compounds with similar structure. Third, the compound of interest and its analogs are examined for similarity in structure and substituents, and an estimate of the rate constant(s) for the untested compound is made by interpolation from the analog data. Some illustrative examples from Kollig et al. (1993) appear below.

13.6.1   Carboxylic Ester Hydrolysis: Butyl Benzyl Phthalate

Butyl benzyl phthalate is a mixed carboxylic acid ester formed by the condensation of phthalic acid with two different alcohols. No literature values could be found nor any directly applicable QSARs. Based on known reactivity of carboxylic acid esters, any acid hydrolysis should be very slow. Neutral hydrolysis, while reported in the literature for certain reactive esters, is not likely for the phthalate esters. The carboxylic acid esters are most labile to alkaline hydrolysis (Mabey and Mill, 1978).

Figure 13.3 shows the proposed hydrolysis pathway for this compound which is comparable to that of the known pathway for bis(2-ethylhexyl)phthalate. There are two possible intermediate half-acids from the competitive hydrolysis. These half-esters can undergo further hydrolysis. Alkaline hydrolysis rate constants for four bis(alkyl)phthalates were retrieved from the literature. Based on analogy with these alkyl diesters, an alkaline hydrolysis rate constant of 1.2×10^5 M^{-1} y^{-1} was estimated for butyl benzyl phthalate. The alkaline hydrolysis rate constants for the intermediates butyl and benzyl hydrogen phthalate were then estimated to be 6×10^4 M^{-1} y^{-1} by assuming one-half the value for the parent compound.

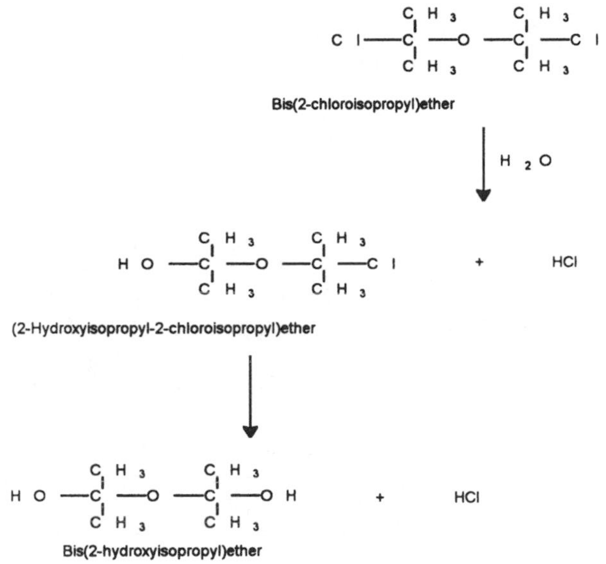

FIGURE 13.4
Proposed neutral hydrolysis pathway for bis(2-chloroisopropyl)ether.

13.6.2 Halohydrin Hydrolysis: Bis(2-chloroisopropyl)ether

Chlorohydrins are compounds characterized by alpha halo-alpha alkoxy groups bound to a common carbon atom. These compounds undergo rapid hydrolysis at this shared carbon atom. Bis(2-chloroisopropyl)ether, a chlorohydrin, has two such carbon atoms, and both react very rapidly with water. In fact, the reactions are so fast that acid and alkaline contributions have not been determined. It is likely, however, that base accelerates the reaction kinetics. The proposed reaction pathway for this compound is based on the reported pathway for bis(chloroethyl)ether (Figure 13.4). The reported rate constant k_n for bis(chloromethyl)ether, of 0.23 sec^{-1} was based on an observed half-life of a few minutes. Similarly, for bis(2-chloroisopropyl)ether, both of the chloro substituents are reactive, and a half-life of a few minutes can be assigned to this compound, as well.

13.6.3 Nitrile/Amide Hydrolysis: Methacrylonitrile

Nitriles are well known to undergo acid and alkaline hydrolysis (Figure 13.5), but there is only one recent report of neutral hydrolysis. Thus, methacrylonitrile, even though it is conjugated with an alkenyl substituent, should undergo both acid-catalyzed and base-mediated hydrolysis via a pathway similar to that for alkyl and aryl nitriles. The amide is a likely intermediate product for nitrile hydrolysis, but it is reactive as well and can undergo acid- and base-mediated hydrolysis to the carboxylic acid and ammonia. The

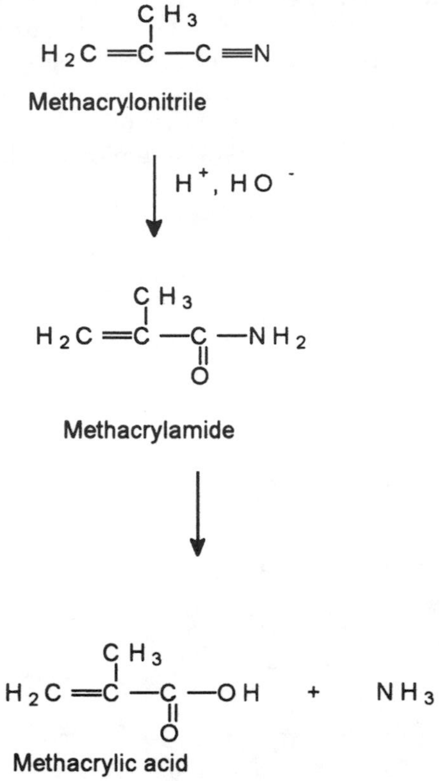

$$H_2C = \underset{\underset{CH_3}{|}}{C} - C \equiv N$$

Methacrylonitrile

$\downarrow$ H^+, HO^-

$$H_2C = \underset{\underset{CH_3}{|}}{C} - \underset{\underset{O}{\|}}{C} - NH_2$$

Methacrylamide

$\downarrow$

$$H_2C = \underset{\underset{CH_3}{|}}{C} - \underset{\underset{O}{\|}}{C} - OH \quad + \quad NH_3$$

Methacrylic acid

FIGURE 13.5
Proposed pathway for acid/base-mediated hydrolysis of methacrylonitrile.

literature contains extensive hydrolysis data, including QSARs for both nitriles and amides, but because of the unsaturation at the number 2 carbon atom in methacrylonitrile, these QSARs are not directly applicable. However, it is believed that the alkenyl substituent should have minimal influence on the acid and base reaction kinetics. Therefore, the rate constants $k_a = 5E2$ M^{-1} sec^{-1} and $k_b = 5.2E3$ M^{-1} sec^{-1} for methyl nitrile are used for the initial hydrolysis to the amide. Similarly, for methacrylamide hydrolysis, the rate constants for methyl amide are recommended: $k_a = 31.5$ M^{-1} sec^{-1} and $k_b = 1.8E-2$ M^{-1} sec^{-1}.

13.6.4   Halogenated Hydrocarbons: 1,2,3-Trichloropropane

Polyhalogenated hydrocarbons can undergo a variety of nucleophilic substitution and elimination reactions. In general, this class of compounds can undergo both neutral and base- mediated reactions, but the saturated compounds cannot undergo acid-mediated reactions. The proposed pathway for 1,2,3-trichloropropane is based on analogy with 1,2-dibromo-3-chloropropane and illustrates the multiple reactions and products that can occur (Figure 13.6). The final products are proposed to be 2-chloro-3-hydroxy-1-propene and glycerol. The route to glycerol proceeds through intermediate haloalcohols and halohydrins. The amount of the elimination product 2-chloro-3-hydroxy-1-propene that is formed will increase at higher pHs.

The disappearance rate constants were estimated to be $k_n = 8.8E-2$ s^{-1} and $k_b = 3.0E-5$ M^{-1} s^{-1}. 2,3-dichloro-1-propanol was estimated to have neutral and base rate constants of $k_n = 0.46$ s^{-1} and $K_b = 1.8E5$ $M^{-1}s^{-1}$. Measured values were found for epichlorohydrin: $k_a = 2.5E4$,

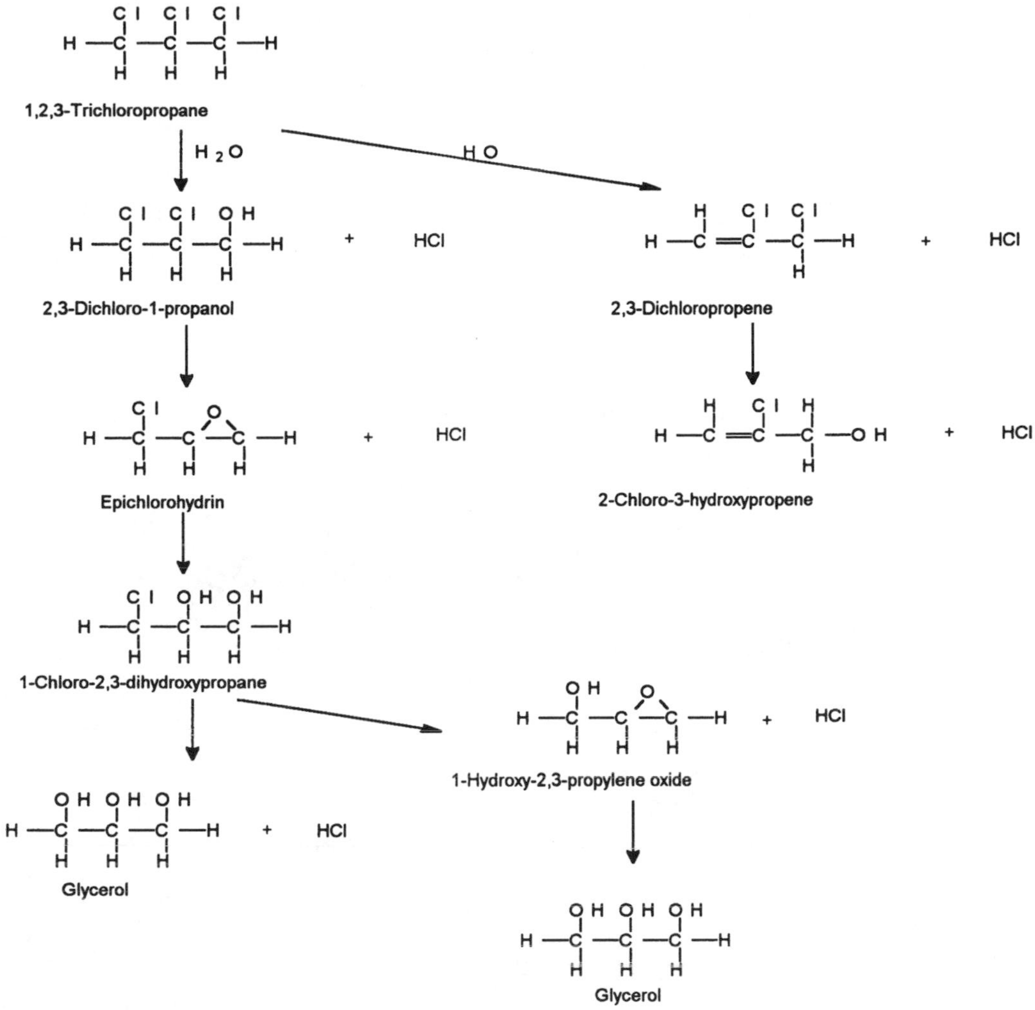

FIGURE 13.6
Proposed hydrolysis pathway for 1,2,3-trichloropropane and its intermediate products..

$k_n = 30.9$ s^{-1}, $k_b = 0$. Measured values were found for 1-chloro-2,3-propylene oxide: $k_a = 7.7E4$, $k_n = 8.9$ s^{-1}, $k_b = 0$. Glycerol is an NLFG compound.

13.6.5 Organophosphate Hydrolysis: Tetraethyl Dithiopyrophosphate

The P-O-P bond of tetraethyl dithiopyrophosphate is very labile to attack by hydroxide even at neutral pH. The resulting O,O-diethylphosphorothioic acid is hydrolyzed to the final products, phosphoric acid and ethanol, by a neutral hydrolysis mechanism (Figure 13.7). The rate constants are $k_a = 0$, $k_n = 84$ Y^{-1} and $k_b = 9E6$ M^{-1} Y^{-1}. The neutral rate constant for O,O- diethylphosphorothioic acid is $k_n = 0.2$ Y^{-1}. The rate constant for O-ethylphosphorothioic acid is $k_n = 1$ Y^{-1}. The rate constant for phosphorothioic acid is $k_n = 3$ Y^{-1}.

13.6.6 Epoxide Hydrolysis: Dieldrin

Dieldrin is an example of a pesticide that contains the epoxy functional group. Dieldrin can exist as both exo and endo isomers (Figure 13.8). Assessment of the hydrolysis of this com-

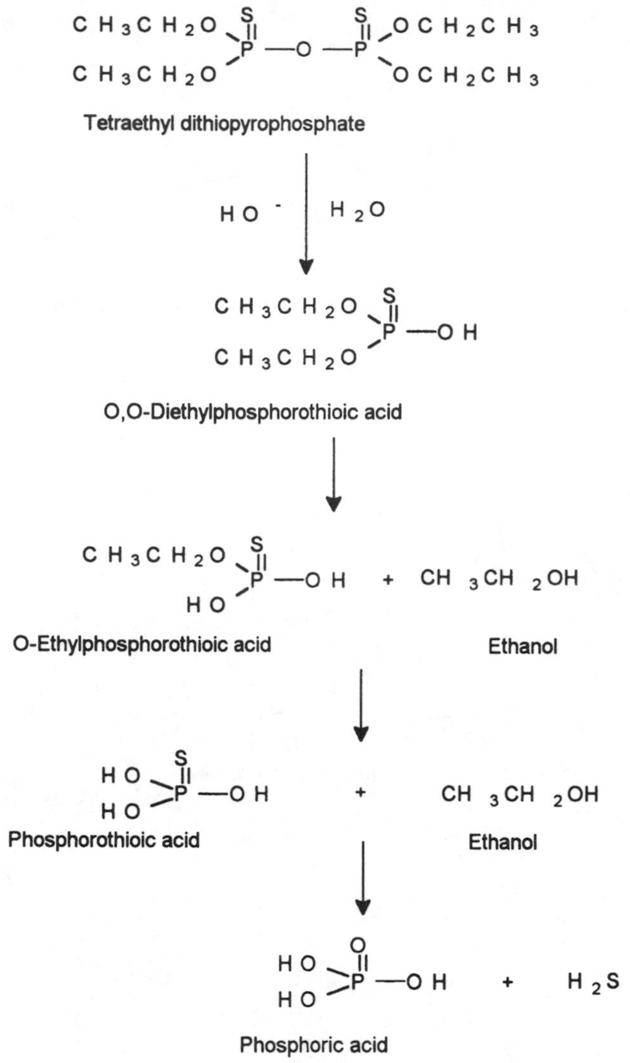

FIGURE 13.7
Reaction pathway for the neutral and alkaline hydrolysis of tetraethyl dithiopyrophosphate.

pound is based on analogy with the reaction of ethylene oxide and water. In this case, however, only a neutral hydrolysis rate constant was estimated, and it is $k_n = 6.3E\text{-}2\ Y^{-1}$.

13.6.7　Sulfonates: Ethyl Methanesulfonate

Alkyl methanesulfonates undergo neutral hydrolysis. The reaction pathway is straightforward, as methylsulfonate is a very good leaving group and nucleophilic attack by water occurs at the methylene group. Nucleophilic substitution results in the formation of the corresponding alcohol (Figure 13.9). Alkyl substituents at the sulfur atom will have relatively minor effects on the reactivity of the esters toward water. Based on reactions of other alkyl sulfonates, the neutral hydrolysis rate constant was estimated to be $k_n = 1.25E3\ Y^{-1}$.

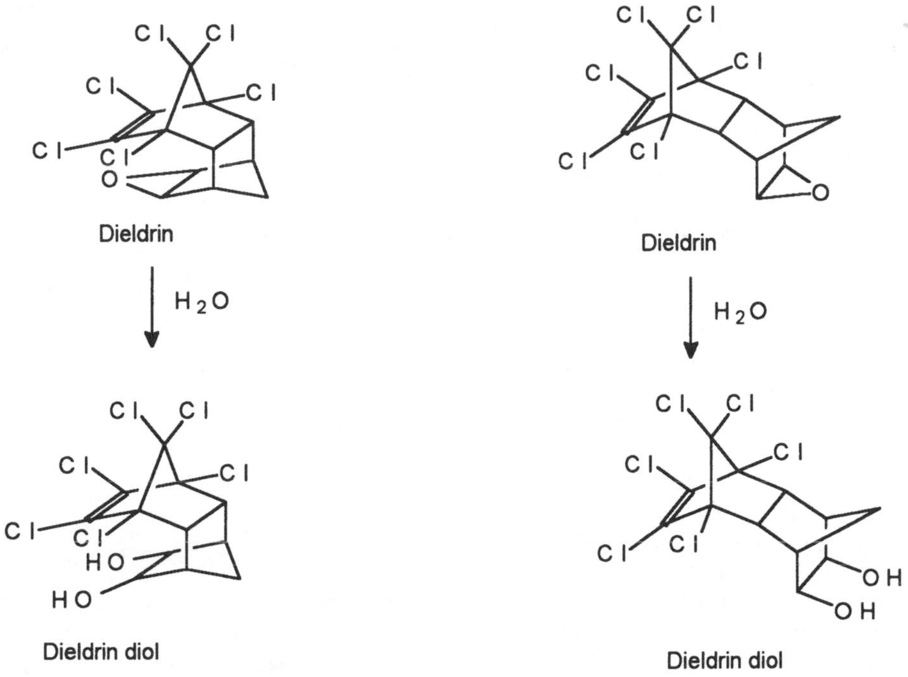

FIGURE 13.8
Products from the hydrolysis of the epoxy group of dieldrin.

13.7 Benchmark Chemicals

Recent years have seen limited advances in formulating quantitative prediction correlations for hydrolysis rate constants. Fortunately, numerous experimental studies provide pH-dependent hydrolysis rate constants for one or more compounds in most classes of organics that might be of environmental concern. Estimation of reactivity by comparison with structural analogs within a given class is often the fastest and most reliable approach.

Consideration of the benchmark chemicals illustrates this approach. Anthracene and 2,6-di-tert-butylphenol have no hydrolyzable functional groups (i.e., are NLFG compounds), hence they cannot undergo hydrolysis. Trichloroethylene hydrolysis has been reported (Jeffers et al., 1989; Jeffers and Wolfe, 1996), but the measured rate constants imply an environmental half-life at pH 7 and 25°C of 100,000 years. Similarly long half-lives have been calculated for other halogenated ethenes, so that, as a class, hydrolysis can be disregarded for these compounds.

For lindane, structural considerations suggest that stepwise 1,2-HCl elimination should occur with relative ease, and that the final product should be 1,3,5-trichlorobenzene. Roberts et al. (1993) included lindane in a QSAR and found close agreement between the predicted and experimentally reported rate constant, with an environmental half-life of about 6 years. The QSAR is useful for compounds that can be analyzed by that approach, especially because the range of reactivity is so vast for the halogenated hydrocarbons. However, the limitations of QSARs are also evident, in that various mechanisms can operate, and a QSAR is developed only for one mechanistic pathway.

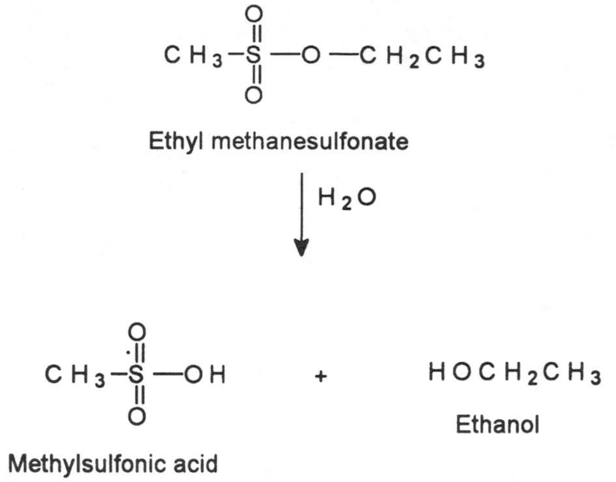

$$CH_3-\overset{\overset{O}{\|}}{\underset{\underset{O}{\|}}{S}}-O-CH_2CH_3$$

Ethyl methanesulfonate

↓ H_2O

$$CH_3-\overset{\overset{O}{\cdot\|}}{\underset{\underset{O}{\|}}{S}}-OH \qquad + \qquad HOCH_2CH_3$$

                                                 Ethanol

Methylsulfonic acid

FIGURE 13.9
Neutral hydrolysis products of ethyl methane sulfonate.

Chlorpyrifos can be fit to a QSAR Wolfe (1980) developed. The calculated as well as observed half-life of about 80 days falls in the middle of the set of diethyl-aryl-phosphorothioates. The reactivity span of these compounds is much lower than that observed for the halogenated alkanes and alkenes, with half-lives generally within the range of several days to several years.

References

Berger, B. and N.L. Wolfe. 1996. Hydrolysis and biodegradation of sulfonylurea herbicides in aqueous buffers and anaerobic water-sediment systems: Assessing fate pathways using molecular descriptors. *Environ. Toxicol Chem.* 15:1599-1507.

Collette, T.W. 1992. Infrared spectroscopy-base property-reactivity correlations for predicting environmental fate of organic chemicals. *Environ. Toxicol. Chem.* 11: 981-991.

Cotham, Jr., W.E. and T.F. Bidleman. 1989. Degradation of malathion, endosulfan, and fenvalerate in seawater and seawater/sediment microcosms. *J. Agric. Food Chem.* 37: 824-828.

Ellington, J.J., F.E. Stancil, W.D. Payne, and C. Trusty. 1986,1987. *Measurement of Hydrolysis Rate Constants.* Vol. I, EPA/600/3-86/043 (1986); Vol. II, EPA/600/3-87/019 (1987).

Fells, I. and E.A. Moelwyn-Hughes. 1958. The kinetics of hydrolysis of methylene dichloride. *J. Chem. Soc.* 1958: 1326-1330.

Haag, W.R. and T. Mill. 1988. Effect of a subsurface sediment on hydrolysis of haloalkanes and epoxides. *Environ. Sci. Technol.* 22: 658-662.

Harris, J.C. 1982. Rate of hydrolysis. In W.J., Lyman, W.F. Reehl, and D.H. Rosenblatt, Eds., *Handbook of Chemical Property Estimation Methods*, pp. 7-1 through 7-48. McGraw-Hill, New York.

Jeffers, P.M., L. Ward, L. Woytowitch, and N.L. Wolfe. 1989. Homogeneous hydrolysis rate constants for selected methanes, ethanes, ethenes, and propanes. *Environ. Sci. Technol.* 23: 965- 969.

Jeffers, P.M. and N.L. Wolfe. 1996a. Hydrolysis and chloride ion exchange rate of methyl bromide in sea water. *Geophys. Res. Lett.* 23: 1773-1776.

Jeffers, P.M. and N.L. Wolfe. 1996b. Homogeneous hydrolysis rate constants – Part II: Additions, corrections and halogen effects. *Environ. Toxicol. Chem.* 15: 1066-1070.

Jeffers, P.M., P. Coty, S. Luczak, and N.L. Wolfe. 1994. Halocarbon hydrolysis rates – A search for ionic strength and heterogeneous effects. *J. Environ. Sci. Health.* A29: 821-831.

Karickhoff, S.W., V.K. McDaniel, C. Melton, A.N. Vellino, D.E. Nute, and L.A. Carreira. 1991. Predicting chemical reactivity by computer. *Environ. Toxicol. Chem.* 10: 1405-1416.

Kollig, H.P., J.J. Ellington, S.W. Karickhoff, B.E. Kitchens, H.P. Kollig, J.M. Long, E.J. Weber, and N.L. Wolfe. 1993. *Environmental Fate Constants for Organic Chemicals Under Consideration for EPA's Hazardous Waste Identification Projects.* EPA/600/R-93/132.

Lacorte, S., S.B. Lartiges, P. Garrigner, and D. Barcelo. 1995. Degradation of organophosphorus pesticides and their transformation products in estuarine waters. *Environ. Sci. Technol.* 29: 431-438.

Lyman, W.J., W.F. Reehl, and D.H. Rosenblatt. 1982. *Handbook of Chemical Property Estimation Methods.* McGraw-Hill, New York.

Mabey, W.R. and T. Mill. 1978. Critical review of hydrolysis of organic compounds in water under environmental conditions. *J. Phys. Chem. Ref. Data* 7: 383-415.

March, J. 1977. *Advanced Organic Chemistry: Reactions, Mechanisms and Structure.* McGraw-Hill, New York.

Masunaga, S. and N. L. Wolfe. 1995. Hydrolysis of para-substituted benzonitriles in water. *Environ. Toxicol. Chem.* 14:1457-1463.

Masunaga, S., N.L. Wolfe, and L.H. Carriera. 1995. Transformation of benzonitriles in anaerobic sediment and in sediment extract. *Environ. Toxicol. Chem.* 14: 1827-1838.

Mill, T., W.R. Mabey, D.C. Bomberger, T.W. Chou, D.G. Hendry, and J.H. Smith. 1982. *Laboratory Protocols for Evaluating the Fate of Organic Chemicals in Air and Water.* EPA/600/3-82/022.

Moelwyn-Hughes, E.A. 1953. The kinetics of hydrolysis. *Proc. Royal Soc. London* A220: 386-393.

Perrin, D.D., B. Dempsey, and E.P. Serjeant. 1981. *pK$_a$ Prediction for Organic Acids and Bases.* Chapman and Hall Ltd., London.

Roberts, A.L., P.M. Jeffers, N.L. Wolfe, and P.M. Gschwend. 1993. Structure-reactivity relationships in dehydrohalogenation reactions of polychlorinated and polybrominated alkanes. *Critical Revs. in Environ. Sci. Technol.* 23: 1-39.

Steinfeld, J.I., J.S. Francisco, and W.L. Hase. 1989. *Chemical Kinetics and Dynamics.* Prentice-Hall, Edgewood Cliffs, NJ.

Well, P.R. 1968. *Linear Free Energy Relationships.* Academic Press, London.

Wolfe, N.L., R.G. Zepp, and D.F. Paris. 1978. Carbaryl, propham, and chlorpropham: a comparison of the rates of hydrolysis and photolysis with the rate of biolysis. *Water Res.* 12: 565-571.

Wolfe, N.L. 1980. Organophosphate and organophosphorothioate esters: application of linear free energy relationships to estimate hydrolysis rate constants for use in environmental fate assessment. *Chemosphere* 9: 571-579.

14

Atmospheric Oxidation

Roger Atkinson

CONTENTS

14.1 The Atmosphere — A Brief Description

The vertical temperature profile of the earth's atmosphere conveniently allows it to be described as comprised of a number of vertical layers. From the earth's surface upward, these are the troposphere, stratosphere, mesosphere, and thermosphere (McIlveen, 1992). Because ~85% of the mass of the atmosphere resides in the troposphere, and most

1-56670-456-1/00/$0.00+$.50
© 2000 by CRC Press LLC

chemicals are emitted directly into the troposphere, this chapter deals only with reactions in that region.

The troposphere extends from the earth's surface to the tropopause at 10–18 km, with the height of the tropopause depending on latitude and season, being highest at the tropics and lowest at the polar regions during wintertime. The troposphere is characterized by decreasing temperature with increasing altitude, from an average of ~290 K at ground level to ~210-220 K at the tropopause. The lowest kilometer or so of the troposphere contains the planetary boundary layer and, in certain locales, inversion layers, which inhibit vertical mixing of chemicals into the "free" troposphere.

Although the atmosphere can be described in terms of a number of vertical layers, up to ~100 km it remains well-mixed, with a composition of 78% N_2, 21% O_2, 0.9% Ar, 0.036% CO_2, varying amounts of water vapor depending on altitude and temperature, and minute amounts of a number of trace gases (McIlveen, 1992). Pressure decreases monotonically with increasing altitude, from an average of 1013 millibar (mb) at the earth's surface to 140 mb at 14 km (the average altitude of the tropopause).

Vertical mixing in the free troposphere is fairly rapid, with a time-scale of ~10–30 days. Within a hemisphere, horizontal mixing in the troposphere is also fairly rapid, with time-scales of a few hours for transport over local distances of <100 km, a few hours to a few days for transport over regional distances of 100–1000 km, and >10 days for transport over long-range or global distances of a few thousand km. The time-scale for transport between the northern and southern hemispheres is ~1 year. The atmospheric lifetime of a chemical therefore determines the distance over which it is transported (i.e., over local, regional, or global distances).

14.2   Theoretical Background

14.2.1   Gas/Particle Partitioning of Chemicals in the Atmosphere

As Chapter 10 discusses in detail, chemical compounds in the atmosphere are partitioned between the gas and particle phases (Pankow, 1987; Bidleman, 1988), and the phase in which a chemical exists in the atmosphere can significantly influence its dominant tropospheric removal process(es) and lifetime (Bidleman, 1988; Atkinson, 1996). Gas/particle partitioning has been conventionally described by the Junge-Pankow adsorption model that depends on the liquid-phase (or sub-cooled liquid-phase) vapor pressure, P_L, at the ambient atmospheric temperature, the surface area of the particles per unit volume of air, θ, and the nature of the particles and of the chemical being adsorbed (Pankow, 1987; Bidleman, 1988). The fraction of the chemical present in the particle phase, ϕ, depends on these parameters through an equation of the form (Pankow, 1987; Bidleman, 1988):

$$\phi = c\phi / (c\phi + P_L) \tag{1}$$

where c is a parameter which depends on the chemical being adsorbed and on the nature of the particle.

Other models treat this partitioning process as absorptions, as shown in Chapter 10. To a first approximation, chemical compounds with liquid-phase vapor pressures of $P_L < 10^{-6}$ Pa ($<10^{-8}$ Torr) at the ambient atmospheric temperature are primarily present in the particle phase, and those with values of $P_L > 1$ Pa ($>10^{-2}$ Torr) at the ambient atmospheric temperature exist essentially totally in the gas-phase (Eisenreich et al., 1981; Bidleman, 1988). Chemicals with intermediate values of P_L are present in both the gas and particle phases and are often termed semi-volatile organic compounds (SOCs). Because P_L varies with temperature, for a given particle surface area a decrease in ambient atmospheric temperature will increase the fraction of the SOC present in the particle phase [Equation (1)].

14.2.2 Wet and Dry Deposition of Gas- and Particle-Phase Chemical Compounds

In addition to photolysis (Chapter 15) and chemical reactions (see the next section), wet and dry deposition also can remove gas- and particle-phase chemical compounds from the troposphere (Eisenreich et al., 1981; Bidleman, 1988). Thus to completely characterize the atmospheric loss processes and overall lifetime of a chemical, we must understand its atmospheric lifetime due to dry and/or wet deposition. Wet deposition refers to the removal of the chemical (or particle-associated chemical) from the atmosphere by precipitation of rain, fog, or snow to earth's surface). Dry deposition refers to the removal of the chemical or particle-associated chemical from the atmosphere to the Earth's surface by diffusion and/or sedimentation.

14.2.1 Wet Deposition

Wet deposition of a chemical compound is characterized in terms of the overall washout ratio, W, which is given by

W = (chemical concentration in aqueous phase)/(chemical concentration in air)

$$= W_g(1 - \phi) + W_p\phi \tag{2}$$

where W_g is the washout ratio for the gas-phase chemical, W_p is the washout ratio for the particle-associated chemical, and ϕ is the fraction of the chemical which is particle associated [see Equation (1)]. For a gas-phase chemical, the partitioning of the chemical between the air and aqueous phases governs wet deposition and the gas-phase washout ratio, W_g, is given by

$$W_g = C_w/C_a = RT/H \tag{3}$$

where C_w is the concentration of the chemical in the aqueous phase, C_a is the concentration of the chemical in air, R is the gas constant, T is the ambient atmospheric temperature (in K), and H is the Henry's law constant [usually estimated as ratio of vapor pressure to aqueous solubility as discussed in Chapter 4 (Eisenreich et al., 1981; Bidleman, 1988).

Values of W_p for particle-associated chemicals are generally in the range $\sim 10^5$ to 10^6 (Eisenriech et al., 1981; Bidleman, 1988). A value of $W = 10^5$ is calculated to result in a residence time for chemicals due to wet deposition in the well-mixed troposphere (with a scale height of 7 km) of ~ 20 days for a constant precipitation rate of 1 m yr^{-1} (the residence time = (height of atmosphere considered)/WJ, where J is the precipitation rate).

14.2.2.2 Dry Deposition

The flux of chemicals to the Earth's surface due to dry deposition, F_{dry}, is given by

$$F_{dry} = V_{dg}C_a + V_{dp}C_p \tag{4}$$

where V_{dg} and V_{dp} are the deposition velocities for gas- and particle-phase chemicals, respectively (and hence V_{dp} is the deposition velocity for the particles with which the chemical is associated), and C_p is the atmospheric concentration of the chemical in the particle phase.

Deposition velocities depend on the atmospheric stability, nature of the surface, nature of the chemical (for a gas-phase chemical), and (for a particle or particle-phase chemical) the size of the particle. For particle deposition, the deposition velocity is a minimum for particles with mean diameter in the range ~0.3–0.5 μm, and it increases with both increasing and decreasing particle size (Eisenriech et al., 1981; Bidleman, 1988).

With a typical value of $V_{dp} = 0.2$ cm s^{-1} for particles of this size range, the calculated residence time of particles and particle-associated chemicals due to dry deposition in the well-mixed troposphere is ~30 days. Modeling of the transport of particle-associated ^{210}Pb leads to an estimated residence time of particles in the atmosphere of ~5–15 days due to combined wet and dry deposition (Balkanski et al., 1993), consistent with the above calculations.

14.2.3   Chemical Transformations and the Presence of Reactive Species in the Atmosphere

Gas- and particle-phase organic compounds can undergo chemical change by a number of routes (Atkinson, 1995, 1996). For gas-phase chemicals, these involve photolysis, reaction with ozone (O_3), reaction with the hydroxyl (OH) radical, and reaction with the nitrate (NO_3) radical (Atkinson, 1995). (For a discussion of photolysis, see Chapter 15).

14.2.3.1   Ozone in the Troposphere

The presence of O_3 in the troposphere is due to downward transport from the stratosphere with dry deposition at the earth's surface (Logan, 1985) and *in situ* photochemical formation and destruction (Logan, 1985; Ayers et al., 1992). Mixing ratios of O_3 in the "clean" remote lower troposphere are in the range $(10–40) \times 10^{-9}$ (Logan, 1985; Oltmans and Levy, 1994).

14.2.3.2 OH Radicals in the Troposphere

The presence of O_3 in the troposphere leads to the formation of OH radicals through the photolysis of O_3 at wavelengths ~290-350 nm to form the electronically excited oxygen atom, $O(^1D)$, which either reacts with water vapor or is deactivated by reaction with O_2 and N_2 to the ground state oxygen atom, (O^3P) (Atkinson, 1995; Atkinson et al., 1997).

$$O_3 + h\nu \rightarrow O_2 + O(^1D) \qquad (\lambda < 350 \text{ nm})$$

$$O(^1D) + H_2O \rightarrow 2 \text{ OH}$$

$$O(^1D) + M \rightarrow O(^3P) + M \qquad (M = N_2, O_2)$$

Because the OH radical is produced photolytically, the OH radical is present at significant concentrations only during daylight hours, and its concentration exhibits a marked diurnal profile (see, for example, Mount et al., 1997), with a maximum concentration at around solar noon (depending on cloud cover) and with low or negligible concentrations at night. The OH radical is the key reactive species in the troposphere, reacting with all organic compounds apart from the chlorofluorocarbons (CFCs) and those Halons which do not contain a hydrogen atom (for example, CF_3Br, CF_2Br_2 and CF_2ClBr).

Based on direct spectroscopic measurements of OH radical concentrations at close to ground level, peak daytime OH radical concentrations are typically $\sim(3-10) \times 10^6$ molecule cm^{-3} (see, for example, Brauers et al., 1996; Mather et al., 1997; Mount et al., 1997). A diurnally, seasonally, and annually averaged global tropospheric OH radical concentration has been derived from the emissions, atmosphere concentrations, and OH radical reaction rate constant for methyl chloroform (CH_3CCl_3), resulting in a 24-hr average OH radical concentration of 9.7×10^5 molecule cm^{-3} (Prinn et al., 1995).

14.2.3.3 *NO₃ Radicals in the Troposphere*

Emissions of NO from combustion processes and soils and *in situ* formation from lightning are followed by reactions leading to formation of the NO_3 radical (Atkinson, 1995; Atkinson et al., 1997).

$$NO + O_3 \rightarrow NO_2 + O_2$$

$$NO_2 + O_3 \rightarrow NO_3 + O_2$$

Because the NO_3 radical photolyzes rapidly (Atkinson et al., 1997), NO_3 radical concentrations are low during daylight hours but can become elevated at night. Measured ground-level NO_3 radical concentrations range up to 1×10^{10} molecule cm^{-3}, and a 12-hr nighttime average concentration of $\sim 5 \times 10^8$ molecule cm^{-3}, uncertain by a factor of ~ 10, has been proposed (Atkinson, 1991).

14.2.4   Lifetimes of Volatile Organic Compounds in the Troposphere

For the majority of gaseous organic compounds, the potential loss or transformation processes in the troposphere are wet and dry deposition, photolysis, reaction with OH radicals, reaction with NO_3 radicals, and reaction with O_3. The overall lifetime, $\tau_{overall}$, of a chemical present in the gas phase is then given by:

$$1/\tau_{overall} = 1/\tau_{wet\ dep.} + 1/\tau_{dry\ dep.} + 1/\tau_{phot} + 1/\tau_{OH} + 1/\tau_{NO3} + 1/\tau_{O3} \qquad (5)$$

where the lifetime due to photolysis is given by $\tau_{phot} = (\kappa_{phot})^{-1}$ and the lifetimes due to chemical reactions with OH radicals, NO_3 radicals, and O_3 are given by, for example, $\tau_{OH} = (k_{OH}[OH])^{-1}$, where k_{OH} is the rate constant for the reaction of the chemical with the OH radical, and [OH] is the ambient tropospheric OH radical concentration. Equation (5) is essentially a statement that the overall loss rate is the sum of the individual loss rates.

Rate constants have been measured for the gas-phase reactions of a large number of organic compounds with OH radicals (Atkinson, 1989, 1994, 1997), NO_3 radicals (Atkinson, 1991, 1994, 1997), and O_3 (Atkinson and Carter, 1984; Atkinson, 1994, 1997). These measured rate constants can be combined with measured or estimated ambient tropospheric concentrations of OH radicals, NO_3 radicals, and O_3 to provide tropospheric lifetimes with respect to the various loss processes (see, for example, Atkinson, 1995).

14.2.5   Reaction Mechanisms for OH Radical, NO₃ Radical, and O₃ Reactions with Organic Compounds

The mechanisms of the gas-phase reactions of OH radicals, NO_3 radicals, and O_3 with organic compounds are discussed in detail elsewhere (Atkinson and Carter, 1984; Atkinson, 1989, 1991, 1994, 1997).

All organic compounds except the CFCs and certain Halons (i.e., saturated compounds containing only carbon, fluorine, chlorine, and/or bromine) react with the OH radical. The reaction pathways involved are discussed in section 14.25.

NO_3 radicals also react with a large number of organic compounds. The NO_3 radical reactions are potentially important tropospheric loss processes for alkenes, other compounds containing unsaturated >C=C< or –C≡C– bonds, organosulfur compounds, phenolic compounds, and certain nitrogen-containing compounds (Atkinson, 1991, 1994). In general, the initial reaction mechanisms for the NO_3 radical reactions parallel the corresponding OH radical reaction mechanisms (Atkinson, 1991, 1994).

O_3 reacts to any significant extent only with organic compounds containing unsaturated carbon–carbon bonds (for example, alkenes, haloalkenes, alkynes, and oxygen-containing compounds with >C=C< bonds) and with certain nitrogen-containing compounds (Atkinson and Carter, 1984; Atkinson, 1994, 1997). These O_3 reactions proceed by initial addition of O_3 to the >C=C< bond(s) (Atkinson and Carter, 1984; Atkinson, 1994, 1997) or by initial "interaction" with the N-atom in certain nitrogen-containing organic compounds which do not contain >C=C< bonds (Atkinson and Carter, 1984).

For the majority of gas-phase organic chemicals present in the troposphere, reaction with the OH radical is the dominant loss process (Atkinson, 1995). The tropospheric lifetime of a chemical is the most important factor in determining the relative importance of transport, to both remote regions of the globe and to the stratosphere, and in determining the possible buildup in its atmospheric concentration. Knowledge of the OH radical reaction rate constant for a gas-phase organic compound leads to an upper limit to its tropospheric lifetime.

To date, OH radical reaction rate constants have been measured for ~500 organic compounds (Atkinson, 1989, 1994, 1997). However, many more organic chemicals are emitted into the atmosphere, or formed *in situ* in the atmosphere from photolysis or chemical reactions of precursor compounds, for which OH radical reaction rate constants are not experimentally available. Thus the need to reliably calculate OH radical reaction rate constants for those organic compounds for which experimental data are not currently available.

14.3 Estimation Methods for OH Radical Reaction Rate Constants

A number of methods for estimating gas-phase OH radical reaction rate constants for organic compounds have been proposed, ranging from estimation methods for single classes of organic compounds to generalized estimation methods (see, for example, Gaffney and Levine, 1979; Heicklen, 1981; Güsten et al., 1981, 1984; Zetzsch, 1982; Klöpffer et al., 1985; Jolly et al., 1985; Walker, 1985; Atkinson, 1986, 1987, 1988; Güsten and Klasinc, 1986; Cohen and Benson, 1987a,b; Dilling et al., 1988; Hodson, 1988; Cooper et al., 1990; Grosjean and Williams, 1992; Klamt, 1993, 1996; Kwok and Atkinson, 1995, Medven et al., 1996). Many of these estimation methods use molecular properties such as ionization energy (Gaffney and Levine, 1979; Güsten et al., 1984; Güsten and Klasinc, 1986; Grosjean and Williams, 1992), NMR chemical shifts (Hodson, 1988), molecular orbital energies (Cooper et al., 1990; Klamt, 1993, 1996; Medven et al., 1996), bond-dissociation energies (Heicklen, 1981) and infrared absorption frequencies (Jolly et al., 1985), or they involve transition state calculations (Cohen and Benson, 1987a,b). Other estimation methods use correlations between gas- and liquid-phase OH radical reaction rate constants (Güsten et al., 1981; Klöpffer et al., 1985; Dilling et al., 1988; Grosjean and Williams, 1992). However, most of these estimation methods are restricted in their use because of the limited database of such molecular properties as ionization potentials, bond-dissociation energies, and infrared absorption frequencies.

One method used often relies on structure-reactivity relationships (Atkinson, 1986, 1987, 1988; Kwok and Atkinson, 1995). This empirical estimation method has been shown to provide good agreement (generally to within a factor of 2) between experimentally measured and calculated room temperature rate constants for ~90% of ~485 organic compounds (Kwok and Atkinson, 1995). It is the basis of the Syracuse Research Corporation's "Atmospheric Oxidation Program" [see Meylan and Howard (1993) for a discussion of an earlier version of this program]. The general approach of this estimation method has been described (Atkinson, 1986, 1987, 1988; Kwok and Atkinson, 1995).

The method is based on the observations that gas-phase OH radical reactions with organic compounds proceed by four reaction pathways, assumed to be additive: H-atom abstraction from C–H and O–H bonds, OH radical addition to >C=C< and –C≡C– bonds, OH radical addition to aromatic rings, and OH radical "interaction" with N-, S-, and P-atoms and with more complex structural units such as –>P=S, >NC(O)S– and >NC(O)O– groups. The total rate constant is assumed to be the sum of the rate constants for these four reaction pathways (Atkinson, 1986). The OH radical reactions with many organic compounds proceed by more than one of these pathways; estimation of rate constants for the four pathways follow. Section 14.3.5 gives examples of calculations of the OH radical reaction rate constants for the "standard" compounds lindane (γ-hexachlorocyclohexane), trichloroethene, anthracene, 2,6-di-*tert*-butylphenol, and chloropyrofos.

14.3.1 H-Atom Abstraction from C–H and O–H Bonds

The calculation of H-atom abstraction rate constants from C–H and O–H bonds is based on the estimation of group rate constants for H-atom abstraction from –CH$_3$, –CH$_2$–, >CH– and –OH groups (Atkinson, 1986). The rate constants for H-atom abstraction from –CH$_3$, –CH$_2$– and >CH– groups depends on the identity of the substituents attached to these groups, with

$$k(CH_3\text{-}X) = k_{prim}\ F(X)$$

$$k(X\text{-}CH_2\text{-}Y) = k_{sec}\ F(X)\ F(Y)$$

$$k(X\text{-}CH\underset{Z}{\overset{Y}{<}}) = k_{tert}\ F(X)\ F(Y)\ F(Z)$$

where k_{prim}, k_{sec} and k_{tert} are the group rate constants for H-atom abstraction from –CH$_3$, –CH$_2$– and >CH– groups, respectively, for a "standard" substituent, and F(X), F(Y) and F(Z) are the substituent factors for the substituent groups X, Y, and Z, respectively. The standard substituent group is chosen to be X (= Y = Z) = –CH$_3$, with F(–CH$_3$) = 1.00 (Atkinson, 1986). It is assumed that the temperature dependence of the substituent factors F(X) can be expressed in the form $F(X) = e^{Ex/T}$, and hence that the pre-exponential factors A (in the Arrenhius expression $k = A\ e^{-B/T}$ or in the alternative three-parameter expression $A\ T^n\ e^{-D/T}$) for H-atom abstraction from C–H bonds are independent of the substituent groups (Atkinson, 1987).

The group rate constants, k_{prim}, k_{sec}, k_{tert} and k_{OH} obtained by Kwok and Atkinson (1995) are (in cm^3 molecule^{-1} s^{-1} units): $k_{prim} = 4.49 \times 10^{-18}$ T^2 e$^{-320/T}$ (1.36×10^{-13} at 298 K); $k_{sec} = 4.50 \times 10^{-18}$ T^2 e$^{253/T}$ (9.34×10^{-13} at 298 K); $k_{tert} = 2.12 \times 10^{-18}$ T^2 e$^{696/T}$ (1.94×10^{-12} at 298 K); and $k_{OH} = 2.1 \times 10^{-18}$ T^2 e$^{-85/T}$ (1.4×10^{-13} at 298 K). Table 14.1 gives the substituent factors, F(X), obtained by Kwok and Atkinson (1995) and Kwok et al. (1996). Ring-strain effects for cycloalkanes with other than 6-member rings are taken into account through ring strain factors F_{ring} (Table 14.1), with only those rings containing the $-CH_2-$ or $>CH-$ groups being considered in accounting for ring strain.

Initial OH radical addition to $-C(O)OH$ groups in carboxylic acids is assumed, with a rate constant at 298 K and atmospheric pressure of air of 5.2×10^{-13} cm^3 molecule^{-1} s^{-1}. Although Hynes and Wine (1991) showed that OH radical reaction with CH_3CN proceeds by both H-atom abstraction and initial OH radical addition at room temperature and atmospheric pressure of air, Kwok and Atkinson (1995) assumed only H-atom abstraction in order to fit the literature rate constants for CH_3CN and C_2H_5CN. For two ether groups bonded to the same carbon atom ($-ORO-$; R = alkyl), Kwok and Atkinson (1995) arbitrarily assumed that $[F(-O-)]^2 = F(-O-)$. The recent kinetic study of Talukdar et al. (1997) for methyl nitrate, ethyl nitrate and 2-propyl nitrate indicates that the OH radical reactions with alkyl nitrates proceed solely by H-atom abstraction, and re-calculated substituent factors for $-ONO_2$, $-CH_2ONO_2$, $>CHONO_2$ and $->CONO_2$ are listed in Table 14.1.

Because of the generally greater reactivity of $>C=C<$, $-C\equiv C-$ and aromatic rings toward OH radical addition, H-atom abstraction from alkyl or substituted alkyl groups in the alkenes, alkynes, and aromatic compounds is generally of minor importance, and Kwok and Atkinson (1995) set the substituent group factors F($>C=C<$), F($-C\equiv C-$) and F($-C_6H_5$) equal to unity.

14.3.2   OH Radical Addition to >C=C< and –C≡C– Bonds

The calculation of rate constants for OH radical addition to $>C=C<$ and $-C\equiv C-$ bonds assumes that the rate constant for OH radical addition to these carbon-carbon unsaturated bonds depends on the number, identity, and position of substituent groups around the $>C=C<$ or $-C\equiv C-$ bond(s). Conjugated double bond systems are dealt with by considering the entire conjugated $>C=C-C=C<$ system as a single unit (Atkinson, 1986), rather than as conjugated $>C=C<$ sub-units as Ohta (1983) did.

Table 14.2 gives the 298 K rate constants for addition to the $CH_2=CH-$, $CH_2=C<$, *cis-* and *trans-* $-CH=CH-$, $-CH=C<$ and $>C=C<$ non-conjugated carbon-carbon double bond units and the various $-C\equiv C-$ and conjugated $>C=C-C=C<$ units, based on the measured rate constants for the alkenes with the appropriate alkyl substitutent groups (Atkinson, 1986; Atkinson and Kwok, 1995). There is no evidence for ring-strain effects on OH radical addition to cycloalkenes.

For organic compounds containing $>C=C<$ bonds with non-alkyl substituent groups, X, group substituent factors C(X) are used [C(alkyl) = 1.00]. Table 14.3 gives the group factors C(X) derived from the available database for haloalkenes and nitrogen- and oxygen-containing organic compounds containing $>C=C<$ bonds (Kwok and Atkinson, 1995).

14.3.3   OH Radical Addition to Aromatic Rings

The rate constants for OH radical addition to aromatic rings are calculated using the correlation between the OH radical addition rate constant and the sum of the electrophilic substituent constants $\Sigma\sigma^+$ (Zetzsch, 1982; Atkinson, 1986). As Zetzsch (1982) discussed, $\Sigma\sigma^+$ is calculated by assuming that steric hindrance can be neglected; $\Sigma\sigma^+$ is the sum of all of the

TABLE 14.1

Substituent Factors F(X) at 298 K.[a]

X	F(X) at 298 K
$-CH_3$	1.00
$-CH_2-$ $>CH-$ $>C<$ }	1.23
$-F$	0.094
$-Cl$	0.38
$-Br$	0.28
$-I$	0.53
$-CH_2Cl$ $-CHCl_2$ $-CHCl-$ $>CCl-$ }	0.36
$-CH_2Br$ $-CHBr-$ }	0.46
$-CCl_3$	0.069
$-CF_3$	0.071
$-CHF_2$	0.13
$-CH_2F$	0.61
$-CF_2Cl$	0.031
$-CFCl_2$	0.044
$-CHF-$	0.21
$-CF_2-$	0.018
$=O$	8.7
$-CHO$ $>CO$ }	0.75
$-CH_2C(O)-$ $>CHC(O)-$ $->CC(O)-$ }	3.9
$-C_6H_5$ $>C=C<$ $-C\equiv C-$ }	~1.0
$-OH$	3.5
$-OR$ (R = alkyl)	8.4
$-OCF_3$ $-OCF_2-$ $-OCHF_2$ $-OCH_2F$ }	0.17
$-C(O)Cl$	0.067
$-OCH_2CF_3$ $-OCH(CF_3)_2$ $-OCHClCF_3$ }	0.44
$-C(O)OR$ (R=alkyl)	0.31
$-OC(O)R$ (R=alkyl)	1.6
$-C(O)OH$	0.74
$-C(O)CF_3$	0.11
$-ONO_2$	0.14
$-CH_2ONO_2$ $>CHONO_2$ $->CONO_2$ }	0.28
$-CN$	0.19
$-CH_2CN$	~0.12
3–member ring	0.020
4–member ring	0.28
5–member ring	0.64
7– and 8–member rings	~1.0

[a] Values of E_x can be calculated from $F(X) = e^{-E_x/T}$.

TABLE 14.2

298 K Group Rate Constants for OH Radical
Addition to >C=C<, –C≡C– and >C=C–C=C<
Structural Units.

Structural Unit	$10^{12} \times k$ (cm^3 molecule^{-1} s^{-1})
CH$_2$=CH–	26.3
CH$_2$=C<	51.4
cis– –CH=CH–	56.4
trans– –CH=CH–	64.0
–CH=C<	86.9
>C=C<	110
H$_2$C=CHCH=CH– $\Big\}$ H$_2$=CHC=CH$_2$	105
H$_2$C=CHCH=C< H$_2$C=CHC=CH– H$_2$C=CCH=CH– –CH=CHCH=CH– H$_2$C=CC=CH$_2$ $\Bigg\}$	142
H$_2$C=CHC=C< H$_2$C=CCH=C< –CH=CHCH=C< H$_2$C=CC=CH– –CH=CCH=CH– $\Bigg\}$	~190
>C=CHCH=C< H$_2$C=CC=C< –CH=CC=CH– –CH=CHC=C< –CH=CCH=C< $\Bigg\}$	~260
HC≡C–	7.0
–C≡C–	27

TABLE 14.3

Group Substituent Factors, C(X),
at 298 K for OH Radical Addition
to >C=C< and –C≡C– Bonds.

Substituent Group	C(X) at 298 K
–F	0.21
–Cl	0.21
–Br	0.26
–CH$_2$Cl	0.76
–CHO	0.34
–C(O)CH$_3$	0.90
–CH$_2$ONO$_2$ $\Big\}$ >CHONO$_2$	0.47
–C(O)OR (R=alkyl)	0.25
–OR (R=alkyl)	1.3
–CN	0.16
–CH$_2$OH	~1.6

substituent constants of the substituent groups attached to the aromatic ring; in general the value of $\Sigma\sigma^+$ depends on the position of OH radical addition to the ring, and the OH radical adds to the position yielding the most negative value of $\Sigma\sigma^+$; and if all positions on the ring are occupied, the *ipso* position is treated as a *meta* position. The correlation between the OH radical addition rate constants and $\Sigma\sigma^+$ for monocyclic aromatic compounds and biphenyls is (Kwok and Atkinson, 1995):

$$\log_{10} k_{add} \text{ (cm}^3 \text{ molecule}^{-1} \text{ s}^{-1}) = -11.71 - 1.34\Sigma\sigma^+ \tag{6}$$

Note that a "maximum rate constant" of $\sim 2 \times 10^{-10}$ cm^3 molecule^{-1} s^{-1} probably should be used when the calculated value exceeds this rate constant.

Electrophilic substituent factors for the $-C_6H_{5-x}Cl_x$ and $-OC_6H_{5-x}Cl_x$ groups, needed to calculate the OH radical addition rate constants for polychlorinated biphenyls (PCBs), polychlorodibenzo-*p*-dioxins (PCDDs), and polychlorodibenzofurans (PCDFs), appear in Atkinson (1996), with discussion of an approach to calculating the rate constants for the PCDDs and PCDFs. The room temperature rate constants for the reactions of the OH radical with phenanthrene and anthracene, recently measured by Kwok et al. (1994, 1997), are lower by factors of 2.5–8 than the previous recommendations and rate data (Biermann et al., 1985; Atkinson, 1989), casting doubt on the previously proposed correlation between the OH radical addition rate constant and ionization potential (Biermann et al., 1985).

Therefore, the estimation of rate constants for the PAH (including for anthracene) is now uncertain. For substituted PAH, we can estimate approximately the enhancement of the rate constant over that for the parent PAH by using the correlation between the rate constant and $\Sigma\sigma^+$ discussed above for the monocyclic aromatic compounds (Atkinson, 1987), with the enhancement factor being $e^{1.34\,\Sigma\sigma^+}$. This approach also may work for heteroatom-containing aromatic compounds, such as pyridines and triazines.

14.3.4 OH Radical Interaction with N-, S-, and P- Containing Groups

The OH radical reactions with a number of nitrogen-, sulfur- and phosphorus-containing organic compounds appear to proceed, at least in part, by an initial addition reaction (Atkinson, 1989, 1994; Kwok et al., 1996), although the products observed may in some cases be those expected from H-atom abstraction. Note that the recent study of Talukdar et al. (1997) indicates that the reactions of the OH radical with alkyl nitrates proceed only by H-atom abstraction, and Table 14.1 gives the applicable substituent group factors for alkyl nitrates.

Rate constants are available for nitrogen-containing compounds (a number of amines and substituted amines [including $CF_3CH_2NH_2$ and 1,4-diazabicyclo[2.2.2]octane], *N*-nitrosodimethylamine, dimethylnitramine, a number of nitroalkanes, four amides, and three carbamates), sulfur-containing organics, three thiocarbamates containing the structural unit $>NC(O)S-$, six compounds containing $->P=O$ and $->P=S$ groups, and four compounds containing $->P(=X)N<$ groups, with X = O or S (Atkinson, 1989, 1994; Kwok et al., 1996; Koch et al., 1996, 1997). The reactions of these compounds apparently proceed in part by initial addition to the N-, S-, or P- atom. Tables 14.4 and 14.5 give group rate constants and substituent group factors, respectively.

Note that these group rate constants and substituent group factors should be used only for homologs of the compounds from which these factors were derived (Kwok and Atkinson, 1995; Koch et al., 1996). For example, the group rate constants $k_{>N-}$, $k_{>NH}$ and k_{-NH2} and the substituent factors $F(-NH_2)$, $F(>NH)$ and $F(>N-)$ are appropriate only for *alkyl*-substituted amines and not for amines containing halogen or other hetero-atoms (in particular, see Koch et al., 1996). For the phosphorothioamidates (with $->P(=S)N<$ structures) and

TABLE 14.4

Group Rate Constants at 298 K for the Reactions of OH Radicals with N-, S- and P-Containing Organic Compounds.

Group (R=alkyl)[a]	$10^{12} \times k$ (cm^3 molecule^{-1} s^{-1})
RNH$_2$	21
R$_2$NH	63
R$_3$N	66
R$_2$NNO	0
R$_2$NNO$_2$	1.3
RNO$_2$	0.13
RSH	32.5
RSR	1.7[b]
RSSR	225
->P=O	0
->P=S	53
>NC(O)S–	11.9
–NHC(O)O–	2.7

[a] These group rate constants are only applicable for alkyl substituents.

[b] Based on the rate constant for the reaction of the OH radical with dimethyl sulfide at 298 K and atmospheric pressure of air (Atkinson, 1994).

TABLE 14.5

Substituent Group Factors F(X) at 298 K for the Reactions of OH Radicals with N–, S–, and P–Containing Organic Compounds.

Substituent Group X	F(X) at 298 K
–NH$_2$	
>NH	
>N–	9.3
>NNO	
>NNO$_2$	
–CH$_2$NO$_2$	0.14
–NO$_2$	0.0
–SH	
–S–	7.8
–SS–	
–OP<–	20.5
–SP<–	
>NC(O)S–	4.1
–SC(O)N<	
–NHC(O)O–	7.5
–OC(O)NH–	4.8

phosphoroamidates (with ->P(=O)N< structures), the P and N portions of the molecules were dealt with as separate ->P(=X)– and >N- units [X = S or O] rather than as a single ->P(=X)N< unit (Goodman et al., 1988). This is probably not appropriate, although an insufficient database exists to derive any meaningful estimation method for this class of compound.

14.3.5 Sample Calculations for OH Radical Reaction Rate Constants

As examples of the calculation of OH radical reaction rate constants using the method discussed above (Kwok and Atkinson, 1995), the OH radical reaction rate constants for lindane [γ-hexachlorocyclohexane; cyclo-(–CHCl–)$_6$], trichloroethene (CHCl=CCl$_2$), 2,6-di-*tert*-butylphenol, and chloropyrofos appear below. As the section dealing with OH radical addition to aromatic rings mentions, at present the rate constant for the reaction of the OH radical with anthracene (and other PAH) cannot be estimated with the method of Kwok and Atkinson (1995). In carrying out these calculations, one first must draw the structure of the chemical (the structures are shown in the appendix to Chapter 1). Then one carries out the calculations for each of the OH radical reaction pathways which can occur for that chemical.

Lindane (γ-hexachlorocyclohexane).

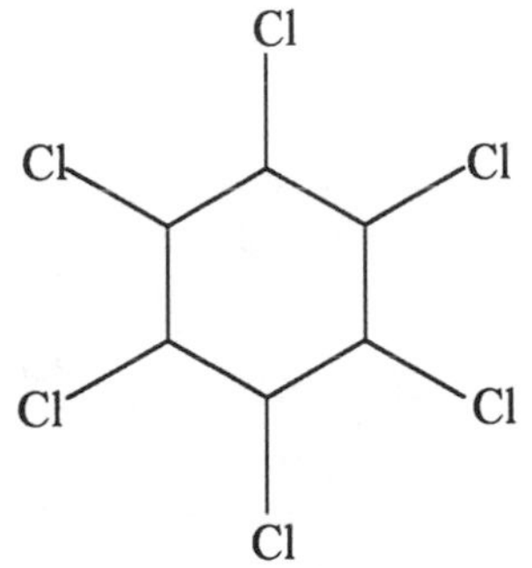

The stereochemical conformation of hexachlorocyclohexane (i.e., whether the Cl atoms are axial or equatorial) has no effect on the estimated rate constant for the OH radical reaction. For the purposes of the calculation, all six carbon atoms are therefore equivalent, with each carbon atom being bonded to two carbon atoms (each with substituent H and Cl atoms) and to one H atom and one Cl atom [i.e., (–CHCl–)$_6$]. The rate constant therefore is given by:

$$k_{total} = 6 \{k_{tert} \ F(-Cl) \ F(-CHCl-) \ F(-CHCl-) \ F(6\text{-member ring})\}$$

$$= 6 \times 1.94 \times 10^{-12} \times 0.38 \times 0.36 \times 0.36 \times 1.00$$

$$= 5.7 \times 10^{-13} \ cm^3 \ molecule^{-1} \ s^{-1}$$

To date, no experimentally measured OH radical reaction rate constant for lindane at room temperature is available in the literature for comparison.

Trichloroethene.

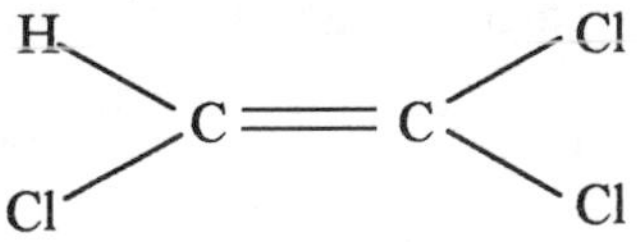

The OH radical reaction occurs at the >C=C< bond, and H-atom abstraction from the vinyl C-H bond is neglected. There are three Cl atom substituents around the >C=C< bond, and the "base" structural unit is the -CH=C< group (Table 14.2). The effects of the three Cl atom substituents are taken into account using the C(-Cl) value from Table 14.3. The rate constant, k_{total}, for trichloroethene is given by:

$$k_{total} = k(\text{OH radical addition to} -CH=C<) \ C(Cl) \ C(Cl) \ C(Cl)$$

$$= 86.9 \times 10^{-12} \times 0.21 \times 0.21 \times 0.21$$

$$= 8.0 \times 10^{-13} \ cm^3 \ molecule^{-1} \ s^{-1}$$

This estimated rate constant is a factor of ~3 lower than the measured value (Atkinson, 1994; Kwok and Atkinson, 1995).

2,6-Di-tert-butylphenol.

OH radical reaction with 2,6-di-*tert*-butylphenol occurs via OH radical addition to the aromatic ring and by H-atom abstraction from the two $-C(CH_3)_3$ substituent groups. H-atom abstraction from the $-OH$ substituent group on the aromatic ring is not considered explicitly in the estimation method of Kwok and Atkinson (1995), although the same rate constant as for H-atom abstraction from an aliphatic $-OH$ group, k_{OH}, could be used (because of the low rate constant for H-atom abstraction from the $-OH$ group, neglecting this pathway has only a minor effect (<1% for 2,6-di-*tert*-butylphenol) on the estimated overall reaction rate constant).

The OH radical reaction rate constant for 2,6-di-*tert*-butylphenol is given by:

$$k_{total} = k(\text{OH radical addition to the aromatic ring})$$

$$+ k(\text{H-atom abstraction from the} -C(CH_3)_3 \text{ groups})$$

The electrophilic substituent constants for a $-C(CH_3)_3$ substituent group are $\sigma^+_{o,p} = -0.256$ and $\sigma^+_m = -0.059$, and for an $-OH$ group $\sigma^+_{o,p} = -0.92$ (Brown and Okamoto, 1958). For 2,6-di-*tert*-butylphenol, the most negative value of $\Sigma\sigma^+ = -1.038$ is obtained for initial OH radical addition at the 4-position. Use of this value of $\Sigma\sigma^+$ together with Equation (6) then leads to a calculated value of $k(\text{OH radical addition to the aromatic ring}) = 4.8 \times 10^{-11} \ cm^3 \ molecule^{-1} \ s^{-1}$. The rate constant for H-atom abstraction from the two $-C(CH_3)_3$ groups is given by:

$$k(\text{H-atom abstraction}) = 2[3\{k_{\text{prim}}\, F(>C<)\}]$$

$$= 2 \times 3 \times 0.136 \times 10^{-12} \times 1.23$$

$$= 1.0 \times 10^{-12}\ \text{cm}^3\ \text{molecule}^{-1}\ \text{s}^{-1}$$

Hence

$$k_{\text{total}} = 4.8 \times 10^{-11} + 1.0 \times 10^{-12}$$

$$= 4.9 \times 10^{-11\cdot}\ \text{cm}^3\ \text{molecule}^{-1}\ \text{s}^{-1}$$

There is no literature rate constant with which to compare this estimated value, although a rate constant of 6.6×10^{-11} cm^3 molecule^{-1} s^{-1} at 296 ± 2 K has been measured for reaction of the OH radical with 2,6-dimethylphenol (Atkinson, 1994), in good agreement with the estimated rate constant of 5.0×10^{-11} cm^3 molecule^{-1} s^{-1}.

Chlorpyrifos.

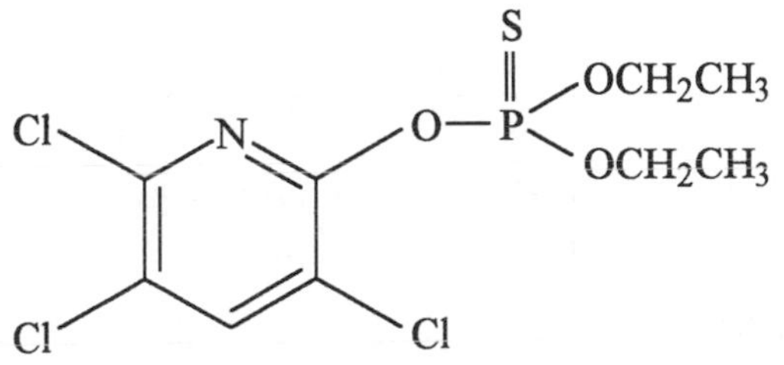

OH radical reaction with chloropyrofos is anticipated to proceed via OH radical addition to the pyridine ring, OH radical "interaction" with the P=S group [with a group rate constant $k_{>P=S}$ (Table 14.4)], and H-atom abstraction from the two –OCH$_2$CH$_3$ groups bonded to the P atom. The total OH radical reaction rate constant is given by:

k_{total} = k(OH radical addition to the pyridine ring)

+ k(OH radical interaction with the P=S group)

+ k(H-atom abstraction from the C–H bonds of the -OCH$_2$CH$_3$ groups)

The rate constant for OH radical addition cannot be calculated because the effect of the –OP(=S)(OCH$_2$CH$_3$)$_2$ substituent is not known. However, because the rate constant for OH radical addition to pyridine is $\sim 3.7 \times 10^{-13}$ cm^3 molecule^{-1} s^{-1} (Atkinson, 1989) and the three Cl atom sustituents will markedly deactivate the ring (Brown and Okamoto, 1958) [as observed, for example, for the OH radical reactions with chlorobenzene, 1,2-, 1,3- and 1,4-dichlorobenzene, and for 1,2,4-trichlorobenzene relative to that for benzene (Atkinson, 1989)], OH radical addition to the pyridine ring is expected to be minor, and its neglect will lead to an estimated lower limit to the total reaction rate constant.

Rate constants for OH radical interaction with the P=S group and H-atom abstraction from the –OCH$_2$CH$_3$ groups are given by:

$$k(\text{OH radical interaction with the P=S group}) = k_{\text{-}>P=S}$$

$$= 5.3 \times 10^{-11} \text{ cm}^3 \text{ molecule}^{-1} \text{ s}^{-1}$$

$$k(\text{H-atom abstraction}) = 2\{k_{\text{prim}} \ F(-CH_2-) + k_{\text{sec}} \ F(-CH_3) \ F(-OP<-)\}$$

$$= 2 \ \{0.136 \times 10^{-12} \times 1.23 + 0.934 \times 10^{-12} \times 1.00 \times 20.5\}$$

$$= 3.9 \times 10^{-11} \text{ cm}^3 \text{ molecule}^{-1} \text{ s}^{-1}$$

Therefore,

$$k_{\text{total}} = k(\text{OH addition to pyridine ring}) + 5.3 \times 10^{-11} + 3.9 \times 10^{-11}$$

$$\geq 9.2 \times 10^{-11} \text{ cm}^3 \text{ molecule}^{-1} \text{ s}^{-1}.$$

No experimental measurement has been reported with which to compare this estimated rate constant.

14.4   Literature Estimation Methods for NO₃ Radical and O₃ Reaction Rate Constants

As section 14.2.5 notes, NO_3 radical reactions are potentially important atmospheric loss processes for alkenes and other organic compounds containing >C=C< bonds, organosulfur compounds, and certain nitrogen-containing compounds (Atkinson, 1991, 1994), while the O_3 reactions are potentially important atmospheric loss processes for alkenes and other organic compounds containing >C=C< bonds (Atkinson and Carter, 1984; Atkinson, 1994). Analogous to the corresponding OH radical reactions, the room temperature rate constants for the reactions of the NO_3 radical and O_3 with alkenes depend on the number and position of the alkyl substituent groups around the >C=C< bond, with the room temperature rate constants generally increasing with the number of alkyl substituents.

Totally analogous to OH radical addition to >C=C< bonds, group rate constants CH_2=CH–, CH_2=C<, *cis-* and *trans-* –CH=CH–, –CH=C< and >C=C< can be derived from the rate constants for the methyl-substituted ethenes; Table 14.6 gives these group rate constants. Again analogous to the corresponding OH radical reactions, no effect of ring-strain appears on the NO_3 radical reaction rate constants for cycloalkenes (Atkinson, 1991) and the group rate constants in Table 14.6 can be used to estimate NO_3 radical reaction rate constants for non-6-member ring cycloalkenes and non-conjugated cyclodienes. In contrast, a marked effect of ring strain appears on the O_3 reaction rate constants for cycloalkenes (Atkinson and Carter, 1984), and the group rate constants in Table 14.6 for the O_3 reactions can be used only for acyclic alkenes and for cycloalkenes with six-member, essentially nonstrained, rings (Atkinson and Carter, 1984).

Analogous to the corresponding OH radical reactions, the effects of non-alkyl substituent groups around the >C=C< bond can be taken into account using substituent group factors $C(X)$, with $C(\text{-alkyl}) = 1.0$ by definition. However, values of $C(X)$ for the NO_3 radical and O_3 reactions have not been tabulated in the literature, in part because of the smaller database for the NO_3 radical and O_3 reactions than for the corresponding OH radical reactions.

TABLE 14.6

298 K Group Rate Constants for NO_3 Radical and O_3 Addition to >C=C< Structural Units[a]

Structural Unit	k (cm^3 molecule^{-1} s^{-1})	
	NO_3 Reaction	O_3 Reaction
CH_2=CH–	1.0×10^{-14}	1.0×10^{-17}
CH_2=C<	3.3×10^{-13}	1.1×10^{-17}
cis– –CH=CH–	3.5×10^{-13}	1.2×10^{-16}
trans– –CH=CH–	3.9×10^{-13}	1.9×10^{-16}
–CH=C<	9.4×10^{-12}	4.0×10^{-16}
>C=C<	5.7×10^{-11}	1.1×10^{-15}

[a] From Atkinson (1991, 1994).

14.5 Limitations of Estimation Methods for OH Radical Reaction Rate Constants

As section 14.3 notes, to date the majority of the estimation methods proposed have limitations due to the unavailability of specific properties of the molecule considered. Furthermore, all of the methods use the measured rate constant database during their development.

For example, the method Atkinson (1986) proposed requires a reliable rate constant database for the various classes of organic compounds in order to derive the various group rate constants and substituent factors. As may be expected, this estimation technique is reasonably reliable when used within its derivation database (Kwok and Atkinson, 1995). Thus, Kwok and Atkinson (1995) observed that the 298 K rate constants were predicted to within a factor of 2 of the experimental values for ~90% of the 485 organic compounds that were considered and for which experimental data were available [and rate constants for alkyl nitrates are still predicted to within a factor of 2 after re-evaluation to take into account the recent kinetic and mechanistic data of Talukdar et al. (1997)].

Similarly, the most recent version of the Atmospheric Oxidation Program (version 1.8) gives agreement of the calculated and experimental room temperature rate constants to within a factor of 2 for 90% of 647 organic compounds, with a standard deviation of 0.242 log units (i.e., a factor of 1.75) (Meylan, 1997). However, extrapolation of this estimation method to organic compounds outside of the database used for its development and testing lacks reliability and is *not* recommended.

The molecular orbital calculation approaches Klamt (1993, 1996) and Medven et al. (1996) developed appear to hold promise as a more "scientific" approach based on sound chemical principles. However, these approaches also require a database for development and presently appear to apply to hydrocarbons (i.e., alkanes, alkenes, and aromatic hydrocarbons) and to fairly simple oxygenated compounds. Hopefully, these molecular orbital calculational method will be further extended and explored in the near future.

14.6 Conclusions

The empirical estimation method of Atkinson (Atkinson, 1986, 1987; Kwok and Atkinson, 1995) allows the 298 K rate constants of ~90% of approximately 500-600 organic compounds to be predicted to within a factor of 2 of the experimental values. Disagreements between calculated and measured rate constants most commonly occur for halogen-containing organic compounds and especially for the haloalkanes, haloalkenes and halogenated ethers, and problems also arise for ethers, in particular for polyethers and cycloethers (Kwok and Atkinson, 1995).

In addition to the uncertainties in correctly estimating reaction rate constants, we must recognize the uncertainties in the ambient atmospheric concentrations of the OH radical as a function of both time and place, since the lifetime, τ_{OH}, of a chemical is given by $\tau_{OH} = (k_{OH}[OH])^{-1}$. Present estimation methods are limited, and most cannot be used with any degree of reliabiity for organic compounds outside the classes of compounds used to develop the particular method. Further studies should be carried out to develop more direct (less empirical) methods for calculating OH radical (and NO_3 radical and O_3) reaction rate constants. Until that time, rate constants should be experimentally measured when possible, recognizing that experimental measurements are currently difficult for low-volatility chemicals.

References

Atkinson, R., and W.P.L. Carter. 1984. Kinetics and mechanisms of the gas-phase reactions of ozone with organic compounds under atmospheric conditions. *Chem. Rev.* 84:437-470.

Atkinson, R. 1986. Kinetics and mechanisms of the gas-phase reactions of the hydroxyl radical with organic compounds under atmospheric conditions. *Chem. Rev.* 86:69-201.

Atkinson, R. 1987. A structure-activity relationship for the estimation of rate constants for the gas-phase reactions of OH radicals with organic compounds. *Int. J. Chem. Kinet.* 19:799-828.

Atkinson, R. 1988. Estimation of gas-phase hydroxyl radical rate constants for organic chemicals. *Environ. Toxicol. Chem.* 7:435-442.

Atkinson, R. 1989. Kinetics and mechanisms of the gas-phase reactions of the hydroxyl radical with organic compounds. *J. Phys. Chem. Ref. Data* Monograph 1:1-246.

Atkinson, R. 1991. Kinetics and mechanisms of the gas-phase reactions of the NO_3 radical with organic compounds. *J. Phys. Chem. Ref. Data* 20:459-507.

Atkinson, R. 1994. Gas-phase tropospheric chemistry of organic compounds. *J. Phys. Chem. Ref. Data*, Monograph 2:1-216.

Atkinson, R. 1995. Gas phase tropospheric chemistry of organic compounds. In *Issues in Environmental Science and Technology*, Issue No. 4, R.E. Hester and R.M. Harrison, Eds. The Royal Society of Chemistry, Cambridge, UK, pp. 65-89.

Atkinson, R. 1996. Atmospheric chemistry of PCBs, PCDDs and PCDFs. In *Issues in Environmental Science and Technology*, Issue No. 6, R.E. Hester and R.M. Harrison, Eds. The Royal Society of Chemistry, Cambridge, UK, pp. 53-72.

Atkinson, R. 1997. Gas-phase tropospheric chemistry of volatile organic compounds. 1. Alkanes and alkenes. *J. Phys. Chem. Ref. Data*, 26:215-290.

Atkinson, R., D.L. Baulch, R.A. Cox, R.F. Hampson, Jr., J.A. Kerr, M.J. Rossi, and J. Troe. 1997. Evaluated kinetic and photochemical data for atmospheric chemistry: Supplement VI. *J. Phys. Chem. Ref. Data*, 26:1329-1499.

Ayers, G.P., S.A. Penkett, R.W. Gillett, B. Bandy, I.E. Galbally, C.P. Meyer, C.M. Elsworth, S.T. Bentley, and B.W. Forgan. 1992. Evidence for photochemical control of ozone concentrations in unpolluted marine air. *Nature* 360:446-448.

Balkanski, Y.J., D.J. Jacob, G.M. Gardner, W.C. Graustein, and K.K. Turekian. 1993. Transport and residence times of tropospheric aerosols inferred from a global three-dimensional simulation of ^{210}Pb. *J. Geophys. Res.* 98:20573-20586.

Bidleman, T.F. 1988. Atmospheric processes. *Environ. Sci. Technol.* 22:361-367.

Biermann, H.W., H. Mac Leod, R. Atkinson, A.M. Winer, and J.N. Pitts, Jr. 1985. Kinetics of the gas-phase reactions of the hydroxyl radical with naphthalene, phenanthrene, and anthracene. *Environ. Sci. Technol.* 19:244-248.

Brauers, T., U. Aschmutat, U. Brandenburger, H.-P. Dorn, M. Hausmann, M. Heßling, A. Hofzumahaus, F. Holland, C. Plass-Dülmer, and D.H. Ehhalt. 1996. Intercomparison of tropospheric OH radical measurements by multiple folded long-path laser absorption and laser induced fluorescence. *Geophys. Res. Lett.* 23:2545-2548.

Brown, H.C., and Y. Okamoto. 1958. Electrophilic substituent constants. *J. Am. Chem. Soc.* 80:4979-4987.

Cohen, N., and S.W. Benson. 1987a. Transition-state-theory calculations for reactions of OH with haloalkanes. *J. Phys. Chem.* 91:162-170.

Cohen, N., and S.W. Benson. 1987b. Empirical correlations for rate coefficients for reactions of OH with haloalkanes. *J. Phys. Chem.* 91:171-175.

Cooper, D.L., N.L. Allan, and A. McCulloch. 1990. Reactions of hydrofluorocarbons and hydrochlorofluorocarbons with the hydroxyl radical. *Atmos. Environ.* 24A:2417-2419.

Dilling, W.L., S.J. Gonsior, G.U. Boggs, and C.G. Mendoza. 1988. Organic photochemistry. 20. A method for estimating gas-phase rate constants for reactions of hydroxyl radicals with organic compounds from their relative rates of reaction with hydrogen peroxide under photolysis in 1,1,2-trichlorotrifluoroethane solution. *Environ. Sci. Technol.* 22:1447-1453.

Eisenreich, S.J., B.B. Looney, and J.D. Thornton. 1981. Airborne organic contaminants in the Great Lakes ecosystem. *Environ. Sci. Technol.* 15:30-38.

Gaffney, J.S., and S.Z. Levine. 1979. Predicting gas phase organic molecule reaction rates using linear free-energy correlations. I. O(^{3}P) and OH addition and abstraction reactions. *Int. J. Chem. Kinet.* 11:1197-1209.

Goodman, M.A., S.M. Aschmann, R. Atkinson, and A.M. Winer. 1988. Atmospheric reactions of a series of dimethyl phosphoroamidates and dimethyl phosphorothioamidates. *Environ. Sci. Technol.* 22:578-583.

Grosjean, D., and E.L. Williams, II. 1992. Environmental persistence of organic compounds estimated from structure-reactivity and linear free-energy relationships. Unsaturated aliphatics. *Atmos. Environ.* 26A:1395-1405.

Güsten, H., W.G. Filby, and S. Schoof. 1981. Prediction of hydroxyl radical reaction rates with organic compounds in the gas phase. *Atmos. Environ.* 15:1763-1765.

Güsten, H., L. Klasinc, and D. Maric. 1984. Prediction of the abiotic degradability of organic compounds in the troposphere. *J. Atmos. Chem.* 2:83-93.

Güsten, H., and L. Klasinc. 1986. Eine voraussagemethode zum abiotischen abbauverhalten von organischen chemikalien in der umwelt. *Naturwissen.* 73:129-135.

Heicklen, J. 1981. The correlation of rate coefficients for H-atom abstraction by HO radicals with C-H bond dissociation enthalpies. *Int. J. Chem. Kinet.* 13:651-665.

Hodson, J. 1988. The estimation of the photodegradation of organic compounds by hydroxyl radical reaction rate constants obtained from nuclear magnetic resonance spectroscopy chemical shift data. *Chemosphere* 17:2339-2348.

Hynes, A.J., and P.H. Wine. 1991. Kinetics and mechanism of the reaction of hydroxyl radicals with acetonitrile under atmospheric conditions. *J. Phys. Chem.* 95:1232-1240.

Jolly, G.S., G. Paraskevopoulos, and D.L. Singleton. 1985. Rates of OH radical reactions. XII. The reactions of OH with c-C_3H_6, c-C_5H_{10}, and c-C_7H_{14}. Correlation of hydroxyl rate constants with bond dissociation energies. *Int. J. Chem. Kinet.* 17:1-10.

Klamt, A. 1993. Estimation of gas-phase hydroxyl radical rate constants of organic compounds from molecular orbital calculations. *Chemosphere* 26:1273-1289.

Klamt, A. 1996. Estimation of gas-phase hydroxyl radical rate constants of oxygenated compounds based on molecular orbital calculations. *Chemosphere* 32:717-726.

Klöpffer, W., G. Kaufmann, and R. Frank. 1985. Phototransformation of air pollutants: Rapid test for the determination of k_{OH}. *Z. Naturforsch.* 40a:686-692.

Koch, R., H.-U. Krüger, M. Eland, W.-U. Palm, and C. Zetzsch. 1996. Rate constants for the gas-phase reactions of OH with amines: *tert*-butyl amine, 2,2,2-trifluoroethyl amine, and 1,4-diazabicyclo[2.2.2]octane. *Int. J. Chem. Kinet.* 28:807-815.

Koch, R., W.-U. Palm, and C. Zetzsch. 1997. First rate constants for reactions of OH radicals with amides. *Int. J. Chem. Kinet.* 29:81-87.

Kwok, E.S.C., W.P. Harger, J. Arey, and R. Atkinson. 1994. Reactions of gas-phase phenanthrene under simulated atmospheric conditions. *Environ. Sci. Technol.* 28:521-527.

Kwok, E.S.C., and R. Atkinson. 1995. Estimation of hydroxyl radical reaction rate constants for gas-phase organic compounds using a structure-reactivity relationship: An update. *Atmos. Environ.* 29:1685-1695.

Kwok, E.S.C., S.M. Aschmann, and R. Atkinson. 1996. Rate constants for the gas-phase reactions of the OH radical with selected carbamates and lactates. *Environ. Sci. Technol.* 30:329-334.

Kwok, E.S.C., R. Atkinson, and J. Arey. 1997. Kinetics of the gas-phase reactions of indan, indene, fluorene, and 9,10-dihydroanthracene with OH radicals, NO_3 radicals, and O_3. *Int. J. Chem. Kinet.*, 29:299-309.

Logan, J.A. 1985. Tropospheric ozone: Seasonal behavior, trends, and anthropogenic influence. *J. Geophys. Res.* 90:10463-10482.

Mather J.H., P.S. Stevens, and W.H. Brune. 1997. OH and HO_2 measurements using laser-induced fluorescence. *J. Geophys. Res.* 102:6427-6436.

McIlveen, R. 1992. *Fundamentals of Weather and Climate*, Chapman & Hall, London.

Medven, Z., H. Güsten, and A. Sabjic. 1996. Comparative QSAR study on hydroxyl radical reactivity with unsaturated hydrocarbons: PLS versus MLR. *J. Chemometrics* 10:135-147.

Meylan, W. 1997. Estimation Accuracy of the Atmospheric Oxidation Program Version 1.8, Syracuse Research Corporation, Syracuse, NY, February.

Meylan, W.M., and P.H Howard. 1993. Computer estimation of the atmospheric gas-phase reaction rate of organic compounds with hydroxyl radicals and ozone. *Chemosphere* 26:2293-2299.

Mount G.H., F.L. Eisele, D.J. Tanner, J.W. Brault, P.V. Johnston, J.W. Harder, E.J. Williams, A. Fried, and R. Shetter. 1997. An intercomparison of spectroscopic laser long-path and ion-assisted in situ measurements of hydroxyl concentrations during the Tropospheric OH Photochemistry Experiment, fall 1993. *J. Geophys. Res.* 102:6437–6455.

Ohta, T. 1983. Rate constants for the reactions of diolefins with OH radicals in the gas phase. Estimate of the rate constants from those for monoolefins. *J. Phys. Chem.* 87:1209-1213.

Oltmans, S.J., and H. Levy II. 1994. Surface ozone measurements from a global network. *Atmos. Environ.* 28:9-24.

Pankow, J.F. 1987. Review and comparative analysis of the theories on partitioning between the gas and aerosol particulate phases in the atmosphere. *Atmos. Environ.* 21:2275-2283.

Prinn, R.G., R.F. Weiss, B.R. Miller, J. Huang, F.N. Alyea, D.M. Cunnold, P.J. Fraser, D.E. Hartley, and P.G. Simmonds. 1995. Atmospheric trends and lifetime of CH_3CCl_3 and global OH concentrations. *Science* 269:187-192.

Talukdar, R.K., S.C. Herndon, J.B. Burkholder, J.M. Roberts, and A.R. Ravishankara. 1997. Atmospheric fate of alkyl nitrates: Part I. Rate coefficients for the reactions of alkyl nitrates with isotopically labelled hydroxyl radicals. *J. Chem. Soc. Faraday Trans.*, 93:2787-2796.

Walker, R.W. 1985. Temperature coefficients for reactions of OH radicals with alkanes between 300 and 1000 K. *Int. J. Chem. Kinet.* 17:573-582.

Zetzsch, C. 1982. 15th Informal Conference on Photochemistry. Stanford, CA, June 27-July 1.

15

Photoreactions in Surface Waters

Theodore Mill

CONTENTS

15.1 Introduction

Photolysis (or photoreaction) can be defined as "any chemical reaction that occurs only in the presence of light." Environmental photoreactions necessarily take place in the presence of sunlight, which has significant photon fluxes only above 295 nm in the near ultraviolet (UV), extending into the infrared region of the electromagnetic spectrum. Environmental photoreactions occur in surface waters, on soil, and in the atmosphere, sometimes rapidly enough to make them the dominant environmental transformation processes for many organic compounds. In the atmosphere, for example, photooxidation, mediated by hydroxyl radical (HO), is the dominant loss process for more than 90% of the organic compounds found there (See Atkinson, 1989, 1994, and Chapter 14).

Photoreactions are often complex reactions that not only control the fate of many chemicals in air and water, but often produce products with chemical, physical, and biological properties quite different from those of their parent compounds: more water soluble, less volatile, and less likely to be taken up by biota. Photooxidation removes many potentially harmful chemicals from the environment, although occasionally more toxic products form in oil slicks and from pesticides (Larson et al., 1977). Biogeochemical cycling of organic sulfur compounds in marine systems involves photooxidation on a grand scale in surface waters, as well as in the troposphere (Brimblecombe and Shooter, 1986).

This chapter discusses environmental photoreactions chiefly in terms of two broad categories of reactions: direct and indirect. A direct photoreaction occurs when a photon is absorbed by a compound leading to formation of excited or radical species, which can react in a variety of different ways to form stable products. In dilute solution, rate constants for these reactions are the products of the rate constants for light absorption and the reaction efficiencies (quantum yields).

An indirect photoreaction occurs when a sunlight photon is absorbed by one compound or group of compounds to form oxidants or excited states, which then react with or transfer energy to other compounds present in the same environmental compartment to form new products. For example, NO_2 and O_3 in air form HO radicals, and humic acids in water form singlet oxygen and oxyradicals, when they absorb sunlight photons. These oxidants react with other chemicals in thermal (dark) reactions, and the rates for these processes follow simple bimolecular kinetics (Mill, 1988; Zepp et al., 1987a; Tratnyek and Hoigné, 1991).

15.2 Direct Photoreactions: Light Absorption and Kinetics

Only a small proportion of synthetic organic compounds absorb UV light in the sunlight region of the spectrum (above 295 nm) and then photolyze at significant rates. Most aliphatic and oxygenated compounds, such as alcohols, acids, esters, and ethers, absorb only in the far UV region, below 220 nm, and simple benzene derivatives with alkyl groups or one heteroatom substituent absorb strongly in only the far and middle UV region. Nitro or polyhalogenated benzenes, naphthalene derivatives, polycyclic aromatics and aromatic amines, nitroalkanes, azoalkanes, ketones, and aldehydes do absorb sunlight between 300 and 450 nm; polycyclic and azoaromatics (dyes) as well as quinones, also absorb visible light, in some cases to beyond 700 nm (Calvert and Pitts, 1967).

The rate of a direct photoprocess depends only on the product of the rate of light (photon) absorption by compound C (I_A) and the efficiency with which the absorbed light is used to effect reaction (quantum yield, Φ):

$$\text{Rate} = \frac{dC}{dt} = \text{Efficiency} \times \text{Photons Absorbed}/\text{time} = \Phi I_A \qquad (1)$$

Under most environmental conditions, chemicals are present in water or air at low concentrations, so that their light absorbing properties lead to simple kinetic expressions for direct photolysis in water (Zepp and Cline, 1977; Mill and Mabey, 1985; Leifer, 1989) and air (Demerjian et al., 1980; Finlayson-Pitts and Pitts, 1986).

The starting point for modeling light absorption is the Beer-Lambert law, which relates the light, $I_{o\lambda}$, incident on a solution or vapor mixture of chemical C at wavelength λ, to the light emergent from the solution, I_λ, according to the equation

$$\text{Absorption } (A_2) = \log(I_{o\lambda}/I_\lambda) = \varepsilon_\lambda[C]\ell \qquad (2)$$

where ε_λ is the extinction coefficient at wavelength λ of the light-absorbing chemical, [C] is the molar concentration of chemical C in solution, and ℓ is the pathlength of the solution in cm. Equation (2) may be rewritten in the exponential form:

$$I_\lambda = I_{o\lambda}10^{-\varepsilon\lambda[C]\ell} \qquad (3)$$

The amount of light absorbed at wavelength λ, $I_{A\lambda}$, is

$$I_{A\lambda} = I_{o\lambda} - I_\lambda = I_{o\lambda}(1 - 10^{-\varepsilon\lambda[C]\ell}) \qquad (4)$$

Natural substances (dissolved organic matter, DOM) also contribute to light absorption, and α_λ is used to account for light attenuation by DOM in natural water. Total light absorption in a natural water follows the relation in Equation (5):

$$I_{TA\lambda} = I_{o\lambda}[1 - 10^{-(\alpha_\lambda + \varepsilon\lambda[C])\ell}] \qquad (5)$$

However, direct photoprocesses are concerned only with the light absorbed directly by the chemical ($I_{A\lambda}$). When $\alpha_\lambda \gg \varepsilon_\lambda[C]$:

$$I_{A\lambda} = I_{o\lambda}\left[1 - 10^{-(\alpha\lambda\ell)}\right]\left\{\frac{\varepsilon_\lambda[C]}{\varepsilon_\lambda[C] + \alpha_\lambda}\right\} \qquad (6)$$

where $\varepsilon_\lambda[C]/(\varepsilon_\lambda[C] + \alpha_\lambda) = F(C)$, the fraction of light at wavelength λ absorbed by C.

Equation (6) is a master equation which can be used to develop expressions for photolysis rates in sunlight. In shallow, clear water where $\alpha_\lambda \gg \varepsilon_\lambda[C]$ and both terms are <0.02, Equation (6) simplifies to:

$$I_{A\lambda} = 2.3\, I_{o\lambda}\varepsilon_\lambda\ell[C] \qquad (7)$$

If $I_{o\lambda}$ is constant over the measurement interval, the rate of photolysis of [C] at a single wavelength becomes a first-order kinetic equation by substituting Equation (7) into Equation (1):

$$-\left(\frac{dC}{dt}\right)_{P\lambda} = \Phi_\lambda \frac{2.3\, I_{o\lambda}}{jD}\varepsilon_\lambda\ell[C] = k_{P\lambda}[C] \qquad (8)$$

where j is a conversion factor (6×10^{20}) for photons to Einsteins (for compatibility with molar concentration units) and D is the depth of the water body in cm (Zepp and Cline, 1977).

The rate constant at each wavelength can be summed over all wavelengths where $\varepsilon > 0$ to give the rate constant in sunlight (k_{PE}), and, assuming that Φ is independent of wavelength:

$$-\left(\frac{dC}{dt}\right)_{PE} = \frac{\Phi 2.3}{jD} \sum \varepsilon_\lambda I_{o\lambda}[C] = k_{PE}[C] \qquad (9)$$

Figure 15.1 shows the overlap of the solar spectrum (L_λ) with the absorption spectrum(ε_λ) of p-nitroanisole (PNA) to create an area defined by $\Sigma I_{o\lambda}\varepsilon_\lambda$, where $I_{o\lambda}$ is expressed in terms of L_λ values (see below). Equation (9) can be used to estimate k_{PE} under environmental conditions from appropriate data.

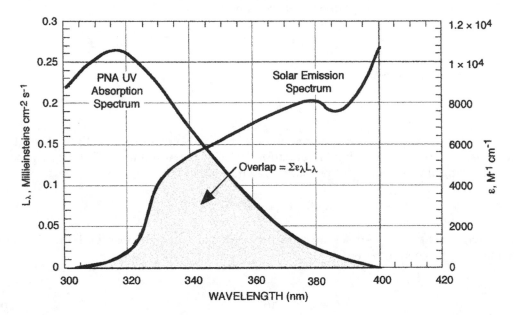

FIGURE 15.1

The overlap of the UV spectrum of p-nitroanisole (PNA) with the solar spectrum at the earth's surface. The overlap of the two curves represents the photoactive region for PNA defined as $\Sigma \varepsilon_\lambda L_\lambda$.

The EPA's GC-SOLAR computer model (EPA, 1982) estimates values of $I_{o\lambda}\ell/D$ as a function of wavelength, time of day, year, and latitude using light flux terms called Z_λ with units of photons $cm^{-2} s^{-1}$. Zepp and Cline (1977) tabulated Z_λ values for mid-day and mid-season at 40°L. GC-SOLAR provides the most reliable method for estimating photolysis rates corresponding to half lives of a few minutes to a few hours.

Mabey et al. (1982) computed diurnal (24-hr) light intensity terms, $L_\lambda = 2.3 \ I_{o\lambda}\ell/D$, for four seasonal dates at specific decadic latitudes in shallow water depths. L_λ values have units of Millieinsteins $cm^{-2} day^{-1}$. L_λ multiplied by ε_λ gives a rate constant averaged over the diurnal changes in sunlight intensity and a 24-hour period and better represents average light intensity over periods of one day or more. Values of L_λ at decadic latitudes from 20° to 70° appear in the EPA OPPTS guidelines for photolysis in water and air (EPA, 1995), in Mill and Mabey (1985), and in Leifer (1988). Table 15.1 lists Z_λ and L_λ values for 40° latitude in summer and winter.

Z_λ and L_λ values in Table 15.1 are averaged over 2-3 nm wavelength bands near 300 nm, where solar intensity is changing rapidly, and then over 10 nm bands beyond 330 nm, where changes are smaller. With substitution of L_λ for $I_{o\lambda}$, Equation (9) then simplifies to:

$$-\left(\frac{d[C]}{dt}\right)_{PE} = \Phi \Sigma \varepsilon_\lambda L_\lambda[C] \qquad (10)$$

TABLE 15.1

Light Intensity Values for 40° Latitude in Summer and Winter.[a]

Wavelength Center, nm	Z_λ (Summer) (Photons cm^{-2} s^{-1})[b]	Z_λ (Winter)[b]	L_λ (Summer) (Millieinsteins cm^{-2} d^{-1})[c]	L_λ (Winter)[c]
297.5	7.16e+11	7.00e+9	6.11e-5	5.49e-7
300.0	2.40e+12	7.33e+10	2.69e-4	5.13e-6
302.5	7.23e+12	3.68e+11	8.30e-4	3.02e-5
305.0	1.81e+13	1.70e+12	1.95e-3	1.19e-4
307.5	3.05e+13	4.50e+12	3.74e-3	3.38e-4
310.0	4.95e+13	8.45e+12	6.17e-3	7.53e-4
312.5	7.17e+13	1.77e+13	9.07e-3	1.39e-3
315.0	9.33e+13	2.71e+13	1.22e-2	2.22e-3
317.5	1.15e+14	3.62e+13	1.55e-2	3.19e-3
320.0	1.35e+14	4.98e+13	1.87e-2	4.22e-3
323.1	2.52e+14	9.06e+13	3.35e-2	8.25e-3
330.0	8.46e+14	3.42e+14	1.16e-1	3.16e-2
340.0	9.63e+14	4.20e+14	1.46e-1	4.31e-2
350.0	1.03E+15	4.49e+14	1.62e-1	4.98e-2
360.0	1.10e+15	4.79e+14	1.79e-1	5.68e-2
370.0	1.22e+15	5.20e+14	1.91e-1	6.22e-2
380.0	1.35e+15	5.62e+14	2.04e-1	6.78e-2
390.0	1.61e+15	8.05e+14	1.93e-1	6.33e-2
400.0	2.31e+15	1.16e+15	2.76e-1	9.11e-2
410.0	3.02e+15	1.54e+15	3.64e-1	1.20e-1
420.0	3.10e+15	1.59e+15	3.74e-1	1.24e-1
430.0	2.98e+15	1.54e+15	3.61e-1	1.20e-1
440.0	3.51e+15	1.84e+15	4.26e-1	1.43e-1
450.0	3.94e+15	2.08e+15	4.80e-1	1.61e-1
460.0	3.98e+15	2.11e+15	4.85e-1	1.64e-1
470.0	4.11e+15	2.19e+15	5.02e-1	1.69e-1
480.0	4.20e+15	2.25e+15	5.14e-1	1.74e-1
490.0	3.96e+15	2.13e+15	4.86e-1	1.65e-1
500.0	4.04e+15	2.18e+15	4.96e-1	1.68e-1

[a] Values in powers of ten: 7.16e+11 = 7.16×10^{11}.
[b] From Zepp and Cline (1977).
[c] From Mill and Mabey (1985).

$$k_{PE} = \Phi \Sigma \varepsilon_\lambda L_\lambda \text{ or } 2.3 \Phi \Sigma \varepsilon_\lambda Z_\lambda / j \quad (11)$$

and

$$-\ln([C]_o / [C_t]) = k_{PE} t \quad (12)$$

Equation (12) is the integrated first order form of Equation (10) in which $[C]_o$ and $[C]_t$ are concentrations at time 0 and t. If L_λ is used in Equation (11), k_{PE} is a 24 hr averaged, near-surface value (day^{-1}); if Z_λ is used, k_{PE} is a midday maximum value (s^{-1}), also near surface (D = 1 cm).

15.2.1 Depth Dependence of k_{PE}

Equation (10) estimates rate constants in shallow surface waters or distilled water where only a small fraction of photons are absorbed, light scattering is minimal, and absorption of light follows the Beer-Lambert relation. Rate constants in deep water, where all photons are absorbed and significant scattering occurs, follow a more complex relation involving the diffuse attenuation coefficient, K_λ. K_λ is a measure of the change in light intensity with depth, at wavelength λ, caused by both absorption and scattering, and is a better measure of transmitted light than α, because α does not account for scattering. The depth-averaged rate constant in deep water when all light is absorbed is:

$$k_{PEavg} = k_{PE(0)}/K_{R\lambda}D \tag{13}$$

and the rate constant at depth D follows the relation (where the near-surface rate constant is $k_{PE(0)}$):

$$k_{PE(D)} = k_{PE(0)}e^{-K_\lambda D} \tag{14}$$

Equation (14) strictly holds only for a narrow wavelength region. Winterle et al. (1987) provide experimental procedures for evaluating wavelength-integrated K_R values in sunlight for several different compounds, using both an actinometer and a radiometer to measure light intensity at several depths and wavelengths.

K_R values over a narrow wavelength region ($K_{R\lambda}$) are compound and water specific and are wavelength dependent. In a moderately eutrophic surface water (Searsville Lake), $K_{R\lambda}$ values varied from 0.12 cm^{-1} at 298 nm to 0.026 cm^{-1} at 400 nm, based on radiometry. Values of $K_{R\lambda}$, integrated over the solar spectral region where $\varepsilon > 0$ for carbazole, varied from 0.072 cm^{-1} in Searsville Lake in June to 0.16 cm^{-1} in an Idaho test pond in September. The value of K_R for p-nitroacetophenone is 0.066 cm^{-1} in Searsville Lake in June. The values of K_R for carbazole and p-nitroacetophenone in Searsville Lake correspond to half of surface photoreaction rates at 9.8 and 11 cm depth, respectively. Based on these studies, the range of K_R values is not large in the near UV, and 0.1 cm^{-1} can be used as an average value of K_R in this region to estimate depth dependence.

15.2.2   Estimation Procedures for k_{PE}

Equation (11) estimates sunlight photoreaction rate constants using computed or tabulated values for Z_λ or tabulated values of L_λ at the appropriate latitude and time of year (Zepp and Cline, 1977; Mill and Mabey, 1985; Leifer, 1988), together with measured values of ε_λ and Φ. Equation (2) can be used to estimate values ε of ε_λ from a uv spectrum measured in water or, if aqueous solubility is low, a polar organic solvent such as acetonitrile or methanol. Absorbance values are converted into ε_λ values at wavelength centers corresponding to those Table 15.1 lists.

If the quantum yield is not known, which is often the case, it can be set equal to 1 and GC-SOLAR or Equation (11) with Z_λ becomes a useful screening tool for estimating the maximum (solar noon) value of k_{PE}. If other transformation processes for the same chemical have rate constants larger than the value calculated for k_{PE} with $\Phi = 1$, then photolysis is not likely to be important, because actual values of Φ are generally less than one and usually less than 0.1 (Calvert and Pitts 1967; Mill and Mabey, 1985; Mill, 1995).

Few reliable procedures are available for estimating Φ even among a closely related group of compounds, because its value depends on competition among several activation and deactivation steps for excited state species, only a few of which are known with any certainty for a few well-studied compounds (Turro, 1967). Mill and Mabey (1985) and Mill (1995) list Φ values for several types of photoreactions in aerated water. Calvert and Pitts (1967) present extensive compilations of Φ values for many photoreactions, but only a few of these measurements were conducted in the presence of air with light >300 nm. Absence of oxygen may cause Φ values to increase because oxygen quenches excited states, and Φ values also may be higher if light much below 300 nm is used.

Peijnenburg et al. (1992) describe a fairly good correlation between Φ values for photo-hydrolysis of a series of halogenated benzenes and their C-Hal bond strengths and steric parameters. This example seems to show that, for closely related compounds, predicting Φ values from structure activity relations may be possible in some cases, but few examples are known.

15.2.3 Estimation of k_{PE} for Benchmark Chemicals

All benchmark chemicals, anthracene (AN), chlorpyrifos (CP), 2,6-di-t-butylphenol (DBP), γ-hexachlorocyclohexane (HCH), and trichloroethylene (TCE) absorb UV light between 200-300 nm, but only anthracene and chlorpyrifos absorb solar photons (> 295 nm) rapidly enough to undergo direct photoreactions at significant rates. Table 15.2 summarizes the UV spectral properties of AN, CP, DBP, and two other compounds, p-nitroanisole (**I**) (PNA) and dinitramide ion (**II**) (DN ion), both of which directly photolyze in sunlight.

TABLE 15.2

UV Spectra of Benchmark Chemicals and PNA and DN ion

Wavelength Center, nm	ε_λ (M^{-1} cm^{-1})				
	AN[a]	CP[b]	DBP[c]	PNA[d]	DN ion[e]
250.0	8890	10	155	11,000	7500
270.0	594	900	490	9,200	5750
290.0	283	3000	282	8,500	5100
297.5	438	2700	14	8,076	4424
300.0	478	2500	5	8,480	4194
302.5	590	1700	0	9,030	3871
305.0	791	1000		9,350	3364
307.5	1025	700		9,790	3180
310.0	1115	400		10,000	2949
312.5	1095	300		10,260	2857
315.0	1190	200		10,350	2719
317.5	1510	150		10,370	2673
320.0	2010	100		10,330	2350
323.1	2425	60		10,120	2595
330.0	2120	0		9,210	2350
340.0	4780			7,165	1982
350.0	3815			4,906	1429
360.0	4045			2,968	876
370.0	3300			1,618	419
380.0	1830			749	166
390.0	45			288	55
400.0	16			87	18
410.0	10			15	5
420.0	7.5			7	7
430.0	0			5	5
440.0				0	0

[a] Anthracene spectrum measured on 2 mM acetonitrile solution.
[b] Chlorpyrifos data estimated from spectrum in Dilling et al. (1984).
[c] Dibutylphenol spectrum estimated from the spectrum of a 10 mM acetonitrile solution of 2,6-di-t-butyl-4-methylphenol.
[d] p-Nitroanisole spectrum measured in 0.1 mM aqueous solution.
[e] Dinitramide ion spectrum measured in 0.1 mM aqueous solution.

All UV spectral data were obtained by measurement using a conventional UV/vis spectrometer (HP8458) with either aqueous or acetonitrile solutions of the benchmark chemicals. The spectrum of DBP was assumed to be the same as the spectrum of 2,6-di-t-butyl-4-methylphenol. Since DBP differs from this compound by only a para methyl group, the assumption of identical spectra is a good approximation and the absence of absorbance above 300 nm quite certain. Values of ε_λ were calculated from the absorption at the wavelength centers λ (shown in Table 15.1, and the concentration using Equation (2).

I (PNA)   II (DN ion)

The preferred method for estimating k_{PE} is to use measured ε_λ and Φ values as inputs to GC-SOLAR, specifying latitude, depth, time of day and time of year. If GC-SOLAR is not available, k_{PE} can be estimated by calculating $\Sigma\varepsilon_\lambda Z_\lambda$ or $\Phi\Sigma\varepsilon_\lambda L_\lambda$ from intensity data in Table 15.1 and ε_λ data (such as that in Table 15.2) for the compound of interest, estimated from its UV spectrum and Equation (2). The calculation of $\Sigma\varepsilon_\lambda L_\lambda$ is easily done with a spreadsheet format in graphical software such as Excel, Cricket Graph or Kaleidagraph. Table 15.3 gives an example of the calculation, using ε_λ data for anthracene from Table 15.2.

UV spectra can be calculated, in principle, from molecular structure, using a combination of molecular mechanics to optimize molecular configuration and quantum chemical methods to estimate excited state energies. The Zindo program was designed to make these calculations and appears to be a promising procedure for screening organic compounds to determine which ones will have significant UV absorption in the solar region. However, the currently available programs estimate spectra for the gas phase and do not account for polar solvent effects on transition energies and cross sections (ε). For the near term, molecular modeling will not substitute for direct measurement of UV spectra, but computational methods change so rapidly that in all likelihood reliable computational estimates of uv spectra of aqueous organic compounds will be possible by 2010.

Table 15.4 lists values of $\Sigma\varepsilon_\lambda L_\lambda$ and k_{PE} for AN, CP, DBP, PNA, and DN ion using the method Table 15.3 illustrates. In most cases L_λ was used, but Z_λ was used to estimate k_{PE} for DN anion because it absorbs light strongly and photolyzes rapidly.

For anthracene, using $\Sigma\varepsilon_\lambda L_\lambda$ ($k_{PE(max)}$) gives a half life of 17 seconds in summer sunlight; if $2.3\Sigma\varepsilon_\lambda Z_\lambda/j$ is used to estimate $k_{PE(max)}$, the half life is 8 seconds. However, since the measured quantum yield is only 0.003 at 366 nm (Zepp and Schlotzhauer, 1979), the calculated value of k_{PE} (which is Φk_{AE}) is closer to 0.48 hr^{-1} (0.003 × 145), corresponding to a half life of 1.5 hours using L_λ or 44 minutes using Z_λ. The measured half life, 45 minutes at solar noon (Zepp and Schlotzhauer, 1979), is in excellent agreement with the calculated value. For a rapidly photolyzing compound, the value of k_{PE} calculated with Z_λ often will be more accurate than with L_λ, because of the strong time-of-day dependence for sunlight intensity.

For chloropyrifos, the 19 d^{-1} value of $\Sigma\varepsilon_\lambda L_\lambda$ or $k_{PE(max)}$ corresponds to a half life of 1.3 hours. The measured quantum yield is 0.0048 at 313 nm (Dilling et al., 1984), giving a calculated value of k_{PE} for CP of 0.091 day^{-1} (19 × 0.0048) or a half life of 7.6 days. The measured half life of 11 days in summer (Meikle et al., 1983) is in fair agreement with this

TABLE 15.3

Calculation of $\Sigma\varepsilon_\lambda L_\lambda$ for Anthracene.[a]

Wavelength Center, nm	L_λ (Summer, 40° Lat) Millieinsteins cm^{-2} d^{-1}	ε_λ M^{-1}cm^{-1}	$\varepsilon_\lambda L_\lambda$ d^{-1}	$\Sigma\varepsilon_\lambda L_\lambda$[b]
297.5	6.11e-5	438	0.03	0.03
300.0	2.69e-4	478	0.13	0.16
302.5	8.30e-4	590	0.49	0.65
305.0	1.95e-3	791	1.54	2.19
307.5	3.74e-3	1025	3.83	6.02
310.0	6.17e-3	1115	6.91	12.93
312.5	9.07e-3	1095	9.93	22.86
315.0	1.22e-2	1190	14.52	37.38
317.5	1.55e-2	1510	23.41	60.79
320.0	1.87e-2	2010	37.59	98.37
323.1	3.35e-2	2425	81.24	179.61
330.0	1.16e-1	2120	245.92	425.53
340.0	1.46e-1	4780	697.88	1123.41
350.0	1.62e-1	3815	618.03	1741.44
360.0	1.79e-1	4045	724.05	2465.49
370.0	1.91e-1	3300	630.30	3095.79
380.0	2.04e-1	1830	373.32	3469.11
390.0	1.93e-1	45	8.69	3477.80
400.0	2.76e-1	16	4.42	3482.22
410.0	3.64e-1	10	3.64	3485.86
420.0	3.74e-1	7.5	2.81	3488.66
430.0		0	0.00	3488.66
				3490.00

[a] Column 2 data are from Table 15.1 and Column 3 data are from Table 15.2.
[b] These values are the running sums of Column 4 values; the last value is $\Sigma\varepsilon_\lambda L_\lambda$.

TABLE 15.4

Estimated and Measured Values of k_{PE} at 40° Latitude, Summer for Benchmark Chemicals and PNA and DN ion

Compound	$\Sigma\varepsilon_\lambda L_\lambda$ (d^{-1})	Φ	$\Phi\Sigma\varepsilon_\lambda L_\lambda$ (k_{PE})	k_{PE} (meas)	$t_{1/2}$ (meas)
Anthracene	3490 (3270)[a]	0.0033	11.5 d^{-1}	22.5 d^{-1a}	44 min[b]
Chloropyrifos	19	0.0048[c]	0.091 d^{-1}	0.063 d^{-1d}	11 days
2,6-Di-t-butylphenol	~0.002	(1)[e]	(~0.002 d^{-1})[e]	—	(346 days)
Hexachlorocyclohexane	0	—	—	—	—
Trichloroethylene	0	—	—	—	—
p-Nitroanisole	5030	0.00028	1.4 d^{-1}	0.99 d^{-1f}	17 hrs
Dinitramide Ion	0.037 s^{-1g}	0.10[h]	0.0037 s^{-1}	0.0033 s^{-1h}	3.5 min

[a] Estimated from $2.3\Sigma\varepsilon_\lambda Z_\lambda/j$.
[b] (Zepp and Schlotzhauer (1979).
[c] Dilling et al, (1984).
[d] Meikle et al. (1983) at pH 5.
[e] Assume $\Phi = 1$ to estimate k_{PE}(max).
[f] Dulin and Mill (1984).
[g] Estimated using $2.3\Sigma\varepsilon_\lambda Z_\lambda/j$.
[h] Su et al. (1995); measured near solar noon.

calculated value. Part of the discrepancy between measured and calculated values of k_{PE} for CP lies in the uncertainty in estimating values of ε from the published spectrum for CP (Dilling et al., 1984); part arises from the methodology Meikle et al. used to measure the sunlight photolysis rate.

The spectrum of PNA exhibits one maximum in the solar region near 320 nm (Table 15.2). The value of $\Sigma\varepsilon_\lambda L_\lambda$ at 40° lat. for PNA is 5030 day^{-1}, corresponding to a 10-second half life in summer sunlight. However, with a measured quantum yield of 2.8×10^{-4} (Dulin and Mill, 1982), the estimated value of k_{PE} is 1.4 day^{-1}, corresponding to a 12-hour half life. The measured half life is 17 hours, again in fair agreement with the estimated value.

DN ion absorbs light in the solar region with a tailing band with a shoulder near 310 nm. However, the quantum yield is 0.1 and does not change with wavelength (Su et. al., 1995). The estimate of k_{AE} for DN ion in summer at 40° lat. gives 0.037 s^{-1}, based on Z_λ:

$$k_{AE} = 2.3\Sigma\varepsilon_\lambda Z_\lambda/j = 2.3(0.016) = 0.037 \text{ s}^{-1} \tag{15}$$

Since the quantum yield is 0.1, k_{PE} is 0.0037 s^{-1} and the half life is 3.1 minutes, compared with the measured half life of 3.5 minutes:

$$k_{PE} = 2.3\Phi\Sigma\varepsilon_\lambda Z_\lambda/j = 2.3(0.1)(0.016) = 0.0037 \text{ s}^{-1} \tag{16}$$

$$t_{1/2} = 0.693/0.0037 = 187 \text{ sec or } 3.1 \text{ min} \tag{17}$$

Dinitramide absorbs sunlight photons more slowly than AN or PNA but uses it much more efficiently than either of the other two compounds, with the result that dinitramide ion photolyzes almost 350 times faster than PNA and fifteen times faster than AN under clear sky conditions.

Since L_λ values are averaged over 24 hours and over each 3-month season, the accuracy of rate estimates depend on the rate of the photolysis process compared with the averaging interval. For example, a compound which photolyzes with a half life of 1 hour will photolyze three times faster at solar noon than at 800 or 1600 hrs. However, if the half life is one week near mid-season, diurnal variations make little difference.

Seasonal variations in light intensity also are important. Average L_λ values for winter are only one fourth to one sixth of summer values, depending on latitude. About one half of total insolation emanates directly from the sun; the other half comes from scattered sky light, so that only heavy cloud cover significantly reduces sunlight intensity. At high latitudes, corrections for reduced photolysis rates under persistent cloud cover can be made by multiplying clear sky rate constants by the average percent reduction clouds caused in light values. The computer model developed for photolysis in northern Europe called ABIWAS specifically takes into account cloud cover (Frank and Klöpffer, 1989).

Reduction in statospheric ozone levels also will shift the solar emission spectral cutoff toward shorter wavelengths, resulting in faster loss rates for any compounds with spectral cutoffs in the same wavelength vicinity. For example, di-t-butylmethyl phenol might then photolyze directly with significant rates in surface waters (see Table 15.2). Indirect photoreactions also will become more rapid for the same reasons.

15.3 Indirect Photoreactions

15.3.1 Photooxidants and Kinetics

Indirect photolysis is most important for compounds that absorb little or no sunlight. Light absorption by chromophores (sensitizers) other than the compound of interest begin the process, forming intermediate (and transient) oxidants or excited states that effect chemical changes in the compound of interest (Atkinson, 1987; Haag and Mill, 1989; Blough and Zepp, 1995; Blough, 1997). Examples of chromophores that serve this purpose are dissolved organic matter (DOM or humic acid) and nitrate ion in water, and ozone and NO_2 in the atmosphere. Transient species formed by indirect photoreactions in water include singlet oxygen and peroxy radicals, both of which are relatively selective and electrophilic. As a result, only electron-rich compounds, such as phenols, furans, aromatic amines, polycyclic aromatic hydrocarbons (PAH), and alkyl sulfides can undergo relatively rapid indirect photoprocesses with these oxidants. Nitroaromatics, though not oxidized, appear to be sensitized by triplet DOM or scavenged by solvated electrons (Mabey et al., 1983). Many of these same compounds (e.g., PAH, nitroaromatics, and aromatic amines), also undergo rapid direct photoreactions (Mill et al., 1981; Zepp and Schlotzhauer, 1979; Behymer and Hites, 1985).

By contrast, HO radical, which dominates tropospheric photochemistry, oxidizes all classes of organic compounds (except perhalogenated compounds), including alkanes, olefins, alcohols, and simple aromatics (Buxton et al., 1988; Atkinson, 1989, 1994). Aqueous HO radical, derived mainly from photolysis of nitrate ion, plays an important role in converting marine DOM to simpler carbonyl compounds, even though the average concentration of HO in marine systems is extremely low ($<2 \times 10^{-18}$ M) (Mopper and Zhou, 1990). HO also appears important in degrading synthetic chemicals in a variety of nitrate-bearing freshwaters, where the HO concentrations appear to be one to two orders of magnitude higher (Zepp et. al., 1987a).

Figure 15.2 illustrates the relationships among redox species formed on insolation (illumination by sunlight) of DOM (Haag and Mill, 1989). In many cases, detailed pathways for forming these oxidants and reductants remain unclear, but identities of several of the transients are fairly well established (Mill, 1989; Blough, 1997). Transient species are transient because they react rapidly with themselves or with a variety of natural organic and metal species in natural waters (Blough, 1997), balancing formation rates to give low average concentrations. Rate equations for indirect photoprocesses depend on bimolecular kinetic equations involving the concentration of each diurnally averaged oxidant concentration $[OX]_{avg}$ and the chemical of interest (C):

$$R_{ox} = -d[C]/dt = \Sigma k_{ox}[Ox]_{avg}[C] = k_{pi}[C] \qquad (18)$$

The kinetic relations controlling $[Ox]_{avg}$ can be formulated in two ways. One way uses photochemical parameters, including light absorption and quantum yields for photooxidant production and rates of transient quenching and loss. The second way uses measured values of ($[Ox]_{avg}$) in insolated surface waters to formulate the kinetic relations shown in Equation (18). The generic time-averaged concentration of a phototransient species ($[Ox]_{avg}$), formed by insolation (ΣL_λ) of a natural sensitizer such as DOM with efficiency Φ where all sunlight is absorbed in the water column of depth D (typical of most natural waters) is given by Equation (19). Loss terms, lumped together as $\Sigma k_d[Q]_i$ may include radical-radical recombinations, excited state deactivation or quenching, and radical –DOM reactions; reaction of the transient with C is usually not important in this equation.

$$[Ox]_{avg} = \Phi\Sigma L_\lambda/D(\Sigma k_d[Q]_i) \qquad (19)$$

Although the value of the term $\Phi\Sigma L_\lambda$ often can be determined, the value of the term $\Sigma k_d[Q]_i$ usually cannot, which is why measured values of $[Ox]_{avg}$ are used in simple second order kinetic relations for estimating the rate of oxidation by each oxidant, and by the sum of the oxidants as Equations (18) through (22) show.

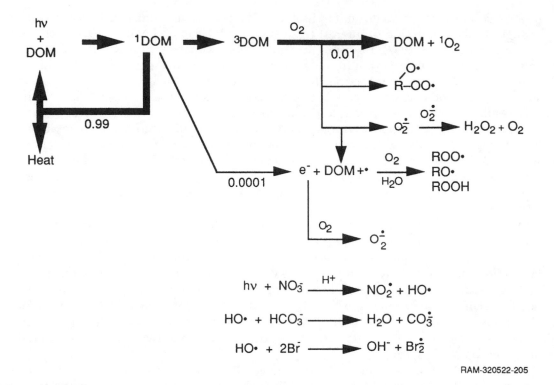

RAM-320522-205

FIGURE 15.2
Photochemical pathways for transient formation in surface waters (Haag and Mill, 1989).

$$-d[C]/dt = k_{ox(i)}[Ox]_{iavg}[C] \tag{20}$$

where $k_{ox(i)}$ is the second order rate constant for the ith oxidant Ox(i) with chemical C. If $[Ox]_{avg}$ is a time-averaged value:

$$\text{Rate} = k'_{OX(avg)}[C] \tag{21}$$

$$\text{Total Rate} = \Sigma k'_{OX(avg)}[C] \tag{22}$$

$$t_{1/2} = 0.693/\Sigma k'_{OX(avg)} \tag{23}$$

Table 15.5 lists concentrations of the major photooxidants in surface waters, diurnally averaged over 24 hours. Note that, even if $k_{ox(i)}$ values are measured or estimated accurately (within a factor two or three), oxidant concentrations in the environment vary widely, and averaged values have a variance of five- to tenfold for any given location. In extreme locations, such as pristine marine waters, or heavily polluted surface waters, oxidant concentrations may be 100 times smaller or larger than the values Table 15.5 lists. Table 15.6 lists rate constants (k_{ox}) for various photooxidants in their reaction with major classes of organic compounds. To estimate the rate of an indirect photoreaction for chemical C (Equation (18)), either a measured or estimated value of k_{ox} is required, specific for each oxidant and for each class of organic compounds. Methods for estimating k_{ox} from molecular structure with structure-activity relationships (SARs) have been developed for many photooxidants and are discussed below.

TABLE 15.5

Time-Average Summer Oxidant
Concentrations in Insolated Surface Waters.[a]

Oxidant	Avg. Concentration, M	Reference
$^1O_2$	3×10^{-14}	f
RO_2	3×10^{-11}	g
HO	$6 \times 10^{-17c} - 1 \times 10^{-18d}$	c,d
DOM^{3b}	6×10^{-14}	h
O_2^-	3×10^{-9}	i
H_2O_2	3×10^{-9}	j
CO_3^-	1×10^{-14e}	e
$e^-(Aq)$	2×10^{-17}	k,l

[a] Values assume clear sky conditions at 40–50°
   latitude with 5 mg/L DOM and 1 mg/L NO_3^-,
   averaged over 24 hr.
[b] Triplet excited states
[c] In freshwater (Zepp et al., 1987a).
[d] In seawater (Mopper and Zhou, 1990).
[e] In Grafensee water with 4 ppm DOC and 1 mM
   CO_3^{2-} (Larson and Zepp, 1988).
[f] Haag and Hoigné (1986).
[g] Mill et al. (1980).
[h] Zepp et al. (1985)
[i] In seawater (Petasne and Zika, 1987).
[j] Cooper et al. (1994).
[k] Fischer et al. (1985).
[l] Zepp et. al. (1987b).

15.3.2   Indirect Photoreaction Estimation Methods

Bimolecular rate constants for oxidation by oxyradicals and singlet oxygen have been measured in numerous studies and have been compiled and used to develop reasonably accurate SARs for photooxidations. Atkinson (1987) developed an additivity SAR to estimate k_{HO}(air) for many kinds of organic structures. Mill (1989) has summarized information on SARs for photooxidants in water and air. See Chapter 14 for a detailed discussion of the procedures used for estimating k_{HO}(air) for most classes of organic compounds.

The most difficult aspect of estimating indirect photoreaction rates is finding a measured value of k_{OX} or estimating k_{OX} for the oxidant and compound of interest. Measured values of k_{OX} are usually much preferred to estimated values, but measured values are available only for a small proportion of organic compounds likely to be found in surface waters. Critical compilations of rate constants for oxidant reactions with organic compounds in water appear in Hendry et al. (RO, RO_2) (1974); Wilkinson et al. ($^1O_2$) (1995); Buxton et al. (HO and e^-Aq) (1988); Hendry and Schuetzle (HO_2) (1976); Neta et al. (RO_2) (1990) and Haag and Yao (HO) (1992).

Environmental oxidants are electrophilic species which effect oxidation by formally transferring an electron or hydrogen atom from a donor species (reductant) or by addition to double bonds to form new O–C bonds. Oxyradicals, including HO, RO, and RO_2 radicals, add to double bonds, transfer H-atoms, and oxidize S and some N atoms (Walling, 1957; Atkinson, 1989, 1994), although only HO radical is sufficiently reactive to add to simple aromatic rings. The high selectivity of RO_2 and $^1O_2$ simplifies requirements for estimating k_{OX} for reaction of these oxidants and a new compound.

Therefore, the first step in estimating an indirect photoreaction rate constant is to characterize the reactivity of all structural units (molecular substructures) in the compound

TABLE 15.6

Rate Constants for Oxidation of Major Classes of Organic
Compounds[a] in Insolated Surface Waters

Class	10^{-9} HO·[b]	RO_2·[c,d,e]	10^{-6} $^1O_2$[f]	$t_{1/2}$[g]
Alkanes	1–3	<0.01	<0.01	44d
Alcohols	2–4	0.1	<0.01	40d
Acids	1–2	<0.1	<0.01	130d
Aromatics	3–6	0.1–0.5	<0.01–10	22d
Olefins	3–20	0.05–1.0	10–100[h]	600h
Phenols	10–20	10^3–10^5	1–20j	13d
Aromatic amines	10–20	10^5	10–300	30d
Furans	10	10^4	100–1000	2.7d
Sulfides	20	<0.1	50[k]	5d

[a] k in units of M^{-1} s^{-1}.
[b] Buxton et al. (1988).
[c] Hendry et al. (1974).
[d] Neta et al. (1990).
[e] t-BuO$_2$· radical.
[f] Wilkinson et al. (1995).
[g] From Equation (21) and [Ox] values in Table 15.5.
[h] Trisubstituted olefins.
[j] Tratnyek and Hoigne (1991).
[k] Monroe (1979).

toward either of these two oxidants, first by a qualitative screening and then by quantitative procedures, illustrated below using the benchmark chemicals. For HO radical oxidations, where most kinds of molecular structures react rapidly, each structural unit can be assigned a quantitative reactivity that, when summed over all units, gives the total molecular reactivity of the compound toward HO. This procedure also is illustrated with the benchmark chemicals. In many cases, the reactivity of one structural unit is known from measurements, but the reactivity of another unit must be estimated from an SAR.

The major difficulty with the measurement databases for these oxidants is that the measured chemicals either have slight solubility in water (>10^{-6} M) or some volatility (vapor pressure > 0.01 torr or > 1 Pa), properties not found for the vast majority of organic chemicals. As a result, databases of k_{HO} (air) and k_{RO2} (air) or k_{RO2} (solvent) are over-represented in low molecular weight or non-polar, simple structures with few representatives of more polar, more complex or higher molecular weight compounds. Databases for k_{HO} (water) and k_{RO2} (water) have the opposite bias with too many polar structures.

Mitigating this situation to some extent are the correlation equations for estimating k_{HO} (water) from k_{HO} (air) or from the Hammett relation (see below), making it possible to tap into the abundant database of measured values for k_{HO} (air) (Atkinson, 1989), the excellent SARs for estimating k_{HO} (air) values (Atkinson, 1987), and the utility of the Hammett equation with its large database of substituent constants (Exner, 1978; Perrin et al., 1981).

15.3.3   SARs For Environmental Oxidants

Structure-activity relations (SARs) rest on the extra-thermodynamic principles first expressed in the Hammett relation for reactions of aromatic (benzene) compounds (Hammett, 1940; Exner, 1972):

$$\log k_x = \rho\sigma_x + \log k_H \tag{24}$$

where k_x is the rate constant for oxidation of a group on an aromatic ring with substituent X, and k_H is for the parent compound without substituent X; σ_x is the substituent parameter which measures the sum of the electronic effects expressed by X (inductive, mesomeric, polar) independent of the reaction type, and ρ is a measure of the sensitivity of a specific reaction to changes in the electronic properties of a series of substituents (Hammett, 1940; Exner, 1972, 1978; Shorter, 1978; Perrin et al., 1981).

The key features of the Hammett relation are that (1) σ is a portable parameter, independent of reaction type and (2) X, as a para- or meta- substituent, is remote from the reaction center and only electronic effects need be accounted for. A negative ρ value indicates an electron-deficient reaction center stabilized by electron donating groups, the most common situation for oxidation reactions.

Taft developed an analogous equation to correlate rate constants for reactions of aliphatic systems in which both steric and electronic effects often are important (Taft, 1953; Shorter, 1972):

$$\log k_x = \rho\sigma_x^* + \delta E_s + \log k_H \tag{25}$$

Taft's σ^* has nearly the same meaning as Hammett's σ (with different numerical values), E_s is the steric parameter for substituent X, and δ is a measure of the sensitivity of the reaction to changes in steric properties near the reaction center (similar to ρ). Extensive compilations of values of σ, σ^* and E_s appear in Perrin et al. (1981) and McPhee et al. (1978), respectively.

SARs for environmental oxidants are empirical correlations of structural parameters with measured rate constants for oxidation by a specific oxidant, for groups of structurally related chemicals in a specific medium. The reliability of any SAR for predicting rate constants for new chemical structures depends on the size and breadth of the database of measured constants originally used to develop the SAR. For HO radical in water, an abundant database of measured values of k_{OH} is available (Farhatziz and Ross, 1977; Buxton et al., 1988; Haag and Yao, 1992) for a wide range of moderately polar compounds. An even larger database of rate constants is available for oxidation by HO in the troposphere (Atkinson, 1989, 1994) and from it a very reliable fragment additivity SAR was developed (Atkinson, 1987). Tropospheric rate data for aliphatic and aromatic compounds can be used to estimate values of k_{OH} in water, using correlation equations (Haag and Yao, 1992; Güsten et al., 1981, 1984; Klöpffer et al., 1985).

Fewer rate constants are available for reactions of oxyradicals (peroxy or alkoxy radicals) or singlet oxygen in water, although large data sets are available for organic solvents (Hendry et al., 1974; Howard 1972; Hendry and Schuetzle, 1976; Neta et al., 1990). Several useful SARs have been developed for reactions of both of these oxidants in organic solvents, and where comparisons with reactions in water are possible, relative reactivities are found to be the same, although rate constants appear to be 2–10 times larger.

Rate constants for superoxide ion (O_2^-) and its conjugate acid HO_2 as oxidant, reductant, and nucleophile have been measured in several solvents (Hendry and Schuetzle, 1976; Sawyer et al., 1978; Bielski et al., 1985), but few SARs have been developed. Moreover, the reactivity of superoxide ion generally is too low for the oxidant to be important in surface waters. Solvated electrons (e^-Aq) also form on insolation of DOM (Fischer et. al., 1985; Zepp et. al., 1988), but its concentration is very low, and target compounds are too few to make $e^-(Aq)$ an important redox agent in surface waters (Buxton et al., 1988). One possible exception is nitroaromatics such as 2,4,6-trinitrotoluene (TNT), which exhibit strong acceleration of photolysis rates in the presence of DOM (Mabey et al., 1983).

Energy transfer can take place between excited DOM (^3DOM) and singlet ground state chemicals to form triplet state chemicals. This process may be another important indirect photolysis process for some classes of chemicals. However, details of the general process and the structural features which control efficiency of energy transfers are poorly understood, other than that the transfer must be exothermic from excited states having triplet energies of no more than 56-60 kcal mole^{-1} (Zepp et al., 1985). The best characterized energy transfer in surface waters is between ^3DOM and $^3O_2$ to form $^1O_2$ (Zepp et al. 1977; Haag and Hoigné, 1986).

15.3.3.1 HO Radical

Atkinson (1987) developed very reliable fragment additivity SARs for estimating $k_{HO}(air)$ from molecular structure using more than 400 compounds in the database (Chapter 14 describes procedures for using these SARs). SARs for HO are based on the premise that rate constants for each of the several different classes of reactions of HO with organic compounds — abstraction of H- atom (k_H), addition to double, triple or aromatic bonds (k_E), and reaction with S or N atoms (k_A) — can be estimated separately and then summed to give the total molecular rate constant, k_{HO}:

$$k_{HO} = k_H + k_E + k_A \tag{26}$$

In addition, most compounds usually have several different types of CH bonds, double bonds, or aromatic rings. For each type of reaction, the effect of substituents at positions α and β to the reaction center must also be taken into account. The detailed calculation of k_{HO} (air) from these SARs has been simplified considerably by availability of a computerized estimation procedure that uses the molecular structure of the compound as the starting point for the calculation (Howard, 1993).

The $k_{HO}(air)$ SAR owes its utility for estimating k_{HO} in water to the correlation between k_{HO} (air) and k_{HO} (water) for many types of aliphatic compounds. Haag and Yao (1992) developed the correlation equation:

$$\log k_{HO}(water) = 1.68 + 0.81 \log k_{HO}(air) \ (in \ M^{-1}s^{-1}) \tag{27}$$

based on 45 compounds in which H-atom abstraction dominates, including alkanes, esters, ethers, epoxides, ketones, nitriles, and amines. This correlation fails for HO radical addition to olefins and aromatics. Rate constants for addition of HO radical to aromatic rings in substituted benzenes or naphthalenes were correlated successfully with the Hammett relation using the σ parameter, summed over all substituents in meta or para positions (Haag and Yao, 1992). This SAR does not use the values of $k_{HO}(air)$:

$$\log k_{HO} (water) = -0.32\Sigma\sigma + 9.83 \ (in \ M^{-1}s^{--1}) \tag{28}$$

Equations (27) and (28) make it possible to estimate k_{HO} (water) values for many more compounds than are listed (Buxton et al., 1988).

15.3.3.2 Oxyradicals

Peroxy radicals(RO_2) react with organic compounds either by H-atom transfer or addition to a double bond. These reactions have rate constants ranging from <0.01 to 300 M^{-1} s^{-1} at 25°C (Howard, 1972; Hendry et al., 1974; Neta et al., 1990) and are rarely important under environmental conditions because of the low average concentration of RO_2 in surface waters (Table 15.5). However, H-atom transfers from phenol OH or aniline NH have large rate constants in the range of 1×10^3 to 1×10^7 M^{-1} s^{-1} (Neta et al., 1990) and can be important contributors to indirect photolysis. Additional reactions of RO$_2$ with some polycyclic aromatics also are very rapid: tetracene, for example, has k_{ox} (RO$_2$) = 2×10^4 M^{-1} s^{-1}, but anthracene's value is only 60 M^{-1} s^{-1} at 60°C (Mahoney, 1965). Values for RO$_2$ addition to benzene or its derivatives are too small to measure.

Halogenation of sp^3 or sp^2 carbon inhibits abstraction and addition reactions by factors of 10 or more, whereas substitution by alkoxy groups or conjugation of two or more double bonds increases rates by factors of 10-100 (Hendry et al., 1974). Thus furans are both conjugated and oxygen substituted, leading to rapid oxidation.

Several large databases of rate constants (Hendry et al., 1974; Howard, 1972) and several correlation equations, facilitate estimating rate constants for RO$_2$ in air or nonpolar solvents, but these data and SARs apply chiefly to hydrocarbons in which H-atom transfers and additions to double bonds are too slow to be important under environmental conditions. SARs for reactions of RO$_2$ radicals with phenols mostly in organic solvents, are listed in Table 15,7, along with the SAR parameters for the Hammett equation (Equation 24). Most of these correlations fit best using σ^+ rather than σ, with ρ values ranging from –1.50 to –0.80.

TABLE 15.7

SARs for Oxidation by RO$_2$ Radicals

Class[a]	RO$_2$ Radical	Solvent	Temp., °C	Rho$^+$ (ρ^+)	Ref.
m- or p- XPhOH	PSO$_2$[b]	Styrene	50	–1.49	c
2,6-(Me)$_2$-4-XPhOH	PSO$_2$	Styrene	65	–1.36	d
2,6-(t-Bu)$_2$-4-XPhOH	PSO$_2$	Styrene	65	–1.11	d
2-t-Bu-4 or 5-XPhOH	PSO$_2$	Styrene	65	–1.46	d
2,6-(t-Bu)$_2$-4-XPhOH	t-BuO$_2$	Isopentane	–37	–1.00	e
2,6-(t-Bu)$_2$-4-XPhOH	t-BuO$_2$	Isopentane	40	–0.8	e
2,6-(t-Bu)$_2$-4-XPhOH	Me$_2$C(CN)O$_2$	Water	25	–0.94	f

[a] X is an alkyl, halogen, nitro or alkoxy substituent.
[b] PSO$_2$ is apolystyrylperoxy radical.
[c] Howard and Ingold (1963a).
[d] Howard and Ingold (1963b).
[e] Howard and Furimsky (1973).
[f] Winterle and Mill (1982).

Hammett coefficients for oxidations of 2,6-di-t-butyl-4-X-phenols by t-BuO$_2$ in isopentane and Me$_2$(CN)CO$_2$ in water are very similar over a temperature range of –40°C to 40°C, suggesting that SARs for other phenols which Table 15.7 lists can be used with moderate reliability to estimate rate constants for oxidations in water.

Blough (1997) suggests that the two most important peroxy radicals in natural waters are acetylperoxy and hydroperoxyl (CH$_3$C(O)O$_2$, HO$_2$). If this is correct, then peroxy radicals play a much more important role in indirect photolysis than indicated above. Acetylperoxy radical is about 10^4 times more reactive than alkylperoxy radicals in H-atom transfer from C-H bonds, (Hendry et al., 1974). Depending on average concentrations of this RO$_2$ radical, making ordinary aliphatic organic compounds much more oxidizable by peroxy radicals and their estimated half lives in these reactions shorter by factors of 100-1000.

15.3.3.3 Singlet Oxygen

Singlet oxygen (^{1}O$_2$) is the first excited state of ordinary triplet oxygen (^{3}O$_2$) and owes its enhanced reactivity to the 24 kcal/mol of excitation energy, coupled with the singlet electronic state, making many kinds of reactions electronically allowable. The range of reactivities for $^1O_2$, combined with its day-averaged surface water concentrations of about 3×10^{-14} M (Table 15.5), limit environmental reactions to those involving furans, dialkyl sulfides, polycyclic aromatics (such as anthracene), pyrroles, thiazoles, oxazoles, eneamines, diazo-compounds, and phenolate ions [Wilkinson et al., 1995; Monroe, 1985].

Some fraction of encounters between organic compounds and $^1O_2$ lead only to quenching to reform $^3O_2$ without chemical transformation (Wilkinson et al., 1995). Some measured rate constants combine reaction and quenching or quenching alone and overestimate reactivity of the chemical toward $^1O_2$ as well as distort the SAR parameters. The Wilkinson et al. compilation (1995) usually indicates which values refer to reaction (k_r) or quenching (k_q) and which are combined values (k_T).

Most SARs for $^1O_2$ reactions are based on reactions in organic solvents (Wilkinson and Brummer, 1981). Foote and Denny (1971) measured reactivities of substituted styrenes toward $^1O_2$ in MeOH and correlated the rate constants with the Hammett equation with ρ = –0.92. Winterle and Mill (1982) measured the oxidation of several of these same styrenes in water and found the same correlation of relative reactivities, but they found absolute reaction constants two to three times larger than in MeOH.

Tratnyek and Hoigné (1991) measured k_r for 22 substituted phenols (mostly m- and p-substituted) in water over a pH range, to evaluate k_r separately for the phenol and phenolate ion. Phenolate ions are 10 to 400 times more reactive toward $^1O_2$. Satisfactory SARs for oxidation of all phenols were developed using Hammett equations with σ^- (Shorter, 1978). Rates for all phenols correlate with σ^- through the relation:

$$\log k_r = 7.1 - 1.0 \Sigma \sigma^-_{o,m,p} \tag{29}$$

with r = 0.84 and n = 19.

By using the fraction of phenol ionized (α) with pH and pK_a, together with rate constants for phenol and phenolate oxidations, it becomes possible to estimate the observed rate constant (k_{obs}) for oxidation of phenols as a function of pH:

$$\alpha = 1/(1 + 10^{pK_a-pH}) \tag{30}$$

where k_r and k_r^- are rate constants for oxidation of the phenol and the anion, respectively

$$k_{obs} = (1-\alpha)\, k_r[PhOH] + \alpha k_r^-[PhO^-] \tag{31}$$

Haag and Mill (1987) reported rate constants and correlation equations for oxidation of a series of phenylazonaphthol dyes, where oxidation of the naphthol and naphtholate ion are accounted for mainly by reaction with singlet oxygen, with some contribution from RO_2. Naphtholate anion is about ten times more reactive than naphthol toward oxidants. The correlation equations for these compounds are similar to those used for phenol oxidations (Tratnyek and Hoigné, 1991):

$$k_{ox} = 1.3 \times 10^7\, (1-\alpha)[Ox] + 1.8 \times 10^8 \alpha[Ox] \tag{32}$$

where [Ox] is the concentration of mixed oxidants, as measured by the rate of oxidation of furfuryl alcohol, a probe molecule that is rapidly oxidized by both RO_2 and $^1O_2$. pK_as for the dyes are calculated readily from Hammett relations (Perrin et al., 1981).

Monroe (1985) lists Hammett parameters for nine singlet oxygen reactions, all of which were conducted in organic solvents. Values of ρ vary from –0.82 to –1.71 with no obvious relation between the value of ρ and the reaction type or class. In most cases, better correlations are found with σ than with σ⁺. Oxidation of thioanisoles in methanol (Monroe, 1979) and chloroform (Kacher and Foote, 1979) had ρ values of –1.6 and –1.67.

Steric effects also can be important in some singlet oxygen reactions. Monroe (1979) reported that, for the series of dialkyl sulfides, R_2S, where R varies from Me to t-Bu and Ph, values of k_{ox} (for reaction) decreased by a factor of 300. Analysis of his data using only the Taft (steric) E_s values for R gave a fair correlation of five sulfides with $\delta = 0.46$ and $r^2 = 0.78$ (Equation (25)).

Table 15.8 compiles SARs for $^1O_2$ reactions with anilines, sulfides, eneamines, furans, phenols and styrenes, all compounds which have high reactivity toward $^1O_2$. Most reactions are in organic solvents. All ρ values are large, negative values indicating significant influence of substitution on the rate at the reaction center.

TABLE 15.8

SARs for Singlet Oxygen Reactions.[a]

Class	Rho (ρ)[b]	Reference
Organic Solvents		
Thioanisoles	–1.6	c
Thioanisoles	–1.67	d
Thiobenzophenones	–1.71	e
Trimethylstyrenes	–0.92	f
Styrylenamines	–0.85	g
Diaryldioxetanes	–0.82	h
Arylfurans	–0.84	i
Di-t-butylphenols	–1.72	j
Water		
Trimethylstyrenes	–0.92	k
o,m,p-Substituted phenols	–1.2[b]	l

[a] For chemical reaction (see text).
[b] Correlation with Hammett σ⁻.
[c] Monroe (1979).
[d] Kacher and Foote (1979).
[e] Rajee and Ramamurthyl (1978).
[f] Foot and Denny (1977).
[g] Wake (1979).
[h] Zalklika et al. (1980).
[i] Young et al. (1972).
[j] Thomas and Foote (1978).
[k] Winterle and Mill (1982).
[l] Tratnyek and Hoigné (1991).

15.3.4 Estimation of $k_{ox}(i)$ for Benchmark Chemicals

15.3.4.1 *Anthracene*

Anthracene is a polycyclic conjugated structure that cannot be fragmented into smaller units. The oxidants each have preferred points of oxidation on the conjugated rings, but the 9 and 10 positions are favored because of symmetry and stabilization of benzylic radicals. Rate constants for oxidation of anthracene by RO_2, $^1O_2$, and HO have been measured in solution or in vapor: $k_{ox}(RO_2)$ is 60 M^{-1} s^{-1} in chlorobenzene at 60°C (Mahoney, 1965);

$k_{ox}(^1O_2)$ is 5×10^{10} M^{-1} s^{-1} for the water soluble 1-sulfonatoanthracene (Wilkinson et al., 1995); k_{ox}(HO) (air) is 6×10^{10} M^{-1} s^{-1} (Atkinson, 1989).

The total indirect oxidation rate constant for anthracene, derived in Table 15.9, is controlled equally by HO and $^1O_2$.

TABLE 15.9

Indirect Photoreaction Rate Constants for Anthracene.

Oxidant	$k_{OX(i)}$(meas.), M^{-1} s^{-1}	$k_{OX(i)}$[Ox]i[a], s^{-1}
RO$_2$	60[b]	2×10^{-9}
$^1O_2$	5×10^8	1.5×10^{-5}
HO	7×10^{10c}	2×10^{-5}
Total Rate Constant		3.5×10^{-5} s^{-1}
Half life		5.5 hr

[a] Average values of Ox(i) are taken from Table 15.5.
[b] Measured value at 60° extrapolated to 25°C by assuming k_{ox}
      = 108.5 e$^{-(E/RT)k}$$_{OX}$ with E = 42 kJ/mole.
[c] Measured in air at 50°C.

15.3.4.2   Chlorpyrifos

The distinctive structure of CP (III), coupled with an absence of any published oxidation rate constants for it, requires that k_{OX} be estimated from additivity of fragment rate constants for the major structural units: 3,5,6-trichloropyridinyl (TCP), diethoxyphosphorothioate (DEP), and the phosphorothioate $\equiv$P=S bond.

III (CP)

Neither RO$_2$ nor $^1O_2$ will react with the pyridine ring or with the –OCH$_2$CH$_3$ units at measurable rates (Hendry et al., 1974; Wilkinson and Brummer, 1981). A limiting rate constant for RO$_2$ reaction with O-CH$_2$CH$_3$ of 0.2 M^{-1} s^{-1} is based on values of k_{OX} for H-atom transfer by t-BuO$_2$ from isopropyl acetate and t-butyl phenyl acetate (Hendry et al, 1974). The estimate assumes the CH groups in the 2 O-ethyl groups have a combined reactivity of (4 × 0.003) + (6 × 0.03) = 0.19 M^{-1} s^{-1}. No data are available for RO$_2$ oxidation rates of the $\equiv$P=S bond in model compounds.

The rate constant for the HO oxidation of trichloropyridinyl group can be estimated from values for pyridine (3 × 10^9 M^{-1} s^{-1}) and 2- and 4-chloropyridines (2 and 3 × 10^9 M^{-1} s^{-1}, respectively) to be close to 1 × 10^9 M^{-1} s^{-1}, allowing for steric effects of the 2-phosphorothioate and 6-chloro groups (Buxton et al., 1988). HO oxidation of >P(O)(OEt)$_2$ groups are modeled well by P(O)(OEt)$_3$ which has k_{HO} = 3 × 10^9 M^{-1} s^{-1}/molec (Buxton et al., 1988). On this basis, the OEt groups in CP should contribute 2 × 10^9 M^{-1} s^{-1} to the overall rate constant. Lastly, HO oxidation of $\equiv$P=S is probably much faster than oxidation anywhere else in CP. Atkinson (1989) lists values of k_{HO}(air) for trimethylphosphate and trimethylphosphorothioate as 4.4 × 10^9 and 4.2 × 10^{10} M^{-1} s^{-1} at 25°C, indicating a factor of ten enhancement in reactivity owing to the $\equiv$P=S group. On this basis $\equiv$P=S oxidation is assigned a fragment value of 3×10^{10} $M^{-1}s^{-1}$. The molecular rate constant for CP is the sum of three fragment constants:

$$k_{HO} = k_{HO}(TCP) + k_{HO}(DEP) + k_{HO}(PS)$$

$$= 1 \times 10^9 + 2 \times 10^9 + 3 \times 10^{10}$$

$$= 3.3 \times 10^{10} \text{ } M^{-1} \text{ } s^{-1}$$

Table 15.10 summarizes the oxidation data for chlorpyrifos.

TABLE 15.10

Indirect Photoreaction Rate Constants for Chloropyrifos.

Oxidant	$k_{OX(i)}$(meas.), M^{-1} s^{-1}	$k_{OX(i)}[Ox]_i$[a]
RO_2	0.19	5.7×10^{-12}
$^1O_2$	$<1 \times 10^4$	$<3 \times 10^{-10}$
HO	3.3×10^{10}	1×10^{-6}
Total Rate Constant		1×10^{-6}
Half life		8 days

[a]See Table 15.9, footnote a.

15.3.4.3 *2,6-Di-t-butyl Phenol*

DBP is oxidized by all three surface water oxidants with medium to large rate constants. RO_2 (t-BuO$_2$) oxidation of DBP was studied in organic solvents to give an extrapolated value of 4×10^3 M^{-1} s^{-1} (Howard and Furimsky, 1973; Neta et al., 1990).

Oxidations of phenols by singlet oxygen are complicated by the differential rates of oxidation of the phenols and the phenolate ions in water, with reactivity ratios for alkylphenols ranging from 20 to 100 (Faust and Hoigné, 1987; Tratnyek and Hoigné, 1991). However, at pH 7, less than 0.1% of alkylphenols are present as anions, leading at most to a 1% contribution of phenolate oxidation to the overall rate. Singlet oxygen oxidation of DBP in n-butanol has a rate constant of 1.6×10^6 M^{-1} s^{-1} (Wilkinson and Brummer, 1995). If the rate increases by a factor of 4 on going from alcohol to water, k_r in water has a value of ~ 6×10^6 M^{-1} s^{-1}, close to the values for non-hindered alkylphenols that Tratnyek and Hoigné (1991) reported. Their SAR for $^1O_2$ oxidation for all phenols (Equation (28)) gives for DBP $k_r = 3 \times 10^6$ M^{-1} s^{-1}, in good agreement.

HO oxidation of DBP has a rate constant close to 1×10^{10} M^{-1} s^{-1}, based on a measured value (in water) for 4-t-butylphenol of 1.9×10^{10} M^{-1} s^{-1} (Buxton et al., 1988). Table 15.11 summarizes these rate constants and the contributions of each oxidation process to the overall oxidation of DBP in a typical surface water. About half the oxidation rate is due to HO, 30% due to $^1O_2$, and 20% due to RO_2.

15.3.4.4 *Hexachlorocyclohexane (Lindane)*

Significant oxidation of HCH occurs only with HO. Haag and Yao (1992) report that aqueous oxidation of HCH with HO has a rate constant of $(0.5-1) \times 10^9$ M^{-1} s^{-1}. No oxidation of HCH is measurable with $^1O_2$. RO_2 oxidation of a >CHCl group has the limiting k_{RO2} value of < 0.005 M^{-1} s^{-1}, making the molecular rate constant $(6 \times 0.005) < 0.03$ M^{-1} s^{-1}, based on

TABLE 15.11

Photooxidation Rate Constants for
Di–T–Butylphenol in Insolated Surface Water.

Oxidant	k_{OX}, M^{-1} s^{-1}	$k_{OX}[Ox]i$, s^{-1a}
RO_2	4×10^3	1×10^{-7}
$^1O_2$	6×10^6	2×10^{-7}
HO	1×10^{10}	3×10^{-7}
Total Rate Constant		6×10^{-7}
Half life		13 days

a   See Table 15.9, footnote a.

the rate constant of 6×10^{-4} M^{-1} s^{-1} for 2-chloropropane (Hendry et al., 1974). Table 15.12 summarizes HCH oxidation data.

TABLE 15.12

Photoxidation Rate Constants for Lindane in
Insolated Surface Waters.

Oxidant	k_{OX}, M^{-1} s^{-1}	$k_{OX}[Ox]i$, s^{-1a}
RO_2	0.03	9×10^{-13}
$^1O_2$	$<1 \times 10^4$	$<3 \times 10^{-10}$
HO	$0.5\text{-}1 \times 10^9$	$1.5\text{-}3 \times 10^{-8}$
Total Rate Constant	—	$1.5\text{-}3 \times 10^{-8}$
Half life	—	270 days

a   See Table 15.9, footnote a.

15.3.4.5   Trichloroethylene

Trichloroethylene is an electrophilic olefin, which is deactivated toward reaction with $^1O_2$; no $^1O_2$ rate constant is reported. RO_2 oxidation is moderately rapid for this olefin because the carbon radical formed by addition of RO_2 is stabilized, but no measured value is reported. An estimate of 50 M^{-1} s^{-1} assumes TCE is six times as reactive as an ordinary terminal olefin (Mill and Hendry, 1980). Aqueous HO oxidation is very rapid with $k_{HO} = 4.9 \times 10^9$ M^{-1} s^{-1} (Buxton et al., 1988). Table 15.13 summarizes rate constants.

TABLE 15.13

Photooxidation Rate Constants for
Trichloroethylene in Insolated Surface Waters.

Oxidant	k_{OX}, M^{-1} s^{-1}	$k_{OX}[Ox]i$, s^{-1a}
RO_2	50b	3×10^{-11}
$^1O_2$	$<1 \times 10^4$	$<3 \times 10^{-10}$
HO	5×10^9	1.5×10^{-7}
Total Rate Constant	—	1.5×10^{-7}
Half life	—	50 days

a   See Table 15.9 footnote a. b. See text.

15.4   Photoreactions in the Gas Phase

Several thousand different synthetic chemicals are released to the troposphere, where the great majority are oxidized by HO (OH in atmospheric chemistry jargon) radicals in a few days or less. If the average value of k_{HO} is 5×10^9 M^{-1}s^{-1} for most organic compounds, and the diurnally averaged tropospheric HO concentration is about 5×10^5 molec cm^{-3} (Nimitz and Skaggs, 1992), then the average half-life in the troposphere for a chemical will be about 50 hrs. To compete with photooxidation, direct photolysis rate constants (k_{PE}) must be $\geq 4 \times 10^{-6}$ s^{-1}. Only a relatively small number of compounds are both volatile and have strong chromophores in the solar region, making it possible for k_{AE} or k_{PE} to compete with photooxidation. However, some volatile compounds that have weak chromophores in the solar region are much less reactive toward HO ($<1 \times 10^9$ M^{-1}s^{-1}), making lower values of k_{PE} competitive with k_{HO}[HO]. These compounds include nitromethane, perfluoronitroalkanes, azoisobutane and azomethane, biacetyl (2,3-butanedione), and all perfluorochlorocarbonyls, such as hexafluoroacetone and hexachloroacetone. Many of these compounds have gas phase quantum yields in excess of 0.1 (Calvert and Pitts, 1967), making photolysis competitive with HO oxidation or (for polyhalogenated compounds) significantly faster.

Anthracene (AN) is the only benchmark chemical that might photolyze rapidly enough in the vapor phase to compete with HO oxidation . The UV spectra of AN in cyclohexane and acetonitrile are similar enough that the light absorption rate for AN vapor in summer sunlight should be similar to the rate in water. If the quantum yield is similar, then AN vapor has a 45 min half-life at solar noon (Table 15.9). The HO rate constant for AN in the vapor phase is 6.7×10^{10} M^{-1}s^{-1} (Atkinson, 1989), giving $k'_{HO} = 6 \times 10^{-5}$ s^{-1} or a 200 minute half life, making direct photolysis four times faster.

15.5 Summary and Conclusions

Rates of direct photoreactions of chemicals in surface waters are controlled by sunlight photon fluxes, the light absorbing properties of the chemicals, and the overall efficiency in using absorbed photon energy. Although light fluxes are well known, light absorption by individual compounds is not easily calculated, and UV spectra must be carefully measured to estimate the rate constants for sunlight absorption (k_{AE}). k_{AE} is also the maximum value of the direct photolysis rate constant, (k_{PE} (max)), thereby enabling one to assess the possible importance of direct photolysis as a competitive loss process in surface waters. The actual value of k_{PE} can be calculated only if the process efficiency or quantum yield (Φ) is known. In practice, measurement of Φ would be made only if the estimated value of k_{PE} (max) is comparable to other loss process rate constants.

Indirect photoreaction rate constants (k_{OX}(total)) can be estimated using accepted values of average photooxidant concentrations ($[Ox]_{avg}$) for RO$_2$, ^{1}O$_2$, and HO and measured values for second-order rate constants (k_{OX}) for these oxidants with the compounds of interest. In many cases, reliable estimates of k_{OX} can be derived from simple SARs, based on the molecular structure of the compound.

Reactivity of organic compounds toward HO is estimated readily in most cases because of the large databases of values measured for reactions in water and air and the correlation equations relating rates of oxidation of aliphatic compounds in air and water, as well as the availability of Hammett SARs for reactions of aromatic compounds. In addition, 90% of HO rate constants in water and air are within $\pm$ 50% of 5×10^9 M^{-1} s^{-1}. Thus, the limiting indirect photoreaction rate constant for almost any compound in freshwater with more than 1 mg/L of NO$_3^-$ will be ~1.5×10^{-7} s^{-1} at 40°–50° latitude in summer, equivalent to a 50-day half life. In low-nitrate waters, the limiting half life will be extended correspondingly.

Rate constants for the more selective oxidants, RO$_2$ and ^{1}O$_2$, require estimates based first on structural analogy and then on SARs, where available. Both RO$_2$ and ^{1}O$_2$ have very limited databases of rate constants for reactions in water, pointing to the need for additional rate measurements in water of RO_2 and $^1O_2$ oxidation reactions with a variety of oxidizable compounds, from which reliable SARs might be developed.

15.6 Symbols Used

λ	=	Wavelength
A_λ	=	Absorbance at wavelength λ
α_λ	=	Light attenuation caused by dissolved organic matter in solution at wavelength λ
ε_λ	=	Molar absorptivity at wavelength λ (= molar extinction coefficient)
λ	=	Light pathlength; the distance traveled by a beam of light passing through the system
D	=	Depth of water body
[C]	=	Molar concentration of chemical C
$-d[C]/dt$	=	Direct photolysis rate for chemical C
k_{PE}	=	Direct photolysis sunlight rate constant in water bodies in the environment
$(k_{PE})_{max}$	=	k_{AE}, the maximum direct photolysis sunlight rate constant in water bodies in the environment
$k_{P\lambda}$	=	Direct photolysis sunlight rate constant at wavelength λ
Φ	=	Sunlight reaction quantum yield
Φ_λ	=	Sunlight reaction quantum yield at wavelength λ
I_A	=	Intensity of absorbed light in photons
$I_{A\lambda}$	=	Intensity of absorbed light at wavelength λ in photons
$I_{o\lambda}$	=	Intensity of light incident on a solution in photons
I_λ	=	Intensity of light transmitted through a solution in photons
L_λ	=	Solar irradiance in water in the units millieinsteins cm^{-2} day^{-1}
Z_λ	=	Light flux at wavelength λ, with units of photons cm^{-2} sec^{-1}
K_R	=	Diffuse attenuation coefficient
K_λ	=	Diffuse attenuation coefficient at wavelength λ
k_d	=	Indirect photolysis rate constant, usually with oxidant specified; e.g., k_{HO} for oxidation by hydroxyl radical

References

Atkinson, R. 1987. A structure-activity relationship for the estimation of rate constants for the gas-phase reactions of OH radicals with organic compounds. *Int. J. Chem. Kinetics* 19:799-828.

Atkinson, R. 1989. Kinetics and mechanisms of the gas phase reaction of hydroxyl radical with organic compounds. *J. Phys. Chem. Ref. Data*, Monograph No. 1.

Atkinson, R. 1994. Tropospheric chemistry of organic compounds. *J. Phys. Chem. Ref. Data* Monograph No. 2.

Behymer, T.D. and R.A. Hites. 1985. Photolysis of polycyclic aromatic hydrocarbons absorbed on simulated atmospheric particulate. *Environ. Sci. Technol.* 19:1004-1006.

Bielski, B.H.J., D.E. Cabelli, R.L. Arudi, and A.B. Ross. 1985. Reactivity of HO_2/O_2^- radicals in aqueous solution. *J. Phys. Chem. Ref. Data* 14:1041-1100.

Blough, N. V. 1997. Photochemistry in the sea-surface microlayer. In P.S. Liss and R.A. Duce, Eds., pp. 383-424. *Sea Surface and Global Change*, Cambridge University Press, London.

Blough, N.V. and R.G. Zepp. 1995. Reactive oxygen species in natural waters. In C.S. Foote, J.S. Valentine, A. Greenberg, and J.F. Liebman, Eds., pp. 280-333. *Active Oxygen in Chemistry*, Chapman and Hall, New York.

Brimblecombe, P. and D. Shooter. 1986. Photooxidation of dimethylsulfide in aqueous solution. *Mar. Chem.* 19:343-353.

Buxton, G.V., C.L. Greenstock, N.P. Helman, and A.B. Ross. 1988. Critical review of rate constants for reactions of hydrated electrons, and hydroxyl radicals in aqueous solution. *J. Phys. Chem. Ref. Data* 17:513-886.

Calvert, J.G. and J.N. Pitts. 1967. *Photochemistry.* John Wiley, New York.

Cooper, W.J., C. Shao, D.R.S. Lean, A.S. Gordon, and F.E. Scully. 1994. Factors affecting the distribution of H_2O_2 in surface waters. *Adv. Chem. Ser.* 237:391-422.

Demerjian, K.L., K.L. Schere, and J.J. Peterson. 1980. Theoretical estimates of actinic flux and photolytic rate constants of atmospheric species. *Adv. Environ. Sci. Technol.* 10:369-392.

Dilling, W.L., L.C. Lickly, T.D. Lickly, and P.G. Murphy. 1984. Organic photochemistry 19. Quantum yields for O,O-diethyl-O-(3,5,6-trichloropyridinyl)phosphorothioate (chloropyrifos) in dilute aqueous solution and its environmental transformation rates. *Environ. Sci. Technol.* 18:540-543.

Dulin, D. and T. Mill. 1982. Development and evaluation of sunlight actinometers. *Environ. Sci. Technol.* 16:815-820.

EPA. 1982. The GC-SOLAR Program for the PC is available from the Environmental Research Laboratory, U.S. EPA, Athens, GA.

EPA. 1995. Draft guidelines: Direct photolysis rate in water by sunlight. OPPTS 835.2210; maximum direct photolysis rate in air from UV/VIS spectroscopy. OPPTS 835.2310. U.S. Environmental Protection Agency, Washington, D.C.

Exner, O. 1972. The Hammett equation – the present position. In N.B. Chapman and J. Shorter, Eds., pp 1-70. *Advances in Linear Free Energy Relationships*, Chapter 1. Plenum Press, New York.

Exner, O. 1978. A critical compilation of substituent constants. In N.B. Chapman and J. Shorter, Eds., pp 439-540. *Correlation Analysis in Chemistry*, Chapter XV. Plenum Press, New York.

Farhataziz and A.B. Ross. 1977. Selected specific rates of reactions of transients from water in aqueous solution. III. Hydroxyl and perhydroxyl radical. National Bureau of Standards NSRDS-NBS59. Washington, DC.

Faust, B.C., and J. Hoigné. 1987. Sensitized photooxidation of phenols by fulvic acids in natural waters. *Environ. Sci. Technol.* 20:957-963.

Finlayson-Pitts, B.J. and J.N. Pitts. 1986. *Atmospheric Chemistry: Fundamentals and Experimental Techniques.* John Wiley, New York.

Fischer, A.M., D.S. Kliger, J.S. Winterle, and T. Mill. 1985. Direct observation of phototransients in natural waters. *Chemosphere* 14:1299-1306.

Foote, C.S. and J.W. Peters. 1971. Chemistry of singlet oxygen. XIV. A reactive intermediate in sulfide photooxidation. *J. Am. Chem. Soc.* 93:3795-4002.

Foote, C.S. and R.W. Denny. 1971. Chemistry of singlet oxygen. XIII. Solvent effects on the reaction with olefins. *J. Am. Chem. Soc.* 93:5168-5172.

Frank, R. and W. Klöpffer. 1989. A convenient model and program for the assessment of abiotic degradation of chemicals in natural water. *Ecotox. Environ. Safety* 17: 323–332.

Gusten, H., W.G. Filby, and S. Schoof. 1981. Prediction of hydroxyl radical reaction rates with organic compounds in the gas phase. *Atmos. Environ.* 15:1763-1765.

Gusten, H., L. Klasinc, and D. Maric. 1984. Prediction of the abiotic degradability of organic compounds in the troposphere. *J. Atmos. Chem.* 2:83-93.

Haag, W.R. and J. Hoigné. 1986. Singlet oxygen in surface waters—Part III: Steady state concentrations in various types of waters. *Environ. Sci. Technol.* 20:341-348.

Haag, W.R. and T. Mill. 1987. Direct and indirect photolysis of water-soluble azodyes. *Environ. Toxicol. Chem.* 6:359-369.

Haag, W.R. and T. Mill. 1989. Survey of sunlight-produced transient reactants in surface waters. Proceedings of a workshop on effects of solar ultraviolet radiation on geochemical dynamics, Woods Hole, MA., pp. 82-88.

Haag, W.R., and C.C.D. Yao. 1992. Rate constants for reactions of hydroxyl radicals with several drinking water contaminants. *Environ. Sci. Technol.* 26:1005-1013.

Hammett, L. P. 1940. *Physical Organic Chemistry,* pp. 184-228. McGraw-Hill Book Co., Inc., New York.

Hendry, D.G. and D. Schuetzle. 1976. Reactions of hydroperoxy radicals. Comparison of reactivity with organic peroxy radicals. *J. Org. Chem.* 41:3179-3182.

Hendry, D.G., T. Mill, L. Piszkiewicz, J.A. Howard, and H.K. Eigenmann. 1974. A critical review of H-atom transfer in the liquid phase. *J. Phys. Chem. Ref. Data* 3:937-978.

Howard, J.A. and K.U. Ingold. 1963a. The inhibited autooxidation of styrene. Part I. *Can. J. Chem.* 41:1744-1753.

Howard, J.A. and K.U. Ingold. 1963b. The inhibited oxidation of styrene. Part II. *Can. J. Chem.* 41:2800-2806.

Howard, J.A. 1972. Absolute rate constants for reactions of oxyl radicals. *Adv. Free Radical Chem.* 4:49-107.

Howard, J.A. and E. Furimsky. 1973. Arrhenius parameters for reactions of tert-butylperoxy radicals with some hindered phenols and amines. *Can. J. Chem.* 51:3738-3745.

Howard, P. 1993. Atmospheric Oxidation Rate Program®, Syracuse Research Corporation, 6225 Running Ridge Road, N. Syracuse, NY 13212-2509.

Kacher, M.L. and C.S. Foote. 1979. Chemistry of singlet oxygen XXVII. Steric and electronic effects on reactivity of sulfides. *Photochem. Photobiol.* 29:765-771.

Klopffer, W., G. Kaufmann, and R. Frank. 1985. Phototransformation of air pollutants: rapid test for the determination of k_{HO}. *Z. Naturforsch* 40A:686-692.

Larson, R.A., L.L. Hunt, and D.W. Blankenship. 1977. Formation of toxic products from a no. 2 fuel oil by photooxidation. *Environ. Sci. Technol.* 11:492-496.

Larson, R.A. and R.G. Zepp. 1988. Reactivity of carbonate radical with aniline derivatives. *Environ. Toxicol. Chem.* 7:265-274.

Liefer, A. 1988. *The Kinetics of Environmental Aquatic Photochemistry.* American Chemical Society, Washington, D.C.

Mabey, W.R., T. Mill, and D.G. Hendry. 1982. Laboratory protocols for evaluating the fate of chemicals in water and air. U.S. Environmental Protection Agency Final Report. EPA-600/3-82-022, Washington, DC.

Mabey, W.R., D. Tse, A. Baraze, and T. Mill. 1983. Photolysis of nitroaromatics in aquatic systems: 2,4,6-trinitrotoluene. *Chemosphere* 12:3-16.

Mahoney, L.R. 1965. Reaction of peroxy radicals with polynuclear aromatic compounds. II. Anthracene in chlorobenzene. *J. Amer. Chem. Soc.* 87:1089-1095.

McPhee, J.A., A. Panaye, and J.E. Dubois. 1978. Steric effects - I. A critical examination of the Taft steric parameter – E_s. *Tetrahedron* 34:3353-3662.

Meikle, R.W., N.H. Kurihara, and D.H. DeVries. 1983. Chloropyrifos: the photodecomposition rates in dilute aqueous solution and on a surface, and the volatilization rate from a surface. *Archiv. Environ. Contam. Toxicol.* 12:189-193.

Mill, T. and D.G. Hendry. 1980. Kinetics and mechanism of free radical oxidation of alkanes and olefins. In C.H. Bamford and C.F. Tipper, Eds., pp. 1-83, *Comprehensive Chemical Kinetics, Vol. 16, Chapter 1.* Elsevier, New York.

Mill, T., D.G. Hendry, and H. Richardson. 1980. Free-radical oxidants in natural waters. *Science* 207:886-889.

Mill, T., W.R. Mabey, B.Y. Lan, and A. Baraze. 1981. Photolysis of polycyclic aromatic hydrocarbons in water. *Chemosphere* 10:1281-1287.

Mill, T. and W. Mabey. 1985. Photodegradation in water. In W.B. Neely and G.E. Blaue, Eds., pp. 175-216, *Environmental Exposure from Chemicals, Vol. I, Chapter 8.* CRC Press, Boca Raton, FL.

Mill, T. 1989. Structure-activity relationships for photooxidation processes in the environment. *Environ. Toxicol. Chem.* 8:31-43.

Mill, T. 1994. Estimation of rates and half lives for direct and indirect photolysis. Summary report, Task I20/25, EPA contract 68-D9-0166.

Monroe, B.M. 1979. Rates of reaction of singlet oxygen with sulfides. *Photochem. Photobiol.* 29:761-767.

Monroe, B.M. 1985. Singlet oxygen in solution. In A.A. Frimer, ed., *Singlet Oxygen,* pp. 177-224. CRC Press, Boca Raton, FL.

Mopper, K. and X. Zhou. 1990. Hydroxyl radical photoproduction in the sea and its potential impact on marine processes. *Science* 250:661-664.

Neta, P., R.E. Huie, and A.B. Ross. 1990. Rate constants for reactions of peroxyl radicals in fluid solutions. *J. Phys. Chem. Ref. Data* 19:413-512.

Nimitz, J.S. and S.R. Skaggs. 1992. Estimating tropospheric lifetimes and ozone-depletion potentials of one- and two-carbon hydrofluorocarbons. *Environ. Sci. Technol.* 26:739-744.

Peijnenburg, W.J.G.M., K.G.M. deBeer, M.W.A. de Haan, and H. A. den Hollander. 1992. Development of a structure-reactivity relationship for the photohydrolysis of substituted aromatic halides. *Environ. Sci. Technol.* 26:2116-2121.

Perrin, D.D., B. Dempsey, and E.P. Serjeant. 1981. *pKₐ Prediction for organic acids and bases.* Chapman and Hall, London.

Petasne, R.G. and R.G. Zika. 1987. Fate of superoxide in coastal sea water. *Nature* 325:516-518.

Rajee, R. and V. Ramamurthyl. 1978. Oxidation of thiones by singlet and triplet oxygen. Tetrahedron Lett. pp 5127-5133.

Sawyer, D.T., M.J. Gibian, and M.M. Morrison. 1978. On the chemical reactivity of superoxide ion. *J. Am. Chem. Soc.* 100:627-633.

Shorter, J. 1972. Separation of polar, steric and resonance effects. In N.B. Chapman and J. Shorter, Eds., Chapter 2., pp. 71-118, *Advances in Linear Free Energy Relationships,* Plenum Press, New York.

Shorter, J. 1978. Multiparameter extensions of the Hammett equation. In N.B. Chapman and J. Shorter, Eds., Chapter 4, pp 119-174. *Correlation Analysis in Chemistry,* Plenum Press, New York.

Taft, R.W. 1953. Correlation of hydrolysis constants for aliphatic esters. *J. Am. Chem. Soc.* 75:4538-4541.

Thomas, J.J. and C.S. Foote. 1978. Chemistry of singlet oxygen. XXVI. photooxidation of phenols. *Photochem. Photobiol.* 27:683-687.

Tratnyek, P.G. and J. Hoigné. 1991. Oxidation of substituted phenols with environment: A QSAR analysis of rate constants for reaction with singlet oxygen. *Environ. Sci. Technol.* 25:1596-1604.

Turro, N.J. 1967. *Molecular Photochemistry.* W.A. Benjamin, Inc., New York.

Wake, R.W. 1979. Sensitized oxidation of aryl enamines. Ph.D. Thesis. University California, Los Angeles, CA.

Walling, C. 1957. *Free Radicals in Solution.* Wiley, New York.

Wilkinson, F., G. Helman, and A. B. Ross. 1981. Rate constants for the decay and reactions of the lowest electronically excited singlet state of molecular oxygen in solution. An expanded and revised compilation. *J. Phys. Chem. Ref. Data* 24: 663-702.

Winterle, J.S. and T. Mill. 1982. In *Validation of estimation techniques for predicting environmental transformation of chemicals. Final report* (Contract 68-01-6269) U.S. Environmental Protection Agency, Washington, DC.

Winterle, J.S., D. Tse, and W.R. Mabey. 1987. Measurement of attenuation coefficients in natural water columns. *Environ. Toxicol. Chem.* 6:663-672.

Young, R.H., R.L. Martin, N.Chinh, C. Mallon, and R.H. Kayser. 1972. Substituent effects om dye-sensitized photooxidation of furans. *Can. J. Chem.* 50:932-939.

Zaklika, K.A., B. Kaskar, and A.P. Schaap. 1980. Mechanisms of photooxygenation. I. Substituent effects of the [2+2] cycloaddition of vinyl ethers. *J. Am. Chem. Soc.* 102:386-396.

Zepp, R.G. and D.M. Cline. 1977. Rates of direct photolysis in aquatic environments. *Environ. Sci. Technol.* 11:359-365.

Zepp, R.G. and N.L. Wolfe, G.L. Baughman, and R.C. Hollis. 1977. Singlet oxygen in natural water. *Nature.* 267:421-423.

Zepp, R.G. and P.F. Schlotzhauer. 1979. In P.W. Jones and P. Leber, Eds., *Polynuclear Aromatic Hydrocarbons.* Ann Arbor Science Publishers, Inc., Ann Arbor, MI.

Zepp, R.G., P.F. Schlotzhauer, and R.M. Sink. 1985. Photosensitized transformations involving energy transfer. *Environ. Sci. Technol.* 19:74-81.

Zepp, R.G., J. Hoigne, and H. Bader. 1987a. Nitrate-induced photooxidation of trace organic chemicals in water. *Environ. Sci. Technol.* 21:443-450.

Zepp, R.G., A.M. Braun, J. Hoigne, and J.A. Leenheer. 1987b. Photoproduction of hydrated electrons from natural organic solutes. *Environ. Sci. Technol.* 21:485-490.

16

Oxidation-Reduction Reactions in the Aquatic Environment

Paul G. Tratnyek and Donald L. Macalady

CONTENTS

1-56670-456-1/00/$0.00+$.50
© 2000 by CRC Press LLC

16.1 Introduction

Oxidation-reduction (redox) reactions, along with hydrolysis and acid-base reactions, account for the vast majority of chemical reactions that occur in aquatic environmental systems (soils, sediments, aquifers, rivers, lakes, and many remediation operations). This chapter provides a survey of the environmental and substrate characteristics that govern redox transformations in aquatic systems, and it suggests methods for estimating the thermodynamic and kinetic properties for redox reactions involving organic contaminants. The scope of this chapter is limited to non-photochemical, abiotic processes; photochemical processes are the focus of chapters 14 and 15, and microbial transformations are discussed in Chapter 12. Chapters focusing on estimation of properties for redox reactions involving inorganic substances have been published previously (1).

The distinction between biotic and abiotic processes is a particularly important issue in defining the scope of this chapter. Living organisms are responsible for creating the conditions that determine the redox chemistry of most aquatic environmental systems. So, in this sense, most redox reactions in natural systems ultimately are driven by biological activity. Once environmental conditions are established, however, many important redox reactions proceed without further mediation by organisms. These reactions are considered to be "abiotic" when it is no longer practical (or possible) to link them to any particular biological activity (2, 3). This distinction is often clear at the conceptual level, even when operational tests (such as comparing the effects of various antimicrobial treatments) give ambiguous results, and thus increasing numbers of studies treat environmental transformations of organic contaminants as abiotic redox reactions. Much work in this area remains to be done, but this chapter attempts to synthesize recent developments into a general framework for estimating the environmental fate of organic chemicals by abiotic redox reactions.

16.2 Background

16.2.1 Redox Reactions Involving Organic Contaminants

16.2.1.1 Assigning Oxidation States

Redox reactions involve oxidation and reduction; they occur by the exchange of electrons between reacting chemical species. Electrons (or electron density) are lost (or donated) in oxidation and gained (or accepted) in reduction. An oxidizing agent (or oxidant), which accepts electrons (and is thereby reduced), causes oxidation of a species. Similarly, reduction results from reaction with a reducing agent (or reductant), which donates electrons (and is oxidized).

To interpret redox reactions in terms of electron exchange, one must account for electrons in the various reacting species. Various textbooks (e.g., 4, 5) provide simple rules, such as the following, for assigning oxidation states for inorganic redox couples:

- For free elements, each atom is assigned oxidation number 0,
- Monoatomic ions have an oxidation number equal to the charge of the ion,
- Oxygen, in most compounds, has the oxidation number –2,
- Hydrogen, in most compounds, has the oxidation number +1,
- Halogens, in most environmentally relevant compounds, have the oxidation number –1.

These rules, however, are not easily applied to organic redox reactions, and this difficulty has led to a steady stream of alternative methods for assigning oxidation states (e.g., 6, 7).

16.2.1.1.1 Simplest Method

For present purposes, familiarity with two methods for assigning oxidation states to organic molecules is sufficient. The first, and easiest where it applies, reflects the qualitative observations from which the historical concepts of oxidation and reduction originated (8):

- Oxidation is the gain of O, Cl, or double bonds, and/or the loss of H.
- Reduction is the gain of H, saturation of double bonds, and/or loss of O or Cl.

Thus, for example, "mineralization" of any hydrocarbon to CO_2 and H_2O involves oxidation (see Chapter 12), and dechlorination of any halogenated compound to hydrocarbon products involves reduction.

16.2.1.1.2 Recommended Method

For more-complex cases, or where a quantitative accounting of oxidation states is needed, the following method is most commonly used (9, 10). For each atom of interest, its oxidation state is assigned the sum of:

- +1 for each bond to a more electronegative atom,
- –1 for each bond to a less electronegative atom, and
- 0 for each bond to an atom of identical electronegativity.

Recall that electronegativities increase across rows and up columns of the periodic table, but the increments are not consistent, so the absolute electronegativities of common elements increase in the order $H < P < C, S, I < Br < N, Cl < O < F$.

Example: Assigning Oxidation Numbers

To illustrate the use of the recommended method, consider the reduction of N, N-nitroso-dimethylamine (NDMA), a mutagenic and carcinogenic contaminant that has been subject to considerable study (11). Reduction can occur at the N-N bond (Equation 1), or at the N-O bond (not shown).

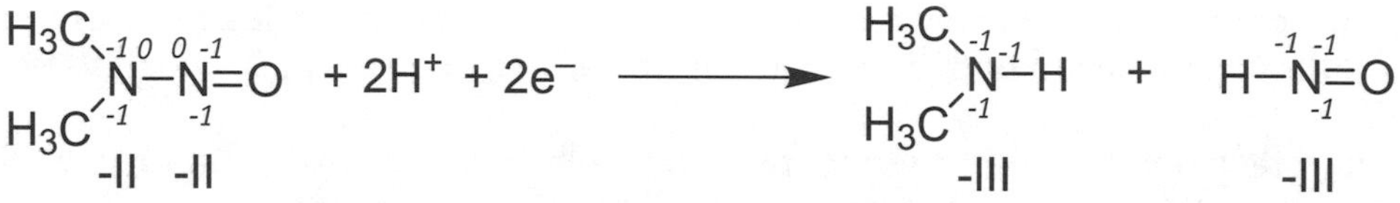

$$
\underset{-II \quad -II}{\overset{H_3C}{\underset{H_3C}{>}} \underset{-1 \ 0 \ 0 \ -1}{N-N=O}} + 2H^+ + 2e^- \longrightarrow \underset{-III}{\overset{H_3C}{\underset{H_3C}{>}} \overset{-1 \ -1}{N}\overset{-1}{H}} + \underset{-III}{\overset{-1 \ -1}{H-N=O}} \quad (1)
$$

For each N atom, the numbers in italics are assigned according to the rules given above. The sum of these values gives the oxidation state for each N, which changes from –II in NDMA to –III in the two products, consistent with a net 2-electron reduction.

In certain cases, these rules, and most other definitions of oxidation and reduction, give counter-intuitive or contradictory results (12). For this reason, in part, few general works on organic reactivity place significant emphasis on reactions classified as oxidations or reductions (major exceptions are 13–17). Environmental chemists, on the other hand, still find it useful to classify organic transformations as oxidations or reductions (e.g., 2, 9, 11, 18, 19) because the environments in which they occur are often distinctive in this regard. The major (abiotic, non-photochemical) oxidation and reduction reactions that influence the environmental fate of organic contaminants are summarized in the two sections that follow.

16.2.1.2 Oxidations

Organic chemicals that are susceptible to oxidation and are of concern from the perspective of contamination and environmental degradation include aliphatic and aromatic hydrocarbons, alcohols, aldehydes, and ketones; phenols, polyphenols, and hydroquinones; sulfides (thiols) and sulfoxides; nitriles, amines, and diamines; nitrogen and sulfur heterocyclic compounds; mono- and di-halogenated aliphatics; linear alkybenzene-sulfonate and nonylphenol polyethoxylate surfactants; and thiophosphate esters. Table 16.1 shows half-reactions for oxidation of some of these chemical groups. See other reviews (9, 18, 19) for additional discussion of the mechanisms of oxidation reactions involving organic substances of environmental interest.

16.2.1.2.1 Example: Oxidation of Phenols

Oxidation of phenols (and anilines) involves free radical reactions that can produce complex mixtures of products, including hydroperoxides and polymers. Simple examples of these two types of products appear below for oxidation of 2,6-di-(t-butyl) phenol (DBP).

TABLE 16.1

Oxidations of Environmentally Relevant Organic Chemicals

Type	Oxidation Half-Reaction
Alkanes to alcohols	$R\text{–}H + H_2O \rightarrow R\text{-}OH + 2H^+ + 2e^-$
Alcohols to aldehydes	$RCH_2OH \rightarrow RCHO + 2H^+ + 2e^-$
Aldehydes to acids	$RCHO + H_2O \rightarrow RCOOH + 2H^+ + 2e^-$
Dehydrogenation	$R_2HC\text{–}CHR'_2 \rightarrow R_2C=CR'_2 + 2H^+ + 2e^-$
Oxidative coupling (Example 2.1.2.1)	$2\ HO\text{–}C_6H_5 \rightarrow HO\text{–}C_6H_4\text{–}O\text{–}C_6H_5 + 2H^+ + 2e^-$
Hydroquinones to quinones (Example 2.1.2.2)	$HO\text{–}C_6H_4\text{–}OH \leftrightarrow O=C_6H_4=O + 2H^+ + 2e^-$
Sulfoxidation	$R\text{–}S\text{–}R' + H_2O \leftrightarrow R\text{-}S(O)\text{–}R' + 2H^+ + 2e^-$
Coupling of thiols	$R\text{–}SH + R'\text{–}SH \leftrightarrow R\text{–}S\text{–}S\text{–}R' + 2H^+ + 2e^-$

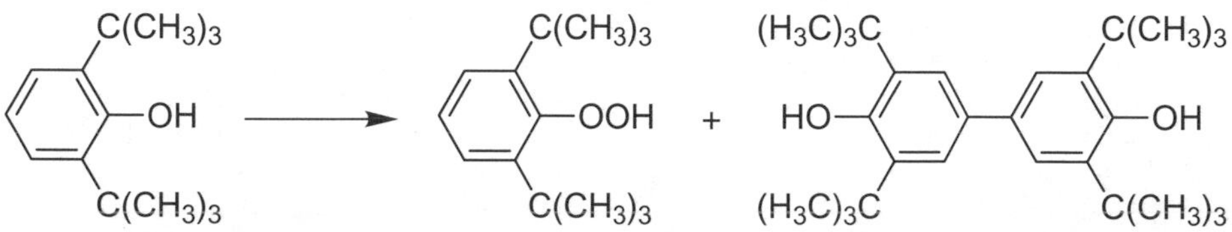

(2)

Further oxidation of these products can result in the consumption of many equivalents of oxidant for each molecule of DBP. This is the chemistry by which antioxidants protect many commercial products from spoilage or material damage by oxidation (20). Antioxidants such as DBP, and the more familiar BHT (butylated hydroxy toluene or 2,6-di-(t-butyl)-4-methyl phenol), are used very widely, so these compounds and their oxidation products are widely distributed in the environment (21).

16.2.1.2.2   Example: Oxidation of Hydroquinones

Oxidation of polyphenolic compounds to their corresponding quinones is another important class of environmental oxidations. When the hydroxyl groups are ortho (vicinal) to one another, the resulting quinone is unstable, which can lead to ring cleavage at the shared C–C bond. In contrast, two hydroxyl groups in para orientation constitute a hydroquinone, which forms a reversible redox couple with the corresponding quinone. The hydroquinone analogue of anthracene, anthrahydroquinone, forms such a redox couple with 9,10-anthraquinone.

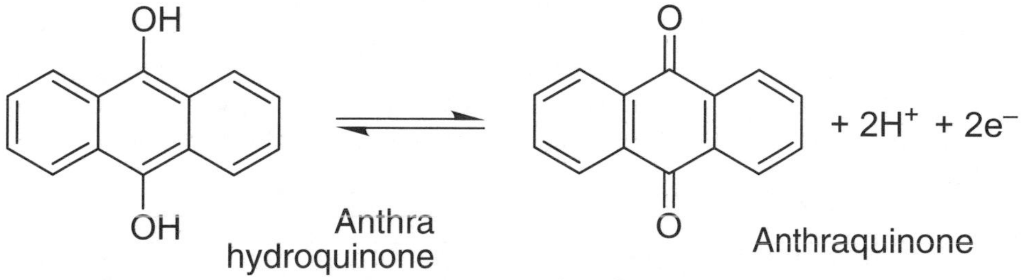

$$+ 2H^+ + 2e^-$$

(3)

Various hydroquinones have been used as model electron donors to study both abiotic degradation pathways (22-25) and microbial respiration (26). However, since quinones rather than hydroquinones are stable under aerobic conditions, the common form of contaminants is quinonoid and the pathway of primary environmental interest is reduction of quinones to the hydroquinone (i.e., the reverse of Equation 3).

16.2.1.3 Reductions

Most interest in reductive transformations of environmental chemicals involves dehalogenation of chlorinated aliphatic or aromatic contaminants and the reduction of nitroaromatic compounds. Other reductive transformations that may occur abiotically in the environment include reduction of azo compounds, quinones, disulfides, and sulfoxides (Table 16.2). See other reviews (2, 9, 11) for additional discussion of the mechanisms of these reactions.

TABLE 16.2

Reductions of Environmentally Relevant Organic Chemicals.

Type	Reduction Half-Reaction
Reductive dehalogenation (Example 2.1.3.1)	$R–X + H^+ + 2e^- \rightarrow R–H + X^-$
Vicinal dehalogenation: (Example 2.1.3.2)	$X–R–R'–X + 2e^- \rightarrow R=R' + 2X^-$
Nitro reduction: (Example 2.1.3.3)	$R–NO_2 + 6H^+ + 6e^- \rightarrow R–NH_2 + 2H_2O$
Azo reduction	$Ar–N=N–Ar' + 4H^+ + 4e^- \rightarrow ArNH_2 + Ar'NH_2$
Disulfides to thiols	$R–S–S–R' + 2H^+ + 2e^- \leftrightarrow R–SH + R'–SH$
Deoxygenation of sulfoxides	$R–S(O)–R' + 2H^+ + 2e^- \leftrightarrow R–S–R' + H_2O$
Nitrosamine reduction (Example 2.1.1.2)	$R_2N–N=O + 2H^+ + 2e^- \rightarrow R_2N–H + HNO$
Quinones to hydroquinones (Example 2.1.2.2)	$O=C_6H_4=O + 2H^+ + 2e^- \leftrightarrow HO–C_6H_4–OH$
Dealkylation	$R–Y–R' + 2H^+ + 2e^- \rightarrow R–YH + R'H$

R and R' = unspecified moieties; Ar = Aryl; X = F, Cl, Br, or I; Y = NH, O, or S.

16.2.1.3.1 Example: Reductive Dehalogenation

Dehalogenation can occur by several reductive pathways. The simplest results in replacement of a C-bonded halogen atom with a hydrogen and is known as *hydrogenolysis* or *reductive dehalogenation*. The process is illustrated for trichloroethene, TCE,

$$(4)$$

where complete dechlorination by this pathway requires multiple hydrogenolysis steps. The relative rate of each step is a critical concern because the steps tend to become slower with each dechlorination (and DCE and VC are at least as hazardous as TCE). Aryl halogens, such as those in the pesticide chlorpyrifos, also are subject to hydrogenolysis, but this reaction rarely occurs abiotically. One notable exception is the rapid abiotic dechlorination of polychlorinated biphenyls (PCBs) by zero-valent iron with catalysis by Pd (27).

16.2.1.3.2 Example: Vicinal Dehalogenation

The other major dehalogenation pathway involves elimination of two halogens, leaving behind a pair of electrons that usually goes to form a carbon-carbon double bond. Where the pathway involves halogens on adjacent carbons, it is known as *vicinal dehalogenation* or *reductive β-elimination*. The major pathway for reductive transformation of lindane involves vicinal dehalogenation, which can proceed by steps all the way to benzene (28). Recently, data has shown that this pathway not only can convert alkanes to alkenes, but can produce alkynes from dihaloalkenes (29).

$$Lindane \xrightarrow[-2Cl^-]{+2e^-} g\text{-}BTC \xrightarrow[-2Cl^-]{+2e^-} \xrightarrow[-2Cl^-]{+2e^-} Benzene \qquad (5)$$

16.2.1.3.3 Example: Nitro Reduction

Reduction of aromatic nitro groups occurs in three steps, via nitroso and hydroxylamine intermediates, to the amine. The amine can go on to form polymeric residues by a mechanism analogous to that for oxidative coupling of phenols, as in Equation 2. Abiotic nitro reduction is well documented for pesticides that contain aromatic nitro groups, such as the phosphorothioate esters methyl and ethyl parathion (22, 30-33).

$$Parathion \xrightarrow[+6e^-]{+4H^+} Aminoparathion + 2OH^- \qquad (6)$$

A great deal of information is also available on the reduction of nitrobenzene, substituted nitrobenzenes, and di- and tri-nitrobenzenes, due to their convenience as model compounds and importance as munitions (e.g., 34-40).

16.2.2 Oxidants and Reductants Relevant to Environmental Systems

The contaminant redox reactions just summarized only occur when coupled with suitable half-reactions involving oxidants or reductants from the environment. In a particular environmental system, these redox agents (along with the physico-chemical factors discussed in section 4.2) collectively determine the nature, rate, and extent of contaminant transformation. Under favorable circumstances, the dominant redox agent(s) can be identified and quantified, thereby providing a rigorous basis for estimating the potential for, and rate of, transformation by abiotic redox reactions.

Such specificity is often possible with systems engineered for contaminant remediation. However, natural systems frequently involve complex mixtures of redox-active substances that cannot be characterized readily. The characterization of redox conditions in complex environmental media is a long-standing challenge to environmental scientists that continues to be an active area of research (Section 3.3).

The remainder of this Section summarizes what is currently known about the identity of oxidants and reductants relevant to environmental systems, in order to provide a basis for estimating rates of contaminant transformations by specific pathways. With respect to natural reductants, however, a great deal remains to be learned, so substantial developments can be expected as new research in this area becomes available.

16.2.2.1 Oxidants

The best opportunities for predicting redox transformations come from engineered systems where a known oxidant is added to achieve contaminant remediation. Well-documented examples include the use of ozone (Example 2.2.1.1) and chlorine dioxide (e.g., 41, 42) in

TABLE 16.3

Environmental Oxidants.

Oxidants	Reduction Half-Reaction
Oxygen (dissolved, ground–state triplet)	$O_2 + 4H^+ + 4e^- \rightarrow 2\ H_2O$
Hydrogen peroxide	$H_2O_2 + 2H^+ + 2e^- \rightarrow 2\ H_2O$
Ozone	$O_3 + 2H^+ + 2e^- \rightarrow O_2 + H_2O$
Hypochlorite	$OCl^- + 2H^+ + 2e^- \rightarrow Cl^- + 2\ H_2O$
Chlorine dioxide	$ClO_2 + e^- \leftrightarrow ClO_2^-$
Ferrate	$FeO_4^{2-} + 8\ H^+ + 3e^- \leftrightarrow Fe^{3+} + 4H_2O$
Permanganate	$MnO_{4-} + 8\ H^+ + 5e^- \leftrightarrow Mn^{2+} + 4H_2O$
Chromate	$HCrO_4^- + 7\ H^+ + 3e^- \leftrightarrow Cr3^+ + 4H_2O$

drinking water treatment. In natural systems, important oxidants are oxides of iron and manganese (43-45), as well as molecular oxygen and various photooxidants (see Chapter 15). Table 16.3 summarizes some of the oxidants responsible for contaminant transformations by abiotic, non-photolytic pathways.

16.2.2.1.1 Example: Oxidation by Ozone

The reactivity of ozone reflects two modes of oxidation: non-selective free radical reactions involving hydroxyl radical, and the selective addition of ozone to form an ozonide intermediate and eventually various carbonyls and carboxylic acids (46). The latter sequence, known as *ozonolysis*, is shown below for anthracene.

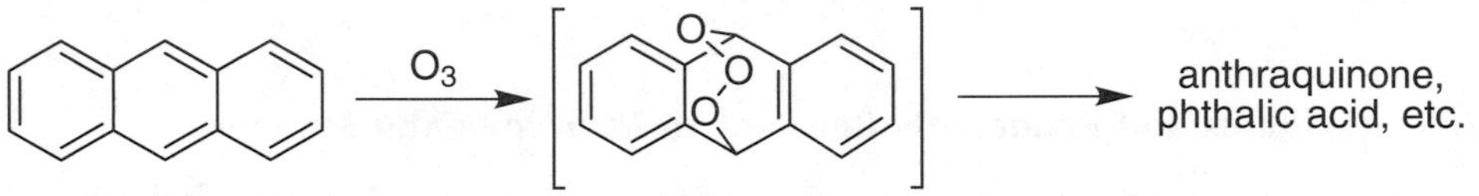

anthraquinone, phthalic acid, etc. (7)

The enormous quantity of research that has been done on environmental effects of ozone reflects its importance in atmospheric chemistry, disinfection, bleaching, and advanced technologies for wastewater treatment (47).

16.2.2.1.2 Oxygen Species

The presence of molecular oxygen, O_2, is used widely as the defining characteristic of "oxidizing" environments, because the overwhelming supply of molecular oxygen makes it the ultimate source of oxidizing equivalents. However, O_2 in its thermodynamic ground-state ($^3O_2$) is a rather poor oxidizing agent and it is not usually the oxidant directly responsible for oxidative transformations of contaminants. Instead, "activated" oxygen species may be involved where they are formed by the action of light on natural organic matter (NOM), peroxides, or various inorganic catalysts (19, 48). Activated oxygen species include singlet oxygen ($^1O_2$), hydroperoxyl radical and superoxide (HO_2/O_2^-), hydrogen peroxide and hydroperoxide anion (H_2O_2/HO_2^-), hydroxyl radical (OH), and ozone (O_3).

16.2.2.1.3 Other Oxidants

Aside from oxygen and the activated oxygen species, there are several other oxidants that cause abiotic oxidation reactions involving environmental contaminants. In engineered systems, these include chlorine (49), chlorine dioxide (50-52), permanganate (53, 54) and ferrate (55, 56). At highly contaminated sites, anthropogenic oxidants such as chromate, arsenate, and selenate may react with co-contaminants such as phenols (57, 58).

In natural anoxic environments, the major alternative oxidants are iron(III) and manganese(IV) oxides and hydroxides. Both are common in natural systems, as crystalline or amorphous particles or coatings on other particles. In the absence of photocatalysis, however, iron and manganese oxides are weak oxidants. As a result, they appear to react at significant rates only with phenols and anilines (45, 59-64).

In the dissolved phase, few alternative abiotic oxidants are available in the natural environment. Nitrate, sulfate, and other terminal electron acceptors used by anaerobic microorganisms are thermodynamically capable of oxidizing some organic contaminants, but it appears that these reactions almost always require microbial mediation.

16.2.2.2 Reductants

Abiotic environmental reductants are not as well characterized as the oxidants because there are fewer remediation applications of reductants, and natural reducing environments are characterized by especially complex biogeochemistry. The most familiar natural reductants are sulfide (present primarily as HS^- and H_2S), Fe(II) and Mn(II), and NOM. Table 16.4 summarizes some of the species that may contribute to abiotic reduction reactions in environmental systems.

TABLE 16.4

Environmental Reductants.

Reductants	Oxidation Half-Reaction
Low molecular weight organics (e.g., oxalate)	$HO_2CCO_2H \rightarrow 2\ CO_2 + 2e^- + 2H^+$
High molecular weight organics (NOM)	$NOM_{red} \leftrightarrow NOM_{ox} + 2e^- + 2H^+$
Dithionite	$H_2O_4^- + 2H_2O \leftrightarrow 2H_2SO_3 + H^+ + 2e^-$
Sulfides (and polysulfides)	$SH^- \leftrightarrow S^0 + H^+ + 2e^-$
Fe(II) at mineral surfaces	$Fe(II)_{surf} \leftrightarrow Fe(III)_{surf} + e^-$
Zero-valent iron	$Fe^0 \leftrightarrow Fe^{2+} + 2e^-$

red = unspecified reduced form; ox = oxidized form; surf = surface species (either adsorbed or part of a mineral lattice, i.e., "structural" (65, 66)).

The transformation of contaminants by sulfur species in anaerobic environments can involve both reduction and nucleophilic substitution pathways. These processes have been studied extensively (67-74), but the complex speciation of sulfur makes routine predictions regarding these reactions difficult.

A similar situation applies for reduced forms of iron (35, 36, 39, 65, 75, 76). As with oxidations, some of the best opportunities for reliably estimating rates of redox transformations are afforded by engineered systems where a reductant of known composition and quantity is added to achieve contaminant remediation. In addition to zero-valent iron, other chemical methods for reduction of contaminants involve dithionite (77-79) and electrolysis (where, in effect, electrons are added directly, e.g., 80, 81).

16.2.2.2.1 Example: Reduction by Zero-Valent Iron

The most established technology for treating contaminants by abiotic reduction reactions relies on zero-valent iron metal (82). In addition to effecting hydrogenolysis (Equation (4)) and reductive elimination (Equation (5)), Fe^0 readily reduces nitro aromatics (Equation (6)), azo dyes, nitrate, chromate, chlorine residual, and some radionuclides. Recently, an investigation of soils contaminated with the herbicide alachlor provided evidence for reductive N-dealkylation (as well as dechlorination) by Fe^0 (83).

$$(8)$$

The redox chemistry of these systems is relatively well defined: contaminant reduction results in oxidative dissolution of Fe^0 by a reaction that is equivalent to corrosion of Fe^0 by organic oxidants. Metals such as Zn and Sn can reduce contaminants by similar reactions.

16.2.2.2.3   NOM as a Reductant

The role of natural organic reductants in environmental systems is even more difficult to characterize than the roles of sulfur and iron because most natural organic matter is of indeterminant composition. To accommodate this, Table 15.4 shows two general categories: high molecular weight organic materials such as humic and fulvic acid, and low molecular weight compounds such as acids, alcohols, etc. Specific examples of the latter include glycolate, citrate, pyruvate, oxalate, and ascorbate (84). These types of compounds have been studied extensively for their role in global cycling of carbon (e.g., 85, 86, 87), but very little work has been done on whether they act as specific reductants of organic contaminants.

  In contrast, the possibility that high molecular weight NOM acts as a reductant in environmental systems is widely acknowledged . Although most evidence for this involves the reduction of metal ions (88-95), several studies have shown that the process extends to various model organic contaminants (24, 40). Presumably, the reducing potential of NOM is due to specific moieties such as complexed metals (96) or conjugated polyphenols (22-24). Often, redox reactions involving these moieties are reversible, which means that NOM may serve as a mediator of redox reactions rather than being just an electron donor (or acceptor).

16.2.3   Mediators and Catalysts

An additional consideration in formulating redox reactions is the possibility of catalysis by substances that mediate the transfer of electrons between the bulk reductant (or oxidant) and the substrate being transformed. Such considerations arise frequently in many areas of chemistry, especially electrochemistry and biochemistry (e.g., 97). In environmental applications, the most common model for mediated electron transfer involves a rapid and reversible redox couple that shuttles electrons from a bulk electron donor to a contaminant that is transformed by reduction.

$$(9)$$

16.2.3.1 Criteria for Mediated Electron Transfer

Demonstrating that a redox transformation of a contaminant involves mediated electron transfer requires meeting several criteria: (i) the overall reaction must be energetically favorable, (ii) the mediator must have a reduction potential that lies between the bulk donor and the terminal acceptor so that both steps in the electron transfer chain will be energetically favorable, and (iii) both steps in the mediated reaction must be kinetically fast relative to the direct reaction between bulk donor and terminal acceptor. Most evidence for involvement of mediators in reduction of contaminants comes from studies with model systems, because natural reducing media (such as anaerobic sediments) consist of more redox couples than can be characterized readily. Although this is an active area of research, we can identify a variety of likely mediator half-reactions (see Table 16.5).

TABLE 16.5

Mediators and Catalysts of Environmental Redox Reactions

Type	Redox Half-Reaction
Hydroquinones/Quinones	$HO-Ar-OH \leftrightarrow O=Ar=O + 2H^+ + 2e^-$
High molecular weight organics (NOM)	$NOM_{red} \leftrightarrow NOM_{ox} + 2H^+ + 2e^-$
Fe(II) at mineral surfaces	$Fe(II)_{surf} \leftrightarrow Fe(III)_{surf} + e^-$
Porphyrins, corronoids, etc.	$Fe(II)_{porphyrin} \leftrightarrow Fe(III)_{porphyrin} + e^-$

Ar = aryl moiety; red = unspecified reduced form; ox = oxidized form; surf = surface.

16.2.3.2 Advantage of the Model

An advantage of the mediator model (Equation 9) is that it can be used to simplify the problem of describing contaminant reduction reactions if the mediator is characterized more easily than the bulk donor. In this case, the bulk donor is best neglected and the problem reduced to the mediator and contaminant half-reactions. The advantage is greatest when a complex microbiological transformation process can be reduced to a reaction with a well defined biogenic mediators, such as quinones (98, 99), porphyrins, or corronoids (100-102).

16.2.3.3 Example: Mediated Reduction of Nitro Compounds

Reduction of nitro aromatic compounds often appears to be a two-step process, in which a mediator is required for facile transfer of electrons from a bulk reductant to the contaminant. A well documented example is the coupling of organic matter oxidation by iron reducing bacteria to "abiotic" nitro reduction by biogenic Fe(II) that is adsorbed to mineral surfaces in a column containing aquifer material (36, 39, 76).

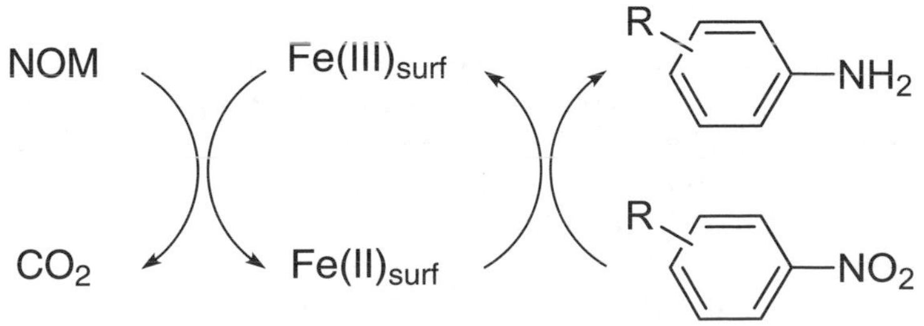

(10)

Although it has long been known that the adsorbed Fe(II) can be an effective reductant, its potential role as a mediator of reductive transformations of contaminants only recently has gained widespread recognition. Of particular interest are its possible roles in "natural attenuation" (65) and remediation technologies where the bulk reductant is dithionite (79) or Fe⁰ (66).

16.2.3.4   NOM as a Mediator

Like the various forms of iron, NOM apparently serves as both bulk reductant and mediator of reduction as well as bulk reductant (recall section 2.2.2). NOM also can act as an electron acceptor for microbial respiration by iron reducing bacteria (26), thereby facilitating the catabolism of aromatic hydrocarbons under anaerobic conditions (103). In general, it appears that NOM can mediate electron transfer between a wide range of donors and acceptors in environmental systems (104, 105). In this way, NOM probably facilitates many redox reactions that are favorable in a thermodynamic sense but do not occur by direct interaction between donor and acceptor due to unfavorable kinetics.

16.3 Methods for Estimating the Thermodynamics of Redox Reactions

16.3.1   Assessing the Energetics of a Transformation Reaction

Once the relevant oxidation and reduction half-reactions have been identified (e.g., from Tables 16.1-5), they can be combined and balanced to determine the overall reaction for any redox transformation. In generalized form, this can be written

$$a_1 O_1 + m_1 H^+ + n_1 e^- \leftrightarrow b_1 R_1 + w_1 H_2 O \tag{11}$$

$$b_2 R_2 + w_2 H_2 O \leftrightarrow a_2 O_2 + m_2 H^+ + n_2 e^- \tag{12}$$

$$n_2 a_1 O_1 + n_1 b_2 R_2 + (n_2 m_1 - n_1 m_2)\, H^+ \leftrightarrow n_2 b_1 R_1 + n_1 a_2 O_2 + (n_2 w_1 - n_1 w_2)\, H_2 O \tag{13}$$

where O and R represent the oxidant and reductant, respectively (106). The sum of the standard reduction and oxidation potentials for the two half-reactions gives the net potential (E^0_{net}), which can be used to assess the thermodynamic feasibility of a particular redox reaction,

$$\Delta G^0 = -nF\,(E^0_{net}) = -nF\,(E^0_{red} + E^0_{ox}) \tag{14}$$

where n equals the number of electrons exchanged in the net reaction, F is the Faraday constant (96,485 J V⁻¹ mol⁻¹), and E^0_{red} and E^0_{ox} are the standard potentials for Equation 11 and Equation (12), respectively. Note that the sign on E^0_{ox} is opposite that of the standard reduction potential for the corresponding reduction half-reaction. Complete redox reactions (i.e., Equation (13)) with positive E^0_{net} (or negative ΔG) can occur spontaneously.

To use Equation (14), it is necessary to have appropriate values of E^0_{red} and E^0_{ox}. Reduction potentials are widely tabulated for the classical "standard" conditions of 25°C and unit activity for all reactants and products (including H⁺, i.e., pH = 0). Tables 16.6 gives selected values of E^0_{red}. However, for the evaluation of energetics under environmental conditions, it is convenient to define a standard state for conditions that more closely approximate those of natural systems. These conditions are usually taken to be $[H^+] = 10^{-7}$ (i.e., pH = 7.0), $[HCO_3^-] = 10^{-3}$ M, $[Cl^-] = 10^{-3}$ M, $[Br^-] = 10^{-5}$M, and $[O] = [R] = 1$M. The environmental literature (9), designates the corresponding standard potential as E^0_w, although the standard state designated by most biochemists as $E^{0'}$ has essentially the same meaning. In Table 16.6, values of E^0_w are given for selected reduction half-reactions. Calculation of the E^0_w value for the overall redox transformation reaction indicates whether the free energy change for the reaction is favorable under typical environmental conditions.

TABLE 16.6

Selected values of E^0 and E^0_w.

Reduction Half–Reaction	E^0	E^0_w
C_2Cl_4 (perchloroethene) + H$^+$ + 2e$^-$ → C_2HCl_3 (trichloroethene) + Cl$^-$	+0.79	+0.58
C_2Cl_6 (hexachloroethane) + 2e$^-$ → C_2Cl_4 (tetrachloroethene) + 2 Cl$^-$	+1.14	+1.14
C_2HCl_3 (trichloroethene) + H$^+$ + 2e$^-$ → $C_2H_2Cl_2$ (cis–1,2–dichloroethene) + Cl$^-$	+0.75	+0.54
C_6Cl_5OH (pentachlorophenol) + H$^+$ + 2e$^-$ → C_6HCl_4OH (2,3,4,6–tetrachlorophenol) + Cl$^-$	+0.66	+0.45
C_6Cl_6 (hexachlorobenzene) + H$^+$ + 2e$^-$ → C_6HCl_5 (pentachlorobenzene) + Cl$^-$	+0.68	+0.47
$C_6H_5-NO_2$ + 6H$^+$ + 6e$^-$ → $C_6H_5-NH_2$ + 2 H$_2$O	+0.83	+0.42
$CCl_3C(O)OH$ (trichloroacetate) + H$^+$ + 2e$^-$ → $CHCl_2C(O)OH$ (dichloroacetate) + Cl$^-$	+0.68	+0.47
CCl_4 (carbon tetrachloride) + H$^+$ + 2e$^-$ → $CHCl_3$ (chloroform) + Cl$^-$	+0.88	+0.67
$CH_3-S(O)-CH_3$ (dimethylsulfoxide) + 2H$^+$ + 2e$^-$ ↔ CH_3-S-CH_3 + H$_2$O	+0.57	+0.16
ClO^- + 2 H$^+$ + 2e$^-$ ↔ Cl$^-$ + H$_2$O	+1.71	+1.30
ClO_2 + e$^-$ ↔ ClO_2^-	+0.95	+0.95
CO_2 + 4H$^+$ + 4e$^-$ → 1/6 $C_6H_{12}O_6$ (glucose) + H$_2$O	–2.01	–0.432
CrO_4^{2-} + 8 H$^+$ + 3 e$^-$ ↔ Cr^{3+} + 4 H$_2$O	+1.51	+0.48
Fe(III)$_{porphyrin}$ + e$^-$ ↔ Fe(II)$_{porphyrin}$	+0.17	+0.06
Fe^{2+} + 2e$^-$ ↔ Fe^0	–0.44	–0.44
Fe_2O_3(s, hematite) + 6 H$^+$ + 2e$^-$ ↔ 2 Fe^{2+} + 3 H$_2$O	+0.66	–0.35
Fe^{3+} + e$^-$ ↔ Fe^{2+}	+0.77	+2.77
FeO_4^{2-} + 8 H$^+$ + 3e$^-$ ↔ Fe^{3+} + 4 H$_2$O	+1.70	+2.59
FeOOH(s, goethite) + 3 H$^+$ + e$^-$ ↔ Fe^{2+} + 2 H$_2$O	+0.67	–0.34
H$^+$ + e$^-$ ↔ 1/2 H$_2$(g)	0.00	–0.41
H_2O_2 + 2 H$^+$ + 2 e$^-$ ↔ 2 H$_2$O	+1.76	+1.35
HCO_3^- + 9 H$^+$ + 8e$^-$ → CH_4(g) + H$_2$O	+0.23	–0.22
IO_3^- + 6 H$^+$ + 5e$^-$ ↔ 1/2 I$_2$(s) + 3 H$_2$O	+1.18	+0.68
MnO_2(s, vernadite) + 4 H$^+$ + 2 e$^-$ ↔ Mn^{2+} + 2 H$_2$O	+1.29	+0.58
MnO_4^- + 4 H$^+$ + 3e$^-$ ↔ MnO$_2$ + 2 H$_2$O	+1.69	+1.14
MnO_4^- + 8 H$^+$ + 5e$^-$ ↔ Mn^{2+} + 4 H$_2$O	+1.51	+0.84
MnOOH(s, manganite) + 3 H$^+$ + e$^-$ ↔ Mn^{2+} + 2 H$_2$O	+1.50	+0.49
NO_3^- + 10 H$^+$ + 8 e$^-$ ↔ NH$_4^+$ + 3 H$_2$O	+0.88	+0.36
NO_3^- + 2 H$^+$ + 2 e$^-$ ↔ NO$_2^-$ + H$_2$O	+0.83	+0.42
NO_3^- + 6 H$^+$ + 5 e$^-$ ↔ 1/2 N$_2$(g) + 3/2 H$_2$O	+1.24	+0.74
O_2(g) + 4 H$^+$ + 4 e$^-$ ↔ 2 H$_2$O	+1.23	+0.81
O_3 + 2H$^+$ + 2e$^-$ ↔ O2 + H$_2$O	+2.08	+1.66
$O=C_6H_4=O$ + 2H$^+$ + 2e$^-$ ↔ HO–C$_6$H$_4$–OH (Catechol)	+0.79	+0.38
$O=C_6H_4=O$ + 2H$^+$ + 2e$^-$ ↔ HO–C$_6$H$_4$–OH (hydroquinone)	+0.70	+0.29
S^0(s) + H$^+$ + 2e$^-$ ↔ SH$^-$	–0.06	–0.27
SO_3^{2-} + 2 H$^+$ + e$^-$ ↔ 1/2 S$_2$O$_4^{2-}$ + H$_2$O	0.42	–0.41
SO_4^{2-} + 9 H$^+$ + 8 e$^-$ ↔ SH$^-$ + 4 H$_2$O	+0.25	–0.21
$(-SCH_2CH(NH_2)COOH)_2$ (cystine) + 2H$^+$ + 2e$^-$ ↔ 2 HSCH$_2$CH(NH$_2$)COOH (cysteine)	+0,02	–0.39

Sources include (5, 8, 84, 107–112).

16.3.1.1 *Example: Energetics of Redox Reactions*

Perhalogenated aliphatic compounds such as hexachloroethane and perchloroethene (PCE) are highly oxidized compounds that are subject to reductive dehalogenation

(Example, 2.1.3.1) with relatively large positive standard potentials. Dihydric phenols such as hydroquinone and catechol are moderately reducing substances that can be oxidized to the corresponding quinones (Example 2.1.2.2). For the case of PCE and catechol, the combination of these reactions gives:

	E^0	E^0_w
C_2Cl_4 (PCE) + H^+ + $2e^-$ $\leftrightarrow$ C_2HCl_3 (TCE) + Cl^-	+0.79	+0.58
$HO\text{-}C_6H_4\text{-}OH$ (Catechol) $\leftrightarrow$ $O=C_6H_4=O$ + $2H^+$ + $2e^-$	–0.79	–0.38
C_2Cl_4 + $HO\text{-}C_6H_4\text{-}OH$ $\leftrightarrow$ C_2HCl_3 + $O=C_6H_4=O$ + H^+ + Cl^-	–0.09	+0.20

Note that the net potential is zero at standard conditions (E^0) and positive at standard aquatic conditions (E^0_w), so the reaction is not favorable except, perhaps, at extremely low pH.

For non-standard conditions, cell (or half-cell) potentials, E, can be calculated with the Nernst equation

$$E = E^0 - (RT/nF) \ln Q \tag{15}$$

where E^0 refers to the standard potential (red, ox, or net), R is the universal gas constant (8.314 J K^{-1} mol^{-1}), T is the absolute temperature (K), and Q is the cell quotient. For the general reduction half-reaction in Equation (11), Q is

$$Q = \{[R_1]^{b_1} / [O_1]^{a_1} [H^+]^{m_1} \} \tag{16}$$

and for the general net redox reaction in Equation 13, Q is

$$Q = \{[R_1]^{n_2 b_1} [O_2]^{n_1 a_2} / [O_1]^{n_2 a_1} [R_2]^{n_1 b_2} [H^+]^{(n_2 m_1 - n_1 m_2)}\} \tag{17}$$

Note that 2.303 RT/F = 0.059 V at 25°C, so Equation 15 can be simplified to

$$E = E^0 - (0.059/n) \log Q \tag{18}$$

and this equation is adequate to relate E^0 to E^0_w for most of the redox couples in Table 16.6.

Systems involving more than one pK_a can become quite complex, in which case it may be useful to compare redox couples graphically in Eh-pH (or Pourbaix) diagrams. These diagrams can be drawn by traditional methods (5, 113-115), obtained from existing compilations (116, 117), or generated with at least one commercially available software package (HSC Chemistry: Outokumpu Research, Pori, Finland). Eh-pH diagrams involving organic substances are not common, but their construction and interpretation are not fundamentally different from those for inorganic substances (8, 118).

16.3.1.2   Example: Effect of pH on Energetics of Redox Reactions

The boundary between all oxidized forms and all reduced forms of a substance can be drawn from Equation (18) by expanding Q (Equation (17)) to include acid/base speciation. Figure 16.1 shows this for five substances that exhibit moderately complex, but well characterized, speciation as a function of pH (uncomplexed Fe(II)/Fe(III), iron porphyrin, juglone, lawsone, and anthraquinone disulfonate). The resulting Eh-pH diagram shows, for example, that the hydroquinone of lawsone is a reductant relative to anthraquinone disulfonate, below pH 7.5, but the relationship is inverted at higher pH. A similar crossing

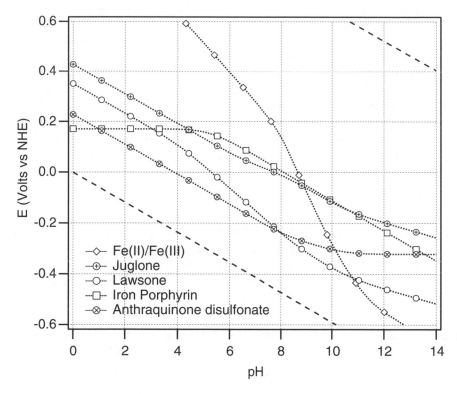

FIGURE 16.1

Eh-pH diagram showing the predominance fields for oxidized (upper right) and reduced (lower left) forms of selected redox-active species. Note that the curves represent totals for each species; i.e., further speciation is not shown. Curves are drawn from variations on Equation (18) for 25°C, using values of E^0 from Table 16.6 and additional constants from various sources (5, 8, 23). Dashed diagonal lines are for the H_2/H_2O (lower) and H_2O/O_2 (upper) couples and together they enclose the conditions over which water is stable.

exists just beloe pH9 for total inorganic ferrous/ferric iron versus both juglone and iron porphyrin.

16.3.2 Estimating Thermodynamic Data for Redox Reactions

The availability of appropriate thermodynamic data for organic redox couples often limits application of the simple formulation presented in section 3.1. This is primarily because few organic substances form reversible redox couples amenable to direct measurement of Nernstian standard potentials.

Two approaches for estimating standard potentials can be used for preliminary assessments of reaction energetics of irreversible redox couples. The first involves measurement of surrogate parameters such as half-wave potentials, and the second involves calculation from free energies of formation.

16.3.2.1 Half-Wave Potentials

Half-wave potentials, $E_{1/2}$, can be obtained from current-potential curves measured using a variety of voltammetric techniques. $E_{1/2}$ is a good approximation of E^0 when the redox couple is reversible and the diffusion coefficients of O and R are equal (119), and a few of the standard potentials in Table 16.6 were obtained in this way. When the redox couple is not reversible, it may still be possible to determine an $E_{1/2}$, but such values cannot be used to derive quantitative estimates of standard potentials without data on electrode kinetics. They can be useful, however, for qualitative or relative assessments of reaction energetics.

The qualitative application is illustrated by the approximate location of E^0_w for the azobenzene/aniline couple on redox ladders constructed by Schwarzenbach et al. (e.g., Figure K.3 in Reference 120). The estimate, around –0.1 V vs. NHE, comes from electrochemical studies that report non-Nernstian dependence of $E_{1/2}$ on pH and additional evidence for the non-reversibility of this reaction (8, 121).

The relative use of $E_{1/2}$ values is exemplified by the ranking of relative reducing potentials for various pesticides Geer et al. developed (28, 122) based on values of $E_{1/2}$ measured at a mercury-coated Pt electrode in dimethyl sulfoxide.

16.3.2.2 Free Energies of Formation

The estimation of standard potentials from other thermodynamic data follows a simple additive procedure. Typically, these calculations are based on published, gas-phase free energies of formation, ΔG^0_f (g), for reactants and products. These gas phase data are adjusted to aqueous phase free energies, ΔG^0_f (aq), using

$$\Delta G^0_f(aq) = \Delta G^0_f(g) + RT \ln H \tag{19}$$

where H is the Henry's constant for each substance. Then the free energies of formation are combined using

$$\Delta G^0 = \sum \Delta G^0_f(products) - \sum \Delta G^0_f(reactants) \tag{20}$$

to give ΔG^0, the free energy of reaction in aqueous solution. Finally, the resulting value can be adjusted to E^0_w using Equation (14) and Equation (18).

One important environmental application of this procedure has been for assessing the energetics of the dehalogenation of chlorinated solvents (110); in fact, all of the values for chlorinated solvents in Table 16.6 were obtained by this method.

Another way to obtain estimates of ΔG^0 is with the group-contribution methods developed by Benson (123) and Mavrovouniotis (124, 125). This approach has been used to extend the list of chlorinated aliphatics for which there are published estimates of E^0_w to include chloroacetates, chloroproprionates, and PCBs (111, 112, 126). A few of these compound also have been included in Table 16.6.

16.3.3 Characterizing the Redox Potential of Environmental Media

Section 5.2 provides the thermodynamic basis for predicting whether or not a specific redox transformation can occur spontaneously in a given environment. The necessary redox half-reactions involving contaminants are usually well characterized because contaminants are the primary motivation for many studies of environmental systems. However, difficulties often arise in selecting the appropriate "environmental" half-reaction with which to balance the overall equation. When an environmental half-reaction cannot be identified, it is tempting to use traditional electrode potential measurements (127, 128) as a generic measure of *in situ* redox conditions. These values (E_{meas}) then might be used as E^0_{red} or E^0_{ox} in Equation 14, to assess the thermodynamic potential of a particular contaminant transformation in a particular environment. However, a number of fundamental difficulties arise with this approach, so we do not recommend the procedure.

16.3.3.1 *Problems with E_{meas} (Not Recommended)*

The problems with using E_{meas} to estimate E^0_{red} or E^0_{ox} are related to the general problem of how redox conditions can be characterized for complex mixtures. Many approaches to this problem have been proposed, but none provide a solution that is both rigorous with respect to chemical fundamentals and practical with respect to application in the field. The issues here are subtle but important.

Several of the key issues are reflected in the debate over the appropriate use of pε to describe redox conditions in natural waters (129-131). The parameter is defined in terms of the activity of solvated electrons in solution (i.e., pε = − log {e^-_{aq}}), but the species e^-_{aq} does not exist under environmental conditions to any significant degree. The related concept of pe (132), referring to the activity of electrons in the electrode material, may have a more realistic physical basis with respect to electrode potentials, but it does not provide an improved basis for describing redox transformations in solution. The fundamental problem is that the mechanisms of oxidation and reduction under environmental conditions do not involve electron transfer from solution (or from electrode materials, except in a few remediation applications). Instead, these mechanisms involve reactions with specific oxidant or reductant molecules, and it is these species that define the half-reactions on which estimates of environmental redox reactions should be based.

Values of E_{meas} (measured at Pt, Au, or carbon electrodes) are of little help because they are not a simple function of the concentrations of specific oxidants and reductants in the solution. Instead, the electrode gives a mixed potential in response to all the redox active species in a given solution, weighted by the sensitivity of the electrode to each species (i.e., the exchange current density). Resolving mixed potentials into concentrations of all the contributing species is not a practical way to characterize the availability of oxidants or reductants to react with a particular contaminant. In fact, the only environmental condition where measured electrode potentials have been related quantitatively to concentrations of redox-active species is in relatively simple systems dominated by relatively high concentrations of dissolved iron (133-136). Inorganic oxidants, such as O_2 and H_2O_2, tend to have small exchange current densities, so their presence is not reliably indicated by electrode potential determinations (137).

16.3.3.2 *Techniques Based on Specific Species (Recommended)*

A related problem associated with efforts to characterize redox conditions of environmental materials is the lack of equilibrium among the chemical constituents of an environmental system (138-141) or between the environmental constituents and a sensor material (142). Thus, even techniques that are based on specific redox active species—such as H_2 (143-146), Hg (147), indicator dyes (148, 149), or other mediators (137)— cannot provide a general characterization of redox conditions. However, we do recommend techniques that quantify the activity of specific oxidants or reductants, because they are necessary for the rigorous application of the approach Section 5.1 describes. Similar considerations apply to the characterization of redox kinetics.

16.4 Methods for Estimating the Kinetics of Redox Reactions

Estimates of the free energy change of redox transformations indicate only which reactions can occur spontaneously, not whether they will occur at appreciable rates in a given environment. For example, reduction of hexachloroethane by water is energetically favorable, and yet the reaction rate is apparently negligible because hexachloroethane is a persistent contaminant in aerobic groundwaters. In fact, in this context, environmental scientists, engineers, and regulators most often need to estimate rates for transformation reactions already known to be possible.

16.4.1   Recommended Method: A Simple Bimolecular Kinetic Model

Estimation of rates for redox reactions in environmental systems requires that the problem be formulated in terms of specific oxidation and reduction half-reactions. In addition, we assume that the rate-limiting step of the transformation mechanism is bimolecular—that is, the slow step requires an encounter (collision) between the electron donor and electron acceptor. Under most conditions found in environmental systems, such reactions exhibit rate laws for the disappearance of a pollutant, P, that are first-order in concentration of P and first-order in the concentration of environmental oxidant or reductant, E,

$$- d[P]/dt = k \, [P] \, [E] \tag{21}$$

where k is the second-order rate constant for the reaction. The major advantage of this approach is that values of k are conventional rate constants in that they should be independent of environmental conditions except for temperature, and, in some cases, ionic strength(150). Quantitative corrections for temperature and ionic strength are discussed in section 4.2.2.

In many cases, the concentration of the environmental oxidant or reductant is effectively constant over the time frame of interest, so Equation (21) can be simplified to a pseudo-first-order rate law

$$- d[P]/dt = k_{obs} \, [P] \tag{22}$$

where the rate constant k_{obs} is the product of k and [E]. From Equation (21) and Equation (22) it is apparent that k_{obs} is defined by

$$k_{obs} = k \, [E]_{ss} \tag{23}$$

where the subscript, ss, indicates the steady-state concentration of E. Thus, k_{obs} (or $t_{1/2}$, from $\ln 2/k_{obs}$) can be calculated for any redox reaction as long as $[E]_{ss}$ can be determined, and k for reaction of P with E is known. Table 16.7 gives selected rate constants for oxidations and Table 16.8, for reductions. Section 5 discusses methods for estimating additional values of k.

16.4.1.1   Example: Kinetics of Oxidation of Aromatics by Ozone

Oxidation by ozone is a homogeneous (solution-phase) reaction, so oxidation rates are readily estimated using Equation (23) and second-order rates constants from the literature (151-154, 159, 160). Thus, for a typical concentration of ozone used in drinking water disinfection operations (10^{-5} M), and the appropriate k for, say, benzene (from 152), we can estimate

$$k_{obs} = 2 \text{ M}^{-1} \text{ s}^{-1} \times 10^{-5} \text{ M} = 2 \times 10^{-5} \text{ s}^{-1} \tag{24}$$

which corresponds to a half-life of 9.6 hours. Direct reaction of aromatic compounds with ozone (i.e., ozonolysis as in example 2.2.1.1) becomes more rapid with increasing numbers of fused rings: e.g., k for naphthalene is 3000 M^{-1} s^{-1} (152). Presumably, anthracene will react with ozone even more rapidly.

TABLE 16.7

Selected Rate Constants for Oxidations of Environmental Contaminants.

Donor, P	Acceptor, E	k (M^{-1} s^{-1})	Source
Alachlor	ozone	3.8 ± 0.4	(151)
Benzene[1]	ozone	2.0 ± 0.2	(152)
Carbon tetrachloride	ozone	<0.005	(152)
Diethylether	ozone	1.1 ± 0.1	(152)
2,6-Dimethylphenol[1]	ozone	1.9×10^4 (k_{ArOH})	(153)
Naphthalene	ozone	$(3.0 \pm 0.6) \times 10^3$	(152)
Phenol[1]	ozone	$(1.3 \pm 0.2) \times 10^3$ (k_{ArOH})	(154)
		$(1.4 \pm 0.4) \times 10^9$ (k_{ArO^-})	
Phenol[1]	chromate	$(2.63 \pm 0.06) \times 10^{-5}$ (k_{rOH})	(58)
Phenol[1]	chlorine dioxide	0.4 ± 0.1 (k_{rOH})	(42)
		$(4.9 \pm 0.5) \times 10^7$ (k_{rO^-})	
Trichloroethene	ozone	$17 \pm 4, 15 \pm 2$	(151, 152)
Trichloroethene	permanganate	6.57×10^{-4}	(155)

[1] Source includes data for other related compounds. For additional data on ozone, chlorine dioxide, and other inorganic radicals see (156). Data on hydroxyl radicals can be found in (157, 158) and chapters 14 and 15.

TABLE 16.8

Selected Rate Constants for Reductions of Environmental Contaminants.

Acceptor, P	Donor, E	k	Source
Trichloroethene	Dithionite	0.15 M^{-1} s^{-1}	(77)
Nitrobenzene[1]	Iron porphyrin	0.96 M^{-1} s^{-1}	(23)
Nitrobenzene[1]	Mercaptojuglone	0.079 M^{-1} s^{-1}	(23)
Hexachloroethane	Mercaptojuglone	0.55 M^{-1} s^{-1}	(25)
Nitrobenzene[1]	Zero-valent iron	3.9×10^{-2} L min^{-1} m^{-2}	(34)
Carbon tetrachloride[2]	Zero-valent iron	0.1 L hr^{-1} m^{-2}	(161, 162)
Trichloroethene	Zero-valent iron	3.9×10^{-4} L hr^{-1} m^{-2}	(161, 162)

[1] Source includes data for other related compounds. Data for hydrated electrons and hydrogen atoms are available in (157).

16.4.1.2 Example: Reduction of Chlorinated Alkenes by Zero-Valent Iron

Reduction by Fe^0 is a surface reaction, so reduction rates are most conveniently estimated from Equation 23 using surface-area normalized values of k_{obs} (k_{SA}). Representative values have been tabulated for a wide range of chlorinated solvents (161, 162). The corresponding value for TCE is $k_{SA} = 3.9 \times 10^{-4}$ L m^{-2} h^{-1} Thus, we can calculate a half-life of

$$t_{1/2} = \ln 2/(k_{SA} \times 3.5 \text{ m}^2 \text{ mL}^{-1} \times 1000 \text{ mL L}^{-1}) = 30 \text{ min} \tag{25}$$

in a treatment zone containing 3.5 m^2 mL^{-1} iron surface area, which is fairly typical of current engineering practice. Note that actual barrier performance varies considerably, but progress has been made in quantifying this uncertainty (163, 164).

16.4.2 Factors that Affect Redox Kinetics

The kinetic model just described is a compromise that affords a realistic possibility of making quantitative estimates with available data and yet preserves a level of deterministic rigor by requiring that the problem be formulated in terms of specific redox-active species.

As discussed in section 3.3, there is no reason to expect that measures of "overall" redox conditions (such as Pt electrode potentials or concentrations of dissolved H_2) will ever provide an improved basis for quantitatively predicting rates of environmental redox reactions. However, extensions and refinements to the simplified bimolecular model can be made when sufficient data are available.

16.4.2.1 Temperature

Temperature affects the rates of redox reactions, just as it does other transformation reactions like hydrolysis. Although a variety of models describe the effect of temperature (165), the approach is to resolve k into a function of temperature with the Arrhenius equation

$$k = Ae^{-Ea/RT} \tag{26}$$

where A is a constant known as the pre-exponential factor, and E_a is the energy of activation. Unfortunately, activation energies are rarely available for redox reactions of environmental interest. The few exceptions include limited data for reduction of chlorinated aliphatics by iron metal (166).

In the absence of compound specific data on temperature effects, Equation 26 can still be useful for approximate corrections using assumed values of E_a. Thus, the rule-of-thumb that reaction rates approximately double for every 10°C increase in temperature, is justified because most reactions of organic substances in solution have an E_a of about 50 kJ/mol. Most reported rate constants probably overestimate environmental rates slightly because the former typically are measured near 25°C, and 15°C is more typical of natural waters.

16.4.2.2 Effect of Ionic Strength

Throughout this chapter we have formulated rate laws in terms of concentrations and ignored activity corrections, as is almost always done in environmental chemistry. However, where ionic strength, I, varies and both reactants are charged, a substantial "primary salt effect" can be expected (167). The effect is described by

$$\log (k/k_0) = 2.34 \, Z_{ox} \, Z_{red} \, I^{1/2} \tag{27}$$

where Z_{ox} and Z_{red} are the charges on each reactant and k_0 is the rate constant extrapolated to zero ionic strength. Although ionic strength effects are likely for, say, the oxidation of pentachlorophenol (which exist mostly as the phenoxide anion at neutral pH) by chromate, there seem to be no documented examples where the effect of ionic strength on kinetics of a redox reaction is significant under environmental conditions.

16.4.2.3 pH

When hydrogen ions are directly involved in the rate-limiting step of a reaction, they usually appear as explicit terms in the rate law. However, the role of hydrogen ions in both halves of the redox reaction must be well-defined before generalizing on the effect of pH. Protonated and deprotonated forms of redox agents react as independent species, so the observed rate constant will vary with pH due to changes in speciation of the reactants. The second-order rate law (Equation (21)) can be modified to take this into account,

$$k_{\text{total}} = \Sigma k_{ij} \, [P_i] \, [E_j] \tag{28}$$

where the subscripts i and j reflect the various degrees of protonation for P and E, respectively.

This approach will be rigorous if a complete speciation calculation is done for P and E (as described in standard textbooks of aquatic chemistry) and all the necessary values of k_{ij} are available. Fortunately, the analysis can usually be simplified to reaction between one or two dominant species, and most of the available rate constants are for these same species.

16.4.2.4 Example: Speciation Effects on Phenol Oxidation

The oxidation of substituted phenols illustrates the importance of including speciation. Dissociation of the phenolic hydroxyl group results in an equilibrium mixture of the parent compound and its dissociated form, the phenoxide (or phenolate) anion. The undissociated phenol and the phenoxide anion react as independent species with very different rate constants, designated k_{ArOH} and k_{ArO^-}. For the oxidation of 4-nitrophenol (pKa = 7.2) by ClO_2, $k_{ArOH} = 1.4 \times 10^{-1}$ M^{-1} s^{-1}, and $k_{ArO^-} = 4.0 \times 10^3$ M^{-1} s^{-1} (42). Estimates of the pH-corrected second-order rate constant, k_{total}, can be made using

$$k_{total} = \{(k_{ArOH} \times f_{ArOH}) + (k_{ArO^-} \times f_{ArO^-})\} \tag{29}$$

where f_{ArOH} is the fraction of the phenol which is in the protonated form,

$$f_{ArOH} = 10^{(pKa-pH)}/(1 + 10^{(pKa-pH)}) \tag{30}$$

and f_{ArO^-} is the fraction in the deprotonated form.

$$f_{ArO^-} = 1/(1 + 10^{(pKa-pH)}) \tag{31}$$

Thus, k_{total} is 25.3 M^{-1} s^{-1} at pH 5 and 3.45×10^3 M^{-1} s^{-1} at pH 8, a 136-fold increase as the speciation shifts from 99% phenol to 86% phenoxide. A similar trend can be expected if the oxidant were O_3. In contrast, the protonated phenol dominates the oxidation rate of most phenols by aqueous chromate (57, 58).

16.4.2.5 Sorption

Sorption to surfaces can have important effects on the rates of contaminant transformation, but these effects may be very different, depending on how the mechanism of sorption (i.e., hydrophobic partitioning, donor-acceptor interactions, or ligand exchange) relates to the mechanism of contaminant transformation (i.e., reaction in solution, reaction at surface sites, etc.). In general, however, the contributions of each compartment can be treated as additive as long as the kinetics of adsorption/desorption are fast, relative to contaminant transformation (168). Just as with the effect of pH (Section 4.2.3), each term is simply the product of the reactant concentrations in the compartment and the corresponding rate constant.

$$k_{total} = \Sigma k_i \, [P_i] \, [E_i] \tag{32}$$

where, in this case, the subscript i reflects the various compartments into which the reactants are distributed: including the solution phase, non-reactive surface sites, and reactive surface sites.

16.4.2.6 Example: Sorption Effects on Reduction of Azo Dyes

Hydrophobic adsorption to sediment particles appears to retard reduction of organic contaminants in anaerobic sediment slurries, so a quantitative kinetic model has been proposed that involves two types of "sites"(168-170),

$$k_{total} = \{(k_{surf} \times f_{surf}) + (k_{soln} \times f_{soln})\} \tag{33}$$

where f_{surf} is the fraction of contaminant adsorbed onto the sediment particles,

$$f_{surf} = \rho K_p/(1 + \rho K_p) \tag{34}$$

f_{soln} is the fraction in the dissolved phase,

$$f_{soln} = 1/(1 + \rho K_p) \tag{35}$$

ρ is the sediment to water mass ratio (effectively, the "concentration" of sediment), and K_p is the coefficient describing equilibrium partitioning between the contaminant and the sediment. Note, that this formulation uses ρ for a particular sediment sample as a surrogate for an unspecified [E]. Only when the model can be rewritten in terms of specific reductants, will it be possible to estimate reduction rates in sediments in general. A slightly more rigorous model involving reactive and non-reactive sites has been applied to dehalogenation kinetics by Fe^0 (171, 172).

16.5 Quantitative Structure-Activity Relationships (QSARs) to Predict Properties

For redox reactions of a series of closely related compounds, redox potentials and rate constants often correlate to descriptor variables that reflect the electron donor or electron acceptor properties of P. Such correlations can be used to derive quantitative structure-activity relationships (QSARs), and these QSARs provide the basis for predicting properties of environmental contaminants that have not previously been measured (173).

16.5.1 Common Descriptor Variables

Commonly used descriptor variables for QSARs involving redox reactions include substituent constants (σ), ionization potential, electron affinity, energy of the highest occupied molecular orbital (E_{HOMO}) or lowest unoccupied molecular orbital (E_{LUMO}), one-electron reduction or oxidation potential ($E^{1'}$), and half-wave potential ($E_{1/2}$). One descriptor variable (D), fit to a log-linear model, is usually sufficient to describe a redox property of P. Such a QSAR will have the form

$$\log R_i = \beta_0 + \beta_1 D_i \tag{36}$$

where R_i is response variable of interest (in this context, usually E_i^0 or k_i), D_i is the descriptor variable, the subscript i distinguishes the congeners that make up the training set of compounds, and the fitted intercept and slope are β_0 and β_1, respectively. Table 16.9 and Table 16.10 summarize QSARs that are currently available for environmental redox reactions.

TABLE 16.9

QSARs for Environmental Oxidation Reactions

Substrates	Oxidant	QSAR Equation[1]	R²(n)	Source
Phenols	ClO_2	$\log k_{ArO-} = 3.2\ \Sigma\sigma_{o,m,p}^- + 8.2$	0.94 (23)	(41, 173)
Phenols	ClO_2	$\log k_{ArO-} = -4.5\ E_{HOMO} - 49$	0.76 (22)	(41, 174)
Phenols[2]	O_3	$\log k/k_0 = -3.1\ \sigma+$	(7)	(152)
Phenols[3]	ClO_2 and $^1O_2$	$\log k(^1O_2) = 5.5 + 0.36\ \log k(ClO_2)$	0.82 (10)	(173, 175)
Phenols	MnO_2	$\log k_{ArOH} = 6.1 - 9.7\ E_{1/2}$	0.85 (10)	(173, 176)
Phenols	MnO_2	$\log k_{ArOH} = -3.7\ E_{HOMO} - 36$	0.86 (9)	(174, 176)
Phenols	Chromate $(HCrO_4^-)$	$\log k_{ArOH} = 6.2 - 17\ E_{1/2}$	0.95 (10)	(58)
Phenols	Chromate $(HCrO_4^-)$	$\log k_{ArOH} = -7.8\ E_{HOMO} - 67$	0.90 (13)	(58, 174)
Phenols	Peroxy-disulfate $(S_2O_8^{2-})$	$\log k_{ArOH} = -7.8\ E_{HOMO} - 67\ ^1$	0.90 (43)	(174, 177)

[1] All k's in $M^{-1}\ s^{-1}$.
[2] For para-substituted phenols only.
[3] For phenoxide anions (ArO^-)

TABLE 16.10

QSARs for Environmental Reduction Reactions

Substrates	Reductant	QSAR Equation	R²(n)	Source
Nitrobenzenes	Mercapto-juglone	$\log k = E^{1\prime}/0.059 + 7.2^1$	0.98 (7)	(23)
Nitrobenzenes	NOM	$\log k = E^{1\prime}/0.059 + 4.4^1$	0.91 (10)	(40)
Halocarbons	Fe^0	$\log k = -3.2\ E_{LUMO} - 3.3^2$	0.85 (11)	(178)

[1] k in $M^{-1}\ s^{-1}$ for pH 7.5 with 5 mM hydrogen sulfide as bulk reductant (25°C), and $E^{1\prime}$ is the one-electron potential in V vs SHE.
[2] k in $L\ hr^{-1}\ m^{-2}$ and E_{LUMO} is the energy of the lowest unoccupied molecular orbital.

16.5.2 Example: Rate Constants for Oxidation of DBP by ClO_2

The antioxidant 2,6-di-(t-butyl)phenol (DBP) has bulky substitutes in both positions ortho to the phenolic moiety, so steric effects are likely to make predictions of oxidation rate constants unreliable. However, the QSAR for k_{ArO-} of substituted phenols reacting with ClO_2 (Table 16.9) has been shown to be relatively robust with respect to ortho effects (179). Assuming additivity of substituent effects, and no steric effects, we can use $\sigma^- = -0.15$ for t-butyl groups (180) to estimate $k_{ArO-} = 1.7 \times 10^7\ M^{-1}\ s^{-1}$ for oxidation of DBP by ClO_2. Then, equations (29), (30) and (31) can be used to compute the effect of pH and estimate k_{total}.

16.5.3 Reliability Limits

Properties estimated by interpolation within the range of conditions over which a QSAR was calibrated should be reliable, but extrapolations beyond this range cannot be made with certainty. A similar restriction applies to experimental variables factored out of the training set data before deriving the QSAR (e.g., the effect of pH on oxidation of phenolic compounds, or the effect of surface area on reductions with Fe^0).

All properties estimated from QSARs should be treated with caution, because most QSARs exhibit outliers due to molecular effects that the correlation model does not take into account, and the occurrence of outliers is not always easy to anticipate. To make the most reliable use of available QSARs such as those in Table 16.9 and Table 16.10, consult the original study from which the QSAR was derived for a complete description of the model's limitations.

16.6 Acknowledgments

The Swiss Federal Institute for Environmental Science and Technology (EAWAG/ETH) provided facilities for DLM during the initial preparation of this manuscript. PGT's recent contributions in this area were supported, in part, by the National Science Foundation (BCS-9212059 and EGB-9708554). Helpful suggestions on this manuscript were provided by T. Mill, M. Sheuer, C. Baker, and D. McCubbary.

16.7 List of Symbols

a_1, a_2	Stoichiometric coefficients for oxidized species
β_0, β_1	Intercept and slope of the linear QSAR
$\sigma, \sigma^-, \sigma^+$	Hammett sigma constants
b_1, b_2	Stoichiometric coeffients for reduce species
ΔG^0	Free energy of reaction
$\Delta G_f^0(aq)$	Free energy of formation in aqueous phase
$\Delta G_f^0(g)$	Free energy of formation in gas phase
D_i	Descriptor variable for QSAR, subscripts i distinguish congeners
E	Potential under non-standard conditions
$E^0_{ox}, E^0_{red}, E^0_{net}, E_i^0, E^0$	Standard potentials (reactants and products at unit activity, hydrogen ion activity of 1, i.e. pH = 0), of reduction, oxidation, net reaction, the ith redox-active species, and any half-reaction
E^0_{ox}	Standard reduction potential at hydrogen ion activity of 10^{-7} (pH = 7.0)
$E^{0\prime}$	One-electron reduction potential
$E_{1/2}$	Half-wave potential
E_{HOMO}	Energy of the highest occupied molecular orbital (HOMO)
E_{LUMO}	Energy of the lowest unoccupied molecular orbital (LUMO)
E_{meas}	Redox potential measured at a Pt, Au, or C electrode
E_j	The jth environmental oxidant or reductant
$[E]_{ss}$	Concentration of E at a steady-state
F	Faraday constant, 96,485 J V^{-1} mol^{-1}
f_{ArO^-}	Fraction in the deprotonated phenoxide form
f_{ArOH}	Fraction in the protonated phenolic form
f_{soln}	Fraction in solution phase
f_{surf}	Fraction sorbed to surfaces
H	Henry's constant
I	Ionic strength
k	Second-order rate constant
Ik_0	Rate constant extrapolated to ionic strength = 0
k_{ArO^-}	Rate constant for the phenoxide (deprotonated) form
k_{ArOH}	Rate constant for the phenolic (protonated) form

k_i	Rate constant for the ith species or phase
k_{ij}	Second-order rate constant for reaction between the ith and jth species
k_{obs}	Observed, pseudo-first-order rate constant
K_p	Equilibrium partition coefficient between aqueous and adsorbed phases
k_{soln}	Rate constant for the dissolve species
k_{surf}	Rate constant for the adsorbed species
k_{total}	Combined second-order rate constant for all forms of a reactant
m_1, m_2	Stoichiometric coefficient for the number of hydrogen ions
n, n_1, n_2	Number of electrons exchanged in the net reaction
O_1, O_2	Oxidized species
P_i	The ith pollutant
pe, pε	$-\log[e^-]$ for electrons of an electrode, $-\log[e^-]$ of solvated electrons
PKa	$-\log$ of the acid dissociation constant
Q	Cell quotient
R	Gas constant ($8.314\ \mathrm{J\ K^{-1}\ mol^{-1}}$)
ρ	Sediment to water mass ratio, i.e., "concentration" of sediment
R_1, R_2	Reduced species
R_i	Response variable of interest (in this context, usually E_i^0 or k_i)
T	Absolute temperature (K)
$t_{1/2}$	Half-life
w_1, w_2	Stoichiometric coefficient for water
Z_{ox}, Z_{red}	Charge on the oxidized and reduced species

References

1. Lyman, W.J., I. Bodek, W.F. Reehl, and D.H. Rosenblatt. 1988. *Environmental Inorganic Chemistry: Properties, Processes, and Estimation Methods*. Pergamon, New York, 1988.
2. Macalady, D.L., P.G. Tratnyek, and T.J. Grundl. 1986. Abiotic reduction reactions of anthropogenic organic chemicals in anaerobic systems. *J. Contam. Hydrol.* 1, 1-28.
3. Wolfe, N.L. and Macalady, D.L. New perspectives in aquatic redox chemistry: Abiotic transformations of pollutants in groundwater and sediments. *J. Contam. Hydrol.* 1992, 9, 17-34.
4. Chang, R. 1998. *Chemistry;* 6th ed. McGraw-Hill, Boston.
5. Pankow, J.F. 1991. *Aquatic Chemical Concepts*. Lewis, Chelsea, MI.
6. Klemm, L.H. 1995. A numerical measure of the degree of oxidation (DOX) of an organicmolecule, with special attention to heteroatoms and heterocyclic compounds. *J. of Heterocycl. Chem.* 32, 1509-1512.
7. Kjonaas, R.A. 1986. Number of oxidations relative to methylene. A convenient method of recognizing and quantifying organic oxidation-reduction. *J. Chem. Ed.* 63, 311-314.
8. Clark, W.M. 1960. *Oxidation-Reduction Potentials of Organic Systems*; Williams & Wilkins, Baltimore.
9. Schwarzenbach, R.P., P.M. Gschwend, and D.M. Imboden. 1993. Chapter 12.4, Oxidation and Reduction Reactions. In *Environmental Organic Chemistry*, ed., pp. 399-435. Wiley, New York.
10. Krikorian, S.E. 1988. Methodology for quantifying functional group oxidation state levels and for analyzing organic reactions for oxidation-reduction behavior. *Am. J. Pharm. Educ.* 52, 177-180.
11. Larson, R.A. and E.J. Weber. 1994. Chapter 3. Reduction. In *Reaction Mechanisms in Environmental Organic Chemistry*, ed., pp. 169-215. Lewis, Chelsea, MI.

12. Sisler, H.H., and C.A. Van der Werf. 1980. Oxidation-reduction. An example of chemical sophistry. *J. Chem. Ed.* 57, 42-44.

13. March, J. 1985. Chapter 19, Oxidations and Reductions. In *Advanced Organic Chemistry: Reactions, Mechanisms, and Structure*; 3rd ed., ed., pp. 1048-1120. Wiley, New York.

14. Wiberg, K.B. 1963. Oxidation-reduction mechanisms in organic chemistry. In *Survey of Progress in Chemistry*, A.F. Scott, ed., Vol. 1, pp. 211-248.

15. Rinehart, K.L., Jr. 1973. *Oxidation and Reduction of Organic Compounds*. Prentice-Hall, Inc., Englewood Cliffs, NJ, 1973.

16. Hudlicky, M. 1984. *Reductions in Organic Chemistry*. Ellis Horwood, Chichester.

17. Hudlicky, M. 1990. *Oxidations in Organic Chemistry*. American Chemical Society, Washington, DC, 1990.

18. Mill, T. 1990. Chemical and photo oxidation. In O. Hutzinger, ed., *The Handbook of Environmental Chemistry: Reactions and Processes*, Vol. 2A, pp. 77-105, Springer-Verlag, Berlin.

19. Larson, R.A. and E.J. Weber. 1994. Chapter 4. Environmental Oxidations. In *Reaction Mechanisms in Environmental Organic Chemistry*, pp. 217-273. Lewis, Chelsea, MI.

20. Simic, M.G. 1991. Antioxidant compounds: an overview. In K.J.A. Davies, ed., *Oxidative Damage & Repair: Chemical, Biological, and Medical Aspects*, pp. 47-56. Pergamon, Oxford.

21. Lopez-Avila, V. and R.A. Hites. 1981. Oxidation of phenolic antioxidants in a river system. *Environ. Sci. Technol.* 15, 1386-1388.

22. Tratnyek, P.G. and D.L. Macalady. 1989. Abiotic reduction of nitro aromatic pesticides in anaerobic laboratory systems. *J. Agric. Food Chem.* 37, 248-254.

23. Schwarzenbach, R.P., R. Stierli, K. Lanz, and J. Zeyer. 1990. Quinone and iron porphyrin mediated reduction of nitroaromatic compounds in homogeneous aqueous solution. *Environ. Sci. Technol.* 24, 1566-1574.

24. Curtis, G.P. and M. Reinhard. 1994. Reductive dehalogenation of hexachlorethane, carbon tetrachloride, and bromoform by anthrahydroquinone disulfonate and humic acid. *Environ. Sci. Technol.* 28, 2393-2401.

25. Perlinger, J.A., W. Angst, and R.P. 1996. Schwarzenbach. Kinetics of the reduction of hexachloroethane by juglone in solutions containing hydrogen sulfide. *Environmental Science and Engineering* 30, 3408-3417.

26. Lovley, D.R., J.D. Coates, E.L. BluntHarris, E.J.P. Phillips, and J.C. Woodward. 1996. Humic-substances as electron acceptors for microbial respiration. *Nature* 382, 445-448.

27. West, O.R., L. Liang, W.L., Holden, N.E. Korte, Q. Fernando, and J.L. Clausen. 1996. Degradation of polychlorinated biphenyls (PCBs using palladized iron, Oak Ridge National Laboratory, *ORNL/TM-13217*.

28. Beland, F.A., S.O. Farwell, A.E. Robocker, and R.D. Geer. 1976. Electrochemical reduction and anaerobic degradation of lindane. *J. Agric. Food Chem.* 24, 753-756.

29. Roberts, A.L., L.A. Totten, W.A. Arnold, D.R. Burris, and T.J. Campbell. 1996. Reductive elimination of chlorinated ethylenes by zero-valent metals. *Environ. Sci. Technol.* 30, 2654-2659.

30. Wolfe, N.L., B.E. Kitchens, D.L. Macalady, and T.J. Grundl. 1986. Physical and chemical factors that influence the anaerobic degradation of methyl parathion in sediment systems. *Environ. Toxicol. Chem.* 5, 1019-1026.

31. Pritchard, P.H., C.R. Cripe, W.W. Walker, J.C. Spain, and A.W. Bourquin, 1987. Biotic and abiotic degradation rates of methyl parathion in freshwater and estuarine water and sediment samples. *Chemosphere* 16, 1509-1520.

32. Butler, L.C., D.C. Staiff, G.W. Sovocool, and J.E. Davis. 1981. Field disposal of methyl parathion using acidified powdered zinc. *Journal of Environmental Sciences and Health* B16, 49-58.

33. Wahid, P.A., C. Ramakrishna, and N. Sethunathan. 1980. Instantaneous degradation of parathion in anaerobic soils. *J. Environ. Qual.* 9, 127-130.

34. Agrawal, A. and T.G. Tratnyek. 1996. Reduction of nitro aromatic compounds by zero-valent iron metal. *Environ. Sci. Technol.* 30, 153-160.

35. Klausen, J., S.P. Trüber, S.B. Haderlein, and R.P. Schwarzenbach. 1995. Reduction of substituted nitrobenzenes by Fe(II) in aqueous mineral suspensions. *Environ. Sci. Technol.* 29, 2396-2404.

36. Haderlein, S.B. and R.P. Schwarzenbach. 1995. Environmental processes influencing the rate of abiotic reduction of nitroaromatic compounds in the subsurface. In J.C. Spain, ed., *Biodegradation of Nitroaromatic Compounds*, pp. 199-225. Plenum, New York.

37. Barrows, S.E., C.J. Cramer, D.G. Truhlar, M.S. Elovitz, and E.J. Weber. 1996. Factors contolling regioselectivity in the reduction of polynitroaromatics in aqueous solution. *Environ. Sci. Technol.* 30, 3028-3038.

38. Schmelling, D.C., K.A. Gray, and P.V. Kamat. 1996. Role of reduction in the photocatalytic degradation of TNT. *Environ. Sci. Technol.* 30, 2547-2555.

39. Rügge, K., T.B. Hofstetter, S.B. Haderlein, P.L. Bjerg, S. Knudsen, C. Zraunig, and T.H. Christensen. 1998. Characterization of predominant reductants in an anaerobic leachate-contaminated aquifer by nitroaromatic probe compounds. *Environ. Sci. Technol.* 32, 23-31.

40. Dunnivant, F.M., R.P. Schwarzenbach, and D.L. Macalady. 1992. Reduction of substituted nitrobenzenes in aqueous solutions containing natural organic matter. *Environ. Sci. Technol.* 26, 2133-2141.

41. Tratnyek, P.G. and J. Hoigné. 1994. Kinetics of reactions of chlorine dioxide (OClO) in water. II. Quantitative structure-activity relationships for phenolic compounds. *Wat. Res.* 28, 57-66.

42. Hoigné, J. and H. Bader. 1993. Kinetics of reactions involving chlorine dioxide (OClO) in water. I. Inorganic and organic compounds. *Wat. Res.* 28, 45-55.

43. Ulrich, H.-J. and A.T. Stone. 1989. Oxidation of chlorophenols adsorbed to manganese oxide surfaces. *Environ. Sci. Technol.* 23, 421-428.

44. Laha, S. and R.G. Luthy. 1990. Oxidation of aniline and other primary aromatic amines by manganese dioxide. *Environ. Sci. Technol.* 24, 363-373.

45. Ukrainczyk, L. and M.B. McBride. 1992. Oxidation of phenol in acidic aqueous suspensions of manganese oxides. *Clays Clay Miner.* 40, 157-166.

46. Hoigné, J. 1988. The chemistry of ozone in water. In S. Stucki, ed., *Process Technologies for Water Treatment*, pp. 121-143. Plenum.

47. Langlais, B., D.A. Reckhow, and D.R. Brink. 1991. *Ozone in Water Treatment: Application and Engineering*; Lewis, Chelsea, MI.

48. Larson, R.A. 1978. Environmental chemistry of reactive oxygen species. *Crit. Rev. Environ. Cntrl.* 8, 197-246.

49. Rosenblatt, D.H. 1977. Chlorine and oxychlorine species reactivity with organic substances. In J.D. Johnson, ed., *Disinfection Water and Wastewater*, pp. 249-276, Ann Arbor Science, Ann Arbor, MI.

50. Noack, M.G. and S.A. Iacoviello 1992. The chemistry of chlorine dioxide in industrial and wastewater treatment applications, *2nd International Symposium, Chemical Oxidation: Technology for the Nineties*, Nashville, TN, *Technomic, Vol. 2*, pp. 1-19.

51. Rav-Acha, C. 1984. The reactions of chlorine dioxide with aquatic organic materials and their health effects. *Wat. Res.* 18, 1329-1341.

52. Gordon, G. and D.H. Rosenblatt. 1972. The chemistry of chlorine dioxide. In S.J. Lippard, ed., *Progress in Inorganic Chemistry*, Vol. 15; pp. 201-286. Wiley, New York.

53. Vella, P.A. and B. Veronda 1993. Oxidation of trichloroethylene: A comparison of potassium permanganate and Fenton's reagent. *3nd International Symposium, Chemical Oxidation: Technology for the Nineties*, Nashville, TN, Technomic, Vol. 3, pp. 62-78.

54. Walton, J., P. Labine, and A. Reidies. 1991. The chemistry of permanganate in degradative oxidations; *1st International Symposium, Chemical Oxidation: Technology for the Nineties*, Nashville, TN, Technomic, Vol. 1, pp. 205-221.

55. Sharma, V.K., J.O. Smith, and F.J. Millero. 1997. Ferrate(VI) oxidation of hydrogen sulfide. *Environ. Sci. Technol.* 31, 2486.

56. Delaude, L. and P. Laszlo. 1996. A novel oxidizing reagent based on potassium ferrate(VI). *J. of Org. Chem.* 61, 6360-6370.

57. Elovitz, M.S. and W. Fish. 1995. Redox interactions of Cr(VI) and substituted phenols: Products and mechanism. *Environ. Sci. Technol.* 29, 1933-1943.

58. Elovitz, M.S. and W. Fish. 1994. Redox interactions of Cr(VI) and substituted phenols: Kinetic investigation. *Environ. Sci. Technol.* 28, 2161-2169.

59. Weerasooriya, S., C.B. Dissanayake, K. Priyadharsanee, and K. Jinadasa. 1993. Chemical decontamination of aniline by redox sensitive mineral surfaces. 1. Kinetic aspects. *Toxicology and Environmental Chemistry* 38, 101-108.

60. Ukrainczyk, L., and M.B. McBride. 1993. The oxidative dechlorination reaction of 2,4,6- trichlorophenol in dilute aqueous suspensions of manganese oxides. *Environ. Toxicol. Chem.* 12, 2005-2014.

61. Ukrainczyk, L. and M.B. McBride. 1993. Oxidation and dechlorination of chlorophenols in dilute aqueous suspensions of manganese oxides – Reaction products. *Environ. Toxicol. Chem.* 12, 2015-2022.

62. McBride, M.B. 1987. Adsorption and oxidation of phenolic compounds by iron and manganese oxides. *Soil Sci. Soc. Am. J.* 51, 1466-1472.

63. Pizzigallo, M.D.R., P. Ruggiero, C. Crecchio, and R. Mininni. 1995. Manganese and iron oxides as reactants for oxidation of chlorophenols. *Soil Sci. Soc. Am. J.* 59, 444-452.

64. Lehmann, R.G., H.H. Cheng, and J.B. Harsh. 1987. Oxidation of phenolic acids by soil iron and manganese oxides. *Soil Sci. Soc. Am. J.* 51, 352-356.

65. Haderlein, S.B. and K. Pecher. 1998. Pollutant reduction in heterogeneous Fe(II)/Fe(III) systems. In D. Sparks and T. Grundl, Eds., *Kinetics and Mechanism of Reactions at the Mineral/Water Interface*, American Chemical Society, Washington, DC.

66. Scherer, M.M., B.A. Balko, and P.G. Tratnyek. 1998. The role of oxides in reduction reactions at the metal-water interface. In D. Sparks and T. Grundl, Eds., *Kinetics and Mechanisms of Reactions at the Mineral-Water Interface*, American Chemical Society, Washington, DC.

67. Barbash, J.E. and M. Reinhard. 1989. Abiotic dehalogenation of 1,2-dichloroethane and 1,2-dibromoethane in aqueous solution containing hydrogen sulfide. *Environ. Sci. Technol.* 23, 1349-1357.

68. Barbash, J.E., and M. Reinhard. 1989. Reactivity of sulfur nucleophiles toward halogenated organic compounds in natural waters. In E.S. Saltzman and W.J. Cooper, ed., *Biogenic Sulfur in the Environment*, Vol. 393, 101-138. American Chemical Society, Washington, DC.

69. Brock, T.D., and K. O'Dea. 1977. Amorphous ferrous sulfide as a reducing agent for culture of anaerobes. *App. Environ. Microbiol.* 33, 254-256.

70. Kriegman-King, M.R., and M. Reinhard. 1992. Transformation of carbon tetrachloride in the presence of sulfide, biotite, and vermiculite. *Environ. Sci. Technol.* 26, 2198-2206.

71. Roberts, A.L., P.N. Sanborn, and P.M. Gschwend. 1992. Nucleophilic substitution reactions of dihalomethanes with hydrogen sulfide species. *Environ. Sci. Technol.* 26, 2263- 2274.

72. Schwarzenbach, R.P., W. Giger, C. Schaffner, and O. Wanner. 1985. Groundwater contamination by volatile halogenated alkanes: Abiotic formation of volatile sulfur compounds under anaerobic conditions. *Environ. Sci. Technol.* 19, 322-327.

73. Yu, Y.S., and G.W. Bailey. 1992. Reduction of nitrobenzene by four sulfide minerals: Kinetics, products, and solubility. *J. Environ. Qual.* 21, 86-94.

74. Haag, W.R., and T. Mill. 1988. Some reactions of naturally occurring nucleophiles with haloalkanes in water. *Environ. Toxicol. Chem.* 7, 917-924.

75. Heijman, C.G., C. Holliger, M.A. Glaus, R.P. Schwarzenbach, and J. Zeyer. 1993. Abiotic reduction of 4-chloronitrobenzene to 4-chloroaniline in a dissimilatory iron-reducing enrichment culture. *Appl. Environ. Microbiol.* 59, 4350-4353.

76. Heijman, C.G., E. Grieder, C. Holliger, and R.P. Schwarzenbach. 1995. Reduction of nitroaromatic compounds coupled to microbial iron reduction in laboratory aquifer columns. *Environ. Sci. Technol.* 29, 775-783.

77. Rodríguez, J.C., and M. Rivera. 1997. Reductive dehalogenation of carbon tetrachloride by sodium dithionite. *Chem. Lett.* 1133-1134.

78. Fruchter, J.S., J.E. Amonette, C.R. Cole, Y.A. Gorby, M.D. Humphrey, J.D. Istok, F.A. Spane, J.E. Szecsody, S.S. Teel, V.R. Vermeul, M.D. Williams, and S.B. Yabusaki. 1996. In Situ Redox Manipulation Field Injection Test Report - Hanford 100-H Area, Pacific Northwest National Laboratory, PNNL-11372, UC-602.

79. Amonette, J.E., J.E. Szecsody, H.T. Schaef, J.C. Templeton, Y.A. Gorby, and J.S. Fruchter. 1994. Abiotic reduction of aquifer materials by dithionite: A promising in-situ remediation technology; In *Proceedings of the 33rd Hanford Symposium on Health & the Environment. In-Situ Remediation: Scientific Basis for Current and Future Technologies*, Vol. 2, pp. 851-881, Battelle Pacific Northwest Laboratories, Pasco, WA.

80. Simonsson, D. 1997. Electrochemistry for a cleaner environment. *Chem. Soc. Rev.* 26, 181-189.

81. Zhang, S.P. and J.F. Rusling. 1993. Dechlorination of polychlorinated biphenyls by electrochemical catalysis in a bicontinuous microemulsion. *Environ. Sci. Technol.* 27, 1375-1380.

82. Tratnyek, P.G. 1996. Putting corrosion to use: Remediation of contaminated groundwater with zero-valent metals. *Chem. Ind. (London)*, 499-503.

83. Eykholt, G.R. and D.T. Davenport. 1998. Dechlorination of the chloroacetanilide herbicides alachlor and metolachlor by iron metal. *Environ. Sci. Technol.* 32, 1482-1487.

84. Stone, A.T., K.L. Godtfredsen, and B. Deng. 1993. Sources and reactivity of reductants encountered in aquatic environments. In G. Bidoglio, ed., *Chemistry of Aquatic Systems: Local and Global Perspectives*, Kluwer, Dordrecht, The Netherlands.

85. Kieber, D.J., J. McDaniel, and K. Mopper. 1989. Photochemical source of biological substrates in sea water: Implications for carbon cycling. *Nature* 341, 637-639.

86. Mopper, K., X. Zhou, R.J. Kieber, D.J. Kieber, R.J. Sikorski, and R.D. Jones. 1991. Photochemical degradation of dissolved organic carbon and its impact on the oceanic carbon cycle. *Nature* 353, 60-62.

87. Sunda, W.G. and D.J. Kieber. 1994. Oxidation of humic substances by manganese oxides yields low-molecular-weight organic substrates. *Nature* 367, 62-64.

88. Wittbrodt, P.R. and C.D. Palmer. 1995. Reduction of Cr(VI) in the presence of excess soil fulvic acid. *Environ. Sci. Technol.* 29, 255-263.

89. Wilson, S.A. and J.H. Weber. 1979. An EPR study of the reduction of vanadium(V) to vanadium(IV) by fulvic acid. *Chem. Geol.* 26, 345-354.

90. Skogerboe, R.K. 1981. Reduction of ionic species by fulvic acid. *Anal. Chem.* 53, 228-232.

91. Matthiessen, A. 1996. Kinetic aspects of the reduction of mercury ions by humic substances. 1. Experimental design. *Fres. J. Anal. Chem.* 354, 747-749.

92. Szilágyi, M. 1971. Reduction of Fe^{3+} ion by humic acid preparations. *Soil Sci.* 111, 233-235.

93. Alberts, J.J., J.E. Schindler, R.W. Miller, and D.E. Nutter, Jr. 1974. Elemental mercury evolution mediated by humic acid. *Science* 184, 895-897.

94. Deiana, S., C. Gessa, B. Manunza, R. Rausa, and V. Solinas. 1995. Iron(III) reduction by natural humic acids: A potentiometric and spectroscopic study. *Eur. J. Soil Sci.*, 46, 103-108.

95. Chen, Y.Z., B.M. Tan, and Z.J. Lin. 1993. A Kinetic Study of the Reduction of Np(VI) with Humic Acid. *Radiochimica Acta* 62, 199-201.

96. Wittbrodt, P.R. and C.D. Palmer. 1996. Effect of temperature, ionic strength, background electrolytes, and Fe(III) on the reduction of hexavalent chromium by soil humic substances. *Environ. Sci. Technol.* 30, 2470-2477.

97. Fultz, M.L. and R.A. Durst. 1982. Mediator compounds for the electrochemical study of biological redox systems: A compilation. *Anal. Chim. Acta* 140, 1-18.

98. Keck, A., J. Klein, M. Kudlich, A. Stolz, H.-J. Knackmuss, and R. Mattes. 1997. Reduction of azo dyes by redox mediators originating in the naphthalenesulfonic acid degradation pathway of *Sphingomonas* sp. strain BN6. *Appl. Environ. Microbiol.* 63, 3684-3690.

99. Kudlich, M., A. Keck, J. Klein, and A. Stolz. 1997. Localization of the enzyme involved in anaerobic reduction of azo dyes by *Sphingomonas* sp. strain BN6 and effect of artificial redox mediators on the rate of azo dye reduction. *Appl. Environ. Microbiol.* 63, 3691-3694.

100. Gantzer, C.J. and L.P. Wackett. 1991. Reductive dechlorination catalyzed by bacterial transition-metal coenzymes. *Environ. Sci. Technol.* 25, 715-722.

101. Chiu, P.C. and M. Reinhard. 1995. Metallocoenzyme mediated reductive transformation of carbon tetrachloride in titanium(III) citrate aqueous solution. *Environ. Sci. Technol.* 29, 595-603.

102. Burris, D.R., C.A. Delcomyn, M.H. Smith, and A.L. Roberts. 1996. Reductive dechlorination of tetrachloroethylene and trichloroethylene catalyzed by vitamin B_{12} in homogeneous and heterogeneous systems. *Environ. Sci. Technol.* 30, 3047-3052.

103. Lovley, D.R. J.C. Woodward, and F.H. Chapelle. 1994. Stimulated anoxic biodegradation of aromatic hydrocarbons using Fe(III) ligands. *Nature* 370, 128-131.
104. Zimmerman, A.P. 1981. Electron intensity, the role of humic acids in extracellular electron transport and chemical determination of pE in natural waters. *Hydrobiologia* 78, 259-265.
105. Schindler, J.E., D.J. Williams, and A.P. Zimmerman. 1976. Investigation of extracellular electron transport by humic acids. In J.O. Nriagu, ed., *Environmental Biogeochemistry, Vol. 1. Carbon, Nitrogen, Phosphorus, Sulfur, and Selenium Cycles*, pp. 109-115, Ann Arbor Science: Ann Arbor.
106. Buvet, R. 1983. General criteria for the fulfilment of redox reactions. In G. Milazzo, and M. Blank, Eds., *Bioelectrochemistry I. Biological Redox Reactions*, pp. 15-50, Plenum, New York.
107. Schwarzenbach, R.P., P.M. Gschwend, and D.M. Imboden. 1993. *Environmental Organic Chemistry*; Wiley, New York, pp. 681.
108. Bratsch, S.G. 1989. Standard electrode potentials and temperature coefficients in water at 298.15 K. *J. Phys. Chem. Ref. Data* 18, 1-21.
109. Compton, R.G. and G.H.W. Sanders. 1996. *Electrode Potentials*; Oxford University, Oxford, pp. 92.
110. Vogel, T.M., C.S. Criddle, and P.L. McCarty. 1987. Transformations of halogenated aliphatic compounds. *Environ. Sci. Technol.* 21, 722-736.
111. Dolfing, J. and B. K. Harrison. 1992. Gibbs free energy of formation of halogenated aromatic compounds and their potential role as electron acceptors in anaerobic environments. *Environ. Sci. Technol.* 26, 2213-2218.
112. Dolfing, J. and D.B. Janssen. 1994. Estimation of Gibbs free energies of formation of chlorinated aliphatic compounds. *Biodegradation* 5, 21-28.
113. Garrells, R.M. and C.L. Christ. 1965. *Solutions, Minerals, and Equilibria*; Harper & Row, New York, pp. 450.
114. Stumm, W. and J.J. Morgan. 1996. *Aquatic Chemistry*; 3rd ed.; Wiley, New York, pp. 1022.
115. Morel, F.M.M. and J.G. Hering. 1993. *Principles and Applications of Aquatic Chemistry*; Wiley, New York, pp. 588.
116. Pourbaix, M. 1966. *Atlas of Electrochemical Equilibria in Aqueous Solutions*. Pergamon, Oxford, pp. 644.
117. Brookins, D.G. 1988. *Eh-pH Diagrams for Geochemistry*. Springer, Berling, pp. 176.
118. Bailey, S.I., I.M. Ritchie, and F.R. Hewgill. 1983. The construction and use of potential-pH diagrams in organic oxidation-reduction reactions. *J. Chem. Soc. Perkin Trans. II* 5, 645-652.
119. Bard, A.J. and L.R. Faulkner. 1980. *Electrochemical Methods. Fundamentals and Applications*; Wiley, New York, pp. 718.
120. Schwarzenbach, R.P., P.M. Gschwend, and D.M. Imboden. 1995. *Environmental Organic Chemistry: Illustrative Examples, Problems, and Case Studies*. Wiley, New York, pp. 376.
121. Thomas, F.G. and K.G. Boto, 1975. The electrochemistry of azoxy, azo and hydrazo compounds. In S. Patai, ed., *The Chemistry of the Hydrazo, Azo and Azoxy Groups*, pp. 443, Wiley, New York.
122. Geer, R.D. 1978. Predicting the anaerobic degradation of organic chemical pollutants in waste water treatment plants from their electrochemical reduction behavior. In Montana University Joint Water Resources Research Center, Bozeman, MT, Completion Report No. 96
123. Benson, S.W. 1976. *Thermochemical Kinetics: Methods for the Estimation of Thermochemical Data and Rate Parameters*; 2nd ed., Wiley, New York, pp. 320.
124. Mavrovouniotis, M.L. 1991. Estimation of standard Gibbs energy changes of biotransformations. *J. Biol. Chem.* 266, 14440-14445.
125. Mavrovouniotis, M.L. 1990. Group contributions for estimating standard Gibbs energies of formation of biochemical compounds in aqueous solution. *Biotechnol. Bioeng.* 36, 1070-1082.
126. Holmes, D.A., B.K. Harrison, and J. Dolfing. 1993. Estimation of Gibbs free energies of formation for polychlorinated biphenyls. *Environ. Sci. Technol.* 27, 725-731.
127. Walton-Day, K., D.L. Macalady, M.H. Brooks, and V.T. Tate. 1990. Field methods for measurement of ground water redox chemical parameters. *Ground Water Monitor. Rev.* 10, 81-89.
128. Langmuir, D. 1971. Eh-pH determination. In R.E. Carver, ed., *Procedures in Sedimentary Petrology*, pp. 597-634. Wiley, New York.
129. Stumm, W. and J.J. Morgan. 1985. On the conceptual significance of pe. *Am. J. Sci.* 285, 856-859.
130. Hostettler, J.D. 1985. On the importance of distinguishing Eh from pe. *Am. J. Sci.* 285, 859-863.

131. Thorstenson, D.C. 1984. The concept of electron activity and its relation to redox potentials in aqueous geochemical systems. U.S. Geological Survey Open-File Report 84-072.
132. Frevert, T. 1979. The pe redox concept in natural sediment-water systems; its role in controlling phosphorus release from lake sediments. *Arch. Hydrobiol. Suppl.* 55, 278-297.
133. Doyle, R.W. 1968. The origin of the ferrous ion-ferric oxide Nernst potential in environments containing dissolved ferrous iron. *Am. J. Sci.* 266, 840-859.
134. Grundl, T. and D. Macalady. 1989. Electrode measurement of redox potential in anaerobic aqueous iron systems. *J. Contam. Hydrol.* 5, 97-117.
135. Grenthe, I., W. Stumm, M. Laaksuharju, A.-C. Nilsson, and P. Wikberg. 1992. Redox potentials and redox reactions in deep groundwater systems. *Chem. Geol.* 98, 131-150.
136. Macalady, D.L., D. Langmuir, T. Grundl, and A. Elzerman. 1990. Use of model-generated Fe^{3+} ion activities to compute Eh and ferric oxyhydroxide solubilities in anaerobic systems. In D.C. Melchior and R.L. Bassett, ed., *Chemical Modeling of Aqueous Systems II*, Vol. 416; pp. 350-367. American Chemical Society, Washington, DC.
137. Breck, W.G. 1972. Redox potentials by equilibration. *J. Mar. Res.* 30, 121-139.
138. Stumm, W. 1966. Redox potential as an environmental parameter; conceptual significance and operational limitations. In O. Jaag, and H. Liebmann, Eds., *Advances in Water Pollution Research*, Vol. 1; pp. 283-308. Water Pollution Control Federation: Washington, DC.
139. Morris, J.C. and W. Stumm. 1967. Redox equilibria and measurements of potentials in the aquatic environment. In W. Stumm, ed., *Equilibrium Concepts in Natural Water Systems*, pp. 270-285. American Chemical Society, Washington, DC.
140. Whitfield, M. 1969. Eh as an operational parameter in estuarine studies. *Limnol. Oceanogr.* 14, 547-558.
141. Spiro, M. 1986. Polyelectrodes: The behaviour and applications of mixed redox systems. *Chem. Soc. Rev.* 15, 141-165.
142. Peiffer, S., O. Klemm, K. Pecher, and R. Hollerung. 1992. Redox measurements in aqueous solutions—A theoretical approach to data interpretation based on electrode kinetics. *J. Contam. Hydrol.* 10, 1-18.
143. Chapelle, F.H., P.M. Bradley, D.R. Lovley, and D.A. Vroblesky. 1996. Measuring rates of biodegradation in a contaminated aquifer using field and laboratory methods. *Ground Water* 34, 691-698.
144. Lovley, D.R. and F.H. Chapelle. 1995. Deep subsurface microbial processes. *Rev. Geophys.* 33, 365-381.
145. Lovley, D.R., F.H. Chapelle, and J.C. Woodward. 1994. Use of dissolved H_2 concentrations to determine distribution of microbially catalyzed redox reactions in anoxic groundwater. *Environ. Sci. Technol.* 28, 1205-1210.
146. Lovley, D.R. and S. Goodwin. 1988. Hydrogen concentrations as an indicator of the predominant terminal electron-accepting reactions in aquatic sediments. *Geochim. Cosmochim. Acta* 52, 2993-3003.
147. Bisogni, J.J., Jr. 1989. Using mercury volatility to measure redox potential in oxic aqueous systems. *Environ. Sci. Technol.* 23, 828-831.
148. Tratnyek, P.G. and N.L. Wolfe. 1990. Characterization of the reducing properties of anaerobic sediment slurries using redox indicators. *Environ. Toxicol. Chem.* 9, 289-295.
149. Lemmon, T.L., J.C. Westall, and J.D. Ingle, Jr. 1996. Development of redox sensors for environmental applications based on immobilized redox indicators. *Anal. Chem.* 68, 947-953.
150. Hoigné, J. 1990. Formulation and calibration of environmental reaction kinetics: Oxidations by aqueous photooxidants as an example. In W. Stumm, ed., *Aquatic Chemical Kinetics: Reaction Rates of Processes in Natural Waters*, pp. 43-70. Wiley-Interscience, New York.
151. Yao, C.C.D. and W.R. Haag. 1991. Rate constants for direct reactions of ozone with several drinking water contaminants. *Water Res.* 25, 761-773.
152. Hoigné, J. and H. Bader. 1983. Rate constants of reactions of ozone with organic and inorganic compounds in water—I. Non-dissociating organic compounds. *Wat. Res.* 17, 173-183.
153. Gurol, M.D. and S. Nekoulnaini. 1984. Kinetic behavior of ozone in aqueous solutions of substituted phenols. *Ind. Eng. Chem. Fundam.* 23, 54-60.

154. Hoigné, J. and H. Bader. 1983. Rate constants of reactions of ozone with organic and inorganic compounds in water—II. Dissociating organic compounds. *Wat. Res.* 17, 185-194.
155. Yan, Y.E. and F.W. Schwartz. 1999. Oxidative degradation of chlorinated ethylenes by potassium permanganate. *J. Contam. Hydro.* 37, 343-365.
156. Neta, P., R.E. Huie, and A.B. Ross. 1988. Rate constants for reactions of inorganic radicals in aqueous solution. *J. Phys. Chem. Ref. Data* 17, 1027-1284.
157. Buxton, G.V., C.L. Greenstock, W.P. Helman, and A.B. Ross. 1988. Critical review of rate constants for reactions of hydrated electrons, hydrogen atoms and hydroxyl radicals (•OH/•O⁻) in aqueous solution. *J. Phys. Chem. Ref. Data* 17, 513-886.
158. Haag, W.R. and C.C.D. Yao. 1992. Rate constants for reaction of hydroxyl radicals with several drinking water contaminants. *Environ. Sci. Technol.* 26, 1005-1013.
159. Hoigné, J., H. Bader, W.R. Haag, and J. Staehelin. 1985. Rate constants of reactions of ozone with organic and inorganic compounds in water—III: Inorganic compounds and radicals. *Wat. Res.* 19, 993-1004.
160. Zheng, Y., D.O. Hill, and C.H. Kuo. 1993. Rates of ozonation of cresol isomers in aqueous solutions. *Ozone Sci. Eng.* 15, 267-278.
161. Johnson, T.L., M.M. Scherer, and P.G. Tratnyek. 1996. Kinetics of halogenated organic compound degradation by iron metal. *Environ. Sci. Technol.* 30, 2634-2640.
162. Tratnyek, P.G. T.L. Johnson, M.M. Scherer, and G.R. Eykholt. 1997. Remediating groundwater with zero-valent metals: Kinetic considerations in barrier design. *Ground Water Monit. Rem.* 108-114.
163. Eykholt, G.R. 1997.Uncertainty-based scaling of iron reactive barriers. In J. Evans and L. Reddi, ed., *In Situ Remediation of the Geoenvironment*, pp. 41-55, American Society of Civil Engineers, New York.
164. Eykholt, G.R. and T.M. Sivavec. 1995. Contaminant transport issues for reactive-permeable barriers. In Y.B. Acar, and D.E. Daniel, Eds., *Geoenvironment 2000, Vol. 2, Characterization, Containment, Remediation, and Performance in Environmental Geotechnics*, pp. 1608-1621. American Society of Civil Engineers, New York.
165. Laidler, K.J. 1990. *Chemical Kinetics;* 3rd ed., pp. 531; McGraw-Hill: New York.
166. Scherer, M.M., J.C. Westall, M. Ziomek-Moroz, and P.G. Tratnyek. 1997. Kinetics of carbon tetrachloride reduction at an oxide-free iron electrode. *Environ. Sci. Technol.* 31, 2385-2391.
167. Gardiner, W.C., Jr. 1972. *Rates and Mechanisms of Chemical Reactions*, pp. 284; Benjamin/Cummings: Menlo Park, CA.
168. Zepp, R.G. and N.L. Wolfe. 1987. Abiotic transformation of organic chemicals at the particle-water interface. In W. Stumm, ed., *Aquatic Surface Chemistry: Chemical Processes at the Particle-Water Interface*, pp. 423-455. Wiley, New York.
169. Weber, E.J. and N.L. Wolfe, 1987. Kinetic studies of the reduction of aromatic azo compounds in anaerobic sediment/water systems. *Environ. Toxicol. Chem.* 6, 911-919.
170. Jafvert, C.T. and N.L. Wolfe. 1987. Degradation of selected halogenated ethanes in anoxic sediment-water systems. *Environ. Toxicol. Chem.* 6, 827-837.
171. Burris, D.R., T.J. Campbell, and V.S. Manoranjan. 1995. Sorption of trichloroethylene and tetrachloroethylene in a batch reactive metallic iron-water system. *Environ. Sci. Technol.* 29, 2850-2855.
172. Allen-King, R.M., R.M. Halket, and D.R. Burris. 1997. Reductive transformation and sorption of cis- and trans-1,2-dichloroethene in a metallic iron-water system. *Environ. Toxicol. Chem.* 16, 424-429.
173. Tratnyek, P.G. 1998. Correlation analysis of the environmental reactivity of organic substances. In D.L. Macalady, ed., *Perspectives in Environmental Chemistry*, pp. 167-194. Oxford, New York.
174. Rorije, E. and J.G.M. Peijnenburg. 1996. QSARs for oxidation of phenols in the aqueous environment, suitable for risk assessment. *J. Chemometrics* 10, 79-93.
175. Tratnyek, P.G. 1995. Correlating oxidation kinetics for organic solutes: A comparison of QSARs for the major aqueous oxidants, *210th National Meeting, Chicago, IL, American Chemical Society, Vol. 35*, No. 2, pp. 400-401.

176. Stone, A.T. 1987. Reductive dissolution of manganese(III/IV) oxides by substituted phenols. *Environ. Sci. Technol.* 21, 979-988.

177. Behrman, E.J. 1963. Studies on the mechanism of the Elbs peroxydisulfate oxidation. *J. Am. Chem. Soc.* 85, 3478-3482.

178. Tratnyek, P.G. and M.M. Scherer. 1998. Kinetic controls on the performance of remediation technologies based on zero-valent iron, *Proceedings of the 1998 National Environmental Engineering Conference: Water Resources in the Urban Environment,* Chicago, IL, American Society of Civil Engineers, pp. 110-115.

179. Tratnyek, P.G., J. Hoigné, J. Zeyer, and R. Schwarzenbach. 1991. QSAR analyses of oxidation and reduction rates of environmental organic pollutants in model systems. *Sci. Total Environ.* 109/110, 327-341.

180. Exner, O. 1978. A critical compilation of substituent constants. In N.B. Chapman and J. Shorter, Eds., *Correlation Analysis in Chemistry: Recent Advances*, pp. 439-540, Plenum, New York.

Part IV

Specific Classes
of
Substances

17

Estimating the Properties of Surface-Active Chemicals

Johannes Tolls and Dick T.H.M. Sijm

CONTENTS

1-56670-456-1/00/$0.00+$.50

17.1 Introduction

We encounter surface active chemicals in many spheres of our lives. They are used in ore flotation, as emulsifiers in oil drilling, as adjuvants in pesticide formulations, in textile processing. Furthermore, they are personal care products and, most importantly, detergents. The latter products account for the largest portion of annual use of surfactants, about 8 million tons of synthetic surfactants and 9 million tons of salts of fatty acids (Granados, 1996). The word surfactant is an amalgam of **sur**face **act**ive ag**ent**. A chemical is surface active when it is enriched at the interface of the solution with adjacent phases. The general structure of surfactants consists of an apolar and a polar moiety, which commonly are referred to as the hydrophobic tail and the hydrophilic headgroup, respectively (Figure 17.1). According to the charge of the headgroup, surfactants can be subdivided into anionic, cationic, nonionic, and amphoteric surfactants, with all but the amphoteric headgroup used in detergents. The hydrophobic tails can be fluorocarbons, methylated silanes, or, most commonly, hydrocarbons. Hence, surfactants are a diverse group of chemicals. The diversity in the headgroups imparts different surfactants with different solute-solvent and solute-sorbent interactions. Table 17.1 explains abbreviations used for surfactants.

After use, surfactants are discharged into the wastewater. Generally, wastewater treatment plants (WWTP) remove a substantial portion of the surfactants by either biodegradation or sorption to settling sewage sludge. Nevertheless, WWTP effluents are a route of entry of surfactants into surface waters. Furthermore, the WWTP sewage sludge can transport surfactants if sprayed onto soils.

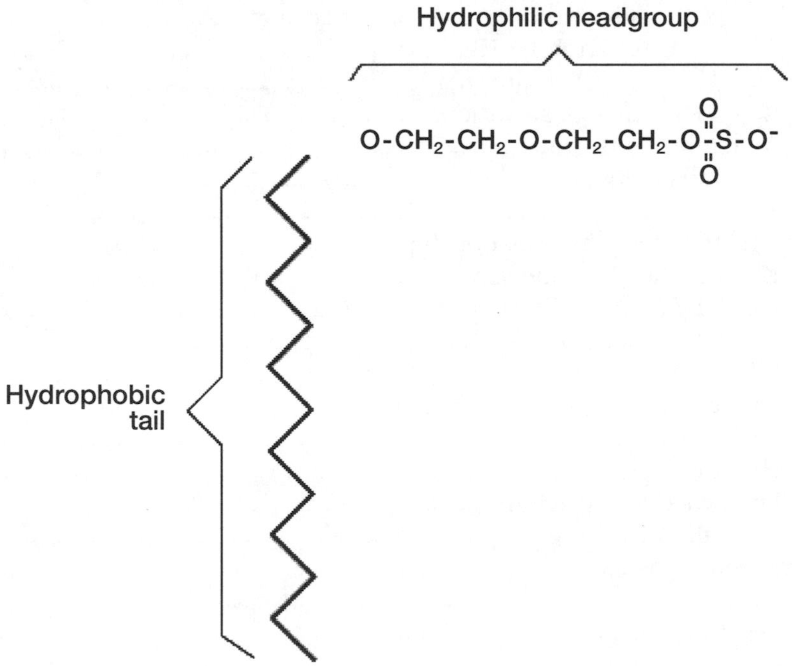

FIGURE 17.1
Structure of $C_{12}EO_3SO_4^-$, an anionic surfactant.

TABLE 17.1

Abbreviations Used for the Structural Moieties in Surfactants

C_n	alkyl chain with n C-atoms
N^+	ammonium N-atom
Me	methyl group
EO_m	oxyethylene chain containing m (C_2H_4O) units
SO_4^-	sulfate ester moiety
SO_3^-	sulfonic acid moiety
LAS	linear alkylbenzenesulfonate*
AP	alkylphenol unit
Pyr	pyridinium moiety
Bz	benzyl moiety
AOS	α-olefin-sulfonate
TPBS	tetrapropylbenzenesulfonate
DEEDMAC	dieethanolester dimethylammonium surfactant used as fabric softeners

* LAS is a mixture of homologs and isomers. If n in C_n is an integer, a mixture consisting of the isomers of one homolog is referred to. If n is a decimal, n gives the average number of C-atoms in the alkyl chain. Pure isomers are referred to if an integer between C_n and LAS, denoting the position of substitution of the sulphophenyl-moiety at the alkyl chain, is included.

The fate of surfactants in soils and surface waters is determined primarily by biodegradation and sorption. Biodegradability depends on the presence of functional groups in surfactants which can be metabolized by microorganisms. Surfactant sorption depends on a) the ability of a surfactant headgroup to undergo interactions with the sorbent and b) the interactions of its hydrophobic tail which is repelled from the water.

The industrially produced surfactants used in detergents are applied in formulations containing mixtures of homologs, isomers, and oligomers. As a result, surfactants occur in the environment as mixtures of chemicals which, within a given class, are closely related. Historically, many properties of surfactants have been determined for such mixtures, most frequently without yielding information for the individual constituents. Therefore, much of the existing data, particularly on environmental behavior, does not lend itself for derivation of structure/property relationships and, in consequence, our selection of data here may appear sparse. For some properties, the scarcity reflects the available data. For other properties, we selected data to illustrate how surfactant structure influences environmentally relevant properties.

However, methods for predicting environmentally relevant properties of surfactants applicable for all surfactant classes are presently not available. Due to the absence of validated estimation methods, this chapter's goal is to supply information necessary to understanding the behavior of surfactants in the environment and to provide data on the relevant properties of surfactants.

17.2 Surface Activity

Due to their structure (Figure 17.1), all surfactants have the tendency to accumulate at interfaces because there the hydrophobic tail can be shielded from interacting with water molecules while the hydrophilic headgroup remains solvated by water molecules. As a result of this orientation, surfactant molecules displace water molecules at the interface. Consequently, the number of hydrogen bonds decreases per unit interface area. This can be measured macroscopically as a decrease in the interfacial tension, which is defined as the work required to increase the interface by unit area.

The solution behavior of surfactants can be illustrated with a curve of the air-water interfacial tension (the so-called surface tension γ_{AW}) vs. surfactant concentration C_w (Figure 17.2). An increase of C_w results in a decrease of γ_{AW} at low C_w until an inflection in the curve occurs. Beyond the inflection region, increasing C_w does not result in a change of γ_{AW}. The inflection indicates saturation of the water-air interface with surfactant molecules. Additional surfactant molecules cannot adsorb to the interface and are forced to remain in the water phase.

The energetically unfavorable interactions of the hydrophobic tails with the water molecules are then minimized by the surfactants forming aggregates with other surfactant molecules. In those aggregates, the hydrophilic headgroups remain solvated by water molecules while the hydrocarbon moieties are shielded from water and create a hydrophobic microenvironment. Examples of these spontaneously formed aggregates are micelles and lamellae. The intersection of the extrapolations of the linear parts of the surface tension curve (Figure 17.2) is the critical micelle concentration (CMC).

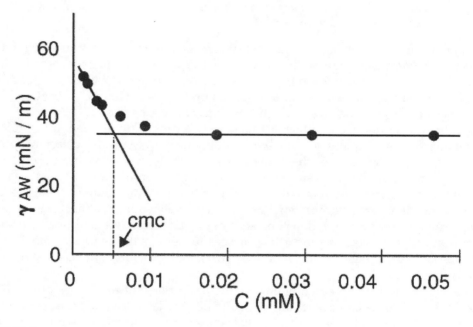

FIGURE 17.2
Idealized curve of the surface tension $\gamma_{A,W}$ against surfactant concentration.

17.2.1   Relationships Between Structure and CMC

17.2.1.1   *Effect of Alkyl Chain Length*

Table 17.2 shows values of the CMC for a series of alkylsulfate surfactants. It reflects the influence of hydrophobicity on the CMC. Within homologous groups of surfactants the CMC decreases with increasing length of the hydrocarbon tail. Regression of log CMC against the number of C-atoms yields a slope of 0.3.

Table 17.3 shows that the slope of this type of plot ranges around 0.3 for all ionic surfactants, while it is approximately 0.5 for nonionic and zwitterionic surfactants. Since it is the logarithm of the CMC which is linearly related to the number of carbon atoms, the slopes

TABLE 17.2

CMC-Values of a Homologous Series of Alkyl Sulfate Surfactants in Distilled Water (Excerpted from Rosen, 1989).

	Counterion	T (°C)	CMC (M)
$C_8SO_4^-$	Na^+	313	$1.4 \cdot 10^{-1}$
$C_{10}SO_4^-$	Na^+	313	$8.2 \cdot 10^{-3}$
$C_{12}SO_4^-$	Na^+	313	$8.6 \cdot 10^{-3}$
$C_{14}SO_4^-$	Na^+	313	$2.2 \cdot 10^{-3}$
$C_{16}SO_4^-$	Na^+	313	$5.8 \cdot 10^{-4}$

TABLE 17.3

Parameters of the Regression Equation Log CMC = $B - An_{C\text{-atoms}}$ for Selected Homologous Series of Surfactants (Excerpted from Rosen, 1989).

Surfactant Series	Counterion	T (K)	B	A
C_nCOO^-	Na^+	293	1.8	0.30
$C_nSO_3^-$	$Na^+ (K)^+$	313	1.5	0.29
LAS	Na^+	328	1.6	0.30
$C_nMe_3N^+$	Br^-	298	2.0	0.32
C_nEO_6		298	1.8	0.49
C_nEO_8		298	1.8	0.50
$C_nMe_2N^+CH_2COO^-$		296	3.1	0.49

indicate that the average decrease in CMC per added C-atom is a factor of 2 and 3 for ionic and nonionic surfactants, respectively.

17.2.1.2 Effect of the Polar Headgroup

Table 17.4 lists values of the CMC for C_{12}-surfactants with different headgroups. Within the nonionic alcohol ethoxylate surfactants, the headgroup area increases with an increasing number of oxyethylene units. The CMC simultaneously increases, illustrating that steric repulsion increases the CMC (Figure 17.3).

Comparing CMCs of a nonionic and an ionic surfactant with approximately equal headgroup area ($C_{12}EO_8$, $C_{12}Pyr$ and $C_{12}EO_2SO_4^-$) makes it apparent that the CMC of the nonionic surfactant is the lowest. It also demonstrates the effect of electrostatic repulsion. While hydrophobic interactions drive micellization, they are counteracted by steric and electrostatic interactions of the headgroups, both of which limit the coverage of the interface with surfactant molecules.

17.2.2 Influence of Environmental Factors on Micellization

Table 17.5 demonstrates the influence of counterions on the CMC of ionic surfactants. An increase of ionic strength decreases CMC. For ionic surfactants, this effect results from the increased concentration of counterions surrounding the ionic surfactant headgroups and thereby shielding the charges of the equally charged headgroups. This causes a decrease in electrostatic repulsion of the headgroups, and this effect is particularly prominent for anionic surfactants in the presence of divalent cations, such as the hardness ions Ca^{2+} and Mg^{2+}.

TABLE 17.4

Polar Headgroup Structure: Influence on CMC and Headgroup Area A for a Series of Dodecyl Surfactants (Excerpted from Rosen, 1989). The Solvent is Distilled Water.

	Counterion	T (°C)	CMC (M)	A Å²
Nonionic				
$C_{12}EO_8$		298	$1.1 \cdot 10^{-4}$	65
C_{12}-β-glucoside		298	$1.9 \cdot 10^{-4}$	
C_{12}-sucrose ester		298	$3.4 \cdot 10^{-4}$	
Influence of number of oxyethylene units				
$C_{12}EO_3$		298	$5.2 \cdot 10^{-5}$	42
$C_{12}EO_5$		298	$6.4 \cdot 10^{-5}$	50
$C_{12}EO_8$		298	$10.9 \cdot 10^{-5}$	65
$C_{12}EO_{12}$		298	$14.0 \cdot 10^{-5}$	
Anionic				
$C_{11}COO^-$	Na^+	298	$12 \cdot 10^{-3}$	47
$C_{12}SO_4^-$	Na^+	298	$12 \cdot 10^{-3}$	53
$C_{12}SO_3^-$	Na^+	298	$8.2 \cdot 10^{-3}$	57
$C_{12}EO_1SO_4^-$	Na^+	298	$3.9 \cdot 10^{-3}$	57
$C_{12}EO_2SO_4^-$	Na^+	298	$2.8 \cdot 10^{-3}$	63
Cationic				
$C_{12}Me_3N^+$	Cl^-	298	$2.0 \cdot 10^{-2}$	38
$C_{12}Pyr$	Cl^-	298	$1.5 \cdot 10^{-2}$	62

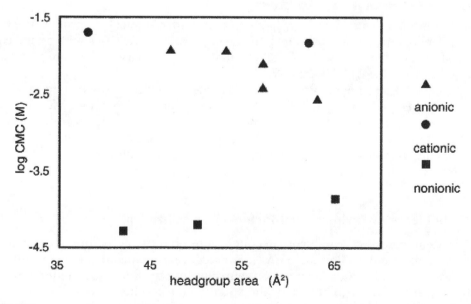

FIGURE 17.3

Plot of CMC vs. headgroup area for nonionic (squares), cationic (spheres), and anionic (triangles) surfactants.

TABLE 17.5

Influence of Counterions and Electrolyte Composition on CMC. If H_2O is Specified as Solvent the Surfactant Salt is the Only Electrolyte Used in the Determination of CMC. $I_{solvent}$ is the Ionic Strength of the Solvent (M).

	Counterion	T	Solvent (°C)	$I_{solvent}$ (M)	CMC (M)
$C_{12}SO_4^-$	Na^+	298	H_2O	0	$8.2 \cdot 10^{-3}$
$C_{12}SO_4^-$	Na^+	298	0.1M NaCl	0.1	$5.6 \cdot 10^{-3}$
$C_{12}SO_4^-$	Na^+	298	0.3M NaCl	0.3	$3.2 \cdot 10^{-3}$
$C_{12}SO_4^-$	Na^+	298	3M urea	0	$9 \cdot 10^{-3}$
$C_{12}SO_4^-$	Li^+	298	H_2O	0	$8.9 \cdot 10^{-3}$
$C_{12}SO_4^-$	K^+	298	H_2O	0	$7.8 \cdot 10^{-3}$
$C_{12}SO_4^-$	Na^+	298	H_2O	0	$8.2 \cdot 10^{-3}$
$C_{12}SO_4^-$	Cs^+	298	H_2O	0	$7.8 \cdot 10^{-3}$
$(C_{12}SO_4)_2^-$	Ca^{2+}	343	H_2O	0	$3.4 \cdot 10^{-3}$
$(C_{12}SO_4)_2^-$	Mg^{2+}	303	H_2O	0	$1.3 \cdot 10^{-3}$
$(C_{12}SO_4)_2^-$	Mn^{2+}	298	H_2O	0	$1.1 \cdot 10^{-3}$
$C_{12}SO_4^-$	$C_{12}Me_3N^+$	298	H_2O	0	$0.05 \cdot 10^{-3}$
$C_{12}EO_2SO_4^-$ [a]	Na^+	298	H_2O	0	$2.9 \cdot 10^{-3}$
$C_{12}EO_2SO_4^-$ [a]	Na^+	298	hard water	$4.6 \cdot 10^{-3}$	$0.6 \cdot 10^{-4}$
$C_{12}Me_3N^+$	Cl^-	298	H_2O	0	$20 \cdot 10^{-3}$
$C_{12}Me_3N^+$	F^-	304.5	0.5M NaF	0.5	$8.4 \cdot 10^{-3}$
$C_{12}Me_3N^+$	Cl^-	304.5	0.5M NaCl	0.5	$3.8 \cdot 10^{-3}$
$C_{12}Me_3N^+$	Br^-	304.5	0.5M NaBr	0.5	$1.9 \cdot 10^{-3}$
$C_{12}Me_3N^+$ [a]	Br^-	298	hard water	$4.6 \cdot 10^{-3}$	$12.6 \cdot 10^{-3}$
$C_{12}Me_3N^+$	NO_3^-	304.5	0.5M NaNO$_3$	0.5	$0.8 \cdot 10^{-3}$
$(C_{12})_2Me_2N^+$	Br^-	298	H_2O	0.5	$0.18 \cdot 10^{-3}$

[a] Data from Rosen et al. 1996. "Hard water" is a salt solution mimicking river water with concentration of hardness ions (1.2mM) that is relatively high for fresh water.

The comparison of CMC data in distilled vs. hard river water shows that the decrease in CMC with hardness has the order anionics >> cationics >> nonionics (Rosen et al., 1996). Hardness increases the dependence of the CMC on alkyl carbon chain length of $C_nEO_mSO_4^-$, indicating that in hard water the influence of additional carbon atoms is the same for $C_nEO_mSO_4^-$ as for C_nEO_m surfactants (Rosen et al., 1996). The influence of ionic strength on micellization of nonionic surfactants is due to a salting out effect of the hydrophobic moiety of the surfactant molecule (Carala et al., 1994).

17.2.3 Prediction of CMC

Recently, Huibers et al. (1996) presented a predictive model using topological predictors, which is capable of predicting CMC values for a wide variety of nonionic surfactants. Zoeller and Blankschtein (1995) attempted to predict CMC based on molecular-thermodynamic theory. Others have tried to predict the influence of temperature (Muller, 1993; Di Toro et al., 1990), of ionic strength (Carale et al., 1994; Di Toro et al., 1990), and of mixing (Zoeller et al., 1995) on the CMC. Combinations of these techniques could make it possible to predict CMCs as a measure of hydrophobicity of surfactants under given environmental conditions.

Within one class of surfactants, relationships between CMC and the LC$_{50}$, an important parameter in ecotoxicology, are often observed (Morall et al., 1997). However, since these relationships are often specific for the set of compounds and the water chemical conditions, extrapolation to other classes of surfactants and/or conditions is quite difficult. Recently, other parameters obtained from the surface tension measurement, such as surface area of the headgroup, as well as the effectivity of the surfactant, defined as the concentration of surfactant required to lower the surface tension by 20Nm^{-1} (Rosen, 1982), have been evaluated with regard to obtaining relationships that are not specific for one class of surfactants (Morall et al., 1997). Readers interested in data on environmental toxicology of surfactants should see the A.D. Little report on toxicology of surfactants (Little, 1991).

17.3   Solubility

Generally, the phase diagram of surfactants has a triple point at which the solid, the dissolved, and the micellar phase coexist. The temperature at this triple point is the Krafft temperature (T_K). For most surfactants, the T_K is below ambient temperature and, therefore, their CMC can be regarded as their molecular solubility. Everything said above about the CMC therefore holds true for aqueous solubility. However, T_K of the cationic surfactants used in fabric softeners is above ambient temperature. The same is true for the individual 2-n-(p-sulfophenyl)-alkanes which are constituents of LAS.

Values of the solubility of surfactants in equilibrium with crystal phase rarely have been measured. In detergent mixtures, other surfactants present in the mixture solubilize the 2-n-(p-sulfophenyl)-alkanes and the fabric softener cationics. A mixing effect which reduces the CMC can explain this.

Most surfactant mixtures deviate negatively from Raoult's Law. The extent of deviation and thus the solubilization is largest for mixtures containing surfactants with different headgroup charges (Scamehorn, 1986) or with significant headgroup size differences (Abe et al., 1992).

Since, in the environment, surfactants always occur in mixtures, the less soluble constituents of the mixtures will be solubilized by the more soluble ones. Therefore one can expect that surfactants will occur in the environment as dissolved species. An exception might be the cationic surfactants used in fabric softeners. It has been speculatedthat one of the representatives $(C_{18})_2Me_2N^+$ occurs in sewage sludge as its LAS salt (Huber, 1984).

Under environmental conditions, the solubility of anionic surfactants, especially of fatty acids, is affected remarkably by the presence of divalent cations. This is reflected in the dependence of the CMC on the counterion. Ca^{2+} and Mg^{2+} tend to precipitate anionic surfactants; however, the solubility products of $Ca(LAS)_2$ are relatively high (>10^{-13} M^{-2}, Matheson et al., 1985), so that precipitation of LAS is unlikely, given that LAS occurs in environmental waters usually at concentrations of less than 1µM.

17.4   Properties Related to Melting Point and Volatilization

Melting point (T_M) data exist for a number of surfactants. Table 17.6 gives a compilation of T_M data. Values of T_M of the linear alcohol ethoxylates range between 289 and 321 K. They increase with increasing length of the alkyl chain as well as the ethoxylate moiety. The values for the ionic surfactants are higher. Table 17.6 shows that the counterion is of influence on the T_M of the salt.

TABLE 17.6

Melting Point Data for Selected Surfactants.

	Counterion	Melting Point (K)	MW (g/mol)	Reference
$C_{10}EO_6$		289.8	422.6	Mulley, 1967
$C_{12}EO_6$		297.8	450.7	"
$C_{14}EO_6$		308.1	478.7	"
$C_{16}EO_6$		309.5	506.8	"
$C_{16}EO_2$		304.8	330.8	"
$C_{16}EO_9$		316.1	638.8	"
$C_{16}EO_{12}$		318.6	770.8	"
$C_{16}EO_{15}$		320.1	902.8	"
$C_{18}\text{-}SO_4^-$	Na^+	462.6-464.1	362.5	Shore and Berger, 1970
	K^+	455.1-456.1	378.5	"
$C_{16}\text{-}Pyr$	Br^-	327.2	402.5	Linfield, 1970
	Cl^-	355.1-357.1	358.0	"
	HSO_3^-	380.1-381.1	403.5	"
	$SO_3\text{-}Tol^-$	396.1-403.1	494.5	"
$C_{18}Me_3N^+$	Br^-	503.1-507.1	392.5	"

For the ionic surfactants, no boiling point data are available, while, for a large variety of nonionic surfactants, boiling point data have been compiled. Table 17.7 presents a selection of the data. The boiling point increases with increasing alkyl chain length, as well as with increasing length of the oxyethylene chain for C_nEO_ms.

TABLE 17.7

Boiling Point Data for Linear
Alcohol Ethoxylate Surfactants
(From Mulley, 1967)

Compound	Boiling Point °C/mm Hg
Influence of alkyl chain length	
$C_{10}EO_6$	200/0.02
$C_{12}EO_6$	205/12
$C_{14}EO_6$	206/0.02
$C_{16}EO_6$	234/0.05
Influence of ethoxylate chain length	
$C_{12}EO_4$	152/0.01
$C_{12}EO_6$	205/12
$C_{12}EO_8$	232/0.01
$C_{12}EO_{12}$	281/0.01

Vapor pressure data are not available for the ionic and nonionic surfactants. Some alcohol ethoxylates have been analyzed by high temperature gas chromatography, but the fact that elution temperatures of the higher ethoxylated AEs are above 520 K on a SE 30 boiling point column (Stancher and Favretto, 1978) indicates that the vapor pressure of these compounds is comparatively low. This is consistent with the high boiling points of these compounds. In addition, since surfactants are rather water soluble, their Henry's law constants can be expected to be very low. Actually, no measured Henry's law constants are available. As a result, evaporation of surfactants can be expected to be negligible.

Since surfactants are ingredients of pesticide formulations, it appears likely that, after they are sprayed, aerosol-associated surfactant molecules occur in the atmosphere. However, to the best of our knowledge, no accounts of atmospheric occurrence of surfactants exist in the literature.

17.5   Octanol Water Partition Coefficient

The 1-octanol water partition coefficient (K_{OW}) is a frequently employed in environmental chemistry and toxicology as a measure of parameter of hydrophobicity.

Table 17.8 gives an overview of experimentally determined values of K_{OW}. An increase of log K_{OW} occurs with increasing carbon chain length. For LAS, this increase is 0.33 log units per C-atom and is unaffected by LAS concentration and water hardness (Kimerle et al., 1975). However, log K_{OW} of C_{12}-LAS varies by two orders of magnitude, depending on the concentration of the surfactant and the electrolyte.

TABLE 17.8

Octanol-Water Partition Ratios for a Selection of Surfactants.

Compound	Log K_{OW}	Reference
C_{10}-LAS	1.08	Kimerle et al. (1975)
C_{12}-LAS	1.89 (0.68[a], 2.68[b])	Kimerle et al. (1975)
C_{14}-LAS	2.29	Kimerle et al. (1975)
C_{12}-SO_4^-	1.60	MedChem database (1985)
C_{12}-Pyr	0.67	Tolls (unpublished)
C_{16}-Pyr	1.43	Tolls (unpublished)
C_{16}-TMA	1.81	MedChem database (1985)
$(C_{16}/C_{18})_2Me_2N^+$	2.69	MedChem database (1985)
C_{12}-Bz-Me_2N^+	0.95	Daoud et al. (1983)
C_{14}-Bz-Me_2N^+	1.81	Daoud et al. (1983)
C_{16}-Bz-Me_2N^+	2.11	Daoud et al. (1983)
$C_{14}EO_7$	2.47	Woltering et al. (1987)

[a]   measured at 10 mg/L LAS, deionized water
[b]   measured at 10 mg/L LAS, 250 ppm hardness

While the variability has been assessed for LAS, little is known for other surfactants. However, the variability of the log K_{OW}-values of LAS indicates that K_{OW} measurements yield conditional partition ratios rather than well defined partition coefficients. On thermodynamic grounds, the partition ratio for ionic surfactants is a parameter depending very much on the electrolyte composition of the aqueous medium and therefore cannot be viewed as a partition coefficient in the sense of Henry's law (Schwarzenbach et al. 1993). In addition, experimental determination of K_{OW} of surfactants could yield erroneous results, due to the emulsifying action of the surfactants in the octanol-water system (Morall et al., 1997).

Summarizing, experimentally determined K_{OW}-values of surfactants, even if determined properly, are reflective of the experimental conditions. Hence, they cannot be used as a general measure of surfactant hydrophobicity as they are for neutral apolar compounds.

Roberts refined the fragment constant method for estimating log K_{OW} for LAS (Roberts, 1989), as well as for C_nEO_ms (Roberts, 1991) by introducing factors which correct for the effective size of the hydrophobic moiety of the surfactant. Kiewiet (1996) employed a thermodynamic model to estimate log K_{OW} of C_nEO_ms. However, the quality of the estimation methods is difficult to evaluate because measured data are not available. This also holds true for estimating K_{OW} by chromatographic methods.

17.6 Biodegradation

Biodegradation is the key process leading to removal of surfactants from the environment. Surfactant biodegradation became an issue with the introduction of tetrapropyleneben-zene sulfonate (TPBS) in detergent formulations. The highly branched alkyl chain of this compound degraded only slowly during wastewater treatment. Since TPBS foams already at concentrations of 1 mg/L TPBS, the consequence of the relatively slow biodegradation became visible as foam on the basins of wastewater treatment plants as well as, in some occasions, on rivers (Swisher, 1987).

Since then, surfactant biodegradation has been investigated intensively (see Swisher (1987) for a thorough account of many aspects of surfactant biodegradation and a large compilation of data.)

Large and consistent data sets are prerequisite for establishing structure activity relation-ships for predicting biodegradability (Mani et al., 1991). However, the existing data are largely inconsistent, and quantitative relationships between surfactant structure and bio-degradability cannot, in general, be derived. Here, we will use some small subsets of data on specific surfactant classes to demonstrate relationships between structure and biodeg-radation. Generally, primary biodegradation alters surfactant molecules in such a way that they lose their surface activity. For LAS this has been observed to result in a reduction of toxicity (Kimerle and Swisher, 1977) and it appears reasonable that this is true for most sur-factants. Therefore, we will focus on primary biodegradation.

An exception to the above rule of the thumb is the alkylphenol ethoxylates. Biodegrada-tion of these compounds yields the relatively stable C_9APEO_m with $0 < m < 3$ ethoxylate units or the respective carboxylates (Ahel, 1989). These ethoxylated products are more hydrophobic and bioaccumulative (Ekelund et al., 1990; Ahel et al., 1993) than the precur-sor surfactant molecules and are also of toxicological concern (Granmo et al. 1989, Jobling et al., 1996).

Biodegradation can be described as a second order reaction (Equation Ia) which, under conditions of constant competent biomass, simplifies to a pseudo first-order reaction (Equation 1b).

$$dC_{surf}/dt = - k_{deg} \cdot C_{surf} \cdot C_{bio} \tag{1a}$$

$$dC_{surf}/dt = - k \cdot C_{surf}, \text{ with } \quad k' = k_{deg} \cdot C_{bio}^{*} \tag{1b}$$

*with k_{deg}, C_{surf}, C_{bio}, and k' being the second-order rate constant for biodegradation, the concentration of the sur-factant and of the biomass, and the pseudo-first-order rate constant.

To obtain a consistent data set for comparison of biodegradability of different surfac-tants, we compiled first-order rate constants k' for disappearance of surfactants mea-sured in river water obtained in river die-away studies, as quantitative measures of biodegradability. In our comparison, we assumed that the solids concentrations in the different experiment were equally low and that surfactant disappearance therefore can be ascribed to biodegradation. We also assumed that the variation in $[C_{bio}]$ in the different studies is negligible. The results are discussed in the following sections.

17.6.1   Structure-Biodegradation Relationships

17.6.1.1   *Rate Constant Differences Among Classes of Surfactants*

Table 17.9 lists some rate constants of primary biodegradation of a number of surfactants, from river die away studies. The anionic surfactants C_{12}-SO_4^- and $C_{12}EO_3$-SO_4^- and the non-ionic surfactants $C_{14}EO_8$, $C_{12}EO_8$, and $C_{12}EO_9$ appear degraded with rate constants ranging between 3 and 24 d^{-1}. In contrast, the biodegradation rate constants for LAS and the cationic surfactants are lower than 1.4 d^{-1}. The same trend has been found in seawater (Vives-Rego et al. 1987), even though the rate constants of 2.3, 0.1, and 0.1, for $C_{12}SO_4^-$, LAS, and $C_{16}Me_3N^+$, respectively, were lower than in river water.

TABLE 17.9

Values of Pseudo First-Order Primary Biodegradation Rate Constants for River Die-Away Studies For Selected Surfactants. The Reaction Types are Ester Hydrolysis (EH), Ether Cleavage (EC), or ω-β-oxidation (ω-β).

	k′ (d^{-1})	Reaction	Reference
DEEDMAC	0.7	EH	Giolando et al. (1995)
$C_{12}NH_3^+$	1.2	ω-β	Ruiz-Cruz (1979)
$C_{12}Me_2NH^+$	0.2	ω-β	Ruiz-Cruz (1979)
$C_{12}Me_3N^+$	0.5	ω-β	Ruiz-Cruz (1979)
$(C_{12})_2Me_2N^+$	0.2	ω-β	Ruiz-Cruz (1979)
$C_{16}Me_3N^+$	0.4	ω-β	Ruiz-Cruz (1979)
$C_{18}Me_3N^+$	0.5	ω-β	Ruiz-Cruz (1979)
$C_{14}EO_8$	3.3	EC	Marcomini and Zanette (1996)
$C_{12}EO_8$	3.70	EC	Cassani et al. (1996)
$C_{12}EO_9$	4.4	EC	Marcomini and Zanette (1996)
$C_{14}EO_3$	3.6	EC	Itrich and Federle (1995)
$C_{14}SO_4^-$	24	EH	Itrich and Federle (1995)
C_{12}-LAS	0.9	ω-β	Itrich and Federle (1995)
C_{10}-LAS	0.47	ω-β	Moreno et al. (1996)
C_{12}-LAS	1.15	ω-β	Cassani et al. (1995)
C_{12}-LAS	0.72	ω-β	Moreno et al. (1996)
C_{14}-LAS	1.4	ω-β	Moreno et al. (1996)

17.6.1.2   *Trends Within Classes of Surfactants*

17.6.1.2.1   *Polar Headgroups*

Generally the biodegradability of anionic surfactants seems to follow the order $C_{12}SO_4^-$ > $C_{12}EO_3SO_4^-$ > $C_{12}SO_3^-$ > C_{12}-LAS (Feighner, 1970). The data for k′ confirm this. $C_{12}SO_4^-$ is hydrolyzed, while primary degradation of $C_{12}EO_3SO_4^-$ occurs mainly via cleavage of an ether bond. $C_{12}SO_3^-$ is converted by oxidation of the sulfonated C-atom and subsequent release of the bisulfite.

Biodegradation of LAS involves ω-oxidation of the alkyl terminal C-atom to an alcohol which is further oxidized via an aldehyde to a carboxylic acid, usually followed by β-oxidation until no further acetyl-CoA can be cleaved off the former surfactant molecule. Since intermediates of the subsequent reactions hardly can be isolated, ω-oxidation appears likely the rate-limiting step in LAS biodegradation.

Within the dodecyl cationic compounds the order of the rate constants is $C_{12}N^+H_3$ > $C_{12}Me_3N^+$ > $(C_{12})_2Me_2N^+$ > $C_{12}Me_2N^+H$. This trend in biodegradability does not reveal any clear indication of an influence of the charge. While the charge of $C_{12}N^+H_3$ and $C_{12}Me_2N^+H$ is pH-dependent, that of $C_{12}Me_3N^+$ and $(C_{12})_2Me_2N^+$ is permanent.

We are not aware of studies on the influence of the pH on the biodegradation of the long chain amines. However, since the pK_A values of the former two compounds are higher than 10, the ionized species will be dominant under most environmentally relevant pH conditions.

Table 17.9 shows that DEEDMAC is the quaternary cationic surfactant with the highest rate constant. Since it is the only onethat is hydrolyzed, the different degradation pathways probably account for the differences.

AEs primarily undergo ω-oxidation of the hydroxylterminus, as well as cleavage of the ether bond between the alcohol and the polyoxyethylene chain (Swisher, 1987). The latter pathway appears the preferred one, since degradation studies show the release of polyethylene glycol (Marcomini and Zanette, 1996; Steber and Wierich, 1985). The rate constants indicate that this reaction is relatively rapid. The increase in the degree of ethoxylation in the remaining AE material during biodegradation experiments indicates that the longer the EO-chain, the lower the rate of primary biodegradation (Marcomini and Zanette, 1996).

17.6.1.2.2   Hydrophobic Moiety

For LAS, the biodegradation rate increases with increasing alkyl chain length. Activated sludge (Bock and Wickbold, 1966) and isolated microorganism (Divo, 1976) data also show that, for a given alkyl chain length the longer the distance between the benzenesulfonate moiety and the ω-carbon atom, the higher the biodegradation rate. This is commonly referred to as "distance principle" (Swisher, 1987).

Within the homologous group of alkyl trimethyl ammonium surfactants, increasing alkyl chain length seems to lead to decreasing biodegradation rate constants. A similar trend appears for the hydrolysis of alkylsulfates by purified enzymes and bacteria (Swisher, 1987). The influence of alkyl chain length has not been investigated systematically for the alcohol ethoxylates. However, comparison of k' values of pure $C_{14}EO_8$ with those for $C_{12}EO_8$ obtained from different experiments, but by the same laboratory, indicate that k' decreases with increasing alkyl chain length.

Desulfonation of LAS under sulfate limiting conditions appears to proceed fastest for shorter alkyl chain lengths and faster for isomers in which the benzenesulfonate moiety is attached to a central position on the alkyl chain (Kertesz et al., 1994). Since in wastewater treatment sulfate is not limiting, this reaction is not relevant in the perspective of the fate of the bulk of LAS. However, the reaction shows that the trend of increasing rate of ω-β-oxidation with increasing alkyl chain length is an exception, not the rule.

While monobranched alkylbenzenesulfonates biodegrade rapidly (Cavalli et al., 1996), multiple branching of the alkyl chain significantly reduces the rate of ω-oxidation. This is why the linear analogs LAS replaced the highly branched TPBS. The alkylphenols used to produce the corresponding C_nPEO_m surfactants also contain branched side chains which are oxidized only at a low rate. Comparing C_nEO_ms with a multi-branched alkyl chain to C_nEO_ms containing one branch and to non-branched C_nEO_ms (derived from primary alcohols) also shows the effect of branching for (Zanette and Marcomini, 1996). In addition, branching also reduces the rate of biodegradation for $C_nSO_4^-$ and $C_nSO_3^-$ surfactants (Swisher, 1987).

17.6.1.3   Further Observations

Generally, values of k' are higher when determined with activated sewage sludge than in river-die away studies, most probably due to the higher biomass. The biodegradation rate observed in a shallow stream biofilm die-away study (k' = 14.7 d^{-1} (Takada et al. 1994)) is higher than that in the river water die-away studies, implying that LAS biodegradation predominantly in the biofilm at the water–solid interface may be more important than in the water column.

17.6.2   Anaerobic Biodegradation

Unfortunately, the data for anaerobic degradation of surfactants is scarce, so that we cannot present rate data. Field observations demonstrate that surfactants such as $C_nSO_3^-$, $(C_n)_2N^+(Me)_2$, and LAS accumulate in anaerobic sediments (Field et al., 1992; Fernandez et al., 1993; Field et al., 1994). This indicates that these surfactants are relatively stable in anaerobic environments.

Hydrolytic reactions and the central fission of ether bonds do not depend on the presence of O_2. They thus can proceed in anaerobic environments, too, as has been demonstrated by, among others, Steber and Wierich (1985) for $C_{13}EO_7$ and Nuck and Federle (1996) for $C_{14}SO_4^-$, $C_{14}E_3SO_4^-$, and $C_{14}EO_6$.

17.6.3   Generalizations

Replacement of alkyl chains by an ester increases the values of k′ in the case of the $(C_n)_2Me_2N^+$ cationic surfactants, indicating that hydrolysis is a more rapid biodegradation pathway than ω-β-Oxidation. However, comparison of the k′-values of $C_{12}SO_4^-$ with that of DEEDMAC demonstrates that other structural moieties also have significant influence on biodegradability. The available data do not allow conclusions as to which factors are involved.

ω-β-oxidation seems to be the slowest biodegradation pathway for surfactants. Surfactants containing ester bonds, such as sulfate esters in the alcohol sulfates and the esters in the so-called esterquats used in fabric softeners, or ether bonds, such as in alcohol ethoxylates of alkyl polyglucosides, are degraded at a higher rate. They are being used to replace less biodegradable surfactants.

Biodegradation in soils is relevant because sewage sludge containing surfactants is used as fertilizer. LAS disappearance half-lives in sludge-amended soils have been measured in the range of 7-22 d (Holt et al., 1989; Waters et al., 1989; Pflugmacher, 1992). Degradation in anaerobic and aerobic soils appears to occur at similar rates (Pflugmacher, 1992). Unfortunately, few data are available on rates of primary biodegradation in soil for surfactants other than LAS.

17.7   Sorption

Since surfactants are designed to be enriched at interfaces, surfactant sorption onto environmental solids should be of major importance particularly when the ratio of water volume to water-solid interface is small. Those conditions exist in wastewater treatment and in soil.

Sorption is measured by recording sorption isotherms, which themselves are a way to express the amount of surfactant sorbed as function of the concentration of the compound in the solution. The Freundlich isotherm (Equation II) is a general sorption isotherm which describes sorption behavior and often is used in studies of surfactant sorption. K_F is the Freundlich sorption coefficient which expresses the affinity of a surfactant for a given solid sorbent. The exponent n is a measure of isotherm non-linearity. For n approaching 1, the Freundlich model of sorption becomes equivalent to a linear sorption model.

$$C_s = K_F \cdot C_w^n \qquad (2)$$

Figure 17.4 shows a surfactant sorption isotherm from low to high (>CMC) concentrations of the surfactant. It can be divided into three parts (Figure 17.5). In Region 1, individual surfactant molecules are in equilibrium with the surfactant molecules adsorbed to the solid sorbent. In Region 2, the surfactant concentration in the water has exceeded the CMC. That is equivalent to saturation of the air/water interface with surfactant molecules. Subsequent addition of surfactant molecules leads to increased sorption due to formation of sorbed surfactant aggregates (Region 2). In Region 3, the aggregates in solution (micelles) are in equilibrium with the sorbed aggregates, the so-called admicelles..

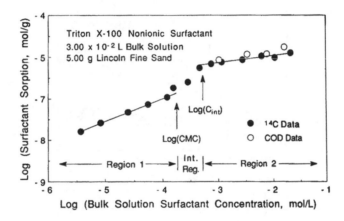

FIGURE 17.4
Sorption isotherm of the nonionic surfactant $C_8APEO_{9.4}$ (Triton X-100) to Lincoln fine sand (Reprinted from Edwards et al., 1994).

Surfactant concentrations as high as this occur under engineered conditions, such as enhanced oil recovery or subsurface remediation. In soil of aquifer remediation schemes, surfactants are used to solubilize and mobilize (West and Harwell, 1992) or retard (Wagner et al., 1994) hydrophobic contaminants. In enhanced oil recovery, surfactants are injected into the drilling mud. The surfactants decrease the interfacial tension of the oil-water interface and emulsify the oil so that it can be pumped with the drilling water. In both enhanced oil recovery and the contaminant mobilization schemes, the surfactant concentrations are so high that sorption of the surfactant to surfaces is negligible. By solubilizing sparingly soluble chemicals, such as oil constituents or organic contaminants, the surfactant micelles increase the mobility of these substances as desired.

Under environmental conditions, C_w is in the low µM-range or below. Table 17.10 is a compilation of data on surfactant sorption for a variety of surfactants to soils, sediments, or minerals, employing information generated at low surfactant concentrations.

17.7.1 Structural Sorption Relationships

17.7.1.1 Sorption Linearity

Many values of n for different surfactants and sorbents are outside the range 0.9 to 1.1, indicating that isotherms are nonlinear and that the affinity of the sorbent for the surfactant is

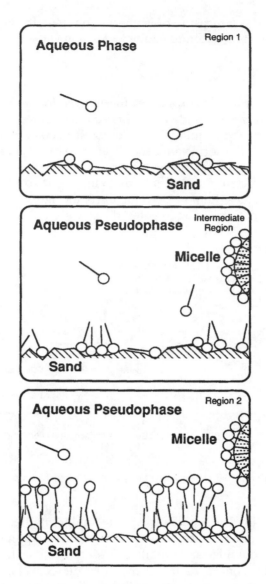

FIGURE 17.5
Schematic of the concentration-dependent processes during surfactant sorption (Reprinted from Edwards et al., 1994).

dependent on C_w. According to Brownawell et al. (1992), isotherm nonlinearity can be ascribed to a) changes of the surface charge with increasing ionic surfactant sorption and b) sorption site heterogeneity with preferential sorption to the energetically favored sites.

Figure 17.6 shows the adsorption behavior of surfactants of three different classes, LAS (anionic), C_{12}-Pyr (cationic), and $C_{13}EO_n$ (nonionic), on EPA-12 sorbent. Table 17.10 presents the corresponding Freundlich parameters. The nonlinearity parameter n is smaller than 1 for all surfactants tested, implying that sorption affinity decreases with increasing surfactant concentrations. This implies that concentration dependency should be taken into account when assessing sorption of surfactants. At the typically low surfactant concentration in environmental waters, the sorption coefficient might be higher than the sorption coefficient measured in the laboratory with higher concentrations of surfactants.

TABLE 17.10

Freundlich Sorption Coefficient K_F and Nonlinearity Parameter n for a Selection of Surfactants for Sorption to Soils, Sediments, and Minerals.

	log K_F	n	Sorbent	Reference
Anionics				
C_{12}-2-LAS	1.7	1[a]	EPA-B1	Hand and Willliams (1987)
C_{12}-6-LAS	1.3	1[a]	EPA-B1	Hand and Willliams (1987)
C_{14}-2-LAS	2.6	1[a]	EPA-B1	Hand and Willliams (1987)
C_{12}-LAS	1.7	1[a]	EPA-B1	Hand and Willliams (1987)
C_{12}-LAS	2.0	1[a]	EPA-5	Hand and Willliams (1987)
C_{12}-LAS	2.7	1[a]	RC4	Hand and Willliams (1987)
C_{12}-LAS	3.2	1[a]	RC3	Hand and Willliams (1987)
C_{12}-LAS	ca 0.1	1	different soils	Ou et al. (1996)
C_{12}-LAS	2.8	1.15	mar.sed	Rubio et al. (1996)
C_{12}-2-LAS	0.98	0.86	EPA12	Brownawell et al. (1992)
C_{12}-LAS	0.6	0.77	soil, clay loam (A)	Abe and Sano (1985)
C_{12}-LAS	1.4	1.20	soil, clay loam (B)	Abe and Sano (1985)
C_{12}-LAS	1.2	1.19	soil, sandy loam	Abe and Sano (1985)
AOS	1.1	1.21	soil, clay loam (B)	Abe and Sano (1985)
$C_{12}SO_4^-$	1.0	1.40	soil, clay loam (B)	Abe and Sano (1985)
$C_{11.0}$-LAS	1.9	1.79	Al_2O_3	Matthijs and De Henau (1985)
$C_{11.0}$-LAS	1.9	1.23	Fe_2O_3	Matthijs and De Henau (1985)
Nonionics				
$C_{10}AE_5$	1.68	1[a]	lake sediment	Kiewiet et al. (1996)
$C_{12}AE_5$	2.86	1[a]	lake sediment	Kiewiet et al. (1996)
$C_{14}AE_5$	3.54	1[a]	lake sediment	Kiewiet et al. (1996)
$C_{16}AE_5$	3.68	1[a]	lake sediment	Kiewiet et al. (1996)
$C_{13}AE_3$	1.90	0.92	EPA12	Brownawell et al. (1992)
$C_{13}AE_6$	1.22	0.79	EPA12	Brownawell et al. (1992)
$C_{13}AE_9$	0.35	0.63	EPA12	Brownawell et al. (1992)
Cationics				
C_{12}-Pyr	0.43	0.54	EPA12	Brownawell et al. (1992)
$C_{16}Me_3N^+$	4.85	?	EPA-18	Larson and Vashon (1983)
$C_{18}Me_3N^+$	5.4	?	EPA-18	Larson and Vashon (1983)
$(C_{18})_2Me_2N^+$	4.1	?	EPA-18	Larson and Vashon (1983)

[a] Isotherm is linear according to specifications of the authors.
? No information on isotherm linearity

17.7.1.2 Effect of Alkyl Chain Length

Table 17.11 details the dependence of log K_F on the number of carbon atoms ($n_{C\text{-atoms}}$) according to the Equation log $K_F = B + A \cdot n_{C\text{-atoms}}$. A varies between 0.33 and 0.46 for LAS and AEs. The rather small variation in A indicates that the increase in sorptivity per added C-atom does not differ substantially between the two classes of surfactants. In addition, Hand and Williams (1987) observed that the sorption coefficient increased by a factor of two when the p-sulfophenyl moiety moved from a central position in the alkyl chain to the 2-position.

This behavior is analogous to the trend observed for the inverse of CMC. Since the inverse of the CMC can be viewed as a measure of hydrophobicity, the relationship between substitution position of the p-sulfophenyl-moiety and the sorption coefficients is indicative of the influence of hydrophobicity on sorption.

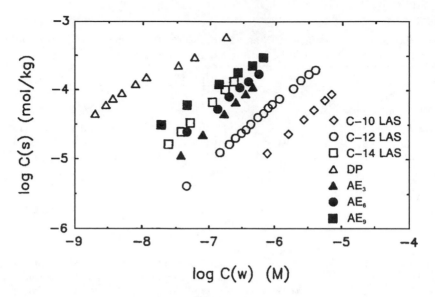

FIGURE 17.6
Sorption isotherms for a variety of structurally different surfactant to EPA B12-sediment (Reprinted from Brownawell et al., 1992, DP = C_{12}-Pyr, AE-3 = $C_{13}EO_3$, AE-6 = $C_{13}EO_6$, AE-9 = $C_{13}EO_9$, C_{10}-LAS = C_{10}-2-LAS, C_{12}-LAS = C_{12}-2-LAS, C_{14}-LAS = C_{14}-2-LAS).

TABLE 17.11

Alkyl Chain Length Dependence of K_F of Surfactants Expressed in the Parameter of the Equation $\log K_F = B + A \cdot n_{C\text{-atoms}}$ Under Different Conditions.

Surfactant	Sorbent	B	A	Reference
LAS	EPA1	−4.0	0.46	Hand and Williams (1987)
LAS	EPA B12	−3.9	0.43	Brownawell et al. (1992)
LAS	marine sed.	−1.2	0.33	Rubio et al. (1996)
AEs	lake sed.	−1.7	0.33	Kiewiet et al. (1996)

17.7.1.3 Influence of Headgroup on Sorption

Figure 17.6 shows that the cationic surfactant C_{12}Pyr sorbs more strongly to the test sorbent EPA-12 than three LAS homologs and three $C_{13}EO_m$ compounds. The same is true for other sorbents. C_{16}Pyr sorbs more strongly to river sediment than $C_{16}Me_3N^+$, and $C_{18}Me_3N^+$ sorbs stronger than $(C_{18})_2Me_2N^+$ (Table 17.10). For all compounds, the electrostatic contribution to sorption can be expected to be the same. Therefore, the examples demonstrate the influence of the headgroup structure on surfactant sorption. In addition, the second example indicates the involvement of sterical factors in cationic surfactant sorption since, even though $C_{18}Me_3N^+$ is much less hydrophobic than $(C_{18})_2Me_2N^+$ and electrostatic forces are equal, the sorption of the latter compound is weaker.

The pronounced sorption nonlinearity for C_{12}Pyr is explained by sorption to heterogeneous cation exchange sites (Brownawell et al., 1992; Xu and Boyd, 1995). With increasing ionic strength and, as a result, increasing saturation of the cation exchange sites, sorption of C_{12}Pyr decreases, confirming the involvement of cation exchange. Cationic surfactants can displace alkali metal ions in the clay interlayers. Therefore, clays provide rather specific sorption sites. The results of biodegradation tests indicate that sorption in clay interlayers renders surfactants unavailable for biodegradation (Games et al. 1982).

Within the anionic surfactants, the order of sorption strength is C_{12}-LAS > $C_{12}SO_3^-$ > olefinsulfonate. Brownawell et al. (1992) and Kiewiet et al. (1996) found that an increasing degree of ethoxylation results in increased sorption and increased nonlinearity. This is in contrast to the behavior expected based on hydrophobicity and implies that the additional ethoxylate moieties increase the capacity of the C_nEO_ms to interact specifically with the solid sorbent. Klumpp and Schwuger (1997) point out that, in nonionic surfactant adsorption on layer silicates, the hydrated ethylene glycol groups can interact with the solid phases via ion-dipole interactions and hydrogen bonding. X-ray measurements demonstrated that nonionic surfactants can intercalate into the clay layer interspaces. Formation of surface complexes with adsorbed monovalent cations also might be involved.

17.7.2 Effect of Sorbent on Surfactant Sorption

The values K_F for C_{12}-LAS for different sorbents indicate that, even when measured in one laboratory, differences of almost 1.5 log units can be observed. The difference between the highest value measured by Hand and Williams (1987) and the value reported by Ou et al. (1996) is three log units. This holds true, too, for the nonionic surfactants, when comparing the data from Brownawell et al. (1992) with those of Kiewiet et al. (1996).

For LAS, an increase of sorption with increasing concentration of Na^+, Ca^{2+}, and H^+ was observed (Brownawell et al., 1992). The pH-dependence is interpreted as an effect of a) the surface potential or b) the pH-dependent surface complex reactions. The magnitude of the effect of Na^+ indicates that Na-ions attenuated the electrostatic repulsion of LAS in the electrical double layer. The effect of Ca^{2+} cannot be explained by nonspecific effects on the electrical double layer. Instead, formation of Ca - LAS surface complexes are postulated and in line with the results of Rubio et al. (1996). For marine sediment the intercept is more than two orders of magnitude higher than for the other two sediments (Table 17.11), indicating that, under conditions of high ionic strength, sorption of LAS is stronger than under freshwater conditions. This could be due to the more effective shielding of the negative charges of LAS under conditions of high ionic strength.

Metal oxides not only provide cation exchange capacity for sorption of cationic surfactants but also possess ligand exchange sites by which these solids can interact specifically with anionic surfactants. Therefore, they can contribute significantly to sorption of anionic surfactants such as LAS (Matthijs and de Henau, 1985; Lajtar et al., 1993).

The correlations of K_F-values with sorbent properties by Hand and Williams (1987) as well as Matthijs and De Henau (1987) give no clear indication which sorbent properties govern surfactant sorption. LAS sorption appears brought about by specific as well as unspecific hydrophobic interactions, and the prevalent mechanism depends on the nature of the sorbent.

Sorption onto sludge particles is an important process in removing surfactants from wastewater (Brunner et al., 1988). The particles in activated sludge are rich in bacterial biomass and, therefore, sorption onto sludge particles brings the degrading organism into contact with the surfactants. Usually, the sludge particles settle in the secondary sedimentation phase and are then further reduced in volume in the anaerobic phase of wastewater treatment. After being removed from the wastewater treatment plant, the sludge can be applied as fertilizer to agricultural fields. Hence, sorption to sludge particles and application of the sludge as fertilizer is an entry path to the terrestrial environment which can result in soil organisms being exposed to surfactants.

TABLE 17.12

Sorption Coefficients for Selected Surfactants to
Activated Sludge.

Compound	Log K_F	Reference
$C_{12}(EO)_5SO_4^-$	3.2	Urano and Saito (1984)
AOS	3	Urano and Saito (1984)
LAS	3.4	Urano and Saito (1984)
$C_9APEO_{10}av$	3.9	Urano and Saito (1984)
$C_{16}EO_6$	4	Urano and Saito (1984)
$C_{12}EO_{10}$	3.4	Kiewiet (1996)
C_{18}-Me_3N^+	4.3–4.7	Games et al. (1982)
C_{16}-Me_3N^+	3.6	Gerike (1982)
$C_{12}Bz(Me)_2N^+$	4	Gerike (1982)
$(C_{10})_2Me_2N^+$	4.2	Gerike (1982)
$(C_{16/18})_2Me_2N^+$	4.8	Gerike et al. (1994)
$(C_{18})_2Me_2N^+$	4.9	Topping and Waters (1982)

Table 17.12 gives a selection of sorption coefficients of surfactants for sorption to activated sludge. In general, it appears that sorption onto sludge particles is higher than onto sediments and soils, with less variation in the sorption coefficients. With regard to surfactant structure, it appears that the cationic surfactants sorb more strongly than either nonionics or anionics. The differences of log K_F-values between anionic surfactants are not very pronounced; however, sorption follows the order AOS < AES < LAS. Sorption coefficients for the nonionic surfactants are somewhat higher than those for the anionic surfactants. Within the group of the cationic surfactants, the differences between C_{18}-TMAC and $(C_{18})_2$-DMAC and between $(C_{10})_2$-DMAC and $(C_{18})_2$-DMAC are less than a factor of 3 and 4, respectively, even though the difference in the number of C-atoms in the alkyl chain is 18 and 16, respectively. This is less than expected from the behavior of monoalkylsubstituted surfactants.

Association of surfactant molecules onto humic substances is also termed sorption and has been given some attention. In the case of cationic alkyl trimethyl and dimethyldioctadecyl ammonium surfactants (Versteeg and Shorter, 1992), the presence of dissolved organic carbon in the water significantly reduced the toxicity of these compounds to fathead minnows. The effect of toxicity reduction increased with increasing hydrophobicity. For the anionic surfactant LAS, this mitigating effect was less pronounced (Traina et al., 1996).

17.7.3 Prediction of Sorption Coefficients

Based on their data for sorption onto a lake sediment, Kiewiet et al. (1996) derived an equation predicting sorption coefficients of C_nEO_ms as a functions of alkyl chain length and the number of oxyethylene units. Di Toro et al. (1990) proposed a model for description of sorption of anionic surfactants which includes sorbent properties (organic carbon content, cation exchange capacity, and particle concentration) and the CMC as a function of the solution properties (ionic strength, temperature). The CMC is used as a relative hydrophobicity parameter. Since the model takes the contribution of electrostatic as well as hydrophobic forces explicitly into account, it is an example of an attempt to model surfactant behavior on the basis of the underlying mechanisms.

17.8 Chemical Reactivity

17.8.1 Hydrolysis

An increasing number of surfactants contains ester bonds, such as the $C_nSO_4^-$, the $C_nEO_mSO_4^-$, the polyol ester surfactants, and the replacements of the dialkyldimethylammonium surfactants, the esterquats. The ester bond is susceptible to base hydrolysis (Holmberg, 1996) in contrast to the acetal bond in, for instance, the alkylpolyglucoside, which is acid-susceptible (Steber et al. 1995).

Data for chemical hydrolysis are not generally available in the open literature. Since surfactants do survive the washing process in the washing machine, often at elevated temperature, it can be concluded that chemical hydrolysis is not a very fast process. Generally, microbial hydrolysis is the prevailing reaction leading to hydrolysis of the parent surfactants under environmental conditions.

17.8.2 Photochemical Reactivity

Generally, surfactants are assumed to be stable towards direct photodegradation, mainly because they lack chromophoric moieties which absorb actinic light. Shortwave UV-radiation may cause photochemical degradation of quaternary ammonium surfactants by (Huber, 1984; Boethling, 1994). C_9APEO_m is degraded by sensitized photolysis, with $^1O_2$ being the reactive species generated by radiation (Ahel et al., 1994).

Secondary C_nEO_m surfactants can be degraded photochemically in an engineered system (Sherrard et al., 1995). An irradiated TiO_2-suspension is used to generate OH-radicals, which attack the tertiary carbon. This degradation pathway is highly nonspecific, and no information is available on the reactivity of other surfactants with photochemically generated oxidative species.

17.9 Bioconcentration

Organisms can accumulate chemicals via either ingestion of food or directly from the environmental medium. The overall process is referred to as bioaccumulation. If accumulation from food results in a higher concentration in the exposed organism on a weight basis than in its food, this is called biomagnification. In the environment, surfactants are present mainly in water rather than in air or soil. Therefore, bioconcentration from the water is the most important process leading to bioaccumulation of surfactants.

The majority of the bioconcentration data is based on experiments with radiolabelled surfactants. Since the measurements of radioactivity were performed without prior separation of metabolites, the data are not specific for the parent surfactants and therefore not a quantitative measure of surfactant bioconcentration (Tolls et al., 1994). The general trend observed in these data is that bioconcentration of surfactant increases with decreasing values of the CMC, indicating that a relationship between surfactant bioconcentration and hydrophobicity exists (Tolls and Sijm, 1995). This trend has been confirmed with parent compound-specific data measured for LAS (Tolls et al., 1997).

Bioconcentration experiments (Tolls, unpublished results; Tolls and Sijm, 1999; Tolls et al., accepted for publication) as well as literature data for indicate that LAS and C_nEO_ms are extensively biotransformed by fish (v. Egmond et al., 1995). Therefore, correlations of the bioconcentration behavior with physical-chemical parameters can be expected only if a) biotransformation occurs at the same rate for all C_nEO_ms, or b) it depends on physical-chemical properties of the surfactants.

17.10 Recommendations

Most of the environmental research on surfactants has focused on their fate in wastewater treatment, where sorption and biodegradation are the processes which efficiently reduce the surfactant load in the effluent. Both processes also dominate the subsequent fate in environmental waters.

Surfactants are a heterogenous group of chemicals. They have in common that, within homologous groups, certain properties vary similarly with the alkyl chain length. In addition, different polar headgroups cause different surfactants to interact with their environment via different mechanisms. Knowledge of these interactions is insufficient for deriving comprehensive structure-property relationships which consider environmental factors, and so validated methods for predicting environmentally relevant properties of surfactants as a compound class are not available.

To estimate properties of surfactants, we recommend obtaining data for the desired property for compounds as closely related structurally as possible. Select data obtained under environmental conditions resembling the ones of interest, and evaluate the data in the light of the relationships this outlined in this chapter. Estimates derived in this way may deviate by about one order of magnitude. Given the large variation in reported surfactant sorption and biodegradation data, such a degree of accuracy is acceptable, since quantitative structure/property relationships are not available.

The data presented here may be helpful in modeling the environmental fate of surfactants. A recent modeling study for LAS (Mackay et al., 1996) has indicated that, due to negligible gas-water exchange, LAS interphase transfer via the gas phase will not occur. Instead, the aqueous phase is the central compartment for environmental fate of surfactants. Besides biodegradation, which reduces the amount of surfactants in the environment, sorption determines the partitioning of the surfactants between the aqueous and the solid phase.

References

Abe, S. and M. Seno. 1985. Adsorption of anionic surfactants on soils. *Nippon Kagaku Kaishi* 5, 814-819.

Abe, M., H. Uchiyama, T. Yamaguchi, T. Suzuki, and K. Ogino. 1992. Micelle formation of pure nonionic surfactants and their mixtures. *Langmuir* 8, 2147-2151.

Ahel, M., J. McAvoy, and W. Giger. 1993. Bioaccumulation of the lipophilic metabolites of nonionic surfactants in freshwater organisms. *Envir. Pollut.* 79, 243-248.

Ahel, M., W. Giger, and M. Koch. 1994. Behavior of alkylphenol polyethoxylate surfactants in the aquatic environment – I. Occurrence and transformation in rivers. *Water Res.* 28, 1143-1152.

Ahel, M., F.E. Scully Jr, J. Hoigne, and W. Giger. 1994. Photochemical degradation of nonylphenol and nonylphenol polyethoxylates in natural waters. *Chemosphere* 28, 1361-1368.

Bock, K.J., and R. Wickbold. 1966. Effects of change to easily biodegradable detergents in a large scale sewage treatment plant and its effluents. *Vom Wasser* 33, 242-253.

Boethling, R.S. 1994. Environmental aspects of cationic surfactants. In J. Cross and E.J. Singer, Eds., *Cationic Surfactants, Surfactant Science Series,* Vol. 53, pp. 95-135. Marcus Dekker, New York.

Brownawell, B.J., H. Chen, J.M. Collier, and J.C. Westall. 1990. Adsorption of organic cations to natural materials. *Environ. Sci. Technol.* 24, 1234-1241.

Brownawell, B.J., H. Chen, W. Zhang, and J.C. Westall. 1992. Adsorption of surfactants. In R.A. Baker, ed., *Organic substances in sediment-water systems, Vol. 2,* p. 127-147, Lewis, New York.

Brunner, P.H., S. Capri, A. Marcomini, and W. Giger. 1988. Occurrence and behaviour of linear alkylbenzensulphonates, nonylphenol, nonylphenol mono- and nonylphenol diethoxylates in sewage and sewage sludge treatment. *Water Res.* 22, 1465-1472.

Carale, T.R., Q.T. Phan, and D. Blankschtein. 1994. Salt effects on intramicellar interactions and micellization of nonionic surfactants in aqueous solutions. *Langmuir,* 10, 109-121.

Cassani, G., M. Lazzarin, G. Nucci, and L. Cavalli. 1996. Biodegradation kinetics of linear alkyl-benzenesulfonates (LAS) and alcohol ethoxylates (AE). Poster Presentation. SETAC 1996, Washington.

Cavalli, L., G. Cassani, M. Lazzarin, C. Maraschin, G. Nucci, and L. Valturta. 1996. Iso-branching of linear alkylbenzene sulphonate (LAS). *Tenside Surf. Det.* 33, 393-398.

Daoud, N.N., N.A. Dickinson, and P. Gilbert. 1983. Antimicrobial activity and physico-chemical properties of some alkyldimethylbenzylammonium chlorides. *Microbios* 37, 73-85.

Di Toro, D., L.J. Dodge, and V.C. Hand. 1990. A model for anionic surfactant sorption. *Environ. Sci. Technol.* 24, 1013-1020.

Divo, C. 1976. Indagine sulla ittiotossicita e sulla biodegradabilita di alchibenzensulfonati di sodio a catena lineare. *Riv. Ital. Sost. Grasse,* 53, 88-93.

Edwards, D.A., Z. Adeel, and R.G. Luthy. 1994. Distribution of nonionic surfactant and phenanthrene in a sediment/aqueous system. *Environ. Sci. Technol.* 28, 1550-1560.

v. Egmond, R., C.S. Newsome, D. Howes, and S.J. Marshall. 1995. Fate of some anionic and alcohol ethoxylate surfactants in *Carassius auratus. Tenside Surf. Detergents* 32, 498-503.

Ekelund, R., Å. Bergman, Å. Granmo, and M. Berggren. 1990. Bioaccumulation of 4-nonylphenol in marine animals – a reevaluation. *Environ. Pollut.* 64, 107-120.

Feighner, G.C. 1970. Alkylarylsulfonates. In W.M. Linfield, ed., *Anionic Surfactants,* Surfactant Science Series Vol. 7. Marcel Dekker, New York.

Fernandez, P., A. Alder, and W. Giger. 1993. Quantitative determination of cationic surfactants in sewage sludges and sediments by supercritical fluid extraction and HPLC applying post-column ion-pair extraction. Presentation at the American Chemical Society Meeting, Denver, CO, March 28-April 2, 1993.

Field, J.A., D.J. Miller, T.M. Field, S.B. Hawthorne, and W. Giger. 1992. Quantitative determination of sulfonated aliphatic and aromatic surfactants in sewage sludge by ion-pair/supercritical fluid extraction and derivatization gas chromatography/mass spectrometry. *Anal. Chem.* 64, 3161-3167.

Field J. A., T.M. Field, T. Poiger, and W. Giger. 1994. Determination of secondary alkane sulfonates in sewage wastewaters by solid-phase extraction and injection-port derivatization gas chromatography/mass spectrometry. *Environ. Sci. Technol.* 28, 496-503.

Games, L.M., J.E. King, and R.J. Larson. 1982. Fate and distribution of a quaternary ammonium surfactant, octadecylammonium chloride, (OTAC), in wastewater treatment. *Environ. Sci. Technol.* 16, 483-488.

Gerike P. 1982. Biodegradation and bioelimination of cationic surfactants. *Tenside Detergents* 19, 162-164.

Gerike P., H. Klotz, J.G.A. Kooijman, E. Matthijs, and J. Waters. 1994. The determination of dihardened tallowdimethyl ammonium compounds (DHTDMAC) in environmental matrices using trace enrichment techniques and high performance liquid chromatography with conductometric detection. *Water Res.* 28, 147-154.

Giolando, S.T., R.A. Rapaport, R. J. Larson, T.W. Federle, M. Stalmans, and P. Mascheleyn. 1995. Environmental fate and effects of DEEDMAC: a new rapidly degradable cationic surfactant for use in fabric softeners. *Chemosphere* 30, 1067-1083.

Granados, J. 1996. Surfactant raw materials: Constant evolution and a solid future, In *Proceedings of the 4th World Surfactant Congress*, Barcelona, Spain, 3–6 June 1996, Vol. A, p. 100.

Granmo, Å., R. Ekelund, K. Magnusson, and M. Berggren. 1989. Lethal and sublethal toxicity of 4-nonylphenol to the common mussel (Mytilus edilus L.) *Environ. Pollut.* 59, 115-127.

Hales, S.G. 1981. Microbial degradation of linear ethoxylate surfactants, Ph.D. Thesis, Univ. Wales, Cardiff.

Hand, V.C., and G. K. Williams. 1987. Structure-activity relationships for sorption of linear alkylbenzenesulfonates. *Environ. Sci. Technol.* 21, 370-373.

Holmberg, K. 1996. Surfactants with controlled half-lives. *Curr. Opinion Coll. Interf. Sci.* 1, 572-579.

Holt, M.S., E. Matthijs, and J. Waters. 1989. The concentration and fate of linear alkylbenzenesulfonate in sludge amended soils. *Water Res.* 23, 749-759.

Hoffmann, H., and G. Pössnecker. 1994. The mixing behavior of surfactants. *Langmuir*, 10, 381-389.

Huber, L.H. 1984. Ecological behavior of cationic surfactants from fabric softeners in the aquatic environment. *JAOCS 61*, 377-381.

Huibers, P.D.T., V.S. Lobanow, A.R. Katrizky, D.O. Shah, and M. Karelson. 1996. Prediction of critical micelle concentration using a quantitative structure-property relationship approach. 1. Nonionic surfactants. *Langmuir*, 12, 1462-1470.

Itrich, N.R., and T.W. Federle. 1995. Primary and ultimate biodegradation of anionic surfactants under realistic discharge conditions in river water. Poster presentation at 2nd SETAC World Congress, 5-9 Nov. 1995, Vancouver, Can.

Jobling, S., D. Sheahan, J.A. Osborne, P. Matthiessen, and J.P. Sumpter. 1996. Inhibition of testicular growth in rainbow trout (Oncorhynchus mykiss) exposed to oestrogenic alkylphenolic chemicals. *Environ. Toxicol. Chem.* 15, 194-202.

Kertesz, M.A., P. Koelbener, H. Stockinger, S. Beil, and A.M. Cook. 1994. Desulfonation of linear alkylbenzenesulfonate surfactants and related compounds by bacteria. *Appl. Environ. Microbiol.* 60, 2296-2303.

Kiewiet, A.T., K.G.M. de Beer, J.R. Parsons, and H.A.J. Govers. 1996. Sorption of linear alcohol ethoxylates on suspended sediments. *Chemosphere* 32, 675-680.

Kiewiet, A.T. 1996. Environmental fate of nonionic surfactants, Ph.D. thesis, Universiteit van Amsterdam.

Kimerle, R.A., R.D. Swisher, and R.M. Schroeder-Comotto. 1975. Surfactant structure and aquatic toxicity. In *Proc. IJC Symposium on Structure Activity Correlations in Studies on Toxicity and Bioconcentration with Aquatic Organisms*, pp. 22-25, March 11-13, 1975.

Kimerle, R.A., and R.D. Swisher. 1977. Reduction of aquatic toxicity of linear alkylbenzenesulfonate (LAS) by biodegradation. *Water Res.* 11, 31-37.

Klumpp, E., and M.J. Schwuger. 1997. Physicochemical interactions of surfactants and contaminants in soil. In M.J. Schick and F.M. Fowkes, Eds., *Surfactants in the Environment, Surfactant Science Series Vol. 63*, pp. 39-64. Marcel Dekker, New York.

Lajtar, L., J. Narkiewicz-Michalek, W. Rudzinski, and S. Partyka. 1993. A new theoretical approach to adsorption of ionic surfactants at water/oxide interfaces: Effects of oxide surface heterogeneity, *Langmuir* 9, 3174-3190.

Larson, R.J., and R.D. Vashon. 1983. Adsorption and biodegradation of cationic surfactants in laboratory and environmental systems. *Dev. Ind. Microbiol.* 24, 425-434.

Linfield, W.M. 1970. Straight chain alkylammonium compounds. In E. Jungermann, ed., *Cationic Surfactants, Surfactant Science Series, Vol. 4*, pp 9-70. Marcel Dekker, New York.

Little, A.D. 1991. *Environmental and human safety of major surfactants, Final Report*, The Soap and Detergent Association, New York.

Mackay, D., C.E. A. di Guardo, S. Paterson, G. Kicsi, C.E. Cowan, and D.M. Kane. 1996. Assessment of chemical fate in the environment using evaluative regional and local-scale models: illustrative application to chlorobenzene and linear alkylbenzene sulfonates. *Environ. Toxicol. Chem.* 15, 1638-1648.

Mani, S.V., D.W. Connell, and R.D. Braddock. 1991. Structure activity relationships for the prediction of biodegradability of environmental pollutants. *Crit. Rev. Environ. Control.* 21, 217-236.

Matheson, K.L., M.F. Cox, and D.L. Smith. 1985. Interactions between linear akylbenzene sulfonates and water hardness ions. I. Effect of calcium ion of surfactant solubility and implications for detergency performance. *JAOCS* 62, 1391-1399.

Marcomini, A., and M. Zanette. 1996. Biodegradation mechanisms of aliphatic alcohol polyethoxylates (AE) under standardized aerobic conditions (modified OECD screening test 301E). *Riv. Ital. Sost. Grasse* 73, 213-218.

Matthijs, E. and H. de Henau. 1985. Adsorption and desorption of LAS, *Tenside Surf. Detergents* 22, 299-304.

Medchem. 1985. MedChem software release CLOGP 3.33. Medicinal Chemistry Project, Pomona College, Claremont, CA.

Morall, S.W., R.R. Herzog, P. Kloepper-Sams, and M.J. Rosen. 1996. Octanol-water partitioning of surfactants and its relevance to toxicity and environmental behavior. In *Proceedings of the 4th World Surfactant Congress*, Barcelona, Spain, 3–6 June 1996, Vol. C, pp. 220.

Morall, S.W., M.J. Rosen, Y.P. Zhu, D.J. Versteeg, and S.D. Dyer. 1997. Physical chemical descriptors for the development of aquatic toxicity QSARs for surfactants, Proceedings of the 7th Internatl. *Workshop on QSARs in Environmental Science*, SETAC Press, USA, in press.

Moreno, A., C. Verge, J. Ferrer, J.L. Berna, and L. Cavalli. 1996. Effect of LAS alkyl chain length & concentration on biodegradation kinetics. Poster Presentation , SETAC 1996, Washington.

Muller, N. 1993. Temperature dependence of critical micelle concentrations and heat capacities of micellization for ionic surfactants. *Langmuir* 9, 96-100

Mulley, B.A. 1967. Synthesis of homogeneous nonionic surfactants, In M.J. Schick, ed., *Nonionic surfactants, Surfactant Science Series, Vol. 1*, pp 421-440, Marcel Dekker, New York.

Nuck, B.A., and T.W. Federle. 1996. Anaerobic die-away test for determining the kinetics of primary and ultimate biodegradartion in digester sludge, Poster presentation at SETAC 17th annual meeting, Washington, 17-21 Novermber, 1996.

Ou, Z., A. Yediler, Y. He, L. Jia, A. Kettrup, and T. Sun. 1996. Adsorption of linear alkylbenzenesulfonate (LAS) on soils. *Chemosphere* 32, 827-839.

Pflugmacher, J. 1992. Abbau von linearen Alkylbenzolsulfonaten, UWSF-Z. Umweltchem. *Ökotoxikol.* 4, 329-332.

Roberts, D.W. 1989. Aquatic toxicity of linear alkyl benzene sulfonates (LAS)—A QSAR analysis. *Communicaciones presentadas a las Jornadas Espanol de la Detergencia* 29, 35-43.

Roberts, D.W. 1991. QSAR issues in aquatic toxicity of surfactants. *Sci. Tot. Environ.* 109/110, 557-568.

Rosen, M.J. 1989. *Surfactants and interfacial phenomena*, 2nd ed. Wiley, New York.

Rosen, M.J., Y.-P. Zhu, and S.W. Morrall. 1996. Effect of hard river water on the surface properties of surfactants. *J. Chem. Eng. Data* 41, 1160-1167.

Rubio, J.A., E. Gonzalez-Mazo, and A. Gomez-Parra. 1996. Sorption of linear alkylbenzenesulfonates (LAS) on marine sediment. *Marine Chem.* 54, 171-177.

Ruiz-Cruz, J., and M.C. Dobarganas-Garcia. 1979. Relation between structure and biodegradation of cationic surfactants in river water. *Grasas Aceites* 30, 67-74.

Scamehorn, J.F. 1986. An overview of phenomena involving surfactant mixtures. In J.F. Scamehorn (ed.), *Phenomena in Mixed Surfactant Systems ACS symposium series 311*, pp 1-22. American Chemical Society, Washington D.C.

Schwarzenbach, R.P., P.M. Gschwend, and D. Imboden. 1993. *Environmental Organic Chemistry.* Wiley, New York, pp. 124-156.

Sherrard, K.B., P.J. Marriott, R.G. Amiet, R. Colton, M.J. McCormick, and G.C. Smith. 1995. Photocatalytic degradation of secondary alcohol ethoxylate:spectroscopic, chromatographic, and mass spectrometric studies. *Environ. Sci. Technol.* 29, 2235-2242.

Shore, S. and D.R. Berger. 1970. Alcohol and ether alcohol sulfates. In W. Linfield, ed., *Anionic Surfactants, Surfactant Science Series, Vol. 7*, 136-217, Marcel Dekker, New York.

Stancher, B., and L. Favretto. 1978. Gas-liquid chromatographic fractionation of polyoxyethylene nonionic surfactants—Polyoxyethylene mono-n-alkyl ethers. *J. Chromatogr.* 150, 447-453.

Steber, J., and P. Wierich. 1985. Metabolites and biodegradation pathways of fatty alcohol ethoxylates in microbial biocenoses of sewage treatment plant. *Appl. Env. Microbiol.* 49, 530-537.

Steber, J, W. Guhl, N. Stelter, and F.R. Schröder. 1995. Alkyl polyglucosides—ecological evaluation of a new generation of nonionic surfactants. *Tenside Surf. Deterg.* 32, 515-521.

Swisher, R.D. 1987. *Surfactant Biodegradation*, 2nd Edition. Surfactant Science Series Vol. 18, Marcel Dekker, New York.

Takada, H.K. Mutoh, N. Tomita, T. Miyadzu, and N. Ogura. 1994. Rapid removal of linear alkylbenzenesulfonates (LAS) by attached biofilm in an urban shallow stream. *Water. Res.* 28, 1953-1960.

Tolls, J., P. Kloepper-Sams, and D.T.H.M. Sijm. 1994. Bioconcentration of Surfactants: a critical review, *Chemosphere* 29, 693-717.

Tolls, J., and D.T.H.M. Sijm. 1995. A preliminary evaluation of the relationships between hydrophobicity and bioconcentration for surfactants. *Environ. Toxicol. Chem.* 13, 1675-1685.

Tolls, J., M. Haller, I. de Graaf, M.A.T.C. Thijssen, and D.T.H.M. Sijm. 1997. LAS bioconcentration: Experimental determination and extrapolation, submitted.

Tolls, J. and D.T.H.M. Sijm. 1999. Bioconcentration and biotransformation of the nonionic surfactant octaethylene glycol monotridecylether in Fathead minnows *(Pimephales promelas)*. *Environ. Toxicol. Chem.* 2689-2696.

Tolls, J. M. Lehmann, and D.T.H.M. Sijm. 2000. In vivo biotransformation of LAS in Fathead minnows *(Pimephales promelas)*. *Environ. Toxicol. Chem.* 20: accepted for publication.

Topping, B.W., and J. Waters. 1982. Monitoring of cationic surfactants in sewage treatment plants. *Tenside Detergents* 19, 164-169.

Traina, S.J., D.C. McAvoy, and D.J. Versteeg. 1996. Association of linear alkylbenzenesulfonates with dissolved humic substances and its effects on bioavailability. *Environ. Sci. Technol.* 30, 1300-1309.

Urano, K., and M. Saito. 1984. Adsorption of surfactants on microbiologies. *Chemosphere* 13, 285-293.

Versteeg, D.J., and S.J. Shorter. 1992. Effect of organic carbon on the uptake and toxicity of quaternary ammonium compounds to the fathead minnow, Pimephales promelas. *Environ. Toxicol. Chem.* 11, 571-580.

Vives-Rego, J., M.D. Vaque, J. Sanchez-Leal, and J. Parra. 1987. Surfactants biodegradation in seawater. *Tenside Surf. Detergents* 24, 20-22.

Wagner, J., H.Chen, B.J. Brownawell, and J.C. Westall. 1994. Use of cationic surfactants to modify soil surfaces to promote sorption and retard migration of hydrophobic organic compounds. *Environ. Sci. Technol.* 28, 231-237.

Waters, J., M.S. Holt, and E. Matthijs. 1989. Fate of LAS in sludge amended soils. *Tenside Surf. Detergents* 26, 129-135.

West, C.C., and J.H. Harwell. 1992. Surfactants and subsurface remediation. *Environ Sci. Technol.* 26, 2324-2330.

Woltering, D.M., R.J. Larson, W.D. Hopping, R.A. Jamieson, and N.T. de Oude. 1987. The environmental fate and effects detergents. *Tenside Surf. Det.* 24, 286-296.

Xu, S., and S.A. Boyd. 1995. Alternative model for cationic surfactant adsorption by layer silicates. *Environ. Sci. Technol.* 29, 3022-3028.

Zoeller, N.J., and D. Blankschtein. 1995. Development of user-freindly computer programs to predict solution properties of single and mixed surfactant systems. *Ind Eng. Chem. Res.* 34, 4150-4160.

Recommended Reading

A short (and somewhat arbitrary) guide to literature on environmental aspects of surfactants:

- Comprehensive compilation of knowledge on surfactant biodegradation.

    de Oude, N.T., ed. 1992. Detergents, *Handbook of Environmental Chemistry*, Volume 3, Part F. Springer, Berlin.

    Swisher, R.D. 1987. Surfactant biodegradation. In *Surfactant Science Series*, Vol. Marcel Dekker, New York.

- Recent review of analytical techniques used in environmental research of surfactants.

Kloster, G. 1997. Analytical Methods for surfactants and complexing agents at concentrations relevant to environmental occurence. In M.J. Schick and F.M. Forbes, Eds., *Detergents in the Environment, Surfactant Science Series 63*, pp. 65-123, Marcel Dekker, New York.

- Compilation and evaluation values of critical micelle concentration of surfactants.

    Mukerjee, P., and K. Mysels. 1967. Critical micelle concentration of aqueous surfactants systems, National Bureau of Standard Reference Data Series. *NSDRS-BS*, 36, National Bureau of Standards, Washington D.C.

- Periodicals dedicated to surfactant science:

    Surfactant Science Series: reviews on all aspects of surfactants, including physical-chemical properties and environmental chemistry.

    Tenside Surf. Deterg.: journal carrying contributions on environmental issues on a regular basis.

    JAOCS: journal carrying contributions on environmental issues on a regular basis.

18

Estimating the Properties of Synthetic Organic Dyes

David G. Lynch

CONTENTS

18.1 Introduction

Dyes differ from pigments in that dyes, but not pigments, are soluble or can be solubilized in the medium in which they are used. Prior to 1856, pigments were exclusively animal, vegetable, and mineral matter. With Perkin's introduction of the first synthetic dye, mauve, to the English textile industry, and the subsequent expansion of the synthetic dye market, the use of natural dyes on textiles was virtually eliminated. In 1993 almost 155 million kilograms of synthetic dyes were manufactured in the U.S. (Table 18.1).

Dyes are significant because of these production volumes and their potential release into the environment from manufacture, processing, and use. Spent dye solutions from textile dyeing operations ultimately are discharged to surface water after some form of wastewater treatment, and dyes not effectively removed from the wastewater may result in environmental and human exposure.

TABLE 18.1

U.S. Dye Production 1993
(USITC 1994).

Dye Class	Production X1000 Kg
Acid	6,335
Basic	5,693
Direct	22,493
Disperse	14,399
Solvent	5,819
Vat	14,044
Other[1]	85,736
TOTAL	154,519

[1] Includes azoic dyes and components, fiber reactive dyes, fluorescent brighteners, F.D.&C. Colors, mordant dyes, sulfur dyes, and miscellaneous dyes.

18.2 Classifications of Dyes

18.2.1 General Use Classes

Acid, basic, and *direct* dyes are all ionic in nature. Acid dyes contain free acid groups which are ionized in the aqueous application medium (dyebath). They generally used to dye polyamine, wool, or silk and are primarily azo, anthraquinone, or triarylmethane structures.

Basic (cationic) dyes contain cationic groups which are ionized in the dyebath. They are applied to acrylics, nylon, polyester, and paper and are most often azo, anthraquinone, triarylmethane, quinoline, thiazine, methine, oxazine, and acridine types.

Direct dyes are applied directly to cellulosic fibers in the presence of electrolytes. Major applications are to cotton, regenerated cellulose, leather, and paper.

Disperse dyes are nonionic and have very low water solubility. They are applied as a dispersion to polyester, nylon, cellulose diacetate, triacetate, and acrylics.

Fiber reactive, sulfur, vat, and mordant dyes all undergo reactions which enhance their attachment to the substrate. Fiber reactive dyes contain a reactive group that forms a covalent bond with a group on the substrate, usually hydroxyl or amine. Fiber reactive dyes are often of the azo or anthraquinone type.

Sulfur dyes are applied to cellulose in an alkaline reducing environment. They become insoluble and attach to fiber after oxidation.

Vat dyes are insoluble in water and applied to cellulose as dispersions. They are reduced to a water soluble leuco form which has an affinity for cellulose in an alkaline dyebath. Oxidation returns vat dyes to an insoluble form and fixes them to cellulose. Vat dyes are often anthraquinone or indigoid structures.

Mordant dyes are applied to wool using the same methods as that for acid dyes. They are then treated with metal salts to form the dye-metal complex on the fiber. The metal salts can be applied to the substrate before, during, or after dyeing. Mordant dyes are generally of the monoazo type. Substitution at both positions ortho to the azo bonds with –OH,

$-OCH_3$, $-OCH_2COOH$, $-COOH$, $-COOC_2H_5$, $-NH_2$, and similar groups is a structural requirement for effective mordant dyeing.

18.2.2 Chemical Classes

The Society of Dyers and Colourists and the American Association of Textile Colorists and Chemists have divided dyes into thirty structural classes in their Color Index. Those of major commercial importance are discussed below.

Azo dyes, characterized by the presence of one or more azo groups (-N=N-), are the most commercially important class of dyes. These compounds are synthesized using a diazotization reaction in which a primary aromatic amine reacts with nitrous acid to form a diazonium salt. The diazonium compound then typically is coupled with phenols, napthols, aromatic amines, heterocycles, or a variety of other compounds containing active methylene groups. Azo dyes are used in acid, direct, disperse, fiber reactive, and mordant applications.

Anthraquinone dyes are characterized by the presence of one or more carbonyl groups in association with a conjugated system. These dyes also may contain hydroxy, amino, or sulfonic acid groups as well as complex heterocyclic systems. Anthraquinones uses include disperse, vat, acid, mordant, and fiber reactive applications.

Azoic components (stabilized diazonium salts) are applied to fibers containing a coupling component, and the color form on the substrate. Azoic components can be used on cotton, polyester, rayon, and cellulose acetate.

Triarylmethane dyes are derivatives of triphenylmethane and diphenylnaphthylmethane. The presence of one or more primary, secondary, or tertiary amino or –OH groups in the para position to the methane carbon determines the dye color. Halogen, carboxyl, or sulfonic acid substituents also may be present on the aromatic rings. Triarylmethane colorant applications include basic, acid, solvent, and mordant dyes. Major uses are in printing inks.

Stilbene dyes contain the stilbene structural moiety ($C_6H_5C=CC_6H_5$) but are synthesized from 4- nitrotoluene-2-sulfonic acid. Direct yellow stilbene dyes for dyeing paper are probably the most commercially important.

Xanthene dyes contain the xanthylium or dibenzo-gamma-pyran nucleus. Amino or hydroxy groups meta to the oxygen are usually present. Colors range from greenish yellow to violet. Xanthene dyes are used for mordant dyeing of cotton and direct dyeing of wool and silk. They also color paper, leather, wood, food, drugs, cosmetics, paints, and varnishes.

Oxazine, azine and thiazine dyes are named for the characteristic heterocyclic ring systems 1,4-oxazine, 1,4-diazine, and 1,4-thiazine. The dyes are generally cationic (basic) or acid dyes. They also can be reduced to colorless forms, then oxidized back to the dye, as in vat dyeing. The dyes also have been used to a limited extent in disperse and fiber reactive applications. They are used as titration indicators and may be applied to acrylic fibers and leather.

Indigoid dyes contain carbonyl groups and are used as vat dyes. They represent one of the oldest known dye classes. Indigo is the blue used for dyeing blue jeans and is the indigoid dye of commercial significance. The cross conjugated system known as the H-chromogen is responsible for its color.

Phthalocyanine dyes were first synthesized by passing ammonia gas into molten phthalic anhydride containing iron filings. The phthalocyanines are stable analogs of the naturally occurring pigments chlorophyll and heme. Adding substituents produces colors ranging from blue to green. Copper phthalocyanines have become the most commercially important members of this group, with major uses in printing inks and paints, because of their strong coloring abilities.

18.3 Environmental Fate of Dyes

18.3.1 Pathways to the Environment

Dye manufacturing, processing, and use generates wastewater and solids. The wastewater typically undergoes biological treatment at the manufacturing or use site or at a municipal wastewater treatment facility before discharge to surface waters. Solids containing waste from dye manufacturing and use are often disposed of in municipal landfills.

Dyes in wastewater released to biological wastewater treatment are subject to removal from the aqueous phase via sorption to sludge, aerobic biodegradation, and anaerobic biodegradation, if anaerobic sludge digestion or other such processes are used, orif transient anaerobic conditions occur elsewhere in the treatment system. Since dyes tend to have extremely low vapor pressures and Henry's law constants, volatilization is generally not a significant process. Those dyes passing through treatment into surface waters are subject to direct and indirect photolysis, aerobic biodegradation in the water column, and anaerobic biodegradation if they enter and are incorporated into deep sediments. Dyes sorbed to sludges and disposed of via landfilling or land spreading are subject to aerobic and anaerobic biodegradation as well as photolysis on soil or sludge surfaces exposed to sunlight.

Dye-containing solid wastes, including cleanup wastes from manufacturing and textile dyeing operations, are likely to enter landfills, where aerobic and anaerobic biodegradation, sorption to soils, and leaching may occur.

18.3.2 Biodegradation

Research to date has focused primarily on azo and anthraquinone dyes, due to their commercial importance. Environmental processes including biodegradation, photolysis, sorption to soils and sediment, and abiotic transformation in sediment/water systems have been studied. The quantity of dyes apparently entering and potentially passing through wastewater treatment systems unaltered has prompted research on the behavior of these chemicals in biological and other types of wastewater treatment systems.

While most dyes appear to resist biodegradation under aerobic conditions, a variety of microorganisms can reduce azo bonds in the absence of oxygen. Pagga and Brown (1986) tested 87 dyes for their biodegradability under aerobic conditions. Most of these dyes showed no change in 28-day aerobic tests. Observed color removal was attributed to sorption to biomass in the test system. Under anaerobic conditions, however, all of the azo dyes showed significant loss of color, indicating disruption of their conjugated structure, most likely by azo reduction. Further studies on degradation products generated in similar experiments conducted under anaerobic conditions associated the appearance of the corresponding aromatic amines with the disappearance of the parent azo dye (Table 18.2). Anthraquinone dyes showed little transformation in the presence or absence of oxygen.

TABLE 18.2
Dyes and Biodegradation Products (Pagga 1986).

Parent Compound	Metabolite
Acid Orange 7	4-aminobenzenesulfonic acid
Acid Yellow 36	3-aminobenzenesulfonic acid N-phenyl-1,4-diaminobenzene
CI 13155	2,5-dichloroaniline
Acid Red 114	4,4-diamino-3,3'-dimethylbiphenyl 4 -methylbenzenesulfonic acid – (4'-aminophenyl) ester

TABLE 18.2 (CONTINUED)
Dyes and Biodegradation Products (Pagga 1986).

Acid Blue 25	1-amino-4-phenylaminoanthraquinone
Acid Yellow 25	3-amino-6-methylbenzene-N-phenylsulfonamide 4-amino-3-methyl-1-(4'-sulphophenyl) pyrazolone
Acid Yellow 151	3-amino-4-hydroxybenzenesulphonamide
Acid Black 24	1-aminonaphthalene-5-sulphonic acid

TABLE 18.2 (CONTINUED)
Dyes and Biodegradation Products (Pagga 1986).

Direct Red 7	4,4'-diamino-3,3'-dimethoxybiphenyl
Direct Blue 14	4,4'-diamino-3,3'-dimethylbiphenyl
Direct Blue 15	4,4'-diamino-3,3'-dimethoxybiphenyl
Direct Yellow 12	1-amino-4-ethoxybenzene 4-amino-2-sulphobenzaldehyde

18.3.3 Aquatic Environment/Sediment Water Systems

Baughman (1995) examined the behavior of nine acid and five direct dyes in suspensions of river and lake sediments representative of waters of the southeastern U.S. piedmont

TABLE 18.2 (CONTINUED)
Dyes and Biodegradation Products (Pagga 1986).

Direct Yellow 50	2-aminonaphthalene-4,8-disulfonic acid
Basic Blue 3	4-(N-diethylamino)-1-hydroxybenzene
Mordant Black 9	3-amino-4-hydroxybenzenesulphonic acid
Mordant Black 11	dihydroxyazo naphthalene dihydroxyamino naphthalene other structural details unknown

region and calculated partition coefficients for dyes that did not undergo transformation (Table 18.3). In general, the sediments did not strongly sorb the dyes.

Kinetic studies in sediment/water systems with Direct Red 2, Acid Black 92, Acid Red 4, Acid Red 18, and Direct Yellow 1 lead to linear and biphasic plots of dye loss over time. For all but Direct Yellow I, dye loss was usually preceded by a lag or adaptation phase. Acid Black 92 and Direct Red 2 were transformed completely in less than 24 and 48 hours, respectively, but Acid Yellow 151 and Direct Yellow 1 showed half-lives of greater than 2 years. The rapid initial drop in concentration of all dyes observed, with the exception of Acid Red 18, was presumed to be due to sorption. Tests to determine the effect of pH on

TABLE 18.3

Sediment Sorption Data for Dyes at 24 hours (Baughman 1995)

Dye	Sediment	$K_p + CV$ ($l\ kg^{-1}$)
Acid Red 4	Oconee River[b]	23 + 29
	Oconee River	206 +46
Direct Yellow 1	Herrick Lake	227 + 18
Acid Yellow 151	Oconee River[b]	65 + 52
	Herrick Lake[b]	95 + 31
Acid Red 18	Herrick Lake	<10
	Oconee River	<5
Acid Black 92	Oconee River[b]	20 + 20

[b] = Boiled

sorption of the dyes showed that, at pH values of > 5.5, the effect of pH on the sorption isotherm was less than a factor of 2 per pH unit. At lower pH levels, the effect of pH was more pronounced.

Thus sorption, followed by intra-particle diffusion of the dyes, causes rapid initial loss followed by slow long term loss. Transformation pathways probably involve azo reduction. Diffusion may limit the dyes transformation rates because of the large size of the molecules.

Weber and Adams (1995) observed the rapid reduction of the azo dye Disperse Blue 79 chemically as well as in three anoxic sediment/water systems. Half lives were on the order of minutes to hours. An initial rapid loss of the dye was followed by a much slower rate of transformation, and, most probably, chemical or cometabolic processes were responsible for transformation.

Baughman (1992) measured the disappearance rate constants for a number of solvent and disperse azo, anthraquinone, and quinoline dyes in anaerobic sediments. The half-lives ranged from 0.1 to 140 days. Product studies of the azo dyes showed that reduction of the azo linkages and nitro groups resulted in the formation of substituted anilines. The 1,4-diaminoanthraquinone dyes underwent complex reactions thought to involve reduction and replacement of amino with hydroxy groups. Demethylation of methoxyanthraquinone dyes and reduction of anthraquinone dyes to anthrones also was observed.

These studies suggest that azo dyes entering anaerobic sediments are likely to undergo relatively rapid chemical/biological reactions resulting in the reductive cleavage of azo bonds and in the reduction of nitro groups, with the corresponding formation of amine products.

18.3.4 Precipitation of Dyes in Natural Waters

Hou (1992) used a simple screening test to determine whether acid and direct dyes precipitate at calcium concentrations typical of hard waters of the SE Piedmont region of the U.S. Of the 52 dyes tested, only three direct dyes (Direct Black 19, Direct Black 22, and Direct Blue 75) and seven acid dyes (Acid Red 88, Acid Red 114, Acid Red 151, Acid Brown 14, Acid Black 24, Acid Orange 8, and Acid Blue 113) precipitated. Although the Ca salts of acid and direct dyes were thought to be the most likely metal salts to precipitate after dye discharge to natural waters, the precipitation is not likely to occur unless dye concentrations exceed 0.02 to 0.6 mg/L, a level far greater than reported concentrations of dyes in surface waters.

18.3.5   Removal of Dyes in Activated Sludge Wastewater Treatment

Shaul et al. (1986) conducted experiments to determine the removal of water soluble dyes in pilot scale activated sludge biological wastewater treatment systems. Adsorption isotherms also were developed using biologically inactive activated sludge as an adsorbent. Overall, the dyes showed a wide range of removal in the pilot scale treatment systems. Failure to account for all of the dye added to the system in the effluent or sorbed to sludge suggested biodegradation as a possible removal process for some simple azo dyes (Table 18.4). Comparison of the adsorption isotherm results with the sorption measured in pilot scale systems showed that, with some exceptions, the adsorption isotherms were reasonably good predictors of the extent of removal by adsorption in pilot scale systems.

TABLE 18.4

Removal of Dyes in Pilot Scale Treatment Systems (Shaul 1986).

Dye	Percent Removed		Percent Apparent Biodegradation	
	1 mg/L	5 mg/L	1 mg/L	5 mg/L
Acid Blue 113	92	83	24	33
Acid Orange 7 (two runs)	83	96	81	96
	75	88	73	87
Acid Orange 10	5	10	0	0
Acid Red 1	12	8	0	0
Acid Red 88	97	>99	96	>99
Acid Red 151	72	89	27	22
Acid Red 337	25	21	0	0
Acid Yellow 151	35	9	0	0
Direct Violet 9	43	81	0	0

Ganesh et al. (1994) conducted aerobic and anaerobic batch sludge digestion experiments to determine the fate of the vinyl sulfonyl fiber reactive azo dyes Reactive Black 5 and Navy 106 in the biological wastewater treatment process. Since reactive dyes not bonded to the substrate during the dyeing process generally undergo hydrolysis before release to wastewater treatment, the authors studied both the parent Reactive Black 5 and its hydrolysis product, using rinse water containing Navy 106 that was collected from a textile plant.

Under aerobic conditions Reactive Black 5 maximum total color removal was 92%. The authors attributed initial rapid loss primarily to adsorption to sludge biomass. Hydrolyzed Reactive Black 5 showed little color removal, adsorption, or degradation. Navy 106 also exhibited a loss of color over time. The initial rapid loss was likely adsorption to sludge, but the subsequent decrease may have been due either to anaerobic transformation of dyes in microanaerobic zones of the aerobic system, or to aerobic microbial degradation.

Under anaerobic conditions, the color levels in the Reactive Black 5 hydrolysis product reactor decreased by 66%. Little or no dye was found in sludge extracts. Similar results were obtained with Navy 106. As much as a 49% decrease in color was observed by the end of day 1 of the experiment, with a change in hue suggesting cleavage of the azo bond.

18.3.6   Photolysis

Adams et al. (1994) studied the photolysis of dyes used to color military smoke grenades, and measured rates of photolysis in natural and simulated sunlight on soil surfaces and as a function of soil depth. At 1 mm soil depth, all the dyes showed some photodegradation. With one exception, the photodegradation kinetics were similar in natural and simulated sunlight. Initial rapid loss was generally observed, followed by a slower rate of loss. This behavior may be attributed to light attenuation effect and/or an inner particle shielding effect.

Because the dyes were undergoing loss at a soil depth greater than that at which direct photolysis would be expected, indirect photolysis of the dyes was suspected as a loss mechanism. The mechanism was thought to be sensitized photooxygenation via reaction with singlet oxygen. Further studies with Disperse Red 9 showed that, in the presence of laboratory air, appreciable degradation of Disperse Red occurred at a depth beyond which direct photolysis would be expected. In contrast, under a nitrogen atmosphere, the dye was photolyzed at the shallower depths associated with direct photolysis.

Haag and Mill (1985) studied the direct and indirect photolysis of a series of commercially important azo dyes in efforts to develop structure/activity relationships. The azo dyes photolyzed more rapidly by indirect processes in natural waters containing humic sensitizers than by direct photolysis, with azo cleavage occurring in almost all cases. Photolysis involves oxidation of the dye by singlet oxygen and possibly other radical oxidants. Photolysis half-lives in surface waters depended on the pH of the water and the pK_a of the dye. The dissociation of –OH groups in o-hydroxyazonaphthyl phenyl azo dyes resulted in an order of magnitude or greater increase in the reactivity of the dye. Chromium complexation greatly increases the stability of azo dyes.

Three equations were derived for predicting indirect photolysis rate constants (s^{-1}) for azo dyes in natural waters

Diarylamine dyes
$$k_p = k_p^o + 1 \times 10^8 \tag{1}$$

Chromium complexed
o,o-dihydroxyazo dyes
$$k_p = k_p^o + 1 \times 10^7 \tag{2}$$

Arylazophenol
$$k_p = k_p^o + \left(1.3 \times 10^7 (1-\alpha) + 1.8 \times 10^8 \alpha\right)[Ox] \tag{3}$$

Arylazonaphthyl
Arylazochromotropic acid
Arylazopyrazolone

where: k_p^o = direct photolysis rate constant, which is normally negligible in humic rich waters
  $[Ox]$ = apparent sum of $^1O_2$ and radical concentrations in sunlight measured using a furan such as FFA
  α = $1/[1 + 10^{(pK_a - pH)}]$ = degree of dissociation
  $t_{1/2}$ = $\ln 2 / k_p$

Table 18.5 shows indirect photolysis half-lives of azo dyes in natural waters at 40°N at the surface and 1 meter depth using measured rate constants and singlet oxygen concentrations. They apply only if singlet oxygen and/or radical oxidation are the only mechanisms for degradation.

See Chapter 15 for a detailed discussion of photolysis and its estimation.

18.3.7 Fate of Dyes in Landfills

Tincher (1988) studied the behavior of dyes under simulated landfill conditions. These results offer limited evidence that, due to one or several factors, dyes disposed of in landfills do not migrate through the soil to a significant extent. Under the anaerobic conditions that can develop in saturated, compacted soils, reduction of azo dyes to their corresponding amines would be expected. In addition, nitro groups, if present, potentially could also

TABLE 18.5

Estimated Indirect Photolysis Half-lives (hours) of Azo Dyes in Natural Waters at 40°N (Haag and Mill, 1985).

Type	Dye	pH 5		pH 7		pH 9	
		Surface	1 Meter	Surface	1 Meter	Surface	1 Meter
o,o-dihydroxy	Acid Alizarin Violet N	5	100	2	40	1	20
	Mordant Blue 9	12	240	4	80	0.6	13
	Palatine Chrome Black 6BN	6	110	0.8	15	0.5	10
o-arylazonaphthol	Orange G	35	700	35	700	30	600
	Sudan I	30	600	30	600	25	500
	Acid Orange 8	11	220	11	220	10	200
	Direct Red 81	100	2000	100	2000	70	1400
	Acid Red 4	25	500	20	400	2	40
weakly H-bonded - OH	Acid Red 29	60	1200	40	800	2	40
	Acid Brown 14	20	400	14	280	2	40
	Acid Yellow 34	40	900	14	280	2	40
diarylamine	Acid Blue 113	5	100	4	70	2	50

Surface $[^1O_2] = 10^{-12}$ M
1 m depth $[^1O_2] = 5 \times 10^{-14}$ M
$t_{1/2}$ (hours)

undergo reduction. The potential for aromatic amines to bind with soil components also is known and that process could be a factor in limiting the mobility of the reduction products. However, factors such as soil properties and oxygen concentrations in the landfill environment should be considered before making broad generalizations.

18.4   Estimation of Key P-Chem Properties

18.4.1   Water Solubility and Octanol/Water Partition Coefficient

Water solubility and the octanol-water partition coefficient (K_{ow}) are fundamental data that can be helpful in predicting the environmental partitioning behavior of chemical substances (see chapters 5 and 7). Dyes in the acid, direct, and basic classes tend to be salts which readily dissociate in water. Their water solubilities typically exceed 100 g/L and their K_{ow} values reflect extremely low solubilities in octanol. Thus, as a general rule, unless another process such as ion exchange or precipitation as an insoluble salt is suspected, these dye classes would be expected to remain in the water column in the aquatic environment and show little affinity for organic matter or biota.

Nonionic dyes of the disperse, vat, sulfur, and solvent classes, in contrast to ionic dyes have significant potential to partition to organic matter and, in some cases, undergo bioconcentration by aquatic organisms.

However, although several methods exist for estimating water solubility based on regression of water solubility against octanol/water partition coefficient, most have used data sets that do not include nonionic dyes.

Baughman and Weber (1991) attempted to improve upon three existing methods for relating water solubility (S) to K_{ow} for disperse dyes. The existing methods were (1) the regression of log K_{ow} against log water solubility, (2) determination of the product of K_{ow} and S, and (3) application of the equation of Yalkowsky and Valvani (1980) using log K_{ow}, log S, entropy of fusion (ΔS_f), and melting point.

The octanol/water partition coefficient and water solubility of crystalline dyes were estimated using equations (4) through (7) below:

$$\log K_{ow} = 1.32 - 0.77 \log S_c \qquad\qquad R^2 = 0.785 \qquad\qquad (4)$$

$$\sigma = 0.67$$

$$n = 20$$

$$\log K_{ow} = 2.16 - 0.77 \log S_c - 0.0049 \,(mp\ -25) \qquad R^2 = 0.80 \qquad\qquad (5)$$

$$\sigma = 0.64$$

$$n = 20$$

$$\log S_c = 0.34 - 1.02 \log K_{ow} \qquad\qquad R^2 = 0.785 \qquad\qquad (6)$$

$$\sigma = 0.76$$

$$n = 20$$

$$\log S_c - 1.27 - 1.03 \log K_{ow} - 0.0050 \,(mp\ -25) \qquad R^2 = 0.797 \qquad\qquad (7)$$

$$\sigma = 0.74$$

$$n = 20$$

S_c = water solubility of crystalline dye in mol/m³
mp = melting point in degrees C
R = correlation coefficient
σ = root mean square deviation

Equations (4) through (7) produce more reliable estimates of K_{ow} and S for hydrophobic dyes than other methods. Isnard and Lambert (1989) calculated root mean square deviations for a dataset of 20 disperse and solvent dyes, using a number of available equations They showed that equations in the form of Equation (4) had root mean square deviations (σ) values of 1.6 to 3.3 log K_{ow} units, regressions in the form of equations (6) and (7) gave σ values ranging from 1.3 to 3.3, and equations similar to Equation (5) had root mean square deviations ranging from 0.57 to 1.4.

Isnard and Lambert derived regression equations relating S and K_{ow} for a database of 300 compounds. Although they determined an R^2 value of 0.93 and σ of 0.47 for K_{ow} estimation and 0.44 for S, σ values increased to 1.5 log units when these equations were used to estimate S or K_{ow} for the Baughman dataset. Thus, the Isnard and Lambert equations may not be well suited for use with dyes of extremely low water solubility, such as the disperse dyes.

The product of S and K_{ow} has been referred to as "pseudooctanolsolubility" (Q). Baughman et al. calculated Q values for 20 hydrophobic dyes using actual and sub-cooled liquid solubilities and measured K_{ow} values. Root-mean-square deviations for log Q values were lowest (0.66) when Q was calculated using solubilities of sub-cooled liquids where ΔS_f and melting point values were known. No additional equations were developed for this approach.

The use of entropies of fusion (ΔS_f) improved both the correlation coefficient and root mean square deviation of the equation below for estimating log K_{ow}:

$$\log K_{ow} = -0.82 \log S_c - 1.05 \times 10^{-4} \Delta S_f (mp - 25) - 0.03 \qquad (8)$$

$$R^2 = 0.836$$

$$\sigma = 0.58$$

$$n = 20$$

Examination of the sources of error in the equations suggests the contribution of a number of factors. The purity of dyes, differences in solubilities of polymorphic forms of dyes, tautomerization, hydrogen bonding, and polarization all can contribute to the difficulty in accurately measuring the S and K_{ow}. The use of an "average" value of ΔS_f in the presence of variability of a factor of two in the values for rigid structures (e.g., anthraquinone dyes) versus more flexible molecules (e.g., azo dyes) may lead to significant errors in estimation.

Baughman concluded that while none of the three methods examined was markedly superior, lack of data limits the use of equations dependent on entropies.

Few reliable measured values for the log K_{ow} and water solubility of hydrophobic dyes have been published. Table 18.6 presents measured values for these properties, Baughman and Perenich (1988).

18.4.2 Acid Dissociation Constant

The acid dissociation constant (pK_a) of an organic compound is useful in assessing its environmental fate. The pK_a value can be used to define the degree of ionization of a compound at a given pH and the potential for sorption to surfaces by cation exchange. The extent to which a compound is sorbed can have a significant effect on its bioavailability, transport, photolysis, and biodegradability.

Organic ions are known to sorb to soils and sediments via cation exchange. This sorption decreases as pH increases above the pK_a. Thus, sorption should be important for a compound with a pK_a below or near the pH range of natural waters (i.e., 5–7). The processes responsible for sorption are not limited to cation exchange, however, and the uncharged portion of the molecule can sorb via hydrophobic partitioning.

Few simple methods exist for estimating pK_a for complex dye structures. However, complex artificial intelligence techniques combining the results of fundamental and empirical approaches have been developed which can predict pK_a values for dye structures to within the experimental error of laboratory measurements.

Hilal (1994) calculated the pK_a values of 214 dye molecules using the SPARC (SPARC Performs Automated Reasoning in Chemistry) computer program. SPARC computational methods use the knowledge base of organic chemistry and conventional Linear Free Energy Relationships (LFER), Structure/Activity Relationships (SAR), and Perturbed Molecular Orbital (PMO) methods.

The root mean square deviation for measured versus estimated pK_a values for 214 azo dyes and related aromatic amines was 0.62 pK_a units. In comparison to experimental errors

TABLE 18.6

Measured Water Solubility and Log Kow for Selected
Dyes (Baughmann and Perenich, 1988).

Dye Name	Structural Type	Water Solubility (μg/L)	Log K_{ow}
Disperse A. 1	mono azo	272.3	
Disperse A. 2	mono azo	287339	
Disperse Blue 23	anthraquinone	782.4	
Disperse Blue 79	mono azo	5.2	4.8
Disperse Orange 3	mono azo	290.4	
Disperse Red 1	mono azo	91060	4.3
Disperse Red 11	anthraquinone	2106480	3.5
Disperse Red 17	mono azo	736160	
Disperse Red 19	mono azo	234.3	
Disperse Red 274			3.8
Disperse Red 5	mono azo	143.83	4.3
Disperse Red 7	mono azo	400.95	
Disperse Violet 1	anthraquinone	228.48	3
Disperse Yellow 1	diphenylamine	7892.5	
Disperse Yellow 42			4.6
Disperse Yellow 54	indigoid	0.74	5
N 1	mono azo	0.69	5.4
N 2	mono azo	74.5	3.4
N 5	mono azo	0.59	5.5
N 7	mono azo	15.7	5.4
N 9	mono azo	276	4
Ra	mono azo	7.9	3
Solvent Yellow 1	mono azo	260040	
Solvent Yellow 2	mono azo	382.5	
Solvent Yellow 58	mono azo	510150	
Solvent Yellow 7	mono azo	21780	

in the measurements of pK_a for some of these dyes of 2 pK_a units, the RMS for estimation of the value appears acceptable for first approximations.

18.4.3   Appropriate Tests

Many laboratory test methods exist that can help in assessing the environmental fate of dyes. It is most important to first identifiy the environmental processes that may affect the dye and the properties of the dye that enhance or mitigate the effect of these processes. Water solubility and log k_{ow} are the primary properties needed for initial assessment. Since the vapor pressures of dyes tend to be extremely low, neither this property, nor Henry's law constant generally needs to be determined.

Useful information regarding the potential for removing dyes in biological wastewater treatment systems can come from a number of tests. Tests of "Ready Biodegradability" are relatively simple to conduct, and a range of methods exist for both poorly water soluble and water soluble dyes. The 28-day tests are relatively stringent, and so, a chemical that passes the criteria to be classified as "readily biodegradable" can be expected to biodegrade rapidly in most aerobic environments. However, most dyes, due to their structural complexity and other factors, will not pass these tests.

Screening tests for anaerobic biodegradation are of particular value for azo dyes. Since disruption of conjugation of the molecule results from reduction of the azo bond, color change or loss of color is an excellent indicator of the initial steps of degradation of these compounds under anaerobic conditions. The ETAD method No. 105 and slight modifications of the EPA method 835.3400, Anaerobic Biodegradability of Organic Chemicals, provide simple methods to determine the potential for dyes to undergo biodegradation in the absence of oxygen.

The potential of dyes to be removed in wastewater treatment via sorption to sludge can be assessed using simple or more-complex approaches. The EPA method 835.1110, Activated Sludge Sorption Isotherm Test, is a simple test designed to allow the measurement of partitioning of a test chemical between sludge and an aqueous test medium. Measured partition coefficients can be used to estimate the removal via sorption to sludge in full scale treatment systems. Laboratory-scale wastewater treatment simulation tests represent a much more complex approach to assessing the treatability of dyes. Batch methods such as the SCAS test (EPA 835.3210) and continuous flow methods, such as the Porous Pot Test (EPA 835.3220), also can give insight into the removal of dyes in biological wastewater treatment plants, with appropriate analyses and analytical methods.

Dyes passing through treatment into surface waters may undergo direct or indirect photolysis. These processes are likely to proceed rapidly for most dyes and can be important in determining their environmental fate. A measurement of UV/Visible absorption spectrum (EPA 830.7050) can provide a basis for assessing photolysis potential. Several methods exist for measuring direct and indirect photolysis in water. These include EPA 835.2210, Direct photolysis rate in water by sunlight, and EPA 835.5270, Indirect photolysis screening test.

Solid dye wastes often are disposed of in landfills. The tendency of a dye to sorb to soil is useful in assessing the potential for landfilled dye wastes to migrate to groundwater. Sediment/soil adsorption/desorption isotherm, soil thin layer chromatography, and soil column leaching studies all can give valuable insight into this behavior. The biodegradation of dyes in soils also can have a significant impact on their fate. The EPA Soil biodegradation test (835.3300) can be conducted under aerobic and anaerobic conditions to examine dye biodegradation potential in upper and lower layers of soils.

References

Adams, R.L., E.J. Weber, and G.L. Baughman. 1994. Photolysis of smoke dyes on soils. *Environmental Toxicology and Chemistry* 13, 889-896.

Baughman, G.L. and T.A. Perenich. 1988. Fate of dyes in aquatic systems: I Solubility and partitioning of some hydrophobic dyes and related compounds. *Environmental Toxicology and Chemistry* 7, 183-199.

Baughman, G.L. 1992. Unpublished data.

Baughman, G.L. and E.J. Weber. 1991. Estimation of water solubility and octanol/water partition coefficient of hydrophobic dyes. Part I: Relationship between solubility and partition coefficient. *Dyes and Pigments* 16, 261-271.

Baughman, G.L. 1995. Fate of azo dyes in aquatic systems. Part 3: The role of suspended sediments in adsorption and reaction of acid and direct dyes. *Dyes and Pigments* 27, 197-210.

Brown D. and B. Hamburger. 1987. The Degradation of dyestuffs: Part III- Investigation of their ultimate biodegradability. *Chemosphere* 16, 1539-1553.

Brown D. and P. Laboureur. 1983. The Degradation of dyestuffs: Part I—Primary biodegradation under anaerobic conditions. *Chemosphere* 122, 397-404.

Ecological and Toxicological Association of the Dyestuffs Manufacturing Industry. Ecological method No. 105. A Screening test for assessing the primary anaerobic biodegradability of water soluble dyestuffs. ETAD, Basel, Switzerland.

Ganesh, R., G.D. Boardman, and D. Michelsen. 1994. Fate of azo dyes in sludges. *Water. Res.* 28, 1367-1376.

Grayson, M., Ed. 1978. *Kirk-Othmer Encyclopedia of Chemical Technology, 3rd Edition.* John Wiley and Sons, New York.

Haag, W., and T. Mill. 1986. Direct and Indirect Photolysis of Azodyes in Water. Data Generation and Development of Structure Reactivity Relationships for Environmental Fate Processes/Properties. EPA Contract No. 68-02-3968.

Hilal, S.H., L.A. Carreira, G.L. Baughman, S.W. Karickhoff, and C.M. Melton. 1994. Estimation of ionization constants of azo dyes and related aromatic amines: environmental implication. *J. Phys. Org. Chem.* 7, 122-141.

Hou, M., and G.L. Baughman. 1992. Predicting the precipitation of acid and direct dyes in natural waters. *Dyes and Pigments* 18, 35-46.

Isnard, P., and S. Lambert. 1989. Aqueous solubility and n-octanol/water partition coefficient correlations. *Chemosphere* 18, 1837-1853.

Pagga, U., and D. Brown. 1986. The Degradation of dyestuffs: Part II – Behavior of dyestuffs in aerobic biodegradation tests. *Chemosphere* 15, 161-166.

Shaul, G.M., R. Lieberman, C. Dempsey, and K. Dostal. 1986. Treatability of water soluble azo dyes by the activated sludge process. *Industrial Wastes Symposia Proceedings. 59th Water Pollution Control Federation Annual Conference, October 5-9.*

Tincher, W.C. 1988. *Dyes in the Environment: Dyeing Wastes in Landfills.* Georgia Institute of Technology, Atlanta, GA.

United States International Trade Commission 1994. *Synthetic Organic Chemicals: United States production and sales, 1993.* USITC Publication 2810.

USEPA. 1995. *OPPTS Test Guidelines Series 835 Fate, Transport and Transformations. Office of Prevention, Pesticides and Toxics.*

Weber, E.J., and R. Adams. 1995. Chemical and sediment mediated reduction of the azo dye Disperse Blue 79. *Environ. Sci. Technol.* 29, 1163-1170.

Yalkowsky, S.H., and S.C. Valvani. 1980. Solubility and partitioning I: solubility of nonelectrolytes in water. *J. Pharm. Sci.* 69, 912.

Yen, C-P.C., T.A. Perenich, and G.L. Baughman. 1989. Fate of dyes in aquatic systems: II Solubility and octanol/water partition coefficients of disperse dyes. *Environmental Toxicology and Chemistry* 8, 981-986.

Appendix: Chemical structures of some dyes mentioned in this chapter.

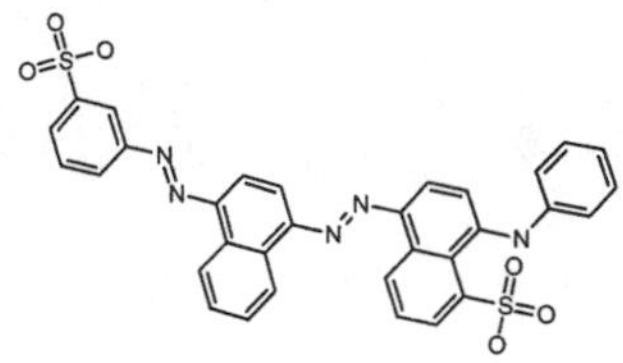

Acid Blue 113

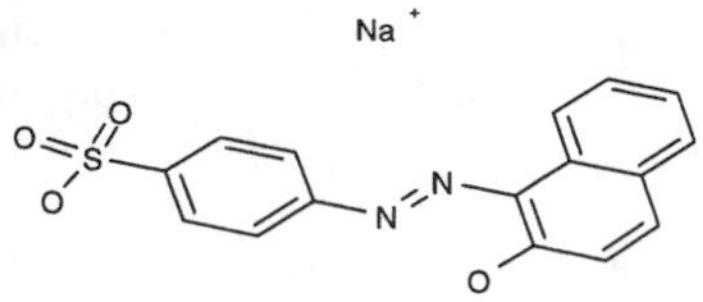

Acid Orange 7

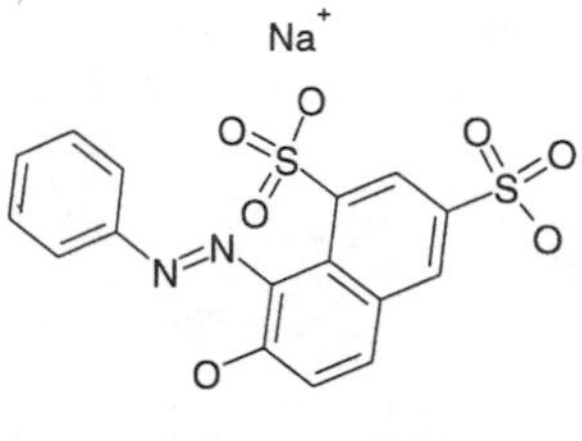

Acid Orange 10

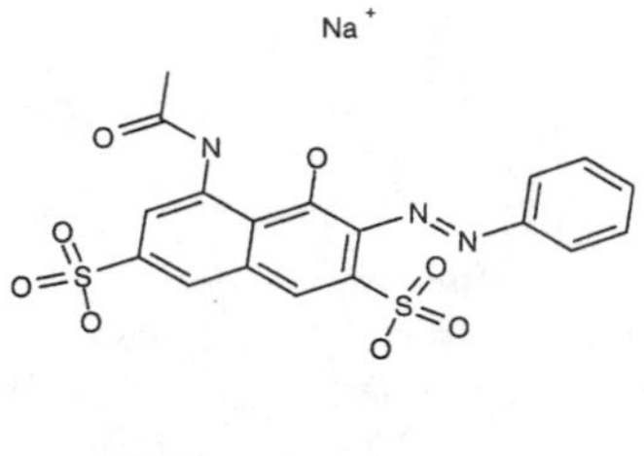

Acid Red 1

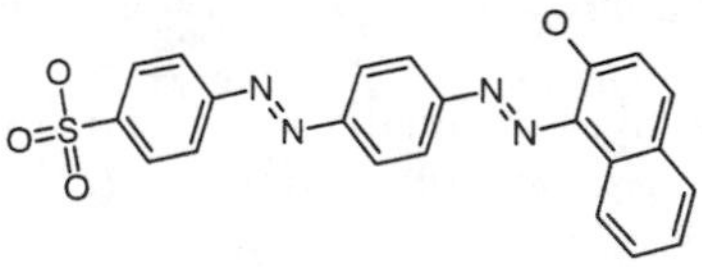

Acid Red 88

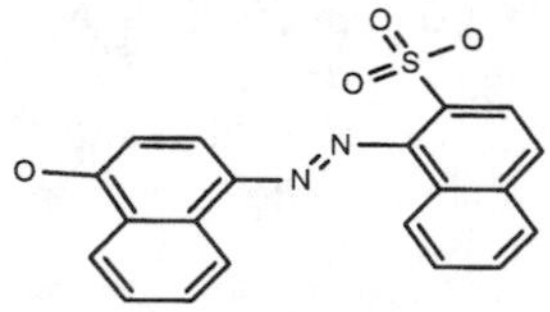

Acid Red 151

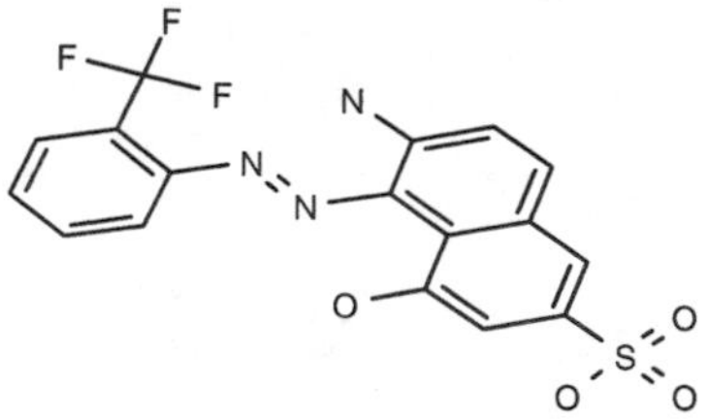

Acid Red 337

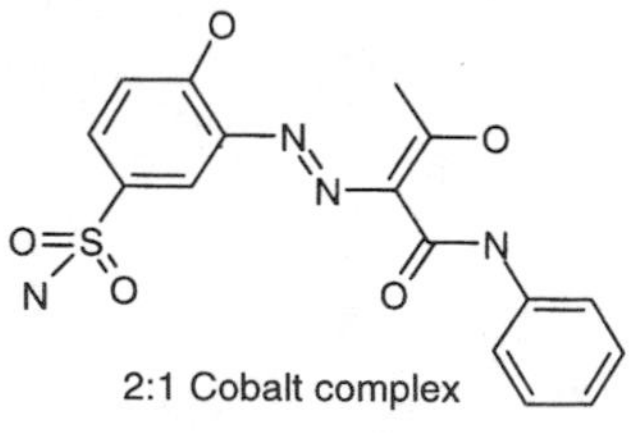

2:1 Cobalt complex

Acid Yellow 151

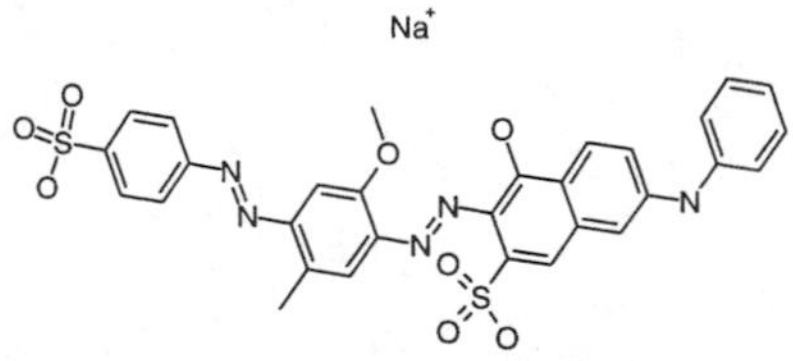

Direct Violet 9

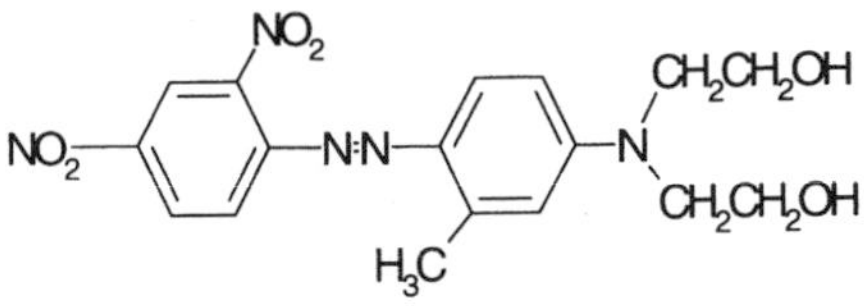

Disperse A. 1

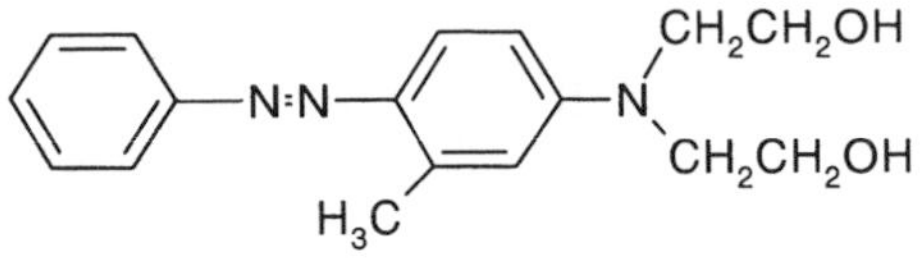

Disperse A.2

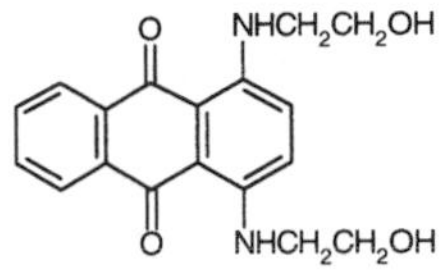

Disperse Blue 23

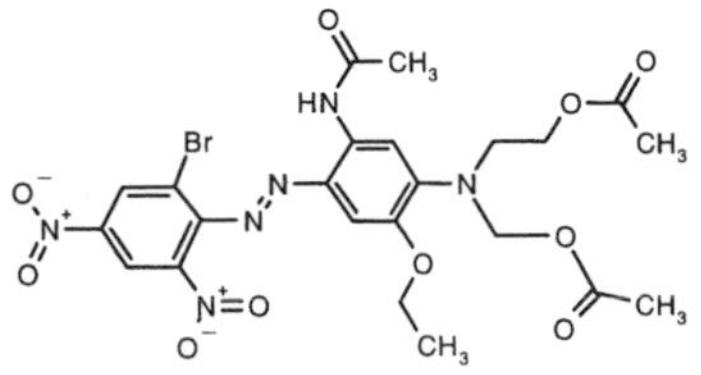

Disperse Blue 79

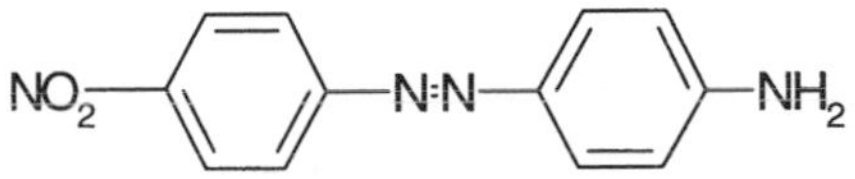

Disperse Orange 3

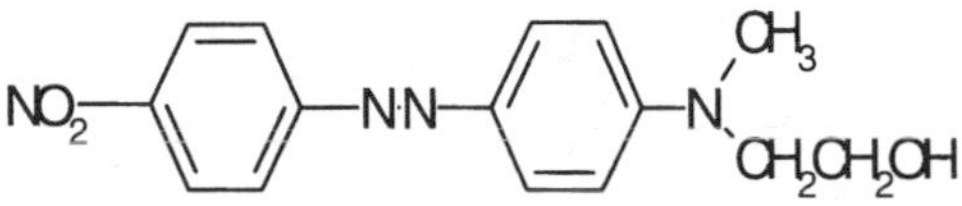

Disperse Red 1

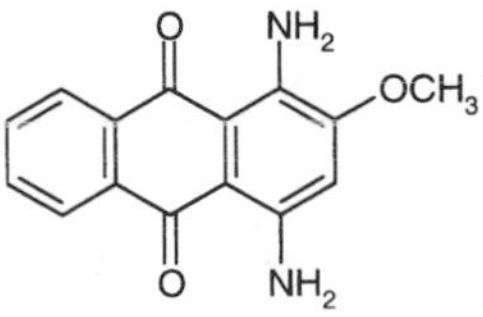

Disperse Red 11

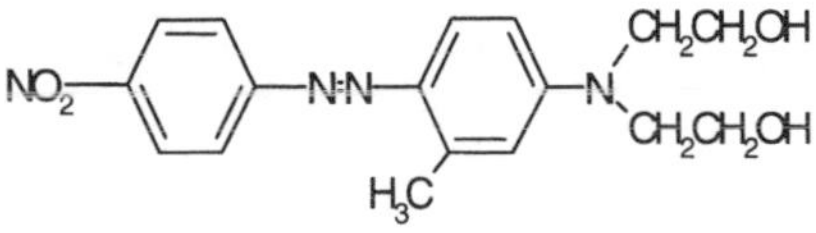

Disperse Red 17

Disperse Red 19

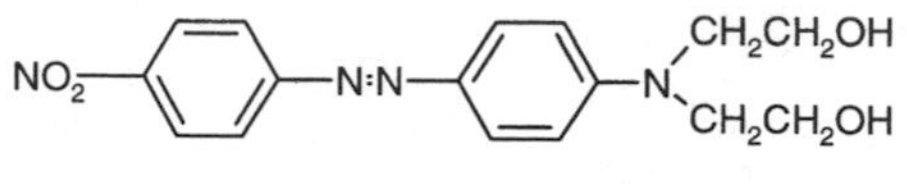

Disperse Red 5

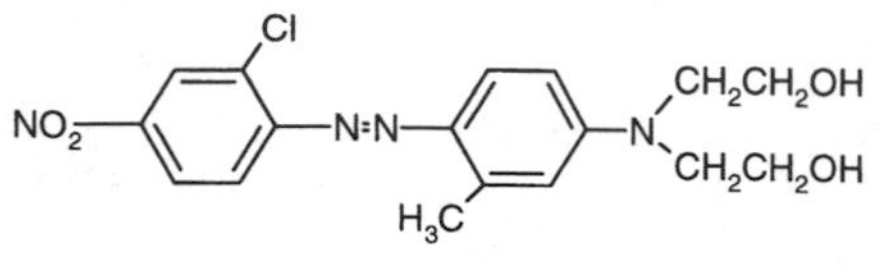

Disperse Red 7

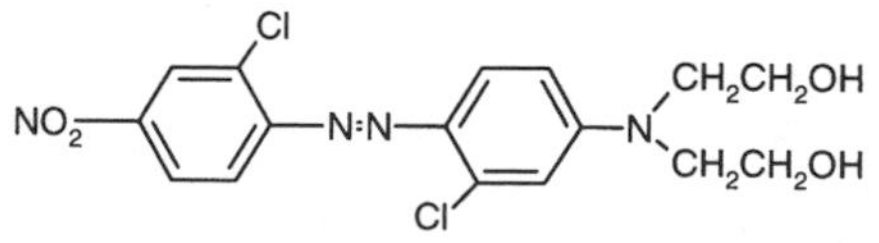

Disperse Violet 1

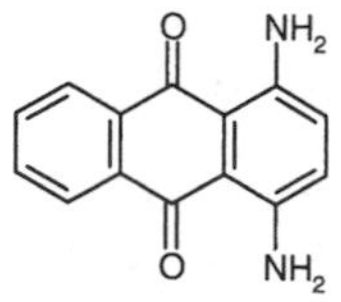

Disperse Yellow 54

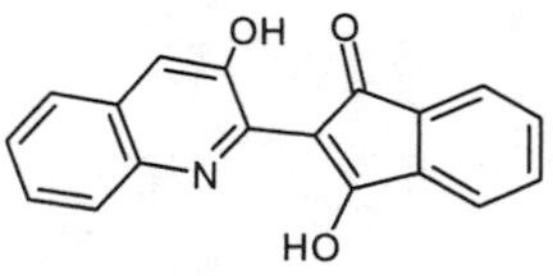

N 1

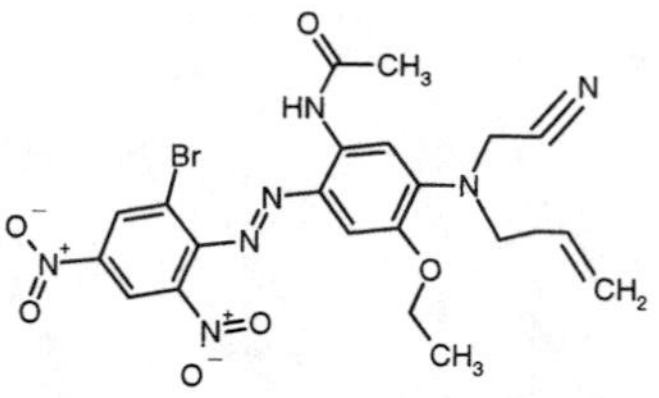

N 2

N 5

N 7

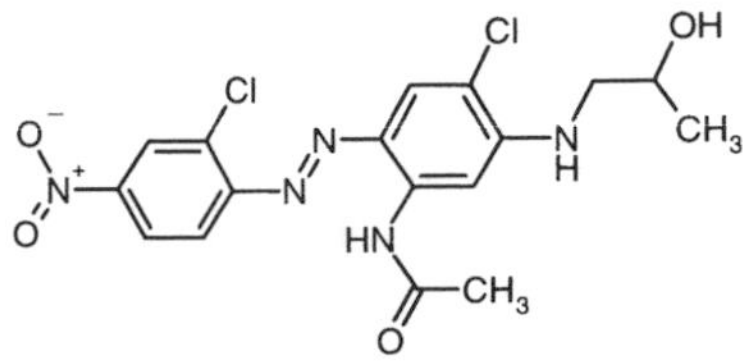

N 9

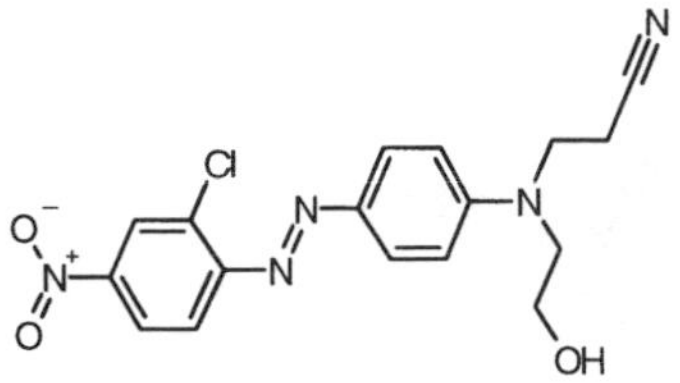

Ra

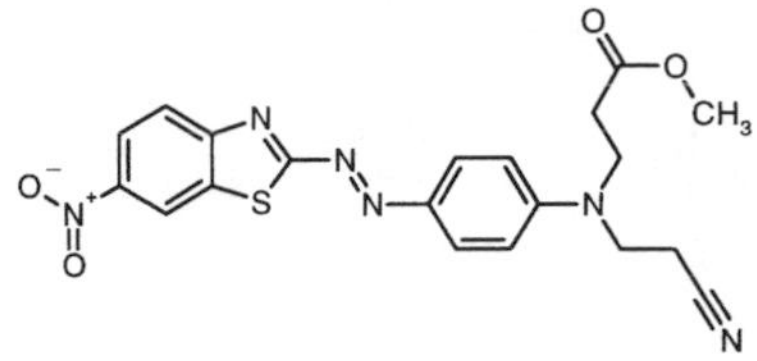

Solvent Yellow 1

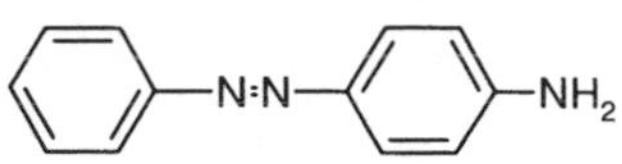

Solvent Yellow 2

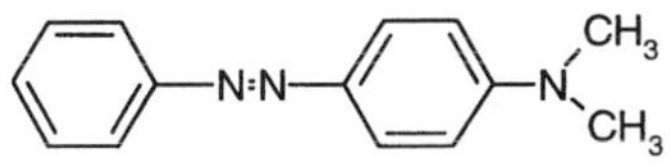

Solvent Yellow 58

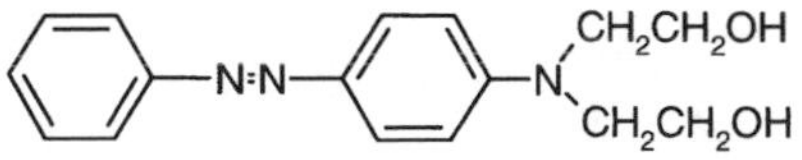

Index

A